Selected Papers of
Leonid V Keldysh

SELECTED PAPERS OF
LEONID V KELDYSH

Editor

Michael V Sadovskii

Russian Academy of Sciences, Russia

World Scientific

NEW JERSEY • LONDON • SINGAPORE • BEIJING • SHANGHAI • HONG KONG • TAIPEI • CHENNAI • TOKYO

Published by

World Scientific Publishing Co. Pte. Ltd.
5 Toh Tuck Link, Singapore 596224
USA office: 27 Warren Street, Suite 401-402, Hackensack, NJ 07601
UK office: 57 Shelton Street, Covent Garden, London WC2H 9HE

Library of Congress Control Number: 2023947078

British Library Cataloguing-in-Publication Data
A catalogue record for this book is available from the British Library.

SELECTED PAPERS OF LEONID V KELDYSH

ISBN 978-981-12-7945-4 (hardcover)
ISBN 978-981-12-7946-1 (ebook for institutions)
ISBN 978-981-12-7947-8 (ebook for individuals)

For any available supplementary material, please visit
https://www.worldscientific.com/worldscibooks/10.1142/13503#t=suppl

Desk Editor: Nur Syarfeena Binte Mohd Fauzi

CONTENTS

PREFACE

Leonid Veniaminovich Keldysh was one of the most prominent and influential Soviet and Russian theoretical physicists, who made most important contributions to condensed matter theory, developing new approaches and methods and discovering beautiful new physical effects later confirmed by experiment.

He was born in Moscow on 7 April 1931 in the family of scientists. His mother Lyudmila Vsevolodovna Keldysh was a leading Soviet mathematician, her brother, an applied mathematician, Mstislav Vsevolodovich Keldysh was one of the leaders of the Soviet space program, later becoming the President of the USSR Academy of Sciences. Leonids's step father was Petr Sergeevich Novikov the full member of the Academy and also a leading mathematician, while Leonid's younger step brother Sergei Petrovich Novikov also becoming a mathematician and Academy member, was later awarded the Fields medal. But Leonid's choice was theoretical physics.

In 1948 Keldysh entered the physical faculty of the Moscow State University, which he graduated in 1954 (attending also the courses at the faculty of mechanics and mathematics for an extra year) and started his work at the Theoretical Physics Department of P.N. Lebedev Physical Institute, which remained his workplace till the end of his life. His scientific supervisor at Lebedev Institute was Vitaly Lazarevich Ginzburg, and Theoretical Department at a time was headed by Igor Evgenievich Tamm (both later becoming the Nobel prize winners). However, since these early years Leonid was essentially the self-made man in science.

In his early works (1957–1958) Keldysh has developed the consistent theory of phonon assisted tunneling in semiconductors, which was immediately recognized by semiconductor community. Most famous work of this period was devoted to calculation of electric field induced shift of absorption edge in semiconductors, what is now called Franz–Keldysh effect. He also developed the original theory of deep levels in semiconductors. One of his most famous works of this period was the 1964 theory of multiple photon ionization of atoms by intensive electromagnetic waves, thereby laying the foundation for the entire field of intense laser radiation interaction with atoms, ions, molecules, and solids. This theory introduced optical tunneling and above-threshold ionization, experimentally observed about 15 years later. The "Keldysh parameter" determines the boundary between multiphoton and tunneling regimes.

Leonid Keldysh came to science during the heroic period, when the quantum field theory methods were successively applied in condensed matter physics. His name is probably best known due to his 1964 work on general diagram technique for non-equilibrium processes. Introducing Green's functions with time-ordering along what in now known as Keldysh contour, he was able to construct the standard Feynman diagrams for these Green's functions at finite temperatures and for general non-equilibrium states. Since then this approach became a standard one, with multiple applications in solid state theory, the theory of quantum liquids, quantum field theory and even quantum cosmology.

Strangely enough, even at that time, ten years after starting his work, he was not yet awarded any scientific degree. However, when he finally submitted the Candidate of Science (PhD) thesis in 1965, he was immediately awarded the degree of Doctor of Science (similar to Habilitation in Germany), and in 1968 he was elected the corresponding member of the USSR Academy of Sciences, becoming the full member (Academician) in 1976.

Since 1964 Keldysh interests moved to the problem of many excitons in semiconductors. In his work with Yu.V. Kopaev he introduced the new concept of excitonic insulator. Actually this was a new mechanism of metal–insulator transitions. In later works by Keldysh and his collaborators it was definitely shown, that there are no superfluidity properties in this model, as was initially suspected by some authors and he moved to the studies of non-equilibrium system of excitons, appearing under intensive laser pumping of semiconductors, where superfluidity of excitons was shown to be possible.

However at that time (1968) Keldysh realized, that in most semiconductors (with multiple bands) the non-equilibrium system of many excitons actually transforms into electron-hole quantum liquid (with excitons destroyed), forming electron-hole drops. Interestingly enough, this idea was expressed only in his summary talk at Moscow International Conference on Semiconductors and was not published anywhere for rather long time. However, it immediately stimulated the experimental studies and electron-hole drops were soon discovered, leading to many further experimental and theoretical works on this new state of matter. Essentially, he supervised these works around the Soviet Union, continuing to introduce new concepts, like phonon "wind" in the system of electron-hole drops.

Since 1965 Keldysh was professor of Moscow State University, where he also headed the chair of quantum radiophysics (1978–2001). He had many PhD students, a number of which later became famous theorists, professors and members of the Russian Academy of Sciences. He was the member of editorial boards of the leading physical journals and the Editor-in-Chief of Physics Uspekhi (2009–2016).

Keldysh was awarded numerous awards, including the Lenin prize (1974), the Hewlett-Packard Prize (1975), Alexander von Humboldt Prize (1994), Rusnanoprize (2009), Eugene Feinberg Memorial Medal (2011), Pomeranchuk Prize (2014) and the Grand Lomonosov Gold Medal of the Russian Academy of Sciences (2015). He was elected the foreign member of the USA National Academy of Sciences (1995) and the Fellow of the American Physical Society (1996).

In the late 1980s Keldysh had to perform different administrative duties, which he actually did not like at all, but considered impossible to reject during this hard period for the Russian Science. These included the heading of Theoretical Physics Department and the directorship of Lebedev Institute (1989–1994) and also the position of Academician–Secretary of the General Physics Department of the RAS (1991–1996). During this period he lived in his own way, not conforming to external circumstances. He was highly independent person and it was impossible to persuade him to take a decision which he disagreed with. He was among the leading RAS members who sharply rejected the Government proposed reform of the Academy in 2013 becoming the member of the influential "July 1 Club" within the RAS, opposing this reform.

Actually his scientific interests were much more wide as compared to what one can judge from his formal list of publications.

Keldysh has authored about 80 scientific papers during his lifetime. Below we present a collection of some of his most important works. Actually the choice of these papers was rather difficult, as well as the task of limiting to a reasonable size of this volume. As an Editor I am the single person responsible for this choice, while I have to thank a number of people with whom I have discussed these matters, especially Sergei Tikhodeev (Moscow State University) and David Khmelnitskii (Cambridge University). The volume is started by memorial paper describing in more details Leonid Keldysh scientific career. The original papers are published in chronological order and an original form, as they appeared in various (English language) editions, without any editorial corrections.

Michael V. Sadovskii

OBITUARY

physica status solidi b

In Memoriam Leonid V. Keldysh

Michael Bonitz, Antti-Pekka Jauho, Michael Sadovskii, and Sergei Tikhodeev*

Leonid V. Keldysh – one of the most influential theoretical physicists of the 20th century – passed away in November 2016. L. V. Keldysh is best known for the diagrammatic formulation of real-time (nonequilibrium) Green functions theory and for the theory of strong field ionization of atoms. Both theories profoundly changed large areas of theoretical physics and stimulated important experiments. Both these discoveries emerged almost simultaneously – like Einstein, also L. V. Keldysh had his *annus mirabilis* – the year 1964. But the list of his theoretical developments is much broader and is briefly reviewed here.

1. Introduction

On November 11, 2016 Leonid Veniaminovich L. V. Keldysh passed away in Moscow. L. V. Keldysh was a Russian theoretical physicist who had a tremendous influence on many fields of physics. In this article we briefly describe L. V. Keldysh's most important contributions and discuss how they influenced and continue to influence modern physics. In particular, we concentrate on his work on nonequilibrium many-particle physics that is related to his discovery of nonequilibrium Green functions theory, see Section 3. At the same time it is of high interest to recall many of his other activities, see Section 4, including his contributions to strong field ionization, Section 4.1, and exciton physics, *cf.* Section 4.2.

L. V. Keldysh's heritage includes 77 scientific publications,[1–77] that reflect the broad range of topics, L. V. Keldysh was interested in and how this interest evolved over time. His two most influential papers on real time Green Functions[12] and strong field ionization[13] were published in 1964 in the *Zhurnal Eksperimentalnoi i Teoreticheskoi Fiziki*[78] and have collected about 3300 and 6000 citations thus far.[79] In fact, these two papers were written almost at the same time. They were submitted by L. V. Keldysh to the journal on April 23 and May 23, 1964, respectively, making the year 1964 L. V. Keldysh's *annus mirabilis*. All articles of L.V. Keldysh are listed in chronological order in the reference section at the end of this paper.

Prof. M. Bonitz
Institut für Theoretische Physik und Astrophysik
Christian-Albrechts-Universität zu Kiel
24098 Kiel, Germany
E-mail: bonitz@theo-physik.uni-kiel.de

Prof. A.-P. Jauho
Department of Micro- and Nanotechnology
Technical University of Denmark
Lyngby, Denmark

Prof. M. Sadovskii
Institute for Electrophysics
RAS Ural Branch
Ekaterinburg 620016, Russia
and
M.N. Mikheev Institute for Metal Physics
RAS Ural Branch
Ekaterinburg 620016, Russia

Prof. S. Tikhodeev
Department of Physics
M.V. Lomonosov Moscow State University
Moscow 119991, Russia
and
A.M. Prokhorov General Physics Institute
Russian Academy of Sciences
Vavilova Street, 38, Moscow 119991, Russia

The ORCID identification number(s) for the author(s) of this article can be found under https://doi.org/10.1002/pssb.201800600.

DOI: 10.1002/pssb.201800600

Figure 1. Leonid V. Keldysh (1931–2016), photograph from around 2010, provided by M. Sadovskii.

ADVANCED SCIENCE NEWS
www.advancedsciencenews.com
physica status solidi b

Also, the activity of L. V. Keldysh in support of Russian science, in general, and the Academy of Sciences, in particular, is documented in seven articles published between 1992 and 1999.[80–86] They are interesting historical documents in their own but also show that L. V. Keldysh was speaking up publicly when he thought this is necessary, often together with other leading Russian colleagues. Some information on the often difficult political environment is contained in the biographical notes in Section 2. Moreover, L. V. Keldysh co-authored a remarkable number of 61 short notes in honor of leading Russian physicists – a special tradition in Soviet and Russian science. These articles include 42 birthday congratulations and 29 obituaries.

Finally, we also include some remarks on L. V. Keldysh's students and L. V. Keldysh's work as a mentor, in Section 5.

2. Biographical Notes

Leonid V. Keldysh was born in Moscow on April 7, 1931 in a family of scientists. His mother, Lyudmila Vsevolodovna Keldysh, was a leading Soviet mathematician, her brother, an applied mathematician, Mstislav Vsevolodovich Keldysh was one of the leaders of the Soviet space program, later becoming the President of the USSR Academy of Sciences.[87] Leonids's step father was Petr Sergeevich Novikov, a full member of the Academy and also a leading mathematician, while Leonid's younger step brother, Sergei Petrovich Novikov, also a mathematician and Academy member, was later awarded the Fields medal. But Leonid's choice was theoretical physics.

In 1948 L. V. Keldysh enrolled in the Physics Department of Moscow State Lomonosov University (MGU), where he graduated in 1954 (attending also courses at the Department of mechanics and mathematics for an extra year). After this he started to work at the Theoretical Physics Department of the P.N. Lebedev Physics Institute (LPI) of the Academy of Sciences, which remained his work place until the end of his life. His scientific supervisor at LPI was Vitaly Lazarevich Ginzburg, and the Theoretical Department at that time was headed by Igor Evgenievich Tamm (both later becoming Nobel prize winners). However, since these early years, Leonid was essentially a self-made man in science.

In his early works (1957–1958) L. V. Keldysh developed a consistent theory of phonon-assisted tunneling in semiconductors which was immediately recognized by the semiconductor community. His most famous work of this period was devoted to the calculation of the electric field-induced shift of the absorption edge in semiconductors, what is now called the Franz–Keldysh effect,[3,88] see Section 4.3.1. In the early 1960s he proposed to use spatial modulation of the lattice to create an artificial band structure.[7] This idea was later realized in semiconductor superlattices. He also developed an original theory of core levels in semiconductors.[9] One of his most famous works of this period was the 1964 theory of tunnel and (multi-)photon ionization of atoms by intense electromagnetic waves[13] that became the starting point for the entire field of intense laser–matter interaction, including atoms, ions, molecules, plasmas, and solids, *cf.* Section 4.1. This field has recently been reviewed in ref. [89], where it is concluded that the success behind the theory is that it precisely fulfills the

Figure 2. Leonid V. Keldysh around 1962, photograph provided by M. Sadovskii.

Figure 3. Leonid V. Keldysh around 1968, photograph provided by M. Sadovskii.

ADVANCED SCIENCE NEWS
www.advancedsciencenews.com
physica status solidi pss b

Figure 4. Leonid V. Keldysh around 1989, photograph provided by M. Sadovskii.

criterion "making things as simple as possible, but not simpler." This feature is characteristic of many of L. V. Keldysh's other influential papers.

Leonid V. Keldysh started working in science during a period when quantum field theory methods were popular and successfully applied in condensed matter physics. Here he made his most famous contribution with his 1964 work on a general diagram technique for nonequlibrium processes.[12] Introducing Green functions with time–ordering along what is today known as the (Schwinger) L. V. Keldysh time contour, he was able to construct the standard Feynman diagrams for these Green functions at finite temperatures and for general nonequilibrium states, see Section 3.

Strangely enough, even at that time, 10 years after starting his work, he had not yet been awarded any higher scientific degree. However, when he finally submitted the Candidate of Science (PhD) thesis, in 1965, he was immediately awarded the degree of Doctor of Science (similar to habilitation in Germany). In 1968 he was elected corresponding member of the USSR Academy of Sciences, becoming a full member in 1976.

Since 1964 L. V. Keldysh's interests moved to semiconductors. In his work with Yu.V. Kopaev[15] he introduced the new concept of an excitonic insulator and to laser excited nonequilibrium exciton systems, exciton superfluidity[20] and their ioinization into an electron–hole quantum liquid of electron–hole droplets, for details see Section 4.2.

Since 1965 L. V. Keldysh was a professor at MGU heading the chair of quantum radiophysics (1978–2001). He had many PhD students, a number of which later became famous theoreticians, professors, and members of the Russian Academy of Sciences, see Section 5. He was member of editorial boards of the leading Russian physics journals and served as Editor in Chief of Physics Uspekhi, from 2009 to 2016. L. V. Keldysh was awarded numerous prizes, including the Lenin prize (1974), the Hewlett–Packard Prize (1975), the Alexander von Humboldt Prize (1994), the Rusnanoprize (2009), the Eugene Feinberg Memorial Medal (2011), the Pomeranchuk Prize (2014), and the Grand Lomonosov Gold Medal of the Russian Academy of Sciences (2015). He was elected foreign member of the US National Academy of Sciences (1995) and became a Fellow of the American Physical Society in 1996.

In the late 1980s L. V. Keldysh had to perform various administrative duties, which he actually did not like at all, but considered impossible to reject during this difficult period for Russian Science. These included the head of the Theoretical Physics Department and the director of the Lebedev Institute (1989–1994) and also the position of a Secretary of the General Physics Department of the Russian Academy of Sciences (1991–1996). During this period he lived in his own way, never conforming to external circumstances. He always was a highly independent person, and it was impossible to persuade him to take a decision he did not agree with. He was among the leading RAS members who strictly rejected the Government–proposed reform of the Academy in 2013, becoming a member of the influential "Club of July 1" within the RAS, opposing this reform.

3. Nonequilibrium (Real-Time or L. V. Keldysh) Green Functions (NEGF)

Judging by its impact on a huge number of fields L. V. Keldysh's Real-time Green functions theory is, without question, his most important discovery, and we address it in some more detail.

3.1. The Story of NEGF

Quantum many-body systems have been described by many different approaches including wave function methods, reduced density operators (quantum BBGKY-hierarchy) of Bogolyubov, Kirkwood, and others, for example,[90] as well as Green functions and Feynman diagrams by Schwinger, Dyson, and Feynman. Following the results for the ground state soon the extension to thermodynamic equilibrium was developed in the 1950s by Matsubara, Kubo as well as Abrikosov, Gorkov, Dzyaloshinski in the U.S.S.R. which led to the concept of imaginary-time Green functions. The idea to rewrite the canonical density operator in thermodynamic equilibrium as a quantum-mechanical evolution operator, but in imaginary time, was then quite popular in a number of fields, including Feynman's path integral concept, so that step was rather natural.

However, the extension of the technique from thermodynamic equilibrium to *arbitrary nonequilibrium situations* is a huge step that is far less straightforward, and it took more time to develop. These developments occured almost independently in the U.S. and in the U.S.S.R. The works in the U.S. were mostly

ADVANCED SCIENCE NEWS
www.advancedsciencenews.com

physica status solidi pss b

due to Martin and Schwinger who derived the generalization of the BBGKY-hierarchy to the case of many-time Green functions,[91] and Baym and Kadanoff who derived and analyzed the generalization of the Boltzmann equation that includes memory effects.[92] These developments were reviewed in detail by Paul Martin and Gordon Baym in their lectures at the first Nonequilibrium Green Functions conference in Rostock, Germany, in 1999, *cf.* refs. [93,94]. The Russian developments in the field of Nonequilibrium Green functions are due solely to Leonid V. Keldysh and were published in his seminal paper[12] where he introduced the "round trip time contour" – a small but ingeneous mathematical trick – that allowed him to rigorously extend Feynman's diagram technique to nonequilibrium. The Russian developments in thermodynamic and Nonequilibrium Green functions were reviewed by Alexei Abrikosov[95] and Leonid V. Keldysh,[66] respectively. The latter article is reprinted as a supplement[96] to this paper.

3.2. The PNGF Conferences and Leonid V. Keldysh

Interestingly, after writing his paper introducing NEGF in 1964, L. V. Keldysh did not actively continue these developments (the same was true for Baym and Kadanoff). So it must have been a surprise for them that they were in 1999 invited to a conference entitled "Kadanoff–Baym equations – Progress and Perspectives for Many-Body Theory," 35 years after the original developments. In fact, in the 1970s and 1980s NEGF were used only by a few groups world wide but the activities increased significantly in the 1990s when NEGF methods were used in semiconductor optics and various groups learned to directly solve the Keldysh–Kadanoff–Baym equations (KBE) on modern computers, following the pioneering work of Danielewicz[97] on nuclear collisions. Not surprisingly, many theorists[98] expected that these equations would lead to breakthroughs in many fields which indeed turned out to be the case, see Section 3.3.

At the same time, the lengthy title of the conference in 1999 reflects some confusion in the community about the different contributions of the American and Russian founders of the theory and about the priorities. Even Baym was under the impression that L. V. Keldysh's work of 1964 was a follow up to their book,[92] as he pointed out in his conference talk in 1999 and in the proceedings.[94] However, this was an incorrect assumption. Not surprisingly, L. V. Keldysh – who could not participate in the 1999 conference – was very upset when he became aware of Baym's article. He then took the opportunity to attend the second conference, "Progress in Nonequilibrium Functions (PNGF) II" in Dresden in 2002 and, in his lecture, to "straighten" things out. For everybody who uses NEGF today or will do so in the years to come, this turned out to be a very lucky case, because L. V. Keldysh summarized in some detail and in his honest style how his ideas emerged and who influenced him. We are lucky that he published his recollections in the conference book,[99] and his article is reprinted as a supplement to this paper. A photo showing Leonid V. Keldysh at the PNGF II together with, among others, Alex Abrikosov, Pawel Danielewicz, and Paul Martin is presented in **Figure 5**.

The success story of nonequilibrium Green functions and the tremendous impact of L. V. Keldysh's paper[12] is clearly reflected in the next meetings of the conference series and their proceedings[100–103] culminating in the present issue of the proceedings of the 2018 conference.

Figure 5. Part of the participants of the Conference "Progress in Nonequilibrium Green Functions II," Dresden, August 2002. Front row from left: Alexei Abrikosov, Leonid Keldysh, Jörn Knoll, Pawel Danielewicz, Hendrik van Hees, Paul C. Martin. Second and third row, among others: Paul Gartner, Egidius Anisimovas, Vladimir Filinov, Robert van Leeuwen, Alexey Filinov, Roland Zimmermann, Antti-Pekka Jauho, and Rolf Binder. Photograph by M. Bonitz (part of the conference photo, from ref. [101]).

ADVANCED SCIENCE NEWS
www.advancedsciencenews.com

physica status solidi b

3.3. Current Research Fields Based on NEGF

During the last three decades NEGF have seen a dramatic increase in attention. This is mostly due to the increase in computing capabilities that have made direct solutions of the Keldysh–Kadanoff–Baym equations possible. Applications have been developed for a large number of fields where many-body effects, correlations and non-quasiparticle behavior are of relevance. This includes transport in metals,[104] semiconductor optics and transport,[105] nanostructures,[106] atoms and molecules,[107,108] plasma physics,[109,110] nuclear matter,[111,112] cosmology,[113,114] transport properties of strongly correlated cold fermionic atoms,[115,116] among others.

4. Other Research Topics of L.V. Keldysh

4.1. Strong Field Ionization

Cited more than 6000 times, L. V. Keldysh's paper[13] presented the first quantum theory of the ionization of an atom by an intense laser field. The paper introduced optical tunneling, multi-photon ionization, and above-threshold ionization, experimentally observed about 15 years later, for example,[117] L. V. Keldysh presented the first nonlinear quantum-mechanical calculation of the ionization probability of an atom in a strong electromagnetic field. Starting from the time-dependent bound state wave function (we follow the notation of ref. [89]) $\Psi_0(t) = \psi_0(\mathbf{r})e^{iI_pt/\hbar}$ he computes the transition probability amplitude of the electron into a time-dependent continuum state in the presence of the field $\Psi_\mathbf{p}(t)$ (i.e., a Volkov state),[118]

$$M(\mathbf{p}) = -\frac{i}{\hbar}\int_{-\infty}^{\infty} dt \langle \Psi_\mathbf{p}(t)|V_{\text{int}}(t)|\Psi_0(t)\rangle \quad (1)$$

where I_p denotes the ionization potential, p the canonical momentum, and V_{int} the interaction potential of the electron with the field. Note that the explicit forms of $\Psi_\mathbf{p}$ and V_{int} depend on the chosen gauge, so the analysis requires some care. Indeed, many suggested modifications or improvements of L. V. Keldysh's work led to gauge-dependent results giving rise to debates in the community, for details see ref. [89]. The momentum distribution of the photoelectrons is then $dW(\mathbf{p}) = |M(\mathbf{p})^2|d^3p$, and the total ionization probability is the momentum integral of dW. Using the dipole approximation for the field and neglecting Coulomb interaction and relativistic effects on $\Psi_\mathbf{p}$ L. V. Keldysh was able to obtain closed expressions for the ionization probability. The result contains an important dimensionless parameter – the "Keldysh parameter,"

$$\gamma = \sqrt{2mI_p}\frac{\omega}{eE_0} \quad (2)$$

which determines the boundary between multiphoton and tunneling regimes. Here ω and E_0 are, respectively, the frequency and amplitude of the exciting electric field. The Keldysh parameter describes the ratio of the characteristic momentum of the electron in the bound state, $\sqrt{2mI_p}$, to the momentum the electron gains from the field, eE_0/ω. For $\gamma < 1$ ($\gamma > 1$), ionization is dominated by the tunnel (multiphoton) mechanism. In case of a monochromatic field and $\gamma > 1$, multiphoton absorption is possible if the atom absorbs at least

$$N_{\min} = \frac{I_p + U_p}{\hbar\omega} + 1$$

photons which includes the average kinetic energy of the free electron in the field ("ponderomotive potential"), $U_p = (eE_0)^2/4m\omega^2$. If the photon number exceeds $N_{\min}$, ionization will lead to distinct peaks in the photoelectron energy spectrum – which has been called "above threshold ionization" – and has been accurately verified experimentally. With the dramatic progress in laser technology and the availability of coherent radiation sources from the infrared range to X-rays, these effects have achieved fundamental importance in countless fields.

L. V. Keldysh's theory triggered a tremendous wave of further improvements of the theory that include Coulomb interaction, relativistic effects, or the field-induced modification of the bound states (Stark effect). The analysis of ionization processes was extended to more complex atoms, molecules, and semiconductors; and similar approaches were developed for relativistic effects such as pair creation (Schwinger mechanism). For additional information and references, the reader is referred to the review [89].

4.2. Excitons and Electron–Hole Systems

L. V. Keldysh made important contributions to semiconductor physics. He was early on interested in the many-exciton problem in semiconductors. In his work with Yu.V. Kopaev[15] he introduced the new concept of an excitonic insulator. Actually this was a new mechanism of a metal–insulator transition. In later works by L. V. Keldysh and his collaborators it was shown conclusively, that there are no superfluidity properties in this model,[27] as was initially suspected by some authors, and he moved to the study of nonequilibrium systems of excitons, appearing under intense laser pumping of semiconductors, where superfluidity of excitons was shown to be possible.[20,119] However at that time (1968) L. V. Keldysh realized, that in most semiconductors (with multiple bands) the nonequilibrium system of many excitons actually transforms into an electron–hole quantum liquid (where excitons are ionized), forming electron–hole droplets. Interestingly enough, this idea was expressed only in his summary talk at the Moscow International Conference on Semiconductors[120] and it immediately stimulated experimental studies, and electron–hole droplets were soon discovered, leading to many further experimental and theoretical works on this new state of matter. Essentially, he supervised these works around the Soviet Union, continuing to introduce new concepts, such as the phonon "wind" in the system of electron–hole droplets.[35] An overview of the field of electron-hole droplets can be found in ref. [121] Electron-hole droplet formation was also verified in *ab initio* quantum Monte Carlo simulations.[122] The problem of limited

ADVANCED SCIENCE NEWS
www.advancedsciencenews.com
physica status solidi b

life time of electron-hole pairs in optically excited semiconductors can be overcome with indirect excitons predicted by Lozovik and Yudson[123] which have interesting superfluidity properties.[124]

4.3. Further Research Results

Even though L. V. Keldysh is mainly famous for real-time Green functions, strong-field ionization and the theory of excitons, he has made important contributions to many other fields.

4.3.1. Franz–Keldysh Effect

It was a natural question to ask whether the Franz–Keldysh effect (the shift of the absorption edge due to an applied static electric field) could be extended to a situation where the absorbing sample was placed in a time-dependent field. Indeed, early theoretical work addressed some aspects of this situation.[125,126] Experimentally, however, sufficiently strong time-dependent fields were not available until the first free-electron lasers started operation. A detailed study was published in ref. [127], where the excitonic absorption of a quantum well system was studied as a function of the frequency of the impinging strong THz field emanating from the Santa Barbara free-electron laser. The theory developed for this situation agreed very well with the observations. The theory combines three concepts in whose development the pioneering ideas by L. V. Keldysh were crucial: strong field effects in semiconductors, excitonic dynamics, and nonequilibrium Green's functions. It is remarkable that all three ingredients originate from the same author.

4.3.2. Transport in Mesoscopic Systems

The scattering theory of transport, developed by Landauer and Büttiker,[128,129] which expresses the conductance of a mesoscopic sample in terms of its transmission properties, is – despite of its huge success and importance – only valid for systems where electron-electron or electron-phonon interactions can be ignored. The Keldysh diagram technique, which allows for a systematic treatment of interactions, is particularly well-suited for deriving extensions of the Landauer-Büttiker formalism. The Keldysh technique, as applied to transport physics, was introduced in the Western literature in an important series of papers by Caroli, Combescot, and co-workers.[130–133] These papers were mainly concerned with tunneling through a single barrier (including interactions with localized states and phonons in the barrier), but a real breakthrough occurred in 1992, when Meir and Wingreen showed[134] that the calculation of the conductance through a quantum dot with arbitrary interactions could be formulated in a similar manner. Literally thousands of papers have examined transport in situations where interactions are important.

One example is the tunneling of electrons between a tip and a metal through a single adsorbed molecule (or atom) in a scanning tunneling microscope. The Keldysh technique provides an elegant way to describe the inelastic stationary electron tunneling with emission and absorption of vibrational excitations of the molecule, interactions with the phonon baths in the substrate and tip, as well as the overheating of the molecule and its resulting motion – hopping or rotation.[135–139] L. V. Keldysh's theory provides the theoretical basis for inelastic tunneling electron spectroscopy, single molecule chemistry and motors, for details see, for example, the text book.[140]

The approach can be generalized to time-dependent situations,[141] or situations where the partitioning of the system into separate leads and a central region must be re-examined.[142] The next level of abstraction can be achieved by formulating the nonequilibrium theory in a field-theory language. This powerful formulation has found a very large number of applications, which are reviewed, for example, in a recent advanced textbook.[143] The field-theory formulation honors L. V. Keldysh by employing many technical terms that commemorate their inventor, for example, the Keldysh rotation, or the Keldysh action.

4.3.3. The Rytova–Keldysh Potential

In 1979 L. V. Keldysh considered the Coulomb interaction in thin semiconductor and semimetal films, and proposed a form for the interaction potential between charged particles in such systems.[41] (Work along similar lines was reported earlier by Rytova).[144] A central theme in condensed matter physics in our millenium is concerned with two-dimensional materials, such as graphene, or transition metal dichalcogenides. The Rytova-Keldysh potential forms an important ingredient in the physics of these materials. Recent developments are reviewed, for example, in ref. [145] where many references to related work can be found.

4.3.4. Stochastic Methods Applied to the Keldysh Contour

The idea of treating quantum many-body systems out of equilibrium on the L. V. Keldysh time contour has been extended to various other methods. A stochastic sampling method of Feynman diagrams was developed by Werner et al. and is known as diagrammatic Monte Carlo, see ref. [146] and references therein. Diagrammatic Monte Carlo extends earlier equilibrium simulations such as the continuous-time quantum Monte Carlo method for fermions[147] to arbitrary nonequilibrium situations. While it formally can treat strongly coupled systems and is successfully used in condensed matter systems and for cold atoms, it suffers from the dynamic fermion sign problem that strongly limits the simulation duration.

5. The Keldysh School

Actually L. V. Keldysh's scientific interests were much broader than one could judge from his list of publications.

ADVANCED SCIENCE NEWS
www.advancedsciencenews.com
physica status solidi pss b

Table 1. List of L. V. Keldysh's PhD (above the line) and master students (below) in chronological order, their year of graduation, and their scientific topics.

Name	Graduation	Research topics
Yu. V. Kopaev	1965	Semimetal-dielectric phase transitions
D. I. Khomskii	1969	Systems with strong electronic correlations
R. R. Guseinov	1971	Electron-phonon interaction in systems with excitonic instabilities
M. V. Sadovskii	1974	Liquid semiconductors, Pseudogap, Disorder and Fluctuation effects on the 1D Peierls transition
A. P. Silin	1975	Condensation of excitons in semiconductors
B. A. Volkov	1976	Electronic properties of semiconductors with structural instabilities
A. V. Vinogradov	1976	Electronic mechanisms of light absorption of dielectrics in the transparency range
E. A. Andryushin	1977	Electron-hole liquid in layered semiconductors
V. S. Babichenko	1977	Electron-hole liquid in strongly anisotropic semiconductors and semimetals
T. A. Onishchenko	1977	Electron-hole liquid in a strong magnetic field
V. E. Bisti	1978	Exciton interactions in semiconductors
S. G. Tikhodeev	1980	Interaction of the electron-hole liquid in semiconductors with deformations
		Nonequilibrium diagram technique for relaxation processes
A. L. Ivanov	1983	Intense electromagnetic wave in a direct-gap semiconductor
I. M. Sokolov	1984	Localization in the Anderson model with correlated site energies, percolation theory
P. I. Arseev	1986	Electrodynamics of rough surfaces of metals and semiconductors
N. S. Maslova	1987	Resonant interaction of light with a system of nonlinear oscillators
		Non-equilibrium transport through correlated systems
N. A. Gippius	1988	Quantum reflection of an exciton from the surface of an electron-hole droplet
		Interaction of electromagnetic radiation with semiconductors
S. S. Fanchenko	1975	Generalized diagram technique of non-equilibrium processes
		The problem of arbitrary initial conditions

See also the list of references at the end of the paper.

This is, in part, reflected in the broad range of topics his PhD students worked on, see **Table 1**. One of us (MS) recalls "I first met him in 1969 when I was a third year student of the Ural State University and attended his lectures on exciton condensation and electron–hole droplets at the famous winter school on theoretical physics 'Kourovka' near Sverdlovsk (now Ekaterinburg). In 1971 I became his PhD student at the Lebedev Institute in Moscow and, to my surprise, he proposed to me a PhD topic related to the construction of the theory of 'liquid semiconductors' – a research field developed previously in the experimental works of the Ioffe–Regel group in Leningrad and still lacking serious theoretical foundation. This reflected L. V. Keldysh's interest in the general theory of electrons in disordered systems, being only developed at that time in the classical works of Neville Mott, Ilya Lifshits, and Philip Anderson. In the following years we tried (in fact more or less in vain!) to construct such a theory. Our main idea was to produce a theoretical model of the pseudogap – a concept introduced by Mott on qualitative grounds to explain electronic properties of amorphous and liquid semiconductors. Here we were successful and formulated an exactly solvable model of the pseudogap, based on the summation of a complete series of Feynman diagrams for a simplified 1D model. Actually L. V. Keldysh declined co-autorship, so these results appeared under my name only, forming the ground for my future work in many years to follow, leading eventually to the studies of the pseudogap problem in high-T_C superconductors. This model was, in fact, a generalization of a similar diagram summation in L. V. Keldysh's studies of doped semiconductors, which appeared only in his dissertation (1965) and was later used or rediscovered by others. These are only few of many examples of his unpublished results. Most of them he was writing in large notebooks at his home, which some of his students were lucky enough to see."

The list of L. V. Keldysh's PhD students and their year of graduation and research topics are presented in Table 1. Many of them became successful scientists themselves. Kopaev, Khomskii, Volkov, Sadovskii, Tikhodeev, Ivanov, Arseev, Maslova, and Gippius later did their habilitation. Gippius, Khomskii, Sadovskii, Tikhodeev, and Vinogradov became professors. Arseev became a corresponding member and Kopaev and Sadovskii full members of the Russian Academy of Sciences.

ADVANCED SCIENCE NEWS
www.advancedsciencenews.com

physica status solidi b

Figure 6. Leonid V. Keldysh around 2006, photograph by Nikolay Gippius.

6. Conclusions

There have been a number of obituaries for L. V. Keldysh in the U.S.[87] and in Russia[148] that have covered various sides of L. V. Keldysh's scientific work and personality. The 2017 special issue of *Physics Uspekhi* (volume 60, issue 11)[148] covers in detail L. V. Keldysh's scientific work. There is no need to reproduce this material here. Instead, we have taken the particular angle of view on L. V. Keldysh that concentrates on his contributions to nonequilibrium many-body physics, in general, and nonequilibrium Green functions, in particular. L. V. Keldysh's single paper on the subject[12] has dramatically changed the whole field providing us and future generations with a strict mathematical basis and an extremely powerful and fully general tool – nonequilbrium diagram technique. This has allowed the NEGF approach to being introduced in an enormously broad range of areas of physics and quantum chemistry, not just as a tool for recovering familiar kinetic equations or deriving improved approximations but, more and more, as a practical tool for quantitative analysis of time-dependent processes on all time scales. Judging by the impressive increase in the number of publications on NEGF, L. V. Keldysh's work has provided a tremendously fertile ground for theory developments in many decades to come.

Supporting Information

Supporting Information is available from the Wiley Online Library or from the author.

Acknowledgements

The authors were grateful to World Scientific Publishing for the permission to reprint L. V. Keldysh's article from the PNGF II proceedings[66] as a supplement to this paper and to D. Semkat for providing the LaTeX source. The authors thank J.-P. Joost for technical assistence with the formatting of this article. APJ is supported by the Danish National Research Foundation, Project DNRF103.

Conflict of Interest

The authors declare no conflict of interest.

Keywords

Keldysh technique, nonequilibrium Green functions, real-time Green functions

Received: October 17, 2018
Revised: January 4, 2019
Published online: March 19, 2019

[1] L. V. Keldysh, Behavior of Non-Metallic Crystals in Strong Electric Fields, *Soviet Phys. JETP-USSR* **1958**, *6*, 763.
[2] L. V. Keldysh, Influence of the Lattice Vibrations of a Crystal on the Production of Electron-Hole Pairs in a Strong Electrical Field, *Soviet Phys. JETP-USSR* **1958**, *7*, 665.
[3] L. V. Keldysh, The Effect of a Strong Electric Field on the Optical Properties of Insulating Crystals, *Soviet Phys. JETP-USSR* **1958**, *7*, 788.
[4] B. Vul, E. Zavaritskaia, L. V. Keldysh, Impurity Conductivity of Germanium at Low Temperatures, *Dokl. Akad. Nauk SSSR* **1960**, *135*, 1361.
[5] L. V. Keldysh, Kinetic Theory of Impact Ionization in Semiconductors, *Soviet Phys. JETP-USSR* **1960**, *10*, 509.
[6] L. V. Keldysh, Optical Characteristics of Electrons with a Band Energy Spectrum in a Strong Electric Field, *Soviet Phys. JETP-USSR* **1963**, *16*, 471.
[7] L. V. Keldysh, Effect of Ultrasonics on the Electron Spectrum of Crystals, *Soviet Phys.-Solid State* **1963**, *4*, 1658.
[8] L. V. Keldysh, Y. Kopaev, The Energy Spectrum of a Degenerate Semiconductor with an Ionic Lattice, *Soviet Phys.-Solid State* **1963**, *5*, 1026.
[9] L. V. Keldysh, Deep Levels in Semiconductors, *Soviet Phys. JETP-USSR* **1964**, *18*, 253.
[10] L. V. Keldysh, G. Proshko, Infrared Absorption in Highly Doped Germanium, *Soviet Phys.-Solid State* **1964**, *5*, 2481.
[11] V. Bagaev, Y. Berozashvili, B. Vul, E. Zavaritskaya, L. V. Keldysh, A. Shotov, Concerning the Energy Level Spectrum of Heavily Doped Gallium Arsenide, *Soviet Phys.-Solid State* **1964**, *6*, 1093.
[12] L. V. Keldysh, Diagram Technique for Nonequilibrium Processes, *Soviet Phys. JETP-USSR* **1965**, *20*, 1018. [*Zh. Eksp. Teor. Fiz.* **1964**, *47*, 1515].
[13] L. V. Keldysh, Ionization in Field of a Strong Electromagnetic Wave, *Soviet Phys. JETP-USSR* **1965**, *20*, 1307. [*Zh. Eksp. Teor. Fiz.* **1964**, *47*, 1945].
[14] L. V. Keldysh, Concerning Theory of Impact Ionization in Semiconductors, *Soviet Phys. JETP-USSR* **1965**, *21*, 1135.
[15] L. V. Keldysh, Y. Kopaev, Possible Instability of Semimetallic State Toward Coulomb Interaction, Soviet Physics Solid State, *Soviet Phys. Solid State, USSR* **1965**, *6*, 2219.
[16] L. V. Keldysh, Superconductivity In Nonmetallic Systems, *Soviet Phys. Uspekhi-USSR* **1965**, *8*, 496.
[17] V. Bagaev, Y. Berozashvili, L. V. Keldysh, Electrooptical Effect in GaAs. *JETP Lett.-USSR* **1966**, *4*, 246.
[18] L. V. Keldysh, T. Tratas, Dynamic Narrowing of Paramagnetic Resonance Lines in a Compensated Semiconductor, *Soviet Phys. Solid State, USSR* **1966**, *8*, 64.
[19] L. V. Keldysh, A. Kozlov, Collective Properties of Large-Radius Excitons, *JETP Lett.-USSR* **1967**, *5*, 190.

ADVANCED SCIENCE NEWS
www.advancedsciencenews.com

physica status solidi pss b

[20] L. V. Keldysh, A. Kozlov, Collective Properties of Excitons in Semiconductors, *Soviet Phys. JETP-USSR* **1968**, *27*, 521. [*Zh. Eksp. Teor. Fiz.* **1968**, *54*, 978].

[21] V. Bagaev, Y. Berozashvili, L. A. Keldysh, Anisotropy of Polarized-Light Absorption Produced in GaAs and CdTe Crystals by a Strong Electric Field, *JETP Lett.-USSR* **1969**, *9*, 108.

[22] L. V. Keldysh, M. Pkhakadze, Conductivity of Semiconductors Under Pinch-Effect Conditions, *JETP Lett.-USSR* **1969**, *10*, 169.

[23] V. Bagaev, T. Galkina, O. Gogolin, L. V. Keldysh, Motion of Electron-Hole Drops in Germanium, *JETP Lett.-USSR* **1969**, *10*, 195.

[24] L. V. Keldysh, O. Konstantinov, V. Perel, Polarization Effects in Interband Absorption of Light in Semiconductors Subjected to a Strong Electric Field, *Soviet Phys. Semicond.-USSR* **1970**, *3*, 876.

[25] L. V. Keldysh, Electron-Hole Drops in Semiconductors, *Soviet Phys. Uspekhi-USSR* **1970**, *13*, 292.

[26] B. Kadomtsev, R. Sagdeev, L. V. Keldysh, I. Kobzarev, On A.A. TYAPKIN's article "Expression of General Properties of Physical Processes in Space-and-Time Metric of Special Theory of Relativity", *Uspekhi Fizicheskikh Nauk* **1972**, *106*, 660.

[27] R. Guseinov, L. V. Keldysh, Nature of Phase-Transitions under Excitonic Instability Conditions of a Crystal Electron Spectrum, *Zh. Eksp. Teor. Fiz.* **1972**, *63*, 2255.

[28] L. V. Keldysh, A. Silin, Electron-Hole Liquids in Semiconductors in Magnetic-Field, *Fiz. Tverdovo Tela* **1973**, *15*, 1532.

[29] L. V. Keldysh, A. Manenkov, V. Milyaev, G. Mikhailova, Microwave Breakdown and Exciton Condensation in Germanium, *Zh. Eksp. Teor. Fiz.* **1974**, *66*, 2178.

[30] L. V. Keldysh, S. Tikhodeev, Absorption of Ultrasound by Electron-Hole Drops in a Semiconductor, *JETP Lett.* **1975**, *21*, 273.

[31] L. V. Keldysh, A. Silin, Electron-Hole Fluid in Polar Semiconductors, *Zh. Eksp. Teor. Fiz.* **1975**, *69*, 1053.

[32] L. V. Keldysh, Phonon Wind and Dimensions of Electron-Hole Drops in Semiconductors, *JETP Lett.* **1976**, *23*, 86.

[33] L. V. Keldysh, T. Onishchenko, Electron Liquid in a Superstrong Magnetic-Field, *JETP Lett.* **1976**, *24*, 59.

[34] E. Andryushin, V. Babichenko, L. V. Keldysh, T. Onishchenko, A. Silin, Electron-Hole Liquid in Strongly Anisotropic Semiconductors and Semimetals, *JETP Lett.* **1976**, *24*, 185.

[35] V. Bagaev, L. V. Keldysh, N. Sibeldin, V. Tsvetkov, Phonon Wind Drag of Excitons and Electron-Hole Drops, *Zh. Eksp. Teor. Fiz.* **1976**, *70*, 702.

[36] V. Bagaev, N. Zamkovets, L. V. Keldysh, N. Sibeldin, V. Tsvetkov, Kinetics of Exciton Condensation in Germanium, *Zh. Eksp. Teor. Fiz.* **1976**, *70*, 1500.

[37] L. V. Keldysh, S. Tikhodeev, Ultrasound Absorption by Electron-Hole Drops in Semiconductor, *Fiz. Tverdogo Tela* **1977**, *19*, 111.

[38] E. Andrushin, L. V. Keldysh, A. Silin, Electron-Hole Liquid and Metal-Dielectric Phase-Transition in Layer Systems, *Zh. Eksp. Teor. Fiz.* **1977**, *73*, 1163.

[39] L. V. Keldysh, Metal-Dielectric Transformation Under Light Action, *Vestnik Moskovskovo Univ. Ser. 3, Fizika Astronomia* **1978**, *19*, 86.

[40] L. V. Keldysh, Coulomb Interaction in Thin Semiconductor and Semimetal Films, *JETP Lett.* **1979**, *29*, 658.

[41] L. V. Keldysh, Polaritons in Thin Semiconducting-Films, *JETP Lett.* **1979**, *30*, 224. [*Pis'ma v ZhETF* **1979**, *29*, 658].

[42] V. Bagaev, M. Bonchosmolovskii, T. Galkina, L. V. Keldysh, A. Poyarkov, Entrainment of Electron-Hole Drops by a Strain Pulse Produced as a Result of Laser Irradiation of Germanium, *JETP Lett.* **1980**, *32*, 332.

[43] E. Andriushyn, L. V. Keldysh, V. Sanina, A. Silin, Electron-Hole Liquid in Thin Semiconducting-Films, *Zh. Eksp. Teor. Fiz.* **1980**, *79*, 1509.

[44] L. V. Keldysh, A. Kechek, On the Dielectric-Constant of The Non-Polar Fluid, *Dokl. Akad. Nauk SSSR* **1981**, *259*, 575.

[45] A. Ivanov, L. V. Keldysh, The Propagation of Powerful Electromagnetic-Waves in Semiconductors under the Resonant Excitation of Excitons, *Dokl. Akad. Nauk SSSR* **1982**, *264*, 1363.

[46] A. Ivanov, L. V. Keldysh, Modification of the Polariton and Phonon-Spectra of a Semiconductor in the Presence of an Intense Electromagnetic-Wave, *Zh. Eksp. Teor. Fiz.* **1983**, *84*, 404.

[47] N. Gippius, V. Zavaritskaya, L. V. Keldysh, V. Milyaev, S. Tikhodeev, Quantum Nature Of the Reflection of an Exciton from the Surface of an Electron-Hole Drop, *JETP Lett.* **1984**, *40*, 1235.

[48] P. Elyutin, L. V. Keldysh, A. Kechek, The Resonance Dielectric Permittivity of Nonpolar Liquids, *Optika i Spektroskopia* **1984**, *57*, 282.

[49] L. V. Keldysh, S. Tikhodeev, High-Intensity Polariton Wave Near the Stimulated Scattering Threshold, *Zh. Eksp. Teor. Fiz.* **1986**, *90*, 1852.

[50] L. V. Keldysh, S. Tikhodeev, Nonstationary Mandelstam-Brillouin Scattering of an Intense Polariton Wave, *Zh. Eksp. Teor. Fiz.* **1986**, *91*, 78.

[51] N. Gippius, L. V. Keldysh, S. Tikhodeev, Mandelstam-Brilloiun Scattering of an Incoherent Polariton Wave, *Zh. Eksp. Teor. Fiz.* **1986**, *91*, 2263.

[52] L. V. Keldysh, Excitons and Polaritons in Semiconductor Insulator Quantum Wells and Superlattices, *Superlattices Microstruct.* **1988**, *4*, 637.

[53] A. Ivanov, L. V. Keldysh, V. Panashchenko, Low-Threshold Exciton-Biexciton Optical Stark-Effect in Direct-Gap Semiconductors, *Zh. Eksp. Teor. Fiz.* **1991**, *99*, 641.

[54] A. Ivanov, L. V. Keldysh, V. Panashchenko, Nonlinear Optical-Response of Interacting Excitons, *Inst. Phys.: Conf. Ser.* **1992**, *126*, 431.

[55] L. V. Keldysh, Excitonic Molecules in Nonlinear Optical-Response, *Phys. Status Solidi B* **1992**, *173*, 119.

[56] L. V. Keldysh, Coherent Excitonic Molecules, *Solid State Commun.* **1992**, *84*, 37.

[57] N. Gippius, T. Ishihara, L. V. Keldysh, E. Muljarov, S. Tikhodeev, Dielectrically Confined Excitons and Polaritons in Natural Superlattices – Perovskite Lead Iodide Semiconductors, *J. Phys. IV* **1993**, *3*, 437. 3rd International Conference on Optics of Excitons in Confined Systems, Univ Montpellier II, Montpellier, France, Aug 30–Sep 02, 1993.

[58] N. Gippius, S. Tikhodeev, L. V. Keldysh, Polaritons in Semiconductor-Insulator Superlattices with Nonlocal Excitonic Response, *Superlattices Microstruct.* **1994**, *15*, 479.

[59] L. V. Keldysh, Correlations in the Coherent Transient Electron-Hole System, *Phys. Status Solidi B* **1995**, *188*, 11. 4th International Workshop on Nonlinear Optics and Excitation Kinetics in Semiconductors (NOEKS IV), Gosen, Germany, Nov 06–10, 1994.

[60] A. Ivanov, H. Wang, J. Shah, T. Damen, L. V. Keldysh, H. Haug, L. Pfeiffer, Coherent transient in photoluminescence of excitonic molecules in GaAs quantum wells, *Phys. Rev. B* **1997**, *56*, 3941.

[61] L. V. Keldysh, Excitons in semiconductor-dielectric nanostructures, *Phys. Status Solidi A* **1997**, *164*, 3. 5th International Meeting on Optics of Excitons in Confined Systems (OECS 5), Göttingen, Germany, Aug 10–14, 1997.

[62] A. Ivanov, H. Haug, L. V. Keldysh, Optics of excitonic molecules in semiconductors and semiconductor microstructures, *Phys. Rep.* **1998**, *296*, 237.

[63] Q. Vu, H. Hang, L. V. Keldysh, Dynamics of the electron-hole correlation in femtosecond pulse excited semiconductors, *Solid State Commun.* **2000**, *115*, 63.

[64] L. V. Keldysh, Biexcitons at high densities, *Phys. Status Solidi B* **2002**, *234*, 17.

[65] F. Klappenberger, K. Renk, R. Summer, L. V. Keldysh, B. Rieder, W. Wegscheider, Electric-field-induced reversible avalanche

ADVANCED SCIENCE NEWS
www.advancedsciencenews.com

physica status solidi pss b

breakdown in a GaAs microcrystal due to cross band gap impact ionization, *Appl. Phys. Lett.* **2003**, *83*, 704.

[66] L. V. Keldysh, in *Progress in Nonequilibrium Green's functions II*. (Eds: M. Bonitz, D. Semkat). World Scientific Publ, Singapore **2003**, pp. 4–17. [reprinted as Supporting Information].

[67] J. Reithmaier, G. Sek, A. Loffler, C. Hofmann, S. Kuhn, S. Reitzenstein, L. V. Keldysh, V. Kulakovskii, T. Reinecke, A. Forchel, Strong coupling in a single quantum dot-semiconductor microcavity system, *Nature* **2004**, *432*, 197.

[68] N. Gippius, S. Tikhodeev, L. V. Keldysh, V. Kulakovskii, Hard excitation of stimulated polariton-polariton scattering in semiconductor microcavities, *Physics Uspekhi* **2005**, *48*, 306.

[69] Y. Osipov, V. Sadovnichii, V. Kozlov, O. Krokhin, N. Zefirov, E. Velikhov, G. Dobrovol'skii, L. V. Keldysh, S. Nikol'skii, Y. Tret'yakov, K. Frolov, V. Khain, E. Chazov, V. Yanin, V. Kabanov, A. Solzhenitsyn, L. Faddeev, A. Andreev, G. Chernyi, V. Lunin, G. Dobrovol'skii, D. Pushcharovskii, V. Stepin, A. Derevyanko, A. Kudelin, R. Nigmatulin, T. Oizerman, N. Dikanskii, N. Plate, V. Kostyuk, V. Urusov, Joint scientific session of the General Meeting of the Russian Academy of Sciences and the Academic Council of Moscow State University named after M.V. Lomonosov, dedicated to the 250th anniversary of Moscow State University, *Her. Russ. Acad. Sci.* **2005**, *75*, 214.

[70] G. Sek, C. Hofmann, J. Reithmaier, A. Loffler, S. Reitzenstein, M. Kamp, L. V. Keldysh, V. Kulakovskii, T. Reinecke, A. Forchel, Investigation of strong coupling between single quantum dot excitons and single photons in pillar microcavities, *Physica E* **2006**, *32*, 471. [12th International Conference on Modulated Semiconductor Structures (MSS12), Albuquerque, NM, July 10–15, 2005.]

[71] S. Reitzenstein, A. Loffler, C. Hofmann, A. Kubanek, M. Kamp, J. Reithmaier, A. Forchel, V. Kulakovskii, L. V. Keldysh, I. Ponomarev, T. Reinecke, Coherent photonic coupling of semiconductor quantum dots, *Opt. Lett.* **2006**, *31*, 1738.

[72] S. Reitzenstein, C. Hofmann, A. Loeffler, A. Kubanek, J. P. Reithmaier, M. Kamp, V. D. Kulakovskii, L. V. Keldysh, T. L. Reinecke, A. Forchel, Strong and weak coupling of single quantum dot excitons in pillar microcavities, *Phys. Status Solidi B* **2006**, *243*, 2224. [8th International Workshop on Nonlinear Optics and Excitation Kinetics In Semiconductors (NOEKS 8), Münster, Germany, February 20–24, 2006.]

[73] L. V. Keldysh, V. D. Kulakovskii, S. Reitzenstein, M. N. Makhonin, A. Forchel, Interference effects in the emission spectra of quantum dots in high-quality cavities, *JETP Lett.* **2006**, *84*, 494.

[74] S. Reitzenstein, A. Loffler, A. Kubanek, C. Hofmann, M. Kamp, J. P. Reithmaier, A. Forchel, V. D. Kulakovskii, L. V. Keldysh, I. V. Ponomarev, T. L. Reinecke, Coherent photonic coupling of semiconductor quantum dots *Opt. Lett.* **2006**, *31*, 1738; Erratum: *Opt. Lett.* **2006**, *31*, 3507.

[75] L. V. Keldysh, Dynamic Tunneling, *Her. Russ. Acad. Sci.* **2016**, *86*, 413.

[76] L. V. Keldysh, Coherent states of excitons, *Physics Uspekhi* **2017**, *60*, 1180.

[77] L. V. Keldysh, Multiphoton ionization by a very short pulse, *Physics Uspekhi* **2017**, *60*, 1187.

[78] Abbreviated ZhETF, the English translation is being published as JETP or Soviet Physics JETP.

[79] According to Google Schloar, December **2018**.

[80] A. Aleksandrov, Z. Alverov, N. Basov, E. Velikhov, A. Gonchar, A. Dynkin, L. V. Keldysh, D. Knorre, V. Kotelnikov, G. Mesyats, Y. Osipov, V. Pokrovskii, B. Saltykov, V. Subbotin, L. Faddeev, E. Chelyshev, V. Shorin, State Research Centers (Discussion in The Russian-Academy-of-Sciences), *Vestnik Rossiskoi Akademii Nauk* **1992**, 14.

[81] L. V. Keldysh, Russian Science at The Approaching Market, *Vestnik Rossiskoi Akademii Nauk* **1992**, 45.

[82] Z. Alferov, V. Ginzburg, V. Goldanskii, L. V. Keldysh, V. Maslov, A. Spirin, V. Keilisborok, Urgent Appeal for Help, *Chem. Eng. News* **1992**, *70*, 2.

[83] A. Gonchar, A. Spirin, Y. Osipov, D. Knoppe, N. Shilo, L. Faddeev, V. Sadovnichii, Z. Alferov, E. Velikhov, V. Subbotin, V. Martynov, V. Kudryavtsev, I. Makarov, E. Chelyshev, A. Prokhorov, M. Styrikovich, V. Orel, V. Sokolov, P. Simonov, L. V. Keldysh, G. Semin, A. Egorov, B. Saltykov, N. Basov, N. Laverov, What doctrine of science advancement is needed by Russia? Discussion in the RAS Presidium, *Vestnik Rossiskoi Akademii Nauk* **1996**, *66*, 16.

[84] Z. Alferov, Y. Osipov, A. Spirin, V. Subbotin, E. Velikhov, N. Laverov, G. Golitsyn, A. Gonchar, I. Makarov, A. Prokhorov, V. Sobolev, L. V. Keldysh, The Ioffe Physico-Technical Institute in the new economic conditions – Discussion in the RAS Presidium, *Vestnik Rossiskoi Akademii Nauk* **1996**, *66*, 491.

[85] V. Ginzburg, L. V. Keldysh, The age qualification in elections to the Academy of Sciences cannot be tolerated, *Vestnik Rossiskoi Akademii Nauk* **1997**, *67*, 321.

[86] A. Boyarchuk, L. V. Keldysh, From a physics laboratory to the General Physics and Astronomy Division, *Uspekhi Fizicheskikh Nauk* **1999**, *169*, 1289.

[87] F. Capasso, P. Corkum, O. Kocharovskaya, L. Pitaevskii, M. Sadovskii, Leonid Keldysh, *Phys. Today* **2017**, *70*, 75, shortened version of the present text.

[88] W. Franz, Einfluss eines elektrischen Feldes aurf eine optische Absorptionskante, *Z. Naturforsch. Teil A* **1958**, *13*, 484.

[89] S. V. Popruzhenko, Keldysh theory of strong field ionization: history, applications, difficulties and perspectives, *J. Phys. B: At. Mol. Opt. Phys.* **2014**, *47*, 204001.

[90] M. Bonitz, *Quantum Kinetic Theory*, 2 ed., Springer, Cham **2016**.

[91] P. C. Martin, J. Schwinger, Theory of many-particle systems. i, *Phys. Rev.* **1959**, *115*, 1342.

[92] L. Kadanoff, G. Baym, *Quantum Statistical Mechanics*. Benjamin, New York **1962**.

[93] P. Martin, in *Progress in Nonequilibrium Green's functions*. (Ed: M. Bonitz), World Scientific Publ, Singapore **2000**, pp. 2–16.

[94] G. Baym, in *Progress in Nonequilibrium Green's functions*. (Ed: M. Bonitz), World Scientific Publ, Singapore **2000**, pp. 17–32.

[95] A. Abrikosov, in *Progress in Nonequilibrium Green's functions II*. (Eds: M. Bonitz, D. Semkat), World Scientific Publ, Singapore **2003**, pp. 2–3.

[96] Supporting Information, online available at https://doi.org/10.1002/pssb.201800600.

[97] P. Danielewicz, Quantum theory of nonequilibrium processes II. Application to nuclear collisions, *Ann. Phys.* **1984**, *152*, 305.

[98] Here we should mention, in particular, W. Schäfer,[100,149] D. Kremp,[150,151] H. Haug,[105] and K. Henneberger in Germany and their schools.

[99] M. Bonitz, D. Semkat (Eds.), *Progress in Nonequilibrium Green's Functions*. World Scientific, Singapore **2003**.

[100] M. Bonitz, A. Filinov (Eds.), Progress in Nonequilibrium Green's Functions III, *J. Phys.: Conf. Ser.* **2006**, *35*.

[101] M. Bonitz, K. Balzer, Progress in Nonequilibrium Green's Functions IV, *J. Phys.: Conf. Ser.* **2010**, *220*, 011001.

[102] R. van Leeuwen, R. Tuovinen, M. Bonitz, Progress in Nonequilibrium Green's Functions V (PNGF V), *J. Phys.: Conf. Ser.* **2013**, *427*, 011001.

[103] C. Verdozzi, A. Wacker, C. O. Almbladh, M. Bonitz, Progress in Nonequilibrium Green's Functions (PNGF VI), *J. Phys.: Conf. Ser.* **2016**, *696*, 011001.

[104] J. Rammer, H. Smith, Quantum field-theoretical methods in transport theory of metals, *Rev. Modern Phys.* **1986**, *58*, 323.

[105] H. Haug, A. P. Jauho, *Quantum Kinetics in Transport and Optics of Semiconductors*. Springer, Berlin, Heidelberg **2008**.

[106] K. Balzer, M. Bonitz, R. van Leeuwen, A. Stan, N. E. Dahlen, Nonequilibrium Green's function approach to strongly correlated few-electron quantum dots, *Phys. Rev. B* **2009**, *79*, 245306.

ADVANCED SCIENCE NEWS
www.advancedsciencenews.com
physica status solidi pss b

[107] N. E. Dahlen, R. van Leeuwen, Solving the Kadanoff-Baym Equations for Inhomogeneous Systems: Application to Atoms and Molecules, *Phys. Rev. Lett.* **2007**, *98*, 153004.
[108] K. Balzer, S. Bauch, M. Bonitz, Time-dependent second-order Born calculations for model atoms and molecules in strong laser fields, *Phys. Rev. A* **2010**, *82*, 033427.
[109] M. Bonitz, T. Bornath, D. Kremp, M. Schlanges, W. D. Kraeft, Quantum kinetic theory for laser plasmas. dynamical screening in strong fields, *Contrib. Plasma Phys.* **1999**, *39*, 329.
[110] D. Kremp, T. Bornath, M. Bonitz, M. Schlanges, Quantum kinetic theory of plasmas in strong laser fields, *Phys. Rev. E* **1999**, *60*, 4725.
[111] B. Schenke, C. Greiner, Nonequilibrium description of dilepton production in heavy ion reactions, *J. Phys.: Conf. Ser.* **2006**, *35*, 398.
[112] H. S. Köhler, Beyond the quasi-particle picture in nuclear matter calculations using Green's function techniques, *J. Phys.: Conf. Ser.* **2006**, *35*, 384.
[113] M. Garny, M. M. Müller, Kadanoff-Baym equations with non-Gaussian initial conditions: The equilibrium limit, *Phys. Rev. D* **2009**, *80*, 085011.
[114] M. Herranen, K. Kainulainen, P. M. Rahkila, Coherent quasiparticle approximation (cQPA) and nonlocal coherence, *J. Phys.: Conf. Ser.* **2010**, *220*, 012007.
[115] N. Schlünzen, S. Hermanns, M. Bonitz, C. Verdozzi, Dynamics of strongly correlated fermions: Ab initio results for two and three dimensions, *Phys. Rev. B* **2016**, *93*, 035107.
[116] N. Schlünzen, M. Bonitz, Nonequilibrium Green Functions Approach to Strongly Correlated Fermions in Lattice Systems, *Contrib. Plasma Phys.* **2016**, *56*, 5.
[117] S. L. Chin, F. Yergeau, P. Lavigne, Tunnel ionisation of Xe in an ultra-intense CO_2 laser field (10^{14} W cm^{-2}) with multiple charge creation, *J. Phys. B: At. Mol. Phys.* **1985**, *18*, L213.
[118] D. Volkov, Über eine Klasse von Lösungen der Diracschen Gleichung [in German], *Z. Phys.* **1934**, *94*, 250.
[119] Here also a less known paper of 1972 on the coherent state of excitons should be mentioned, that was recently reprinted in Ref. [76] with comments by M. Sadovskii. There Keldysh derived what is now called the Gross-Pitaevkii equation for a coherent excitonic state in an external electromagnetic field.
[120] L. V. Keldysh, Concluding remarks, in: *Proceedings of the IXth International Conference of Semiconductor Physics*, Moscow 23–29 July 1968, Nauka, Leningrad **1969**, pp. 1303–1312.
[121] (a) L. V. Keldysh, C. D. Jeffries (eds.), *Electron–Hole Droplets in Semiconductors*, North-Holland, Amsterdam **1983**; (b) S. G. Tikhodeev, The electron-hole liquid in a semiconductor, *Soviet Phys. Uspekhi* **1985**, *28*, 1.
[122] M. Bonitz, D. Semkat, A. Filinov, V. Golubnychyi, D. Kremp, D. O. Gericke, M. S. Murillo, V. Filinov, V. Fortov, W. Hoyer, S. W. Koch, Theory and simulation of strong correlations in quantum Coulomb systems, *J. Phys. A: Math. Gen.* **2003**, *36*, 5921.
[123] Y. E. Lozovik, V. I. Yudson, Feasibility of superfluidity of paired spatially separated electrons and holes – new superconductivity mechansim, *JETP Lett.* **1975**, *22*, 274.
[124] J. Böning, A. Filinov, M. Bonitz, Crystallization of an exciton superfluid, *Phys. Rev. B* **2011**, *84*, 075130.
[125] Y. Yacoby, High-frequency Franz-Keldysh effect, *Phys. Rev.* **1968**, *169*, 610.
[126] Y. T. Rebane, Semiconductor and dielectric energy-band reconstruction in a field of intensive weak self-absorbing light waves, *Fiz. Tverd. Tela* **1985**, *27*, 1364.
[127] K. B. Nordstrom, K. Johnsen, S. J. Allen, A. P. Jauho, B. Birnir, J. Kono, T. Noda, H. Akiyama, H. Sakaki, Excitonic dynamical Franz-Keldysh effect, *Phys. Rev. Lett.* **1998**, *81*, 457.
[128] R. Landauer, Spatial varaion of currents and fields due to localized scatterers in metallic conduction, *IBM J. Res. Dev.* **1957**, *1*, 223.
[129] M. Büttiker, 4-terminal phase-coherent conductance, *Phys. Rev. Lett.* **1986**, *57*, 1761.
[130] C. Caroli, R. Combescot, P. Nozieres, D. Saint-James, Direct calculation of tunneling current, *J. Phys. C, Solid State Phys.* **1971**, *4*, 916.
[131] C. Caroli, R. Combescot, D. Lederer, P. Nozieres, D. Saint-James, Direct calculation of tunneling current. 2. Free electron description, *J. Phys. C, Solid State Phys.* **1971**, *4*, 2598.
[132] R. Combescot, Direct calculation of tunneling current. 3. Effect of localized impurity states in barrier, *J. Phys. C, Solid State Phys.* **1971**, *4*, 2611.
[133] C. Caroli, D. S. James, R. Combescot, P. Nozieres, Direct calculation of tunneling current. 4. Electron-phonon interaction effects, *J. Phys. C, Solid State Phys.* **1971**, *5*, 21.
[134] Y. Meir, N. S. Wingreen, Landauer formula for the current through an interacting electron region, *Phys. Rev. Lett.* **1992**, *68*, 2512.
[135] S. Tikhodeev, M. Natario, K. Makoshi, T. Mii, H. Ueba, Contribution to a theory of vibrational scanning tunneling spectroscopy of adsorbates. Nonequilibrium Green's function approach, *Surf. Sci.* **2001**, *493*, 63.
[136] T. Mii, S. G. Tikhodeev, H. Ueba, Spectral features of inelastic electron transport via a localized state, *Phys. Rev. B* **2003**, *68*, 205406.
[137] P. I. Arseyev, N. S. Maslova, Tunneling current induced phonon generation in nanostructures, *Pis'ma v ZhETF* **2006**, *84*, 99. [*JETP Lett.* **2006**, *84*, 93].
[138] S. G. Tikhodeev, H. Ueba, How Vibrationally Assisted Tunneling with STM Affects the Motions and Reactions of Single Adsorbates, *Phys. Rev. Lett.* **2009**, *102*, 246101.
[139] Y. E. Shchadilova, S. G. Tikhodeev, M. Paulsson, and, H. Ueba, Rotation of a Single Acetylene Molecule on Cu(001) by Tunneling Electrons in STM, *Phys. Rev. Lett.* **2013**, *111*, 186102.
[140] H. Ueba, S. G. Tikhodeev, B. N. J. Persson, in: *Current-Driven Phenomena in Nanoelectronics*. Pan Stanford Publishing Pte. Ltd, Singapore **2011**, pp. 26–89, chap. 2: Theory of inelastic tunneling current-driven motions of single adsorbates
[141] A. P. Jauho, N. S. Wingreen, Y. Meir, Time-dependent transport in interacting and noninteracting resonant-tunneling systems, *Phys. Rev. B* **1994**, *50*, 5528.
[142] G. Stefanucci, C. O. Almbladh, Time-dependent partition-free approach in resonant tunneling systems, *Phys. Rev. B* **2004**, *69*, 195318.
[143] A. Kamenev, *Field Theory of Non-Equilibrium Systems*. Cambridge University Press, **2011**.
[144] N. S. Rytova, Coulomb interaction of electrons in a thin film, *Doklady Akademii Nauk SSSR* **1965**, *163*, 1118.
[145] A. V. Rodina, A. L. Efros, Effect of Dielectric Confinement on Optical Properties of Colloidal Nanostructures, *J. Exp. Theor. Phys.* **2016**, *122*, 554.
[146] P. Werner, T. Oka, and, A. J. Millis, Diagrammatic Monte Carlo simulation of nonequilibrium systems, *Phys. Rev. B* **2009**, *79*, 035320.
[147] A. N. Rubtsov, V. V. Savkin, and, A. I. Lichtenstein, Continuous-time quantum Monte Carlo method for fermions, *Phys. Rev. B* **2005**, *72*, 035122.
[148] Special Issue: *Physics Uspekhi* **2017**, *60*, 1065.
[149] W. Schäfer, M. Wegener, *Semiconductor Optics and Transport Phenomena*. Springer, Berlin, Heidelberg **2002**.
[150] T. Bornath, W. Kraeft, R. Redmer, G. Röpke, M. Schlanges, W. Ebeling, M. Bonitz, A tribute to Dietrich Kremp, *Contrib. Plasma Phys.* **2017**, *57*, 434.
[151] D. Kremp, M. Schlanges, W. Kraeft, *Quantum Statistics of Nonideal Plasmas*. Springer, Berlin, Heidelberg **2005**.

SOVIET PHYSICS JETP VOLUME 6 (33), NUMBER 4 APRIL, 1958

BEHAVIOR OF NON-METALLIC CRYSTALS IN STRONG ELECTRIC FIELDS

L. V. KELDYSH

P. N. Lebedev Physics Institute, Academy of Sciences, U.S.S.R.

Submitted to JETP editor April 30, 1957

J. Exptl. Theoret. Phys. (U.S.S.R.) **33**, 994-1003 (October, 1957)

An expression is obtained for the number of electron-hole pairs generated in a semiconductor by a uniform electric field. The derivation is made for an arbitrary crystal. The result differs from the usually-employed formula in that it contains an explicit angular dependence and a slightly different dependence on the field. A particularly essential fact is that in the absence of electron-phonon collisions and collisions between electrons themselves the magnitude of the effective potential barrier is determined not by the width of the forbidden band, but by the lower edge of optical absorption (the internal photoeffect), which, as a rule, is considerably greater. This circumstance should lead to an essential increase in the critical fields.

THE presence of a strong electrical field ($E \sim 10^5$ v/cm), as is known, produces in semiconductors additional carriers, the number of which increases sharply with increasing field. The most probable mechanisms causing this fact are, first, shock ionization and second, direct knocking out of valence electrons by the field in the conduction band. This latter mechanism, analogous in a certain sense to cold electron emission from a metal surface, was first considered by Zener[1] in the quasi-classical approximation, which is natural for such a problem. The best expressions obtainable by this method for the number of electrons n passing into the conduction band per unit volume per unit time, is given apparently in the work by McAfee et al.[2] and has the form

$$n = N \frac{eEd}{2\pi\hbar} \exp\left\{-\frac{\pi}{2e\hbar E} \sqrt{2m^*} \Delta^{3/2}\right\}, \tag{1}$$

where N is the number of valence electrons per unit volume, d the crystal lattice period, m^* the effective mass of the electron, Δ the width of the forbidden zone, and e the electron charge.

This formula is somewhat indefinite, since the values of the effective mass of the electron in the valence band and in the conduction band are in general different and it is not clear exactly which value is contained in the exponent. In addition, expression (1) was obtained by solving the unidimensional problem. As will be shown below, the correct allowance for the three-dimensionality leads to certain quite substantial qualitative changes. Finally, a particularly important point, no account is taken in the derivation of Eq. (1) of the scattering of electrons by thermal lattice vibrations, which, as will be shown below, is of decisive significance for this problem.

We shall calculate in this work the probability of the passage of a valence electron into the conduction band by a method already used for this purpose by Houston[3] and representing essentially a method commonly used in perturbation theory to calculate the transition probability per unit time. The entire analysis will be carried out in the so-called single-electron approximation, i.e., the interaction between electrons will be disregarded, with the exception of that portion of the interaction included in the general self-consistent field of the crystal.

The Hamiltonian of the system has in this case the following form:

$$\hat{H} = \hat{H}_{0e} + \hat{H}_{0L} + \hat{H}_{eL} + e\mathbf{Er}; \tag{2}$$

$$\hat{H}_{0e} = \frac{1}{2m}\left(\frac{\hbar}{i}\nabla\right)^2 + W(\mathbf{r}); \quad \hat{H}_{0e}\psi_{0j}(\mathbf{p}, \mathbf{r}) = \varepsilon_j(\mathbf{p})\psi_{0j}(\mathbf{p}, \mathbf{r}); \tag{3}$$

$$\hat{H}_{0L} = \frac{1}{2}\sum_{|\mathbf{k}| \leqslant k_m} \hbar\omega_{\mathbf{k}}(b_{\mathbf{k}}^{+}b_{\mathbf{k}} + b_{\mathbf{k}}b_{\mathbf{k}}^{+}); \tag{4}$$

$$\hat{H}_{eL} = i\sqrt{\frac{\nu}{VN}} \sum_{|\mathbf{k}|<k_m} \alpha(\mathbf{k}) \{b_{\mathbf{k}}^{+} e^{-i\mathbf{k}\mathbf{r}} - b_{\mathbf{k}} e^{i\mathbf{k}\mathbf{r}}\}. \quad (5)$$

In these formulas $\mathbf{r}$ is the electron radius vector, $W(\mathbf{r})$ is the periodic potential of the crystal, $\mathbf{p}$ and $\mathbf{k}$ are the electron quasi-momentum and the phonon wave vector, $\omega_{\mathbf{k}}$ and $\epsilon_j(\mathbf{p})$ are functions that determine the dependence of the phonon frequency on the wave vector and of the electron energy on the quasi-momentum in the band with index j, $b_{\mathbf{k}}^{+}$ and $b_{\mathbf{k}}$ are the creation and annihilation operators of a phonon with wave vector $\mathbf{k}$, V is the normalizing volume, ν the number of valence bands, and finally $\psi_{0j}(\mathbf{p}, \mathbf{r})$ the electron wave function in the following Bloch form

$$\psi_{0j}(\mathbf{p}, \mathbf{r}) = \exp(i\mathbf{p}\mathbf{r}/\hbar)\, u_j(\mathbf{p}, \mathbf{r}), \quad (6)$$

where $u_j(\mathbf{p}, \mathbf{r})$ is the periodic solution of the equation

$$-\frac{\hbar^2}{2m}\nabla^2 u_j(\mathbf{p}, \mathbf{r}) - \frac{i\hbar}{m}\mathbf{p}\nabla u_j(\mathbf{p}, \mathbf{r}) + \left\{W(\mathbf{r}) + \frac{p^2}{2m} - \varepsilon_j(\mathbf{p})\right\} u_j(\mathbf{p}, \mathbf{r}) = 0. \quad (7)$$

The interaction Hamiltonian (5) should generally speaking also contain functions of the Bloch type instead of the plane waves. However, in the calculation of the matrix elements for transitions between states (6) allowance for this circumstance gives only a corrective factor on the order of unity, which is insignificant, since we do not specify the form of $W(\mathbf{r})$ anyhow, nor do we consequently specify $\psi_{0j}(\mathbf{p}, \mathbf{r})$.

No assumptions whatever are made concerning the form of the function $\alpha(\mathbf{k})$, with the exception of the obvious property $\alpha(\mathbf{k}) \sim k^{1/2}$ for small k.

It is well known that the presence of a homogeneous electric field $\mathbf{E}$ leads to a uniform increase in the quasi-momentum of the electron with time in accordance with the law $\mathbf{p} = \mathbf{p}_0 - e\mathbf{E}t$. In other words, if at the instant $t = 0$ the electron is described by a wave function $\psi_{0j}(\mathbf{p}_0, \mathbf{r})$ then in subsequent instants its wave function will have in the zeroth approximation the form

$$\psi_j(\mathbf{p}_0, \mathbf{r}, t) = \exp\left\{-\frac{i}{\hbar}\int_0^t \varepsilon_j(\mathbf{p}_0 - e\mathbf{E}x)\, dx\right\} \psi_{0j}(\mathbf{p}_0 - e\mathbf{E}t, \mathbf{r}). \quad (8)$$

By virtue the periodic dependence (with the period of the reciprocal lattice) of $\epsilon_j(\mathbf{p})$ and $\psi_{0j}(\mathbf{p}, \mathbf{r})$ on the quasi-momentum $\mathbf{p}$, a uniform increase in the latter means that the electron vibrates within the confines of a single band, if the field is directed along one of the principal crystallographic axes (with a period $2\pi\hbar/eEd$ for the case of a simple cubic lattice). If the field is not aligned with any of the reciprocal-lattice vectors, the motion can have a complex aperiodic character, but is still confined to the same band. The exact wave function should, naturally, contain also terms connected with the transitions into other bands, but these terms will obviously be small. It is therefore natural to seek a solution of the Schrödinger equation

$$i\hbar\, \partial\Psi/\partial t = \hat{H}\Psi \quad (9)$$

in the form of a superposition of products of functions of the form (8) (which already include the fundamental effect of the field — uniform acceleration) and of the phonon wave functions

$$\Psi = \sum_{j, \mathbf{p}_0, [N_{\mathbf{k}}]} c_j([N_{\mathbf{k}}], \mathbf{p}_0, t) \exp\left\{e\mathbf{E}\int_0^t \gamma_j(\mathbf{p}_0 - e\mathbf{E}x)\, dx\right\} \psi_j(\mathbf{p}_0, \mathbf{r}, t) \prod_{\mathbf{k}} (b_{\mathbf{k}}^{+})^{N_{\mathbf{k}}} \Phi_0. \quad (10)$$

The symbol $[N_{\mathbf{k}}]$ denotes the set of occupation numbers $N_{\mathbf{k}}$ corresponding to all possible values of $\mathbf{k}$; Φ_0 is the wave function of the lowest energy state of the lattice; summation over $\mathbf{p}_0$ is carried out over all physically-different states, i.e., over the volume of the first Brillouin zone

$$\gamma_j(\mathbf{p}) = \int_{\Omega_0} u_j^{*}(\mathbf{p}, \mathbf{r})\, \mathrm{grad}_{\mathbf{p}} u_j(\mathbf{p}, \mathbf{r})\, d\tau. \quad (11)$$

The integral in (11) is taken over the volume of the elementary cell. The quantity (11) is pure imaginary, since by virtue of the condition

$$\int_{\Omega_0} u_j^{*}(\mathbf{p}, \mathbf{r})\, u_j(\mathbf{p}, \mathbf{r})\, d\tau = 1 \text{ we have } \mathrm{Re}\,\gamma_j(\mathbf{p}) = 0.$$

The advisability of separating out the exponential factor in (10) will be seen from the following. Substituting this expansion into (9), we obtain in the usual manner a system of equations for the coefficients:

$$i\hbar\frac{\partial c_j([N_{\mathbf{k}}],\mathbf{p}_0,t)}{\partial t}=i\hbar\sum_{j'\neq j}\frac{\mathbf{E}\mathbf{J}_{jj'}(\mathbf{p}_0-e\mathbf{E}t)}{\varepsilon_j(\mathbf{p}_0-e\mathbf{E}t)-\varepsilon_{j'}(\mathbf{p}_0-e\mathbf{E}t)}c_{j'}([N_{\mathbf{k}}],\mathbf{p}_0,t)\,Q^0_{jj'}(\mathbf{p}_0,t)+i\sqrt{\frac{\nu}{VN}}\sum_{j',\,|\mathbf{k}'|\leqslant k_m}$$

$$\times\{M_{jj'}(\mathbf{p}_0-e\mathbf{E}t,\ \hbar\mathbf{k}')\sqrt{N_{\mathbf{k}'}}\,c_{j'}([N_{\mathbf{k}}]-1_{\mathbf{k}'},\ \mathbf{p}_0+\hbar\mathbf{k}',\ t)\,Q^+_{jj'}(\mathbf{p}_0,\ \mathbf{k}',\ t)$$

$$-M_{jj'}(\mathbf{p}_0-e\mathbf{E}t,\ -\hbar\mathbf{k}')\sqrt{N_{\mathbf{k}'}+1}\,c_{j'}([N_{\mathbf{k}}]+1_{\mathbf{k}'},\ \mathbf{p}_0-\hbar\mathbf{k}',\ t)\,Q^-_{jj'}(\mathbf{p}_0,\ \mathbf{k}',\ t)\}, \quad (12)$$

where

$$M_{jj'}(\mathbf{p},\ \hbar\mathbf{k})=\alpha(\mathbf{k})\int_{\Omega_0}u^*_j(\mathbf{p},\ \mathbf{r})\,u_{j'}(\mathbf{p}+\hbar\mathbf{k},\ \mathbf{r})\,d\tau,\quad \mathbf{J}_{jj'}(\mathbf{p})=\frac{ie\hbar}{2m}\int_{\Omega_0}\{u_{j'}(\mathbf{p},\ \mathbf{r})\nabla u^*_j(\mathbf{p},\ \mathbf{r})-u^*_j(\mathbf{p},\ \mathbf{r})\nabla u_{j'}(\mathbf{p},\ \mathbf{r})\}\,d\tau,$$

$$Q^{\pm}_{jj'}(\mathbf{p},\ \mathbf{k},\ t)=\exp\Big\{\frac{i}{\hbar}\int_0^t[\varepsilon_j(\mathbf{p}-e\mathbf{E}x)-\varepsilon_{j'}(\mathbf{p}-e\mathbf{E}x\pm\hbar\mathbf{k})\pm\hbar\omega_{\mathbf{k}}]\,dx$$

$$-e\mathbf{E}\int_0^t[\gamma_j(\mathbf{p}-e\mathbf{E}x)-\gamma_{j'}(\mathbf{p}-e\mathbf{E}x\pm\hbar\mathbf{k})]\,dx\Big\},\quad Q^0_{jj'}(\mathbf{p},\ t)\equiv Q^+_{jj'}(\mathbf{p},\ 0,\ t).$$

In the derivation of (12) use is made of the identity

$$\mathbf{J}_{jj'}(\mathbf{p})/e\{\varepsilon_j(\mathbf{p})-\varepsilon_{j'}(\mathbf{p})\}=\int_{\Omega_0}u^*_j(\mathbf{p},\ \mathbf{r})\,\mathrm{grad}_{\mathbf{p}}u_{j'}(\mathbf{p},\ \mathbf{r})\,d\tau.$$

The diagonal term in the first sum over j' drops out virtue of the presence of an exponential factor in the expansion (10). The symbols $[N_{\mathbf{k}}]\pm 1_{\mathbf{k}'}$ denote that in the set of numbers $[N_{\mathbf{k}}]$ the quantity $N_{\mathbf{k}'}$ is replaced by $N_{\mathbf{k}'}\pm 1$, and all the remaining quantities remain the same.

To obtain an expression analogous to (1) we must discard from the system of equations (12) those terms containing collisions between electrons and phonons. Taking it into account that at $t = 0$

$$c_j(\mathbf{p},\ 0)=\delta_{jV}$$

(the indices V and c are necessary in what follows to denote quantities pertaining to the valence and conduction bands respectively), introducing by way of a new variable the vector $\mathbf{p} = \mathbf{p}_0 - e\mathbf{E}t$, and resolving this vector into a component $p_{\parallel}$ parallel to the field and a vector $\mathbf{p}_{\perp}$ perpendicular to the field, we obtain

$$c_c(\mathbf{p}_0,\ t)=\int_{p_{0\parallel}}^{p_{0\parallel}-eEt}\frac{\mathbf{E}\mathbf{J}_{Vc}(\mathbf{p})}{\varepsilon_c(\mathbf{p})-\varepsilon_V(\mathbf{p})}\exp\Big\{i\int_{p_{0\parallel}}^{p_{\parallel}}[\varepsilon_c(\mathbf{p}')-\varepsilon_V(\mathbf{p}')]\frac{dp'_{\parallel}}{e\hbar E}$$

$$+\int_{p_{0\parallel}}^{p_{\parallel}}\mathbf{n}\gamma_{Vc}(\mathbf{p}')\,dp'_{\parallel}\Big\}\frac{dp_{\parallel}}{eE},\quad \gamma_{Vc}(\mathbf{p})\equiv\gamma_c(\mathbf{p})-\gamma_V(\mathbf{p}). \quad (13)$$

Here $\mathbf{n}$ is a unit vector in the direction of the field.

For simplicity and clarity let us first calculate the integral (13) for the simplest case, when the field is directed along one of the principal crystalline axes of a simple cubic lattice with a period d, and then generalize the results to include the case of a lattice of any symmetry and an arbitrarily oriented field.

In this particular case, as indicated above, the motion of the electron in the band is periodic with a period $2\pi\hbar/eEd$. A natural characteristic of the infiltration is therefore the probability of passing through the conduction band during one period

$$D_0(\mathbf{p}_{\perp})=\Big|\int_{-\pi\hbar/d}^{\pi\hbar/d}\frac{\mathbf{n}\mathbf{J}_{Vc}(\mathbf{p})}{e[\varepsilon_c(\mathbf{p})-\varepsilon_V(\mathbf{p})]}\exp\Big\{i\int_{p_{0\parallel}}^{p_{\parallel}}\frac{\varepsilon_c(\mathbf{p}')-\varepsilon_V(\mathbf{p}')}{e\hbar E}\,dp'_{\parallel}+\int_{p_{0\parallel}}^{p_{\parallel}}\mathbf{n}\gamma_{Vc}(\mathbf{p}')\,dp'_{\parallel}\Big\}\,dp_{\parallel}\Big|^2. \quad (14)$$

At a fixed value of $\mathbf{p}_{\perp}$ the functions $\epsilon_c(\mathbf{p})$ and $\epsilon_V(\mathbf{p})$ are different branches of the same infinitely-valued analytic function $\epsilon(\mathbf{p})$ of complex variable $p_{\parallel}$, since they represent different roots of the same eigenvalue problem. Since these functions are close to each other on the real axis, there should be located somewhere near the real axis in the complex plane a branch point $p_{\parallel} = q$, in which

$$\varepsilon_c(q)=\varepsilon_V(q).$$

Obviously, q depends on $p_\perp$. In the vicinity of the point q the branching is into two bands and $\epsilon_c(p_\parallel) - \epsilon_V(p_\parallel) \sim \sqrt{p_\parallel - q}$. The possibility of branching into a larger number of bands,* as well as the multiplicity of the inverse function $p_\parallel(\epsilon)$ in the vicinity of the point q (which corresponds to $\epsilon_c(p_\parallel) - \epsilon_V(p_\parallel) \sim (p_\parallel - q)^{(2n+1)/2}$) would appear only accidently, under particular selection of the potential $W(\mathbf{r})$, and will therefore not be considered below.

By using the general properties of solutions of second-order differential equations it is possible to show (see Supplement) that in the vicinity of the point q the quantity $n\gamma_{Vc}(p_\parallel)$ behaves as $(p_\parallel - q)^{-1/2}$ and the factor ahead of the exponent in (14) has at the same point a simple pole with a universal value of the residue $i/4$. The latter circumstance makes it possible to calculate the integral in Eq. (14). For this purpose we introduce into (14) a new variable $y = y(p_\parallel)$, satisfying the following conditions: (1) $dy(p_\parallel)/dp_\parallel$ is real and positive for all real $p_\parallel$ and has the same period as the reciprocal lattice; (2) $dy(p_\parallel)/dp_\parallel$ behaves in the vicinity of the point $p_\parallel = q$ as $(p_\parallel - q)^{-1/2}$ and has neither zeros nor singularities lying closer to the real axis than q.

It is easy to check that the integrand in (14) is single-valued in the vicinity of the point $y = y(q)$ and has at this point a pole with residue $i/2$. An example of such a function is the integral

$$\int_0^{p_\parallel} \frac{dp'_\parallel}{\varepsilon_c(\mathbf{p}) - \varepsilon_V(\mathbf{p})}.$$

Shifting the contour of integration in (14) to the upper half of the y plane, we obtain

$$D_0(\mathbf{p}_\perp) = \pi^2 \left| \exp\left\{\int_0^q \left[i\,\frac{\varepsilon_c(\mathbf{p}) - \varepsilon_V(\mathbf{p})}{e\hbar E} + n\gamma_{Vc}(\mathbf{p})\right] dp_\parallel\right\}\right|^2 \qquad (15)$$

with accuracy to terms that are exponentially small compared with fundamental term. Equation (15) discards, in addition, terms connected with the limits of integration, and therefore not increasing with time.†

The probability of passage after l periods is,

$$D_l(\mathbf{p}_\perp) = D_0(\mathbf{p}_\perp)\left|\sum_{r=0}^{l} e^{irs_0}\right|^2 = \frac{\sin^2[(l+1)s_0/2]}{\sin^2(s_0/2)} D_0(\mathbf{p}_\perp),$$

where

$$s_0(\mathbf{p}_\perp) = \int_{-\pi\hbar/d}^{\pi\hbar/d} \left\{i\,\frac{\varepsilon_c(\mathbf{p}) - \varepsilon_V(\mathbf{p})}{e\hbar E} + n\gamma_{Vc}(\mathbf{p})\right\} dp_\parallel$$

is a rapidly oscillating function of the field ($s_0 \sim 1/E$).

However, taking it into account that s_0 is a function of $\mathbf{p}_\perp$ and averaging over the narrow region $\Delta\mathbf{p}_\perp$, we obtain

$$D_l(\mathbf{p}_\perp) = lD_0(\mathbf{p}_\perp). \qquad (16)$$

Equations (14)—(16) were obtained, as already remarked, for a field directed along one of the principal crystallographic axes of a simple cubic lattice, when the motion of the electron has a simple peri-

*A distinction must be made between the branching of $\epsilon(p_\parallel)$ for fixed $p_\perp$, considered here, and the frequently-encountered band degeneracy by virtue of the crystal symmetry. In the latter case the equivalent states belonging to the various zones correspond to different directions of the quasi-momentum, i.e., to different $p_\perp$.

†A formula analogous to (14) was also used by Franz.[4] However, he did not take into account thereafter the presence of a pole in the integrand, and as a result his final results differed strongly from (15) and the equations that follow. These include the value of the factor ahead of the exponent at the point q, which as shown above, is infinite. If this factor is nevertheless replaced by the most sensible value $\sim d/2\pi\hbar$, the resultant values are several orders of magnitude smaller than (15), and contain a factor ahead of the exponent that is dependent on E ($\sim E^{4/3}$).

odic character. In general, however, these cannot be used, since the motion is generally speaking aperiodic and the probability of infiltration changes from period to period. But this circumstance, which is significant in the calculation of the probability of passage of each electron, does not play any role for the total effect when the valence electrons fully fill the band, since the place of the departing electron is taken during the next period by another electron, which will infiltrate with the same probability. Still another difference lies in the fact that the time interval, to which formula (15) pertains, is no longer equal to $2\pi\hbar/eEd$, but is determined by the length, divided by eE, of the segment of the straight line lying within the first Brillouin zone parallel to the field and passing through point p_0, viz.: $T(p_0) = \Delta p_{\parallel}(p_0)/eE$. But even this fact is of no significance for the summary effect, because all the electrons having initial states along this segment have an equal probability of passage, and their number is proportional to $\Delta p_{\parallel}(p_0)/2\pi\hbar$. The number of electrons passing in the conduction band per unit time is proportional to $\Delta p_{\parallel}(p_0)/T(p_0)2\pi\hbar = eE/2\pi\hbar$ and is independent of $\Delta p_{\parallel}$, i.e., of the form of the Brillouin band. The equations that will be obtained below are therefore valid for lattices of all symmetries and for all field orientations.

The total number of electrons passing in the conduction band per unit time and per unit volume is given by the expression

$$n = \frac{eE}{2\pi\hbar}\int D_0(p_\perp)\frac{d^2p_\perp}{(2\pi\hbar)^2} = \pi^2\frac{eE}{2\pi\hbar}\int\left|\exp\left\{\int\limits_0^{q(p_\perp)}\left[i\,\frac{\varepsilon_c(p)-\varepsilon_V(p)}{eE\hbar}+n\gamma_{Vc}(p)\right]dp_\parallel\right\}\right|^2\frac{d^2p_\perp}{(2\pi\hbar)^2}. \quad (17)$$

The principal contribution to the integral (17) is made by the narrow region near that value of $p_\perp$, which corresponds to the minimum of the function $\epsilon_c(p) - \epsilon_V(p)$. Actually, for all specified values of $p_\perp$ this function reaches a certain relative minimum $\epsilon_{min}(p_\perp)$ as $p_\parallel$ is varied. The greater $\epsilon_{min}(p_\perp)$, the farther is the branch point $q(p_\perp)$ from the real axis. At values of $\epsilon_{min}(p_\perp)$ that are not too large, $q(p_\perp) \sim \epsilon_{min}^{1/2}(p_\perp)$. The exponent therefore contains a large negative quantity, proportional to $\epsilon_{min}^{3/2}(p_\perp)$ and consequently, the integrand has a sharp maximum in the region where the function $\epsilon_c(p) - \epsilon_V(p)$ has an absolute minimum.

In the vicinity of this point we cannot restrict ourselves to an ordinary quadratic expansion, and must use a somewhat more accurate expression, taking into account the presence of a branch-point surface.

$$\varepsilon_c(p) - \varepsilon_V(p) = \varepsilon_0\Big[1 + \sum_{i,k}(p_i - p_{0i})(p_k - p_{0k})/m_{ik}\varepsilon_0\Big]^{1/2}. \quad (18)$$

On the other hand, the quantity

$$\int\limits_0^{q(p_\perp)} n\gamma_{Vc}(p)\,dp_\parallel$$

can be considered constant within this region with a sufficient degree of accuracy, $-in\gamma_0 Vc \lesssim 1$.

Elementary integration leads then to the following final result:

$$n = 2\pi\nu\left(\frac{eE}{2\pi\hbar}\right)^2\left(\frac{m_1m_2m_3}{m_\parallel^3}\frac{m_\parallel}{\varepsilon_0}\right)^{1/2}\exp\left\{-\frac{\pi}{2e\hbar E}\sqrt{m_\parallel}\,\varepsilon_0^{3/2} + n\gamma_{0Vc}\right\}, \quad (19)$$

where $m_\parallel^{-1} = \sum\limits_i [(\cos^2\gamma_i)/m_i]$, m_i^{-1} are the principal values of the tensor m_{ik}^{-1} [see (18)], and γ_i are the angles between the directions of the field and the principal axes of this tensor, which in general do not coincide with the principal crystal axes.

This formula can be presented in a clearer form if one introduces the "average lattice period" d using the equation $d^3 = \Omega_0 = \nu/N$, where Ω_0 is the volume of the elementary cell. Then

$$n = N\frac{\pi}{2}\frac{eEd}{2\pi\hbar}\left(\frac{m_1m_2m_3}{m_\parallel^3}\right)^{1/2}\frac{eEd}{(\pi\hbar/d)(\varepsilon_0/m_\parallel)^{1/2}}\exp\left\{-\frac{\pi}{2e\hbar E}\sqrt{m_\parallel}\,\varepsilon_0^{3/2} + n\gamma_{0Vc}\right\}. \quad (20)$$

This expression differs from (1) in that the factor in front of the exponent is dependent on the field. The exponential term contains an explicit angular dependence, which can appear also in crystals of cubic symmetry in the presence of degeneracy of the valence band or of the conduction band.

A more substantial difference lies, however, in the meaning that must be ascribed to the quantities ϵ_0 and $m_{\|}$. As can be seen from the above derivation, ϵ_0 is the width of the forbidden band and m_{ik} is the effective mass of the electron (more accurately, the reduced effective mass of the electron and hole) only in that case when the highest states of the valence band and the lowest state of the conduction band correspond to the same value of the quasi-momentum. In practice this never happens, and consequently ϵ_0, coinciding with the red boundary of light absorption for a given crystal, is always substantially greater than the width of the forbidden band. By virtue of this, the values of the critical fields, corresponding to a noticeable infiltration, should be considerably greater than those usually expected on the basis of Eq. (1).

If, however, there exists some interaction that changes the quasi-momentum of the electron, a transition is possible from the highest state of the valence band to the lowest state of the conduction band.

Such interactions may be collisions between electrons and electrons or between electrons and phonons or impurity atoms. The field dependence of the probability of an infiltration involving these processes is given by the same exponential factor as in (19), except that the width of the forbidden band enters in place of ϵ_0. In view of the very strong dependence of (19) on ϵ_0, the role of these processes can turn out to be decisive.

Let us note also that while formula (17) is quite rigorous, formulas (19) and (20) are obtained under the assumption that ϵ_0 is small compared with the widths of the valence and the conduction bands, and therefore expansion (18) is valid everywhere within the confines of the forbidden band. For the opposite case, corresponding to the appproximation of strongly bound electrons, it is also possible to obtain a relatively simple expression. In this approximation, as is known,[5] it is possible to retain only the first terms in the Fourier expansion

$$\varepsilon_c(\mathbf{p}) - \varepsilon_V(\mathbf{p}) = I_0\left[1 + \sum_{\mathbf{g}} \alpha_{\mathbf{g}} e^{i\mathbf{pg}/\hbar}\right], \tag{21}$$

where $\mathbf{g}$ are the vectors of the crystal lattice. In this case $I_0 \approx \epsilon_0 \approx \Delta$ and $\alpha_{\mathbf{g}} \ll 1$. Then

$$n \sim N \frac{eEd}{2\pi\hbar} \frac{eEd}{I_0} \exp\left\{-\frac{I_0}{eEd}\left[q_0 - \sum_{\mathbf{g}} \frac{\alpha_{\mathbf{g}}\hbar}{\mathbf{ng}}\left(\exp\left\{\frac{i\mathbf{p}_m - q_0\mathbf{n}}{\hbar}\mathbf{g}\right\} - \exp\left\{i\frac{\mathbf{p}_m\mathbf{g}}{\hbar}\right\}\right)\right]\right\}, \tag{22}$$

where q_0 is determined by

$$\sum_{\mathbf{g}} \alpha_{\mathbf{g}} \exp\left\{\frac{i\mathbf{p}_m - q_0\mathbf{n}}{\hbar}\mathbf{g}\right\} + 1 = 0.$$

It is easy to verify that the fundamental term in the exponent is always of the order $-(I_0/eEd)\ln(1/\alpha)$, where α is the ratio of the widths of the allowable and forbidden bands and consequently, the transmission coefficient diminishes somewhat slower with increasing ϵ_0 than called for formula (19). In this approximation, the exponential factor (22) is close to that obtained by Feuer[6] for the unidimensional case.

It must be emphasized that the entire above analysis of the problem of production of electron-hole pairs by the electric field starts out with the far-reaching assumptions on which the band theory of solids is based. It is assumed, in particular, that both the electron states (states of the conduction band) and hole states (valence band) can be obtained by solving a certain single-electron problem with a specified periodic potential, and that the electrons and holes created do not interact with each other. As pointed to the author by L. D. Landau, it would be logical to consider the electrons and holes as two different branches of the excitation spectrum of the crystal, without making any further detailed assumption concerning their nature, and also to take it into account that they are created not free, but interact with each other in accordance with the Coulomb law, which may lead to certain changes in the factor in front of the exponent in (19).

It can be shown that such an analysis will lead in practice to the above results if the Coulomb interaction between the electron and hole are disregarded. There are grounds for hoping that allowance for the latter does not change strongly the final derivations, since the value of the Born parameter $e^2/\mu\hbar v$ (where μ is the dielectric constant and v the relative velocity of the electron and hole), a parameter characteristic of this problem, is of the same order of magnitude as the square root of the ratio of the electron and hole bound-state energy to the width of the forbidden band, and is always small.

In conclusion I thank Professor V. L. Ginzburg and Academician L. D. Landau who made many valuable comments when evaluating the results of this work.

SUPPLEMENT

Let us investigate the behavior of the function

$$J_{cV}(\mathbf{p})/e\,[\varepsilon_c(\mathbf{p})-\varepsilon_V(\mathbf{p})]\equiv\int_{\Omega_0} u_c^*(\mathbf{p},\mathbf{r})\,\mathrm{grad}_p u_V(\mathbf{p},\mathbf{r})\,d\tau. \tag{1S}$$

in the complex plane. For simplicity all further arguments will concern the unidimensional case. Equation (1S), which takes place on the real axis, cannot be extended to the entire complex plane, since its right half contains an essentially non-analytic operation, complex conjugation. We can, however, use the circumstance that on the real axis $u_j^*(p, x) = u_j(-p, x)$ and rewrite (1S) as

$$\frac{J_{cV}(p)}{e\,[\varepsilon_c(p)-\varepsilon_V(p)]}\equiv\int_{\Omega_0} u_c(-p, x)\frac{\partial u_V(p, x)}{\partial p}\,dx. \tag{1'S}$$

In such a form it is possible to continue this equation analytically into the region of complex values of p. Analogously, an analytic continuation of the orthogonality and normalization conditions of the functions $u_j(p, x)$ in the p plane leads to the equation

$$\int_{\Omega_0} u_j(-p, x)\,u_{j'}(p, x)\,dx=\delta_{jj'}. \tag{2S}$$

To investigate the properties of (1'S) in the vicinity of the energy branch point q it is advisable to change to a new independent (generally speaking, complex) variable, the energy ϵ. Corresponding to each value of ϵ are two values of p of opposite sign, and consequently two functions $u_j(\pm p(\epsilon), x) = u_j^{\pm}(\epsilon, x)$. The function $p(\epsilon)$ is regular in the vicinity of the point q. The index indicating the number of the zone can be omitted hereinafter, since the states of the different zones correspond to different energies. On the other hand, at the point where the energies of both zones coincide, i.e., at the branch point, the functions $u_c^{\pm}(\epsilon, x)$ and $u_V^{\pm}(\epsilon, x)$ tend to values that differ only by a factor that is independent of x. Actually, were this not so, the initial Schrödinger equation would have at this energy value four linearly-independent solutions, an impossibility for a second-order equation.

If we detour the branch point ϵ_q in the ϵ plane, then by virtue of the uniqueness of the solution of the differential equation with the given boundary condition (i.e., with p specified), we should obtain a function that differs from the original one only by a multiplying factor (which in general depends on ϵ). Consequently, $u^{\pm}(\epsilon, x)$ can be represented in the form $C^{\pm}(\epsilon)v^{\pm}(\epsilon, x)$ where the $v^{\pm}(\epsilon, x)$ are unique functions of ϵ in the vicinity of the point ϵ_q. Furthermore, they can be considered regular at this point, for if they had, for example, a pole of the mth order, the factor $(\epsilon-\epsilon_q)^{-m}$ could be taken out and included in $C^{\pm}(\epsilon)$.

Let us denote by ϵ_1 and ϵ_2 the two values of energy, corresponding to the same value of the quasi-momentum. As p approaches q, both these quantities tend to ϵ_q, and then, to accuracy within terms of higher order, we have $\epsilon_2-\epsilon_q=\epsilon_q-\epsilon_1$. Then

$$v^{\pm}(\varepsilon_2, x)=v^{\pm}(\varepsilon_1, x)+2(\varepsilon_q-\varepsilon_1)\,\partial v^{\pm}(\varepsilon_1, x)/\partial\varepsilon+\ldots. \tag{3S}$$

Let us insert this expansion into the normalization condition for $v^{\pm}(\epsilon_2, x)$ and take account at the same time of the orthogonality of $v^{+}(\epsilon_2, x)$ and $v^{-}(\epsilon_1, x)$

$$1=C^+(\varepsilon_2)C^-(\varepsilon_2)\int_{\Omega_0} v^-(\varepsilon_2, x)v^+(\varepsilon_2, x)\,dx=C^+(\varepsilon_2)C^-(\varepsilon_2)\,2(\varepsilon_q-\varepsilon_1)\int_{\Omega_0} v^-(\varepsilon_2, x)\frac{\partial v^+(\varepsilon_1, x)}{\partial\varepsilon}dx+\ldots. \tag{4S}$$

Let us now compare the right half of (4S) with the quantity of interest to us

$$\int_{\Omega_0} u_c(-p, x)\,\frac{\partial u_V(p, x)}{\partial p}dx=C^-(\varepsilon_2)\,C^+(\varepsilon_1)\frac{d\varepsilon_1}{dp}\int_{\Omega_0} v^-(\varepsilon_2, x)\frac{\partial v^+(\varepsilon_1, x)}{\partial\varepsilon}dx. \tag{5S}$$

As a result we obtain the relation

$$\int_{\Omega_0} u_c(-p, x)\,\frac{\partial u_V(p, x)}{\partial p}dx=\frac{d\varepsilon_V(p)}{dp}\frac{C^+(\varepsilon_V)}{C^+(\varepsilon_c)}\frac{1}{2(\varepsilon_q-\varepsilon_V)}=\frac{C^+(\varepsilon_V)}{C^+(\varepsilon_c)}\frac{1}{4(p-q)}. \tag{6S}$$

It is obviously always possible to choose the functions so as to obtain on the real axis $C^+(\epsilon) = C^-(\epsilon)$. Then this equality is always retained, and in the vicinity of the branch point it follows from this equality that $C^+(\epsilon) = C^-(\epsilon) \sim (\epsilon - \epsilon_q)^{-1/2}$ and, consequently

$$\lim_{p\to q} \frac{C^+(\varepsilon_V)}{C^+(\varepsilon_c)} = \pm i.$$

At first glance it may appear that this deduction is not unique, since $C^{\pm}(\epsilon)$ are determined on the real axis with an accuracy to within an arbitrary phase factor. However, strictly speaking, we are interested not in the quantity (1S), but in its products by the factor

$$\exp\left\{\int_0^p \gamma_{Vc}(p')\,dp'\right\},$$

which is contained in the intergrand of (14) and which, as can be readily verified from the definition of $\gamma_{Vc}(p)$, is in general independent of this phase factor.

We thus have near the branch point

$$\int_{\Omega_0} u_c(-p, x) \frac{\partial u_V(p, x)}{\partial p} dx = \pm \frac{i}{4(p-q)} + \ldots \qquad (7S)$$

From the definition of the quantity $\gamma_{Vc}(p)$ it is seen that the expansion in powers of $\epsilon - \epsilon_q$ contains no even terms, and therefore the expansion begins with the term $(\epsilon - \epsilon_q)^{-1} \sim (p-q)^{-1/2}$.

[1] C. Zener, Proc. Roy. Soc. A145, 523 (1934).

[2] McAfee, Ryder, Shockley, and Sparks, Phys. Rev. 83, 650 (1951). See also Vol'kenshtein, J. Tech. Phys. (U.S.S.R.) 9, 171 (1939).

[3] W. V. Houston, Phys. Rev. 57, 184 (1940).

[4] W. Franz and L. Tewordt, Halbeiterprobleme 3, 1 (1956).

[5] R. E. Peierls, Quantum Theory of Solids, Oxford, 1955.

[6] P. Feuer, Phys. Rev. 88, 92 (1952).

Translated by J. G. Adashko
197

SOVIET PHYSICS JETP VOLUME 34(7), NUMBER 5 NOVEMBER, 1958

THE EFFECT OF A STRONG ELECTRIC FIELD ON THE OPTICAL PROPERTIES OF INSULATING CRYSTALS

L. V. KELDYSH

P. N. Lebedev Physics Institute, Academy of Sciences, U.S.S.R.

Submitted to JETP editor November 1, 1957; resubmitted February 20, 1958

J. Exptl. Theoret. Phys. (U.S.S.R.) **34**, 1138-1141 (May, 1958)

The absorption coefficient is calculated for a crystal placed in a uniform electric field, for frequencies at which the crystal does not normally absorb in the absence of a field. It is shown that there is a shift in the red absorption limit toward the longer wavelengths, by an amount which may reach hundreds of angstroms for reasonable field strengths.

It is known that a uniform electric field greatly alters the electron states in a crystal. Strictly speaking, there is no stationary state at all under these conditions. Instead of the Bloch functions $\psi_j(\mathbf{p}, \mathbf{r})$ for the electrons, we have functions of the form

$$\psi_j(\mathbf{p}_0, \mathbf{r}, t) = \exp\left\{-\frac{i}{\hbar}\int_0^t \varepsilon_j(\mathbf{p}_0 - e\mathbf{E}x)\,dx\right\}\psi_j(\mathbf{p}_0 - e\mathbf{E}t, \mathbf{r}). \quad (1)$$

even in the zero-order approximation. Here $\epsilon_j(\mathbf{p})$ are functions which define the dependence of electron energy on the quasi-momentum in the j-th band; $\mathbf{E}$ is the electric field strength.

These functions satisfy the time-dependent Schrödinger equation for the model under consideration (an electron in a periodic field plus a uniform electric field), accurately up to the exponentially small terms corresponding to the "leakage" of electrons from one band to another under the influence of the field, and describe a uniform acceleration of the electron which is in the state $\psi_j(\mathbf{p}_0, \mathbf{r})$ at the instant $t = 0$.

In what follows we shall express corrections to the wave function for the system which arise from interaction with light, using an expansion in terms of the functions (1). We could have used for this purpose any other system of functions satisfying the above-mentioned Schrödinger equation, so long as this system is complete and orthogonal at all instants of time (since we shall be considering only those expansion coefficients that change with time). However, the system (1), in addition to its physically descriptive nature, has the further advantage that it is obviously complete and orthogonal. As a matter of fact, although each of the functions varies continuously with time, at every instant the system as a whole is identical with the complete system of Bloch functions for the crystal. The particular choice of functions in which to carry out the expansion is quite immaterial to the final result, since the number of quanta absorbed (or, the number of photoelectrons, which is the same thing) depends on the total sum of the squares of the moduli of the coefficients in the expansion, and therefore does not depend on the choice of the base functions.

The probability that a quantum with frequency ω and polarization vector $\mathbf{e}$ will be absorbed by an electron in the state $\psi_v(\mathbf{p}_0, \mathbf{r}, t)$ is

$$w(\mathbf{p}_0, \omega, t) = \left(\frac{e}{m}\right)^2 \frac{2\pi\hbar}{\omega}\left|\int_0^t dt \int d\tau \psi_c^*(\mathbf{p}_0, \mathbf{r}, t)\, e^{-i\omega t}\mathbf{e}\nabla\psi_v(\mathbf{p}_0, \mathbf{r}, t)\right|^2, \quad (2)$$

where the indices v and c refer to the valence and conduction bands respectively; e and m are the charge and mass of electron.

In the general case, in the absence of an electric field, the integrand varies harmonically with time; this implies that the energy must remain constant. The total probability of absorption of a photon $\hbar\omega$ per unit time per unit volume is then given by the expression

$$W(\omega) = \int \frac{dw(\mathbf{p}, \omega, t)}{dt}\frac{d^3p}{(2\pi\hbar)^3} = \left(\frac{e}{m}\right)^2\frac{2\pi\hbar}{\omega}\int |\mathbf{eM}_{vc}(\mathbf{p})|^2\,\delta\{\varepsilon_c(\mathbf{p}) - \varepsilon_v(\mathbf{p}) - \hbar\omega\}\frac{d^3p}{(2\pi\hbar)^3}, \quad (3)$$

$$\mathbf{M}_{vc}(\mathbf{p}) = \int \psi_c^*(\mathbf{p}, \mathbf{r})\nabla\psi_v(\mathbf{p}, \mathbf{r})\,d\tau, \quad (4)$$

from which it can be seen, in particular, that quanta with frequencies less than $\omega_0 = \epsilon_0/\hbar$, where $\epsilon_0 = \min\{\epsilon_c(\mathbf{p}) - \epsilon_v(\mathbf{p})\}$, are not absorbed at all. To be specific, we shall assume that the transition (4) is not forbidden by symmetry considerations, i.e., $\mathbf{M}_{vc}(\mathbf{p}_m) \neq 0$, where $\mathbf{p}_m$ is the value of the quasi-momentum corresponding to the absolute minimum in the function $\epsilon_c(\mathbf{p}) - \epsilon_v(\mathbf{p})$. This case corresponds to the steepest absorption edge, and is therefore particularly suitable for observations. In fact, near this edge the expression (3) can be put into an explicit form by noting that in this region the function $\epsilon_c(\mathbf{p}) - \epsilon_v(\mathbf{p})$ can be written as

$$\varepsilon_c(\mathbf{p}) - \varepsilon_v(\mathbf{p}) = \varepsilon_0 + \sum_{i,k}(p_i - p_{im})(p_k - p_{km})/2m_{ik}. \quad (5)$$

Then

$$W(\omega) = \frac{1}{\pi}\frac{e^2}{\hbar c}\frac{c}{\omega}\left[\frac{m_1 m_2 m_3}{m^3}\frac{2(\hbar\omega - \varepsilon_0)}{m}\right]^{1/2} |\mathbf{e}\mathbf{M}_{vc}(\mathbf{p}_m)|^2, \quad (6)$$

where m_i^{-1} is the principal value of the tensor m_{ik}^{-1}.

The absorption coefficient is thus seen to vary as $(\omega - \omega_0)^{-1/2}$. In the particular case $\mathbf{M}_{vc}(\mathbf{p}_m) = 0$, the coefficient increases much more slowly,[2] as $(\omega - \omega_0)^{3/2}$.

In the presence of an electric field, expression (2) no longer contains delta functions, and therefore the probability of absorbing a quantum of frequency less than ω_0 is not zero. The probability of photon absorption in this case must be calculated in a different manner. We shall first carry out this derivation under the simplifying assumption that the field is directed along one of the axes of a simple cubic lattice with period d. Then because $\epsilon_j(\mathbf{p})$ and $\psi_j(\mathbf{p}, \mathbf{r})$ are periodic functions of the quasi-momentum, the electron undergoes a periodic motion within the j-th band, with the period $T = 2\pi\hbar/eEd$. In this case it is natural to take the absorption probability for one period of this oscillation, $w(\mathbf{p}_0, \omega, 2\pi\hbar/eEd)$, as the characteristic absorption. In order to calculate this quantity, we transform (2) into the new variables of integration $p_x = p_{0x} - eEt$ (the X axis is assumed to be in the field direction) and note that the factor

$$\exp\left\{\frac{i}{e\hbar E}\int[\varepsilon_c(\mathbf{p}) - \varepsilon_v(\mathbf{p}) - \hbar\omega]\,dp_x\right\}$$

occurring in the integrand has a saddle point $p_x = q(p_y, p_z)$ in the complex p_x plane, determined by the conditions

$$\varepsilon_c(q, p_y, p_z) - \varepsilon_v(q, p_y, p_z) - \hbar\omega = 0.$$

Making use of this circumstance, we obtain

$$w\left(\mathbf{p}_0, \omega, \frac{2\pi\hbar}{eEd}\right) = \left(\frac{e}{m}\right)^2 \frac{2\pi\hbar}{\omega}\frac{2\pi\hbar}{eE}|\mathbf{e}\mathbf{M}_{vc}(q, p_{0y}, p_{0z})|^2$$

$$\times\left|\frac{\partial[\varepsilon_c(\mathbf{p}) - \varepsilon_v(\mathbf{p})]}{\partial p_x}\right|^{-1}_{p_x=q}$$

$$\times\left|\exp\left\{\frac{i}{e\hbar E}\int^{p_x=q}[\varepsilon_c(\mathbf{p}') - \varepsilon_v(\mathbf{p}') - \hbar\omega]\,dp'_x\right\}\right|^2. \quad (7)$$

Multiplying this expression by the number of oscillations of the electron per unit time, $eEd/2\pi\hbar$, and integrating over all $\mathbf{p}_0$, we obtain the absorption probability for a photon of frequency ω per unit volume per unit time. Electrons with this same p_{0y} and p_{0z}, but different p_{0x}, will be found to carry out exactly the same motions, but with a shift in time. Hence integration over p_x is equivalent to a simple multiplication by $1/d$, the number of states with a given p_{0y} and p_{0z}. Using the expansion (5), and bearing in mind that, for the almost trivial case under consideration, $m_{ik}^{-1} = m^{*-1}\delta_{ik}$, we arrive at the following final result (for frequencies $\omega < \omega_0$):

$$W(\omega) = \frac{e^2}{\hbar c}\frac{c}{\omega}\left(\frac{m^*}{m}\right)^2\sqrt{2\frac{\varepsilon_0 - \hbar\omega}{m^*}}\frac{(e\hbar E)^2}{m^*(\varepsilon_0 - \hbar\omega)^3}|\mathbf{e}\mathbf{M}_{vc}(\mathbf{p}_m)|^2$$

$$\times\exp\left\{-\frac{4\sqrt{2m^*}}{3e\hbar E}(\varepsilon_0 - \hbar\omega)^{3/2}\right\}. \quad (8)$$

It is not difficult to repeat all the steps outlined above for more general cases as well — lattices with any symmetry and with arbitrary field directions, in which the electron motion, generally speaking, is aperiodic (see, for instance, Keldysh[3]). The number of photons absorbed per unit volume in unit time is then given by the following expression (under the condition $\omega < \omega_0$)

$$W(\omega) = \frac{e^2}{\hbar c}\frac{c}{\omega}\sqrt{\frac{m_1 m_2 m_3}{m^3}\frac{2(\varepsilon_0 - \hbar\omega)}{m}}\frac{(e\hbar E)^2}{m_\parallel(\varepsilon_0 - \hbar\omega)^3}|\mathbf{e}\mathbf{M}_{vc}(\mathbf{p}_m)|^2$$

$$\times\exp\left\{-\frac{4\sqrt{2m_\parallel}}{3e\hbar E}(\varepsilon_0 - \hbar\omega)^{3/2}\right\},$$

$$m_\parallel^{-1} = \sum_i \cos^2\gamma_i/m_i, \quad (9)$$

where m_i^{-1} are the diagonal terms of the tensor m_{ik}^{-1}, and γ_i are the angles between the field and the principal axes of the tensor m_{ik}^{-1}. This formula applies when $\sqrt{m_\parallel}\,(\epsilon_0 - \hbar\omega)^{3/2}/e\hbar E \gtrsim 1$. For $\omega > \omega_0$ and $\sqrt{m_\parallel}\,(\hbar\omega - \epsilon_0)^{3/2}/e\hbar E \gtrsim 1$, formula (6) is still valid.

Comparison of these two expressions shows that in an electric field $\mathbf{E}$ there is a fundamental change in the frequency dependence of the absorption coefficient near the threshold, which can be integrated between known limits to give a shift of the absorption edge toward the red by a distance of the order of

$$\Delta\omega_E = \frac{1}{\hbar}\left[(eE)^2\frac{\hbar^2}{m_\parallel}\right]^{1/3} = \frac{1}{\hbar}\left[(eEd)^2\frac{\hbar^2}{m_\parallel d^2}\right]^{1/3}. \quad (10)$$

For crystals whose absorption edge is in the visible region of the spectrum, in electric fields E of the order of 10^5 v/cm, this shift amounts to hundreds of angstroms, if we assume that $m_\parallel \sim m = 10^{-27}$ g. However, it is very seldom that either an electron or a hole will have an effective mass much less than m, which would lead to an increased $\Delta\omega_E$, since $m_\parallel$ is the resultant effective mass of the electrons and holes, and is therefore determined by the smallest of these masses. Thus the most favorable case for observing this effect is one in which the effective mass is small and the forbidden zone is not too wide (of the order of 1 — 2 ev), so that the relative value of the shift $\Delta\omega_E/\omega_0$ is not too small. The origin of this effect is analogous to the self-ionization which causes widening in the lines of atomic spectra. This case was considered by Lanczos, in whose work[4] it is shown that spectral lines which are separated from the series limit by a frequency $\Delta\omega$ widen and merge into a comlex spectrum when the applied field satisfies the condition $\Delta\omega_E \sim \Delta\omega$. The basic qualitative difference between the two cases is that in crystals this shift has nothing at all to do with the existence of any discrete lines corresponding to bound states of electrons or holes, much less to any broadening of such lines. Furthermore, the cases most suitable for the observation of the effect are those in which such states are completely absent, i.e., when the field E is so strong that $\hbar\Delta\omega_E$ is greater than the binding energy, and consequently bound states are practically non-existent. Under the opposite conditions,[5] the picture is similar to the one put forward by Lanczos.

Obviously there will be an analogous shift of the lower threshold which corresponds to absorption with the formation of phonons.

[1] V. W. Houston, Phys. Rev. **57**, 184 (1940).

[2] G. Dresselhaus, Phys. Rev. **105**, 135 (1957).

[3] L. V. Keldysh, J. Exptl. Theoret. Phys. (U.S.S.R.) **33**, 994 (1957), Soviet Phys. JETP **6**, 763 (1958).

[4] C. Lanczos, Z. Physik **62**, 518 (1930).

[5] E. F. Gross, Progress Phys. Sciences **63**, 575 (1957).

Translated by D. C. West
234

SOVIET PHYSICS JETP VOLUME 37 (10), NUMBER 3 MARCH, 1960

KINETIC THEORY OF IMPACT IONIZATION IN SEMICONDUCTORS

L. V. KELDYSH

P. N. Lebedev Physics Institute, Academy of Sciences, U.S.S.R.

Submitted to JETP editor March 23, 1959

J. Exptl. Theoret. Phys. (U.S.S.R.) **37**, 713-727 (September, 1959)

The effect of impact ionization processes on the distribution function for electrons and holes in a strong electric field is studied. It is shown that the energy dependence of the impact ionization probability near the threshold is essentially different for crystals with small and high dielectric constants; the solution of the kinetic equation is considered in both these cases. Expressions are obtained for the equilibrium number of carriers in a strong field, the impact-ionization coefficient, the critical field, etc. The dependence of the breakdown field on temperature, on specimen thickness, and on the electron-lattice interaction law is found. The connection of the expressions obtained with the known breakdown criteria of Fröhlich and Hippel is established. It is shown that increasing the electric field causes a decrease in the recombination speed, as a result of which the equilibrium number of carriers starts growing as the field increases long before the appearance of impact ionization.

THE electric breakdown of semiconductors apparently takes place as a result of the unlimited growth of carrier concentration with increasing field strength.[1] In a stationary state the number of carriers, n, is determined by the relationship

$$n\{\overline{w_i(E,T)}-\overline{w_r(n,E,T)}\}+n_0(E,T)=0, \qquad (1)$$

where $\overline{w_i(E, T)}$ and $\overline{w_r(n, E, T)}$ are the impact ionization and recombination probabilities averaged over the distribution function, $n_0(E, T)$ is the number of carriers of a given type created in unit volume of the semiconductor in unit time by thermal ionization and direct field extraction of valence electrons into the conduction band, E is the field strength and T the temperature.

With increasing field, as will be shown below, $\overline{w_r}$ decreases but $\overline{w_i}$ grows rapidly and, consequently, n increases. In the field E_c for which $\overline{w_i}=\overline{w_r}$ the carrier concentration tends to infinity, which is the breakdown criterion for the case given. Thus, quantitative consideration of the behavior of a semiconductor in the pre-breakdown region, as well as a study of the mechanism of breakdown itself, requires the solution of the kinetic equation taking into account the processes of impact ionization and recombination. This is the aim of the present work.

Following the usual method,[2] it is not difficult to obtain the following system of equations for determining the symmetric $f_0(\epsilon)$ and antisymmetric $f_1(\epsilon)$ parts of the distribution function, $f(\mathbf{P})$

$$f(\mathbf{P})=f_0(x)+\frac{e\mathbf{EP}}{m}f_1(x)+\ldots$$
$$=f_0(x)-\frac{e\mathbf{EP}}{\varepsilon_i m}\frac{\tau(x)}{1+w_i(x)\tau(x)}\frac{df_0(x)}{dx}, \qquad (2)$$

$$\left[\eta(x)+\left(\frac{E}{E_i}\right)^2\frac{\lambda^2(x)\delta}{x\delta(x)}\frac{1}{1+w_i(x)\tau(x)}\right]\frac{df_0(x)}{dx}$$
$$+f_0(x)-\frac{\tau(x)}{x^{3/2}\delta(x)}\frac{S(x)}{N_i}=0, \qquad (3)$$

$$\frac{1}{N_i x^{1/2}}\frac{dS(x)}{dx}-[w_i(x)+w_r(x)]f_0(x)+n_0(x,E,T)$$
$$+N_i\int_1^\infty w_i(x,x')f_0(x')x'^{1/2}dx'=0, \qquad (4)$$

where $x=\epsilon/\epsilon_i$; $\epsilon=P^2/2m$ is the energy; $\mathbf{P}$ is the momentum; ϵ_i is the threshold ionization energy; τ^{-1} is the frequency of collision with phonons; $w_i(x)$ and $w_r(x)$ are the total probabilities of impact ionization and recombination; $w_i(x, x')$ is the probability of the creation by ionization of a carrier with energy x by a carrier with initial energy x', $\lambda(x)=l(x)/l(1)$; $l(x)=P\tau(x)/m$ is the mean free path; $S(x)$ is the carrier current through the surface $\epsilon(\mathbf{P})=x\epsilon_i$, caused by the field and phonon interactions; $\frac{3}{2}N_i$ is the total number of states with energy $\epsilon<\epsilon_i$;

$$\delta(x)=4m\int_0^{2P}B(q)\hbar\omega_q q\,dq\Big/\int_0^{2P}B(q)(1+2N_q)q^3dq$$
$$\sim\frac{\hbar\omega_P}{\varepsilon(N_P+1/2)}, \qquad (5)$$

$$\eta(x) = \int_0^{2P} B(q)(\hbar\omega_q)^2(2N_q+1)\,q\,dq \Big/ 2\varepsilon_i \int_0^{2P} B(q)\,\hbar\omega_q\,q\,dq,$$

$$\tau^{-1}(x) = \frac{Vm}{4\pi\hbar^4 P^3}\int_0^{2P} B(q)(2N_q+1)\,q^3\,dq, \quad E_i = \frac{\sqrt{3\delta}\,\varepsilon_i}{el(\varepsilon_i)}. \quad (6)$$

Here, q, ω_q and N_q are the momentum, frequency, and number of phonons; B_q is the square of the matrix element of the interaction of an electron with a phonon; V is the normalization volume; E_i is the field in which the mean energy of the carriers becomes of the order ϵ_i; $\delta = \delta(1)$. The small value of δ, the average fraction of the energy lost by an electron in one collision with the lattice, is a condition for the applicability of the approximation considered. For the parameter values of interest to us, $\delta \sim 10^{-2}$. The small quantity $\eta(x)$ we will neglect henceforth. For those valence crystals in which the electrons interact mainly with acoustic phonons $B_{ac}(q) \sim q$ and $\hbar\omega_q^{ac} = cq$, where c is the velocity of sound. When the interaction is with optical phonons $B_{op}(q) = \text{const.}$ and $\hbar\omega_q^{op} = \hbar\omega_0 = \text{const.}$

From (5) and (6) it follows that

$$\delta_{ac}(x) = 4mc^2/kT, \qquad \lambda_{ac}(x) = 1,$$

$$E_i^{ac} = \sqrt{12mc^2/kT}\,\varepsilon_i/el,$$

$$\delta_{op}(x) = \hbar\omega_0/\varepsilon(N_P + 1/2), \qquad \lambda_{op}(x) = 1,$$

$$E_i^{op} = \sqrt{3\hbar\omega_0\varepsilon_i/(N_P + 1/2)(el)^2}. \quad (7)$$

Before proceeding to the solution of Eqs. (2) – (4), we make some remarks on the choice of the probabilities $w_r(x)$, $w_i(x)$ and $w_i(x, x')$. The effect of recombination on the form of the distribution function is insignificant in view of the inequality $w_r\tau \ll 1$, which is well fulfilled. Therefore, the corresponding term in (4) can be considered as a small contribution and the fact that it is in general nonlinear has no effect. In the majority of cases, however, an important part is played by the so-called "radiationless" recombination associated with carrier capture into local states. Its probability for a sufficiently large number of carriers can be considered as independent of concentration. Such a process can only take place for very slow carriers ($\epsilon \lesssim \hbar\omega_m$, where ω_m is the maximum lattice vibration frequency), which in fact are not included in the conditions considered, since for them $\delta(x)$ is not small. Therefore, it is most natural to include radiationless recombination, not in (4), but in the boundary condition for $S(x)$ at $x = 0$. The last two terms in (4) can be taken into account in a similar way. In fact, the probability of creation of a carrier with energy ϵ by thermal ionization and by the field decreases exponentially with increase of ϵ. Likewise, the number of carriers with energies essentially exceeding the ionization threshold is exponentially small. Therefore, both $n_0(E, T)$ and $w_i(x, x')$ cause the creation of only very slow carriers $x \ll 1$, and we include them only in the boundary condition

$$\frac{S(0)}{N_i} - w_r^0 f_0(0) + \frac{n_0(E,T)}{N_i}$$

$$+ \sum_{e,h} N_i \int_0^\infty x^{1/2}dx \int_1^\infty w_i^{(1,2)}(x, x')\, f_0(x')\, x'^{1/2}\, dx' = 0,$$

$$w_r^0 = \int_0^\infty w_r(x)\, x^{1/2}\, dx, \quad (8)$$

where the summation takes into account the presence of two carrier types and the indices 1 and 2 refer to the creation probability of carriers of the same or opposite charge.

The quantity $w_i(x)$, as shown in Appendix 1, can increase near the threshold either linearly or quadratically, depending upon whether the value of the dielectric constant of the crystal is small or large, i.e.,

$$w_i(x) = p(x-1)^j k_j(x)/\tau(x), \qquad j = 1, 2,$$

$$k_j(x) = 1 + \sum_{n=1}^{\infty} k_{jn}(x-1)^n. \quad (9)$$

The dimensionless quantity p thus defined is rather large ($p \sim 10^2$).

Equations (3) and (4) take an essentially different form in the regions $x < 1$ and $x > 1$. It is natural, therefore, to solve them in each of these regions separately and then match the solutions obtained at $x = 1$. Below the ionization potential $w_i(x) = 0$, $w_r(x)$ is a small correction, and all the remaining terms in the right half of (4) do not depend on the value of $f_0(x)$ in this region. Equations (3) and (4) are consequently integrated in the general form. However, on the basis of the remarks made above, we will use a simplified form of the solution in which the value $\eta(x)$ is neglected, and $w_r(x)$, $w_i(x, x')$, and $n_0(x, E, T)$ are taken into account only in the boundary condition (8).

$$S(x) = \text{const} \equiv -\frac{N_i\delta}{\tau(1)}\sigma(E)\, f_0(1),$$

$$f_0(x) = f_0(1)\exp\left\{\left(\frac{E_i}{E}\right)^2 \int_x^1 \frac{x'\delta(x')}{\lambda^2(x')\delta}\, dx'\right\}$$

$$\times\left\{1 + \left(\frac{E_i}{E}\right)^2 \sigma(E)\int_x^1 \exp\left[-\left(\frac{E_i}{E}\right)^2 \int_{x'}^1 \frac{t\delta(t)}{\lambda^2(t)\delta}\, dt\right]\frac{dx'}{x'\lambda(x')}\right\}.$$

(10)

The constants of integration $f_0(1)$ and $\sigma(E)$ must be determined from the boundary conditions. The value of $\sigma(E) \equiv -S(1)\tau(1)/N_i\delta f_0(1)$, giving the mean probability of impact ionization, is simply determined, as will be shown below, by solving the equation in the region $x > 1$. Knowing this value is sufficient to determine all the characteristics of the semiconductor in the strong electric field. In fact, we show below that in the region $x > 1$ the distribution function falls practically to zero for $x - 1 \gtrsim (\delta/p)^{1/4}\sqrt{E/E_i} = \alpha$. Consequently, the contribution of this region to all the observed quantities (number of carriers, conductivity, mean energy) is of the order $\alpha \ll 1$. In other words, all these quantities can be evaluated using the function (10) by averaging over the region $x < 1$. The single quantity completely determined by the distribution of carriers over the ionization potential is the mean probability of impact ionization. But it is equal at the same time to $-S(1) = N_i\delta\sigma(E)f_0(1)/\tau(1)$, which is easy to see by integrating (4) from 1 to ∞, and omitting the last three terms, which are insignificantly small in this region. Thus, all the information which we must obtain from the solution of the kinetic equation in the region $x > 1$ is included in the function $\sigma(E)$, proportional to the ratio $S(x)/f_0(x)$ at $x = 1$.

The total number of carriers n and the mean impact ionization and recombination probabilities are determined from the following relationships

$$n = N_i\int_0^\infty f_0(x)\,x^{1/2}dx = f_0(1)\,N_i \exp\left[\left(\frac{uE_i}{E}\right)^2\right]\left\{\varphi_{1/2}^{(1)}\left(\frac{E}{E_i}\right)\right.$$

$$\left.+\left(\frac{E_i}{E}\right)^2\sigma(E)\exp\left[-\left(\frac{uE_i}{E}\right)^2\right]\varphi_{1/2}^{(2)}\left(\frac{E}{E_i}\right)\right\}, \qquad (11)$$

$$\overline{w_i(E)} = \frac{S(1)}{n} = \frac{\delta}{\tau(1)}\sigma(E)$$

$$\times\frac{\exp[-(uE_i/E)^2]}{\varphi_{1/2}^{(1)}(E/E_i) + (E_i/E)^2\sigma(E)\exp[-(uE_i/E)^2]\,\varphi_{1/2}^{(2)}(E/E_i)}, \qquad (12)$$

$$\overline{w_r(E)} = \frac{N_i}{n}w_r^0 f_0(0)$$

$$= w_r^0\frac{1 + (E_i/E)^2\sigma(E)\exp[-(uE_i/E)^2]\,\zeta(E/E_i)}{\varphi_{1/2}^{(1)}(E/E_i) + (E_i/E)^2\sigma(E)\exp[-(uE_i/E)^2]\,\varphi_{1/2}^{(2)}(E/E_i)}, \qquad (13)$$

$$u^2 = \int_0^1 \frac{x\delta(x)}{\lambda^2(x)\,\delta}dx,$$

$$\zeta(z) = \int_0^1 \exp\left[\frac{1}{z^2}\int_0^x \frac{x'\delta(x')}{\lambda^2(x')\,\delta}dx'\right]\frac{dx}{x\lambda(x)}, \qquad (14)$$

$$\varphi_k^{(1)}(z) = \int_0^1 \exp\left[-z^{-2}\int_0^x \frac{x'\delta(x')}{\lambda^2(x')\,\delta}dx'\right]x^k dx,$$

$$\varphi_k^{(2)}(z) = \int_0^1 \frac{dx}{x\lambda(x)}\int_0^x \exp\left[-z^{-2}\int_x^{x'} \frac{x''\delta(x'')}{\lambda^2(x'')\,\delta}dx''\right]x'^k dx'. \qquad (15)$$

The integral $\zeta(z)$ in general diverges at the lower limit. This is associated with the fact that we have considered in (8) the number of recombinations as proportional to $f_0(0)$. In fact, as follows from the reasons given, this number is determined by the mean value of $f_0(E)$ in the region of very small energies ($\epsilon \lesssim \hbar\omega_m$). Consideration of this fact should lead to the exclusion of the integral for $x \sim \hbar\omega_m/\epsilon_i$. The exact value of this limit for calculating $\zeta(z)$ has no significance, since the divergence of the integral is logarithmic.

Using the expressions (11) – (13) and the evident relationships

$$-S(1) = N_i\int_1^\infty w_i(x)\,f_0(x)\,x^{1/2}dx,$$

$$N_i\int_0^\infty w_i^{(1)}(x, x')\,x^{1/2}\,dx$$

$$= 2N_i\int_0^\infty w_i^{(2)}(x, x')\,x^{1/2}\,dx = 2w_i(x'), \qquad (16)$$

the boundary condition (8) can be rewritten in a form completely analogous to relation (1) (index e refers to electrons and h to holes),

$$[\overline{w_{re}(E)} - \overline{w_{ie}(E)}]\,n_e(E) - \overline{w_{ih}(E)}\,n_h(E) - n_{0e}(E.T) = 0,$$

$$[\overline{w_{rh}(E)} - \overline{w_{ih}(E)}]\,n_h(E) - \overline{w_{ie}(E)}\,n_e(E) - n_{0h}(E,T) = 0. \qquad (17)$$

If w_r^0 can be taken as independent of n, then (see reference 4)

$$n_e(E) = \frac{n_{0e}(E,T)}{\overline{w_{re}(E)}}\,\frac{1 + (n_{0h}(E,T)/n_{0e}(E,T) - 1)\,r_h(E)}{1 - r_e(E) - r_h(E)},$$

where

$$r(E) = \frac{\overline{w_i(E)}}{\overline{w_r(E)}}. \qquad (18)$$

We study the behavior of this expression on increasing the field. In the region $E \ll E_i$, we have $r(E) \ll 1$ and (18) takes the usual form $n(E) = n_0/\overline{w_r(E)}$. Thus, the growth of n is determined mainly by the function $\varphi_{1/2}^{(1)}(E/E_i)$ [we notice that $\sigma(E)$ becomes of the order of unity only when E is comparable with E_i]. In particular, for the cases mentioned above, involving acoustical and optical phonons in valence crystals, n is proportional to $E^{3/2}$ and E^3 respectively. This behavior is occasioned by the fact that when the field becomes sufficiently large ($E \gtrsim 10^3$ v/cm), the mean electron energy starts to increase, the relative number of slower electrons decreases, and there is an associated decrease in the recombination velocity. When the field approximates to $E_i \sim 10^5$ v/cm, $\overline{w_i(E)}$ begins to increase rapidly and at a field E_c, determined by the condition

$$r_e(E_c) + r_h(E_c) = 1, \tag{19}$$

breakdown occurs. As a rule, the values of E_i are different for electrons and holes, therefore one of the quantities $r(E)$ is much greater than the other and the condition of breakdown takes the form $r(E_c) = 1$. The carrier type for which the value E_i is smaller is taken here. A direct comparison of (12) and (13) shows that $r(E_i) \sim \delta/w_r^0\tau \gg 1$ and, consequently, $E_c < E_i$. But the field dependence of all the quantities entering into these formulae is small compared with the exponential, and in the zero-order approximation we have

$$E_c = uE_i \ln^{-1/2}\left\{\frac{\sigma(E_c)\delta}{w_r^0\tau(1)}\left[1 + \left(\frac{E_i}{E_c}\right)^2 \sigma(E_c)\exp\left[-\left(\frac{uE_i}{E_c}\right)^2\right] \times \zeta\left(\frac{E_c}{E_i}\right)\right]^{-1}\right\} \approx uE_i \ln^{-1/2}\frac{\delta}{w_r^0\tau}. \tag{20}$$

The value of w_r^0 is of the order $(\hbar\omega_m/\epsilon_i)^{3/2}/\tau_r$ where τ_r is the recombination lifetime of carriers when the field is absent. Taking this lifetime as of the order of a microsecond, and $\tau \sim 10^{-12}$ sec, we obtain the following estimate of the critical field $E_c \approx E_i u/5$ (see reference 5).

If the region in which the field acts is sufficiently small ($\lesssim 1$ cm), then the lifetime of carriers is determined not by recombination but by their departure from this region. The carrier concentration depends in this case on the distance t from the boundary of the specimen and the number of carriers n_0 flowing through this boundary in unit time[1] ($n_{0h} = 0$):

$$n_e(E, t) = \frac{n_{0e}}{v_{ed}}\exp[\varkappa_e(E)t] \times \frac{\varkappa_e(E)\exp[-\varkappa_e(E)L - \varkappa_h(E)t] - \varkappa_h(E)\exp[-\varkappa_h(E)L - \varkappa_e(E)t]}{\varkappa_e(E)\exp[-\varkappa_e(E)L] - \varkappa_h(E)\exp[-\varkappa_h(E)L]}. \tag{21}$$

Here v_d is the drift velocity of the carriers in the field E, L is the dimension of the region, and $\kappa(E)$ is the so-called impact-ionization coefficient.

$$\varkappa(E) \equiv \frac{\overline{w_i(E)}}{v_d} = \frac{\sqrt{\;}}{l(1)}\frac{E_i}{E}\sigma(E) \times \frac{\exp[-(uE_i/E)^2]}{\varphi^{(1)}(E/E_i) + (E_i/E)^2\sigma(E)\exp[-(uE_i/E)^2]\varphi^{(2)}(E/E_i)}, \tag{22}$$

$$\varphi^{(1)}(z) = \int_0^1 \exp\left[-z^{-2}\int_0^x \frac{x'\delta(x')}{\lambda^2(x')\delta}dx'\right]\frac{d}{dx}[x\lambda(x)]\,dx,$$

$$\varphi^{(2)}(z) = \int_0^1 \frac{dx}{x\lambda(x)} \times \int_0^x \exp\left[-z^{-2}\int_x^{x'} \frac{x''\delta(x'')}{\lambda^2(x'')\delta}dx''\right]\frac{d}{dx'}[x'\lambda(x')]\,dx'. \tag{23}$$

For the particular cases corresponding to (7)

$$\varphi_{ac}^{(1)}(z) = \frac{\sqrt{\pi}}{2z'}\Phi(z'), \quad \varphi_{ac}^{(2)}(z) = \frac{\sqrt{\pi}}{4z'}\int_0^{z'^2} e^t\Phi(\sqrt{t})\frac{dt}{t},$$

$$z' = \frac{1}{\sqrt{2}z}, \qquad \varphi_{op}^{(1)}(z) = z^2(1 - \exp[-z^{-2}]),$$

$$\varphi_{op}^{(2)}(z) = z^2\{\mathrm{Ei}(z^{-2}) + 2\ln z - C\}, \tag{24}$$

where $\Phi(z)$ is the error integral, $\mathrm{Ei}(z)$ is the exponential integral function, and C is Euler's constant ($C \approx 0.577$).

The breakdown field for the limited region is determined according to (21) by the condition

$$[\varkappa_e(E_c) - \varkappa_h(E_c)]L = \ln[\varkappa_h(E_c)/\varkappa_e(E_c)] \tag{25}$$

and is thus a function of the ratio $\sqrt{3\delta}\,L/l(1)$. As long as this ratio is small we have

$$E_c(L) = uE_i \ln^{-1/2}\left\{\frac{\sqrt{3\delta}L}{l(1)}\frac{E_i}{E_c}\frac{\sigma(E_c)}{\ln[\varkappa_h(E_c)/\varkappa_e(E_c)]} \times \left[\varphi^{(1)}\left(\frac{E_c}{E_i}\right) + \left(\frac{E_i}{E_c}\right)^2\sigma(E_c)\exp\left[-\left(\frac{uE_i}{E_c}\right)^2\right]\varphi^{(2)}\left(\frac{E_c}{E_i}\right)\right]^{-1}\right\} \approx uE_i \ln^{-1/2}\frac{\sqrt{3\delta}L}{l(1)}. \tag{26}$$

The temperature dependence of E_c is determined mainly by the quantity $E_i \sim \sqrt{\delta}/l(1)$. For acoustical phonons $E_c \sim \sqrt{T}$, for optical $E_c \sim \coth^{1/2}(\hbar\omega_0/kT)$.

In conclusion, we make a series of remarks on the connection of the parameter we have introduced, u, with the known breakdown criteria of Fröhlich and Hippel. Since $\lambda(1) = 1$ the integrand of u^2 in (14) can be either of the order of unity if $\lambda(x)$ does not increase with increasing energy, or increases sufficiently slowly, or much greater than unity, if the mean free path increases rapidly with growth of x. In the first case, which apparently, obtains always in valence crystals, breakdown occurs in fields of the order E_i, i.e., when the mean carrier energy becomes of the order ϵ_i. In form this condition agrees with Fröhlich's criterion,[6] although the primary idea of this criterion was somewhat different. In the second case, which has been well studied and of which ionic crystals are an example, the integrand of u^2 in (14) attains a maximum for small energies and, therefore,

$E_c \sim uE_i \gg E_i$. In ionic crystals, for energies $\epsilon \gtrsim \hbar\omega_0$, the mean free path $\lambda(x)$ is proportional to x, and consequently $u \sim [\epsilon_i/\hbar\omega_0]^{1/2}$. The breakdown field is determined by the relationship

$$E_c \sim uE_i \ln^{-1/2}\frac{\delta}{w_r^0\tau(1)} \sim \frac{\hbar\omega_0}{el(\hbar\omega_0)}\ln^{-1/2}\frac{\delta}{w_r^0\tau}. \quad (27)$$

In other words, breakdown occurs in fields for which $eE_c l \sim \hbar\omega_0$ for carriers with energy $\epsilon \sim \hbar\omega_0$, which agrees qualitatively with Hippel's criterion.[9]

Starting from (5) and (6), it is not difficult to verify that Fröhlich's criterion is applicable to crystals in which $B(q)$ for $q \to 0$ increases less rapidly than $[q(1+2N_q)]^{-1}$, and Hippel's criterion applies in the opposite case.

We proceed now to the solution of our basic problem — finding the distribution function in the region $x > 1$ where the process of impact ionization is important. It will be convenient here to introduce new units of energy and current

$$y = \frac{x-1}{\alpha_j}, \qquad s(y) = \frac{\tau(1)}{\delta}\gamma_j S(x)/N_i, \quad (28)$$

$$\alpha_j = (\delta E^2/pE_i^2)^{1/(2+j)}, \qquad \gamma_j = (E_i/E)^2\alpha_j, \qquad \beta_j = p\alpha_j^j. \quad (29)$$

The system (3) and (4), expressed in these variables, takes the following form:

$$\frac{ds(y)}{dy} - \frac{xk_j(x)}{\lambda(x)}y^j f_0(y) = 0, \quad (30a)$$

$$\frac{x\lambda(x)}{1+\beta_j k_j(x)y^j}\frac{df_0(y)}{dy} + \gamma_j\frac{x^2\delta(x)}{\lambda(x)\delta}f_0(y) - s(y) = 0. \quad (30b)$$

An essential fact, on which all further discussion is based, is the smallness of α, evident directly from (29). In the most interesting region of the field $\alpha \lesssim 0.1$. The functions $f_0(x)$ and $S(x)$ are essentially different from zero only in the narrow region $x-1 \lesssim \alpha$, outside which they fall off exponentially. In this region the coefficients of (30), in the arguments of which the substitution of y for x has not been made, are very slowly changing functions of y and with great accuracy can be taken as the first terms of corresponding series of degree $x-1 = \alpha y$. In the zero order approximation in α, which we mainly use, all these are unity. Further, in this approximation, (30) can be solved exactly only in particular — although perhaps the most interesting — cases. We therefore now describe a general method allowing us to determine with adequate accuracy the quantity $\sigma(E)$ of direct interest to us.

Equations (30) for any values of the parameters have two linearly independent solutions; one exponentially decreasing at infinity, the other increasing. Apparently, only the first of these is physically permissible. It is determined to within an arbitrary multiplier, but the ratio of $s(y)$ to $f_0(y)$ at any point, including at $y = 0$, is strictly defined. Therefore, the requirement of a solution decreasing at infinity is equivalent to the problem of determining the value of $\sigma(E)$. We eliminate from (30) the function $f_0(y)$. Then for the current $s(y)$ we obtain a second-order equation

$$\frac{d}{dy}\left\{\frac{\lambda(x)}{xk_j(x)y^j}e^{\gamma_j F(y)}\frac{ds(y)}{dy}\right\} - \frac{1+\beta_j k_j(x)y^j}{x\lambda(x)}e^{\gamma_j F(y)}s(y) = 0,$$

$$F(y) = \int_0^y \frac{x'\delta(x')}{\lambda^2(x')\delta}[1+\beta_j k_j(x')y'^j]\,dy'. \quad (31)$$

The quantity $\sigma(E)$ which is sought is derived from its solution in the following manner:

$$\sigma(E) = -\frac{1}{\gamma_j}\lim_{y\to 0}\left\{\frac{xk_j(x)y^j}{\lambda(x)}\frac{s(y)}{ds(y)/dy}\right\} \quad (32)$$

and is a function of the parameters β_j and γ_j. Instead, to find this function, we invert the problem, take fixed values of σ, and seek the inverse function $\beta_j = \beta_j(\sigma, \gamma_j)$. In these circumstances (31) appears as a typical eigenvalue problem; the given boundary conditions at $y = 0$ [Eq. (32)] and at infinity $[s(y) \to 0]$ require the finding of the value of the parameter β_j, for which the equation has a nontrivial solution. Solving then the expression obtained for σ, we find the relationship of importance to us $\sigma = \sigma(\beta_j, \gamma_j) = \sigma(E)$. The eigenvalue of interest to us must, apparently, be the smallest, since, from its physical meaning, the function $f_0(y)$ cannot tend to zero anywhere except at infinity. But, from (30a) and the conditions $xk_j(x)/\lambda(x) > 0$ and $s(\infty) = 0$ it follows that $s(y)$ also has no zeros in the interval $(0, \infty)$. These properties, by virtue of the oscillation theorem, are possessed only by the eigenfunction corresponding to the lowest eigenvalue.

One of the most accurate and at the same time simplest ways of finding the lowest eigenvalue is the variational method. Equation (31) with the boundary condition (32) is equivalent to the following variational problem:[10]

$$-\beta_j = \min\frac{\int_0^\infty\left\{\frac{\lambda(x)}{xk_j(x)y^j}\left[\frac{ds(y)}{dy}\right]^2 + \frac{s^2(y)}{x\lambda(x)}\right\}e^{\gamma_j F(y)}dy - \frac{s^2(0)}{\sigma\gamma_j}}{\int_0^\infty\frac{k_j(x)}{x\lambda(x)}e^{\gamma_j F(y)}y^j s^2(y)\,dy} \quad (33)$$

with the additional condition that only functions satisfying (32) are permissible. The eigenvalue is obtained by this method rather accurately even when the variational functions are only rough approximations to the true solution. In the problem considered, however, one can also hope to obtain a reasonable approximation for the function (although this is not a necessity), since its quantitative behavior is very simple. Directly from (31) it is apparent that $s(y)$ is a smooth function monotonically decreasing with increase of y.

Evaluation of the integrals entering into (33) and the subsequent solution requires as a rule rather cumbersome expressions. Therefore we will utilize this method only in cases when an accurate solution cannot be found.

We proceed now to the actual solution of (30) in different cases.

I. $j = 1$. As already remarked above, this case corresponds to semiconductors with not very large dielectric constants ($\mu \sim 1$). In the null approximation with respect to α, the system (30) takes the following form

$$\frac{ds(y)}{dy} - yf_0(y) = 0,$$

$$\frac{1}{1+\beta_1 y}\frac{df_0(y)}{dy} + \gamma_1 f_0(y) - s(y) = 0. \quad (34)$$

The integration of these equations is carried out in Appendix II and leads to the following results:

$$f_0(y) = \frac{\text{const}}{\sqrt{z}}\left\{W_+(z^2) + \rho W_-(z^2)\right\}\exp\left\{-\frac{1}{4}\beta_1\gamma_1\left(y+\frac{1}{2\beta_1}\right)^2\right\}, \quad (35)$$

$$s(y) = -\frac{1}{\sqrt{\beta_1}}\left\{\frac{W_+(z^2)-\rho W_-(z^2)}{W_+(z^2)+\rho W_-(z^2)} - \frac{\sqrt{\beta_1}\gamma_1}{2}\right\}f_0(y), \quad (36)$$

where

$$z = \beta_1^{1/4}\left\{y + \frac{1}{2\beta_1}\left(1+\frac{\beta_1\gamma_1^2}{4}\right)\right\}, \qquad \rho = \frac{1}{4}\beta_1^{-3/4}\left(1-\frac{\beta_1\gamma_1^2}{4}\right),$$

$W_\pm(x) = W_{\rho^2\pm\frac{1}{4},\frac{1}{4}}(x)$ is the so-called Whittaker function. From (35) and (36) it follows that

$$\sigma(E) = \frac{1}{\sqrt{\beta_1}}\left\{\frac{W_+(z_0^2)-\rho W_-(z_0^2)}{W_+(z_0^2)+\rho W_-(z_0^2)} - \frac{\sqrt{\beta_1}\gamma_1}{2}\right\},$$

$$\text{where } z_0 = 2\rho\,\frac{4+\beta_1\gamma_1^2}{4-\beta_1\gamma_1^2}. \quad (37)$$

The various limiting cases of these formulae are also treated in Appendix II. Here we only remark that the quantity $\beta_1\gamma_1^2 = \delta(E_i/E)^2$ for the fields of interest to us is small, which allows formulae (35) — (37) to be greatly simplified.

II. $j = 2$. The case of large values of the dielectric permittivity. In Appendix I it is shown that μ can be considered as large if the condition

$$\mu \gg e^2/\hbar\sqrt{2\alpha\varepsilon_i/m}, \quad (38)$$

is satisfied, where e is the electronic charge. By inspection this criterion can be interpreted as follows; the quantity μ is considered large in crystals in which the binding energy of Coulombic levels $me^4/\mu^2\hbar^2$ is smaller than the width of the region $\alpha\epsilon_i$ where impact ionization takes place. In the semiconductors of most interest — germanium and silicon — the binding energy of the Coulomb levels is about 10^{-2} ev, and the quantity $\alpha\epsilon_i \sim 0.1$ ev. It is natural, therefore, to suppose that they belong to the case $j = 2$. The intermediate case when

$$w_i(x) = \frac{p}{\tau(1)}\,(x-1) + c_1(x-1)^2] \text{ and } \alpha c_1 \sim 1,$$

also leads to the case $j = 2$ by the transformation $y' = y + 1/2\alpha c_1$ and, therefore, will not detain us.

Putting $j = 2$ in the original system (30) gives

$$\frac{ds(y)}{dy} - y^2 f_0(y) = 0, \qquad \frac{1}{1+\beta_2 y^2}\frac{df_0(y)}{dy} + \gamma_2 f_0(y) - s(y) = 0. \quad (39)$$

It is not possible to find the general solution of these equations for arbitrary values of the parameters. We proceed, therefore, in the following manner: we divide the integral of possible values of the field into two partially overlapping regions, in one of which $\beta_2 \ll 1$, and in the other $\gamma_2 \ll 1$. In the first region, which is apparently the one of greater interest, we find an analytical solution of (39), and in the second utilize the general method described above for determining the function $\sigma(E)$.

We will start with a proof that these regions do, in fact, overlap, and thus the combination of the solutions we obtain contains the solution of the problem for any values of the parameters E and p. To do this, we remark that the product $\beta_2^{3/2}\gamma_2 \sim \sqrt{p\delta^2} \ll 1$, and therefore for all values of E one of the quantities β_2 and γ_2 must be small. The regions overlap when $\beta_2 \sim \gamma_2 \sim (p\delta^2)^{1/5} < 1$, although the margin in the latter inequality is not very large. The considerations given have an obvious physical significance. The energy relaxation time due to collisions with phonons is much greater than the momentum relaxation time. In ionization collisions these times are of the same order. Therefore, if the ionization processes play an important role in establishing momentum equilibrium ($\beta_2 \gtrsim 1$), then the energy relaxation is determined only by them ($\gamma_2 \ll 1$). On the other hand, if phonons make a significant contribution to establishing energy equilibrium ($\gamma_2 \gtrsim 1$), then the momentum loss in ionization for time τ is insignificantly small ($\beta_2 \ll 1$). The region $\beta_2 \ll 1$ is of the greatest interest, since it corresponds to a field $E < E_i$, and in general this condition, as was shown above, is satisfied even by breakdown fields. We commence the discussion with this case.

a) $\beta_2 \ll 1$. Region of comparatively weak fields.

In the zeroth approximation with respect to α_2 and β_2 (the method used, taking into account the corresponding corrections, is given in Appendix III) a second-order equation for $f_0(y)$ can be obtained from (39):

$$d^2 f_0(y)/dy^2 + \gamma_2 df_0(y)/dy - y^2 f_0(y) = 0. \tag{40}$$

By a series of elementary substitutions this can be reduced to the Whittaker equation and its solution takes the following form:

$$f_0(y) = \text{const}\, y^{-1/2} \exp\{-{}^1/_2 \gamma_2 y\}\, W_{-(\gamma_2/4)^2,\, 1/4}(y^2). \tag{41}$$

Using the known behavior of the Whittaker function, it is not difficult to discover the behavior of $f_0(y)$ at zero and at infinity:

$$y > 1:\; f_0(y) = \text{const}\, y^{-(\gamma_2/4)^2 - 1/2} \exp\{-{}^1/_2 (y^2 + \gamma_2 y)\}, \tag{42}$$

$$y < 1:\quad f_0(y) \approx \text{const}\left\{\frac{\Gamma(^1/_2)}{\Gamma(^3/_4 + \gamma_2^2/16)}\left(1 - \frac{1}{2}\gamma_2 y\right) + \frac{\Gamma(-^1/_2)}{\Gamma(^1/_4 + \gamma_2^2/16)}\, y + O(y^2)\right\}. \tag{43}$$

With the help of the last formulae and the second of Eqs. (39) the quantity $\sigma(E)$ of interest to us can be determined:

$$\sigma(E) = -\frac{1}{\gamma_2}\left\{\frac{d}{dy}\ln f_0(y) + \gamma_2\right\}_{y=0} = \frac{2}{\gamma_2}\frac{\Gamma(^3/_4 + \gamma_2^2/16)}{\Gamma(^1/_4 + \gamma_2^2/16)} - \frac{1}{2}. \tag{44}$$

In the limiting cases of small $(\gamma_2 \gg 1)$ and large $(\gamma_2 \ll 1)$ fields we have

$$\sigma(E) = \begin{cases} 2/\gamma_2^4 = (2p/\delta)(E/E_i)^6, & \gamma_2 \gg 1 \\ \dfrac{2}{\gamma_2}\dfrac{\Gamma(^3/_4)}{\Gamma(^1/_4)} = 2\dfrac{\Gamma(^3/_4)}{\Gamma(^1/_4)}\left(\dfrac{p}{\delta}\right)^{1/4}\left(\dfrac{E}{E_i}\right)^{3/2}, & \gamma_2 \ll 1. \end{cases} \tag{45}$$

The distribution function (41) in the latter case $(\gamma_2 \ll 1)$ is close to that obtained by Heller.[11]

b) $\gamma_2 \ll 1$. Region of very strong fields. In the zero approximation with respect to α_2 and γ_2 we must solve the following variational problem:

$$-\beta_2 = \min \frac{\int_0^\infty \{[y^{-1} ds/dy]^2 + s^2(y)\}dy - s^2(0)/\gamma_2\sigma}{\int_0^\infty y^2 s^2(y)\, dy}$$

$$\lim_{y\to 0}\left\{\frac{1}{y^2 s(y)}\frac{ds(y)}{dy}\right\} = -\frac{1}{\gamma_2\sigma}. \tag{46}$$

The detailed method of solution is given in Appendix III. Here we confine ourselves to a summary of the results. The connection between the quantity $g = (\gamma_2\sigma)^{-2/3}$ and β_2 is given in parametric form by the two relationships

$$\beta_2 = g\{g^2\psi_1(\nu) - \psi_2(\nu)\}, \qquad g^2 = \frac{d\psi_2(\nu)}{d\nu} \Big/ \frac{d\psi_1(\nu)}{d\nu}, \tag{47}$$

where

$$\psi_1(\nu) = \frac{\nu \sin \pi\nu}{\pi}\frac{\Gamma^3(\nu)\,\Gamma(4\nu)}{\Gamma^2(2\nu)\Gamma(3\nu)},$$

$$\psi_2(\nu) = \left[\frac{\nu\sin\pi\nu}{3\pi}\Gamma^2(\nu)\right]^{3/4}\frac{\Gamma(4\nu)}{\Gamma(8\nu/3)}\frac{\Gamma(\nu/3)\Gamma^2(4\nu/3)\,\Gamma(7\nu/3)}{\Gamma(\nu)\,\Gamma^2(2\nu)\,\Gamma(3\nu)}. \tag{48}$$

The corresponding function $s(y)$ takes the form

$$s(y) = z^\nu K_\nu(z), \qquad z = \left[\frac{2^{2\nu}\nu\sin\pi\nu}{3\pi\sigma}\Gamma^2(\nu)\right]^{1/2\nu} y^{3/2\nu}, \tag{49}$$

where $K_\nu(z)$ is the Macdonald function. ν varies from $\frac{3}{4}$ to $\frac{1}{2}$. In the limiting case when $\beta_2 \to 0$, ν tends to $\frac{3}{4}$ and the solution (49) agrees with that which is obtained from (41) in the limit as $\gamma_2 \to 0$. Thus, the solutions we have obtained in fact join up in the region where the conditions $\beta_2 \ll 1$ and $\gamma_2 \ll 1$ are simultaneously satisfied. In the other limiting case when $\nu \to \frac{1}{2}$, g tends to infinity. Thus, $\beta_2 \sim g^3 \sim (\gamma_2\sigma)^{-2}$. In fact, for very large fields

$$\sigma(E) = \sqrt{\psi_1(^1/_2)/\delta}\; E/E_i. \tag{50}$$

APPENDIX I

The probability of the creation of electrons with momenta $\mathbf{p}_1$ and $\mathbf{p}_2$ and a hole with momentum $\mathbf{p}_3$, due to an ionizing collision of an electron with an original momentum $\mathbf{p}_0$, can always be written in the form

$$(2\pi/\hbar)\,|M(\mathbf{p}_0;\, \mathbf{p}_1,\, \mathbf{p}_2)|^2\, \delta[\varepsilon_e(\mathbf{p}_0) - \varepsilon_e(\mathbf{p}_1) - \varepsilon_e(\mathbf{p}_2) - \varepsilon_h(\mathbf{p}_3)] \times \delta[(\mathbf{p}_0 - \mathbf{p}_1 - \mathbf{p}_2 - \mathbf{p}_3)/\hbar].$$

Hence, the following expression is obtained for the total ionization probability:

$$w_i(\varepsilon_0) = \frac{2\pi}{\hbar}V^3 \times \int |M(\mathbf{p}_0;\, \mathbf{p}_1,\, \mathbf{p}_2)|^2\, \delta[\varepsilon_0 - \varepsilon_e(\mathbf{p}_1) - \varepsilon_e(\mathbf{p}_2) - \varepsilon_h(\mathbf{p}_3)] \times \delta\left(\frac{\mathbf{p}_0 - \mathbf{p}_1 - \mathbf{p}_2 - \mathbf{p}_3}{\hbar}\right)\frac{d^3p_1 d^3p_2 d^3p_3}{(2\pi\hbar)^9}. \tag{A.1}$$

All the conservation laws can be satisfied only for sufficiently large values of $\mathbf{p}_0$. The ionization threshold is determined by the condition

$$\varepsilon_i \equiv \varepsilon_e(\mathbf{p}_i) = \varepsilon_{min}(\mathbf{p}_i) \equiv \min\{\varepsilon_e(\mathbf{p}_1) + \varepsilon_e(\mathbf{p}_2) + \varepsilon_h(\mathbf{p}_i - \mathbf{p}_1 - \mathbf{p}_2)\}.$$

The condition of a minimum in the right-hand side of the equality means that at the threshold $\nabla\epsilon_e(\mathbf{p}_1) = \nabla\epsilon_e(\mathbf{p}_2) = \nabla\epsilon_h(\mathbf{p}_3) = \mathbf{v}$, i.e., the speeds of all the final particles are equal. Close to the threshold the argument of the energy δ-function can be developed in a power series of the departure of the momenta from their values $\mathbf{p}_m(\mathbf{p}_0)$ determined from the minimum condition written down above. The coefficients of the corresponding quadratic form, after transforming to principal axes, we will label $m_k^{*-1}(\mathbf{p}_0)$. After introducing new variables of integration according to the formulae

$$p_k - p_{km}(\mathbf{p}_0) = \sqrt{2m_k^*(\mathbf{p}_0)\,[\varepsilon_e(\mathbf{p}_0) - \varepsilon_{min}(\mathbf{p}_0)]}\;\pi_k$$

etc., (A.1) acquires the following form:

$$w_i(\mathbf{p}_0) = \frac{2\pi}{\hbar} \frac{m_1^*(\mathbf{p}_0)\, m_2^*(\mathbf{p}_0) m_3^*(\mathbf{p}_0)}{(2\pi)^3 (2\pi\hbar)^6} [\varepsilon_e(\mathbf{p}_0) - \varepsilon_{min}(\mathbf{p}_0)]^2$$

$$\times V^3 \int |M(\mathbf{p}_0;\ \mathbf{p}_1,\ \mathbf{p}_2)|^2 \delta\left(1 - \sum_{k=1}^{6} \pi_k^2\right) d^6\pi =$$

$$= \frac{1}{\hbar} \frac{m_1^*(\mathbf{p}_0)\, m_2^*(\mathbf{p}_0)\, m_3^*(\mathbf{p}_0)}{(2\pi)^2 (2\pi\hbar)^6} V^3 \overline{|M(\mathbf{p}_0)|^2} [\varepsilon_e(\mathbf{p}_0) - \varepsilon_{min}(\mathbf{p}_0)]^2. \tag{A.2}$$

In the Born approximation $M(\mathbf{p}_0, \mathbf{p}_1, \mathbf{p}_2)$ is simply the matrix element of the corresponding interaction energy. Since the momenta which are exchanged by the particles participating in the reaction are of the order $\sqrt{m\epsilon_i}$, the collision parameters making the principal contribution to ionization at the threshold are of the order $\hbar/\sqrt{m\epsilon_i}$. In this region the interaction potential must be of Coulomb order e^2/r, since the polarization of the medium only occurs at large distances. Therefore

$$M \sim \frac{e^2}{\hbar} \sqrt{m\varepsilon_i} \frac{\hbar^3}{V (m\varepsilon_i)^{3/2}} = \frac{1}{V} \frac{e^2\hbar^2}{m\varepsilon_i},$$

so that

$$w_i(\mathbf{p}_0) \sim \frac{e^4 m}{\hbar^3} V \varepsilon_i^{-2} [(\nabla\varepsilon_e(\mathbf{p}_i) - \mathbf{v})(\mathbf{p}_0 - \mathbf{p}_i)]^2 \approx \frac{e^4 m}{\hbar^3} V \left(\frac{\varepsilon_0 - \varepsilon_i}{\varepsilon_i}\right)^2. \tag{A.3}$$

This expression is almost the same as the result obtained by Tevordt[12] from an exact analysis of a somewhat simplified model. In deriving (A.3) we have neglected the difference between the slowly varying quantities $m_k^*(\mathbf{p}_0)$ and their values at the ionization threshold, and have replaced all the m_k^* by some mean value m. Also we took the speed of the final particles v as small compared with the speed of the primary $\nabla\epsilon_e(\mathbf{p}_0)$.

When the Born approximation is inapplicable, the ionization cross-section for slow electrons differs from the Born multiplier $|\psi(0, 0)|^2$, where $\psi(\mathbf{r}_1, \mathbf{r}_2)$ is the wave function of the final state describing the motion of the two electrons relative to the hole.[13] When a long-range Coulomb interaction is present, this multiplier tends to infinity as $(\epsilon - \epsilon_i)^{-1}$.[14*] The evaluation of the matrix element given above is then correct only in the region $\epsilon - \epsilon_i \gtrsim e^4 m/\mu^2\hbar^2$ (criterion for applicability of the Born approximation). The dielectric permitivity μ enters into this criterion because the growth of $|\psi(0, 0)|^2$ is determined by the long-range part of the Coulomb interaction. In the region of small energies

$$w_i(\mathbf{p}_0) \sim \frac{e^4 m}{\hbar^3} V \left(\frac{\varepsilon_0 - \varepsilon_i}{\varepsilon_i}\right)^2 \frac{e^4 m}{\mu^2\hbar^2(\varepsilon_0 - \varepsilon_i)} = \frac{\varepsilon_i}{\hbar} \left(\frac{e^4 m}{\mu\hbar^2\varepsilon_i}\right)^2 V \frac{\varepsilon_0 - \varepsilon_i}{|\varepsilon_i}. \tag{A.4}$$

The energies of interest to us are of the order $\epsilon_i + \alpha\epsilon_i$. Consequently, if μ is large enough so that $\mu^2\alpha\epsilon_i\hbar^2/e^4 m \gg 1$ the Born approximation is applicable for them and Formula (A.3) can be used. In cases where $\mu \sim 1$, the situation is completely analogous to that which exists in gases and the ionization cross section close to the threshold increases linearly.

*The results of Geltman[14] cannot be considered as strictly proven, since one of the terms entering into the interaction of the final particles was considered as a small perturbation. More convincing from our point of view is the fact that the experimentally measured ionization cross section in gases close to the threshold depends linearly on energy.

APPENDIX II

We transform the system (34) by introducing new variables according to the formulae

$$z = \beta_1^{1/4}\left[y + \frac{1}{2\beta_1}\left(1 + \frac{\beta_1\gamma_1^2}{4}\right)\right], \quad \rho = \frac{1}{4}\beta_1^{-3/4}\left(1 - \frac{\beta_1\gamma_1^2}{4}\right),$$

$$\chi_1(z) = \left\{\left(1 - \frac{\sqrt{\beta_1}\gamma_1}{2}\right) f_0(y) + \sqrt{\beta_1} s(y)\right\}$$

$$\times \exp\left[\frac{1}{4}\beta_1\gamma_1\left(y + \frac{1}{\beta_1}\right)^2\right];$$

$$\chi_2(z) = \left\{\left(1 + \frac{\sqrt{\beta_1}\gamma_1}{2}\right) f_0(y) - \sqrt{\beta_1} s(y)\right\}$$

$$\times \exp\left[\frac{1}{4}\beta_1\gamma_1\left(y + \frac{1}{\beta_1}\right)^2\right]. \tag{A.5}$$

The functions $\chi_{1,2}(z)$ then satisfy the following equations:

$$d^2\chi_{1,2}(z)/dz^2 + (4\rho^2 \mp 1 - z^2)\chi_{1,2}(z) = 0, \tag{A.6}$$

which easily lead to the Whittaker equation. The corresponding solutions are given by (35) — (37). Here we retain in the analysis some limiting cases. It was remarked above that in the region of fields of interest to us the quantity $\beta_1\gamma_1^2 = \delta(E_i/E)^2$ is small compared with unity. Therefore we will retain only terms of the order $\sqrt{\beta_1}\,\gamma_1$, and will neglect terms of the order $\beta_1\gamma_1^2$.

a) $\rho \ll 1$. Region of large fields. In this case the quantity $z_0 = 2\rho(4 + \beta_1\gamma_1^2)/(4 - \beta_1\gamma_1^2)$ is also small. In the zero approximation in ρ^2 the solution has the following form:

$$f_0(y) \approx z^{-1/2} \exp\left\{-\frac{1}{4}\beta_1\gamma_1\left(y + \frac{1}{\beta_1}\right)^2\right\} W_{1/4,\,1/4}(z^2)$$

$$\approx \exp\left\{-\frac{\sqrt{\beta_1}}{2}\left(y + \frac{1}{2\beta_1}\right)^2 - \frac{1}{4}\beta_1\gamma_1\left(y + \frac{1}{\beta_1}\right)^2\right\}, \tag{A.7}$$

$$s(y) = -\{1 - \sqrt{\beta_1}\gamma_1/2\} f_0(y)/\sqrt{\beta_1},\ \sigma(E) = 1/\sqrt{\beta_1}\gamma_1 - {}^1/_2. \tag{A.8}$$

b) $\rho \gg 1$. Region of relatively small fields. For this case it is most convenient to start directly from (A.6). We transform afresh to the independent variable $y = (4\rho)^{1/3}(z - 2\rho)$ and neglect the small quantity $\rho^{-4/3}y^2$. The functions $\chi_{1,2}(y)$ then satisfy the equation

$$d^2\chi_{1,2}(y)/dy^2 - (y \pm \sqrt{\beta_1})\chi_{1,2}(y) = 0. \tag{A.9}$$

Consequently,

$$\chi_{1,2}(y) = (y \pm \sqrt{\beta_1})^{1/2} K_{1/3}[{}^2/_3 (y \pm \sqrt{\beta_1})^{3/2}], \tag{A.10}$$

$$f_0(y) \approx \sqrt{y} K_{1/3}({}^2/_3\, y^{3/2}) \exp[-{}^1/_4\, \beta_1\gamma_1 (y + 1/\beta_1)^2] + O(\beta_1),$$
$$s(y) = -[\sqrt{y} K_{2/3}({}^2/_3\, y^{3/2}) / K_{1/3}({}^2/_3\, y^{3/2}) - \gamma_1/2] f_0(y); \tag{A.11}$$

$$\sigma(E) = -3^{1/3}\Gamma({}^2/_3)/\gamma_1\Gamma({}^1/_3) - {}^1/_2. \tag{A.12}$$

If γ_1 is not small, then

$$\sigma(E) = -{}^1/_2 [K_{2/3}(\gamma_1^3/12)/K_{1/3}(\gamma_1^3/12) - 1]. \tag{A.13}$$

This solution is easily obtained also from (34) under the condition $\beta_1 \ll 1$.

APPENDIX III

1. The evaulation of the corrections to the solution of (30) for $j = 2$ which are proportional to β_2, α_2, and α_2^2, can be performed in the following way. We introduce the new independent variable

$$z = \left\{2\int_0^y y \sqrt{\frac{k_2(x)\,|1 + \beta_2 k_2(x) y^2|}{\lambda^2(x)}}\,dy\right\}^{1/2}$$

and then by substitution

$$f_0(y) = \left(\frac{dz}{dy}\right)^{-1} \exp\left\{-\frac{1}{2}\int_0^z \left[\gamma_2 \frac{x^2\delta(x)}{\lambda(x)\delta} + \frac{d}{dy}\left(\frac{x\lambda(x)}{1 + \beta_2 k_2(x) y^2}\right)\right] \frac{\lambda(x)}{x k_2(x)} \frac{dz}{dy}\,dz\right\} \varphi(z) \tag{A.14}$$

arrive at the equation in the normal form

$$d^2\varphi(z)/dz^2 - \{{}^1/_4\, \gamma_2^2 \Psi(z) + z^2\}\varphi(z) = 0. \tag{A.15}$$

We develop $\Psi(z)$ in a power series of z: $\Psi(z) = \Psi(0) + \Psi_1(z) + \Psi_2(z^2) + \ldots.$ In essence this series is an expansion of $\Psi(z)$ in powers of α_2 and β_2: $\Psi \sim \alpha_1$, Ψ_2 contains terms proportional to α_2^2 and β_2, etc. By the transformation

$$z' = \left(1 + \frac{1}{4}\gamma_2^2\Psi_2\right)^{1/4}\left(z + \frac{1}{2}\gamma_2^2 \frac{\Psi_1}{4 + \gamma_2^2\Psi_2}\right)$$

Equation (A.15) leads to the previous form. We limit ourselves here to this preliminary treatment and shall not proceed to explicit expressions for $\Psi(z)$, Ψ_0, Ψ_1 and Ψ_2 in view of their cumbersomeness.

2. As the variational function for the problem formulated in Eq. (46), we will choose the function $s(y) = z^\nu K_\nu(z)$, where $z = (\xi y)^{3/2\nu}$. This function has the correct behavior at zero, $s(y) - s(0) \sim y^3$, and decreases monotonically with increase of y, i.e., it satisfies the basic qualitative requirements for $s(y)$. Also in the limiting cases of small ($\nu \to \frac{3}{4}$) and large ($\nu \to \frac{1}{2}$) values of β_2, it gives an accurate solution of (39) for $\gamma_2 = 0$. Of the two parameters ξ and ν, only one is disposable by virtue of the additional condition (46). We will take ν as the independent variable. The parameter ξ is expressed in terms of ν and σ in the following way:

$$\xi^3 = (2^{2\nu}/3\pi\sigma)\,\nu\Gamma^2(\nu)\sin\pi\nu. \tag{A.16}$$

The evaluation of all the integrals entering into (46) is most conveniently carried out using the known integral forms of the Macdonald functions

$$K_\nu(z) = \frac{1}{2} z^\nu \int_0^\infty t^{-\nu-1} \exp\left\{-\frac{t}{2} - \frac{z^2}{2t}\right\} dt,$$

$$\int_0^\infty z^\mu K_\nu(z) K_{\nu'}(z)\,dz = \frac{2^{\mu-2}}{\Gamma(\mu+1)} \Gamma\left(\frac{\mu+\nu+\nu'+1}{2}\right)\Gamma\left(\frac{\mu-\nu+\nu'+1}{2}\right) \times \Gamma\left(\frac{\mu+\nu-\nu'+1}{2}\right)\Gamma\left(\frac{\mu-\nu-\nu'+1}{2}\right). \tag{A.17}$$

As a result, (46) is reduced to

$$-\beta_2 = g \min\{-g^2\psi_1(\nu) + \psi_2(\nu)\}, \tag{A.18}$$

where the functions $\psi_1(\nu)$ and $\psi_2(\nu)$ are determined by (48). The condition for the minimum of this expression leads to the connection (47) between β_2 and g.

[1] S. L. Miller, Phys. Rev. **99**, 1234 (1955). A. P. Shotov, J. Tech. Phys. (U.S.S.R.) **28**, 437 (1958), Soviet Phys.–Tech. Phys. **3**, 413 (1958).

[2] B. I. Davydov, JETP **7**, 1069 (1937). W. Franz, Z. Physik **113**, 607 (1939); Encycl. Phys. **17**, 155 (1956). R. Stratton, Proc. Roy. Soc. **A242**, 355 (1957).

[3] W. Shockley and W. T. Read, Phys. Rev. **87**, 835 (1952).

[4] N. L. Pisarenko, Izv. Akad. Nauk SSSR, Ser. Fiz. **5**, 631 (1938).

[5] F. Seitz, Phys. Rev. **76**, 1376 (1949).

[6] H. Fröhlich, Proc. Roy. Soc. **A160**, 230 (1937).
[7] B. I. Davydov and I. M. Shmushkevich, JETP 10, 1043 (1940).
[8] V. A. Chuenkov, Usp. Fiz. Nauk **54**, 185 (1954).
[9] A. von Hippel, J. Appl. Phys. **8**, 815 (1937).
[10] R. Courant and D. Hilbert, Methods of Mathematical Physics, Interscience, 1943. Chapter IV.
[11] W. R. Heller, Phys. Rev. **84**, 1130 (1951).
[12] L. Tewordt, Z. Physik **138**, 499 (1954).
[13] L. D. Landau and E. M. Lifshitz, Квантовая механика (Quantum Mechanics), Pergamon, 1958 Sec. 118.
[14] S. Geltman, Phys. Rev. **102**, 171 (1956).

Translated by K. F. Hulme
138

SOVIET PHYSICS JETP VOLUME 18, NUMBER 1 JANUARY, 1964

DEEP LEVELS IN SEMICONDUCTORS

L. V. KELDYSH

P. N. Lebedev Physics Institute, Academy of Sciences, U.S.S.R.

Submitted to JETP editor February 19, 1963

J. Exptl. Theoret. Phys. (U.S.S.R.) **45**, 364-375 (August, 1963)

The problem of local impurity levels having a binding energy comparable with the forbidden band is solved in the two-band approximation, in which the electron and hole bands and their interaction are taken into account. If the two bands are not degenerate, the electron and hole motion can be described by equations of the Dirac type. In the case of triple degeneracy of the valence band, a set of equations is obtained to relate the motion of a scalar particle (electron) with that of a vector particle (hole). Analysis of the equations permits one to explain qualitatively such features of impurity centers with deep levels as attraction to a charged center of carriers with the same charge, capture of both electrons and holes by the same center, etc. The existence of two types of holes (heavy and light) is found to be important for the recombination process: trapping levels correspond mainly to heavy hole states, while large cross sections are due to small light-hole masses.

THE development of the effective-mass method and the detailed clarification of the energy spectra of typical semiconductors have made it possible to solve in exhaustive manner the problem of shallow impurity levels[1-3], i.e., levels with very low binding energies, on the order of 10^{-2} eV, formed near impurity atoms having one less valence electron than the atom of the main substance (for example, group III or V elements in Ge). These states are completely determined by the effective charge of the center, i.e., the dielectric constant of the medium, and by the structure of the bottom of that energy band near which they are formed, since their distances from the other bands are on the order of 1 eV, i.e., much larger than the binding energy. They are described quantitatively by an equation (or system of equations) of the Schröd-inger type with Coulomb potential, and, generally speaking, by anisotropic effective masses characteristic of the given band. The small effective masses and large dielectric constants typical of all semiconductors give rise to low binding energies and to Bohr orbits with tremendous effective radii (compared with the lattice period).

In contradistinction from the shallow levels, it is customary to classify levels as deep if the binding energies are several electron volts, i.e., comparable with the width of the forbidden band, which is formed as a rule near multiply-charged impurity centers, vacancies, etc. Their nature is by far less clear. It is only known that their role in semiconductor physics is very significant: they are usually recombination centers or traps, and consequently determine the lifetimes of the non-equilibrium carriers.

It is customary to point to two main difficulties in the theoretical description of these levels. First, their radius is much smaller than the radius of the shallow levels, so that there are no grounds for assuming a Coulomb field for the center with correction for the dielectric constant and for using the effective-mass approximation. Second, their distance from both bands—valence and conduction—is of the same order, and they cannot therefore be regarded as belonging to either band, unlike the shallow levels, which are determined by a Shröd-inger equation containing the effective mass or other parameters of either the valence or the conduction band.

In the present paper we present, on the basis of a two-band approximation similar to that used by Kane to explain the band structure of InSb[4], an analysis of the deep-level problem, in which at least the second of the foregoing difficulties is resolved. As to the first, from our point of view it has no basic significance at all, since the radii of the bound states deep in the forbidden band, although smaller than the radii of the shallow levels, are nevertheless several times larger than the crystal lattice constant d. Indeed, by virtue of the uncertainty relations the radius a and the energy ϵ of the bound state satisfy the relation $a\sqrt{m\epsilon} \gtrsim \hbar$, where $\hbar$ is Planck's constant and m is some average effective mass for the electron and hole.

Hence

$$a/d \gtrsim \sqrt{(\hbar/d)^2/m\varepsilon} \gtrsim \sqrt{(\hbar/d)^2/m\Delta} \equiv \sqrt{1/\lambda} \gg 1, \qquad (1)$$

where Δ is the half-width of the forbidden band. The last inequality of (1) follows essentially from the definition of the semiconductor, since it requires that the allowed band be much wider than the forbidden band. If we use values typical of semiconductors such as $d = 3 \times 10^{-8}$, $m = 10^{-28}$ g, and $\Delta = 1$ eV, we get $a/d \gtrsim 3$.

The course of the potential in this region can be calculated quite correctly within the framework of the approximation under consideration. Indeed, the polarization at distances $\sim a$ from the center is produced essentially by the valence electrons, bound in the lattice with energies $\lesssim \hbar^2/ma^2 < \Delta$, i.e., lying at distance of the order of Δ from the edge of the band. But such states are known to be well described by the Kane model. Deeper electrons contribute to the polarization cloud at smaller distances $\sim d$ and can be taken into account in the region of interest to us by using some fixed value of the dielectric constant.

In this paper, however, we shall not calculate this potential, but confine ourselves to an analysis of the effects connected with the interactions between the valence and conduction bands at a specified potential, particularly a Coulomb potential, with effective charge Ze, taking into account both the charge of the center and the effective dielectric constant. Even this allows us to explain several qualitative peculiarities of the deep levels, which are utterly unintelligible within the framework of the single-band scheme. Of course, no quantitative meaning should be ascribed to the obtained results, both in view of the arbitrariness in the choice of the potential and in view of the extreme idealization of the band structures which will be employed below. The small parameter characterizing the omitted term is the quantity λ introduced in (1), equal to the ratio of the width of the forbidden band to the distance to the other bands, the influence of which we neglect in most cases. If the potential is more precisely defined, successive neglect of terms of order λ can guarantee an accuracy on the order of 10%.

Before we proceed to a detailed analysis of various specific models, we present now a highly schematized qualitative analysis which permits, however, to obtain in lucid form all the main results and estimate the degree to which they are general. We shall assume here, as in what follows, that the minima of both bands lie at $\mathbf{k} = 0$. Then if the momentum k is small compared with the limiting value $\hbar/d$ but not small compared with $\sqrt{m\Delta}$, the function $\epsilon(\mathbf{k})$, which describes the dependence of the energy on the momentum in both bands, is determined uniquely by the following requirements[5].

1. Its value in the electron and hole bands is given by two branches of one analytic function, since both are obtained by solving the same Schrödinger problem.

2. As k^2 approaches zero, $\epsilon_e(\mathbf{k}) \approx \Delta + k^2/2m_e$ in the electron band and $\epsilon_h(\mathbf{k}) \approx -\Delta - k^2/2m_h$ in the hole band.

This function is of the form

$$\varepsilon(\mathbf{k}) = k^2/2M + \sqrt{\Delta^2 + k^2\Delta/m},$$
$$1/M = 1/2m_e - 1/2m_h, \quad 1/m = 1/2m_e + 1/2m_h. \qquad (2)$$

Higher orders of k^2 are omitted both under and outside the radical sign, in view of the assumed smallness of $(kd/\hbar)^2$. In the energy region under consideration they are of the order of λ. An exact calculation for the two-band model[4] yields the same result. It shows also that the mass M, which is outside the radical sign, is brought about by the interaction with other bands, while m is connected with the branching under consideration, i.e., the interaction between the electron and hole bands, by virtue of which $m/M \sim \lambda$. The term in front of the radical can therefore be omitted henceforth. Experimental data on semiconductors with narrow forbidden bands confirm qualitatively that the masses of the electrons and light holes are of the same order (in the case of Ge we have in mind, of course, the electrons not in the main lateral minima on the (111) axis, but in the additional minimum at $k = 0$).

In accordance with the general principles of quantum mechanics, the Schrödinger equation (at least the symbolic one) for a system described by (2) should be in the form

$$\{\Delta\sqrt{1 - \hbar^2\nabla^2/m\Delta} + V(\mathbf{r})\}\,\psi(\mathbf{r}) = \varepsilon\psi(\mathbf{r}), \qquad (3)$$

where $V(\mathbf{r})$ is the external potential and ∇ the differentiation operator. Let us apply to this equation the operator $\Delta\sqrt{1 - \hbar^2\nabla^2/m\Delta} - V(\mathbf{r})$. Then

$$\Big\{-\frac{\hbar^2}{2m}\nabla^2 + \frac{\varepsilon}{\Delta}V(\mathbf{r}) - \frac{1}{2\Delta}V^2(\mathbf{r})$$
$$+ \frac{1}{2}\Big[\sqrt{1 - \frac{\hbar^2}{m\Delta}\nabla^2},\ V(\mathbf{r})\Big] - \frac{\varepsilon^2 - \Delta^2}{2\Delta}\Big\}\,\psi(\mathbf{r}) = 0. \qquad (4)$$

Expression (4) has the form of the Schrödinger equation for an ordinary particle with mass m, but has several characteristic features. The role of the effective energy is played in it by the quantity $(\epsilon^2 - \Delta^2)/2\Delta$, which goes over in the vicinity of each band into the energy reckoned from the bottom of this band. The potential $V(\mathbf{r})$ has ϵ/Δ

as a coefficient, i.e., it reverses sign with ϵ: if it is attractive for electrons, it is repulsive for holes, and vice versa. Finally, the most interesting circumstance is the appearance of an additional potential $V^2(\mathbf{r})/2\Delta$, which is attractive for both electrons and holes, regardless of the sign of the initial potential. At small distances it always predominates, for $V(r)$ tends to infinity as r tends to zero independently of the character of the screening. This fact explains why the same center can capture both electrons and holes. Capture of the type of carrier which is repelled at large distances calls in this case for activation energy.

In addition, an increase in the degree of singularity of the attractive potential can cause the particle to fall into the center, i.e., recombination, if $V^2(\mathbf{r})$ increases more rapidly than $1/r^2$ with increasing r, or exactly as $1/r^2$ but with a coefficient larger than $\hbar^2\Delta/8m$ [6]. Thus, it is precisely the Coulomb potential $V \sim 1/r$ which serves as the dividing line between the potentials for which bound levels can exist in the forbidden band and the potentials for which instantaneous recombination occurs. From this point of view, the impurities existing in the neutral state only are those for which the potential increases more rapidly than $(\hbar/2r)\sqrt{\Delta/m}$ in the given crystal even for single ionization; singly-charged centers are those for which this situation arises in the case of double ionization, etc. Finally, the usual recombination center should have in the initial state a potential that increases slowly in the sense indicated above, so that it can capture at a stationary level an electron which falls into the center after the capture of a hole, i.e., after the charge increases by unity.

It is easy to understand why the Coulomb potential is the recombination limit. To demonstrate this we repeat the usual arguments showing that for a free particle this role is played by the potential $1/r^2$. Allowing for the uncertainty relation $\Delta k \Delta r \gtrsim h$, the total energy

$$\varepsilon = k^2/2m + V(\mathbf{r}) \gtrsim \hbar^2/2mr^2 - |V(r)| \tag{5}$$

is bounded from below if $V(\mathbf{r})$ increases more slowly than $1/r^2$. In the opposite case the particle falls into the center, i.e., ϵ can tend to $-\infty$. This argument is essentially based on the fact that the kinetic energy increases like $k^2 \gtrsim \hbar^2/r^2$. But it is seen from (2) that if we neglect the term $k^2/2M$, as we have done, the kinetic energy increases in our case in proportion to $|\mathbf{k}|$ when $\mathbf{k}$ is large, so that in a relation of the type (5) even a Coulomb increase is sufficient to cause recombination. An account of the $k^2/2M$ term limits this recombination at very large momenta, i.e., very short distances, which are of no interest to us.

We note finally that Eq. (4) contains one more term—the commutator of kinetic into potential energy. It is equivalent to some effective spin-orbit coupling and behaves at small distances like $\nabla V(\mathbf{r})\nabla$. We shall not analyze it in detail at present, since it has a somewhat different form in real models, which we now proceed to consider.

The main difference between these models and the schematic equation (3) lies in the fact that in these models we deal with a multi-component ψ-function (number of components equal to the number of bands considered), and Eq. (3) is then replaced by a system of first-order equations. Following the method of Luttinger and Kohn [1], we represent the solution of the complete Schrödinger equation

$$\left\{-\frac{\hbar^2}{2m_0}\nabla^2 + W(\mathbf{r}) + V(\mathbf{r})\right\}\psi(\mathbf{r}) = \varepsilon\psi(\mathbf{r}) \tag{6}$$

in the form

$$\psi(\mathbf{r}) = \sum_{n\mathbf{k}} c_n(\mathbf{k})\, e^{i\mathbf{kr}/\hbar} u_{n0}(\mathbf{r}). \tag{7}$$

Here m_0 is the mass of the free electron $W(\mathbf{r})$ the periodic potential of the crystal lattice, $V(\mathbf{r})$ the potential of the impurity center, and u_{n0} the purely periodic (i.e., corresponding to $\mathbf{k} = 0$) solutions of (6) without the external potential $V(\mathbf{r})$. We assume for simplicity that the minima of both the electron and hole bands lie at $\mathbf{k} = 0$. The eigenvalues corresponding to u_{n0} are denoted ϵ_{n0}. We can then readily obtain for the functions

$$\varphi_n(\mathbf{r}) = \sum_{\mathbf{k}} c_n(\mathbf{k})\, e^{i\mathbf{kr}/\hbar} \tag{8}$$

the following system of equations:

$$\left\{-\frac{\hbar^2}{2m_0}\nabla^2 + \varepsilon_{n0} - \varepsilon\right\}\varphi_n(\mathbf{r}) - \sum_{n'}\left\{\frac{i\hbar\,\mathbf{p}_{nn'}}{m_0}\nabla + \widetilde{V}_{nn'}(\mathbf{r})\right\}\varphi_{n'}(\mathbf{r}) = 0, \tag{9}$$

where

$$\widetilde{V}_{nn'}(\mathbf{r}) = \frac{1}{(2\pi)^3}\int e^{i\mathbf{k}(\mathbf{r}-\mathbf{r}')/\hbar}\, u^*_{n0}(\mathbf{r}')\, V(\mathbf{r}')\, u_{n'0}(\mathbf{r}')\, d\mathbf{k}\, d\mathbf{r}', \tag{10}$$

$$\mathbf{p}_{nn'} = -i\hbar\int u^*_{n0}(\mathbf{r})\,\nabla u_{n'0}(\mathbf{r})\,d\mathbf{r}, \quad \mathbf{p}_{nn} = 0, \quad \mathbf{p}_{nn'} = \mathbf{p}^*_{n'n}. \tag{11}$$

In the last formula the integral is taken over the volume of the unit cell, for which the functions u_{n0} are normalized. By virtue of the orthogonality of this system of functions, the potential $\widetilde{V}_{nn'}(\mathbf{r})$ reduces to

$$\widetilde{V}_{nn'}(\mathbf{r}) = V(\mathbf{r})\,\delta_{nn'} \tag{12}$$

accurate to terms of order λ, if $V(\mathbf{r})$ changes noticeably over distances of order $\hbar/\sqrt{m\Delta}$, as we assume.

Let us now break down the entire infinite set of functions φ_n into two groups. The first, designated by indices c and v, will include those numbers n for which u_{n0} correspond to the bottom of the conduction or valence band. The second group includes all the remaining ("far") bands. They will be designated by the index j ($j \neq c, v$). The possible degeneracy of the c and v bands will be taken into account by means of a supplementary index, which will not be written out for the time being, for it is best to introduce it in different fashion for different specific cases. Using standard perturbation theory with respect to $\mathbf{kp}_{nn'}$,[1] we can eliminate the φ_j from the equations for $\varphi_{c,v}$. The system (9) then reduces to

$$\left\{\frac{\hbar^2}{2m_0} D^{cc}_{\alpha\beta} \nabla_\alpha \nabla_\beta + V(\mathbf{r}) + \Delta - \varepsilon\right\} \varphi_c(\mathbf{r}) - \left\{\frac{i\hbar}{m_0} \mathbf{p}_{cv}\nabla - \frac{\hbar^2}{2m_0} D^{cv}_{\alpha\beta}\nabla_\alpha\nabla_\beta\right\} \varphi_v(\mathbf{r}) = 0,$$

$$\left\{\frac{\hbar^2}{2m_0} D^{vv}_{\alpha\beta} \nabla_\alpha \nabla_\beta + V(\mathbf{r}) - \Delta - \varepsilon\right\} \varphi_v(\mathbf{r}) - \left\{\frac{i\hbar}{m_0} \mathbf{p}_{vc}\nabla - \frac{\hbar^2}{2m_0} D^{vc}_{\alpha\beta}\nabla_\alpha\nabla_\beta\right\} \varphi_c(\mathbf{r}) = 0, \tag{13}$$

$$D^{cc}_{\alpha\beta} = -\delta_{\alpha\beta} + 2\sum_j \frac{(p_{cj})_\alpha (p_{jc})_\beta}{m_0(\varepsilon_{j0} - \Delta)},$$

$$D^{cv}_{\alpha\beta} = 2\sum_j \frac{(p_{cj})_\alpha (p_{jv})_\beta}{m_0(\varepsilon_{j0} + \Delta)} \quad \text{etc.} \tag{14}$$

The indices α and β run through the three values x, y, and z; the energy is reckoned from the center of the forbidden band the width of which is 2Δ. We emphasize once more that generally speaking $\varphi_c(\mathbf{r})$ and $\varphi_v(\mathbf{r})$ have in the case of band degeneracy several components. All the quantities $D^{cc}_{\alpha\beta}$, $D^{cv}_{\alpha\beta} = (D^{vc}_{\alpha\beta})^*$, and $D^{vv}_{\alpha\beta}$ are square matrices.

At the energies $\epsilon \sim \Delta$ of interest to us, all the momenta are of order $\sqrt{m\Delta}$, and all terms of the type $D_{\alpha\beta}\nabla_\alpha\nabla_\beta$ can be omitted, with accuracy of order λ. In other words, the effective mass is brought about in the case of small Δ only by the $\mathbf{kp}$ interaction of the neighboring bands[4]. An exception are the bands for which all the matrix elements, of momentum $\mathbf{p}$, coupling them to the other c and v bands, vanish, Singular cases of this type, as is well known, are heavy holes. We therefore begin the analysis with an artificial model, in which heavy holes play no role whatever.

This situation can be attained in the Kane model[4] if it is assumed that the spin-orbit splitting energy is $\Lambda \gg \Delta$, but has a sign opposite that of the real semiconductors InSb, Ge, etc. The upper valence band is then the band with total momentum $j = 1/2$, which is doubly degenerate in the spin. Exactly the same result is obtained if the usual sign of Λ is retained, but it is assumed that the triple representation corresponds to the conduction band, while the valence band is not degenerate. To be specific, we shall consider in what follows the first of these variants. Then there are two functions $\varphi^{\pm}_c$ corresponding to spin projections $+1/2$ and $-1/2$, along the momentum axis, and six functions $\varphi^{\pm}_{v\alpha}$, which transform like x, y, and z with respect to the index α. Accordingly

$$(p_{cv\alpha})_\beta = p\delta_{\alpha\beta}. \tag{15}$$

It is more convenient, however, to go over to a different representation of the valence functions, in which the total momentum j is diagonal, and with it the spin-orbit interaction[4,7]:

$$j = \tfrac{1}{2}, \quad \chi^{\pm}_1 = \pm\left[\varphi^{\pm}_{vz} - (\varphi^{\mp}_{vx} \pm i\varphi^{\mp}_{vy})\right]/\sqrt{3}, \tag{16}$$

$$j = \tfrac{3}{2}, \quad \chi^{\pm}_{3h} = \left[\varphi^{\pm}_{vx} \pm i\varphi^{\pm}_{vy}\right]/\sqrt{2},$$

$$\chi^{\pm}_{3l} = \left[2\varphi^{\pm}_{vz} + (\varphi^{\mp}_{vx} \pm i\varphi^{\mp}_{vy})\right]/\sqrt{6}. \tag{17}$$

We now turn again to (13), eliminating from the second equation the spin-orbit term $(1/2 - j)\Lambda$, which was not considered so far. The analysis is best continued in the $\mathbf{k}$-representation, introducing a special system of coordinates in which the z axis is directed along the vector $\mathbf{k}$. It is then seen directly that the wave function χ_{3h} of the heavy holes is not connected at all with the others, and the equation for it can be left out. The remaining equations assume the form

$$[\varepsilon - \Delta - V(-i\hbar\nabla_{\mathbf{k}})]\,\varphi_c(\mathbf{k}) - (p/\sqrt{3}m_0)\,k\,[\sqrt{2}\chi_{3l}(\mathbf{k}) + \sigma_z\chi_1(\mathbf{k})] = 0,$$

$$[\varepsilon + \Delta - V(-i\hbar\nabla_{\mathbf{k}})]\,\chi_1(\mathbf{k}) - (p^*/\sqrt{3}m_0)\,k\sigma_z\varphi_c(\mathbf{k}) = 0,$$

$$[\varepsilon + \Delta + \Lambda - V(-i\hbar\nabla_{\mathbf{k}})]\,\chi_{3l}(\mathbf{k}) - \sqrt{2/3}\,p^*k\varphi_c(\mathbf{k})/m_0 = 0, \tag{18}$$

where σ_z is the third Pauli matrix.

In the energy region $\epsilon \sim \Delta$ we get from the third equation

$$\chi_{3l}(\mathbf{k}) \approx \sqrt{\frac{2}{3}}\frac{p^*k}{m_0\Lambda}\varphi_c(k) \sim \frac{\Delta}{\Lambda}\varphi_c(\mathbf{k}), \tag{19}$$

and if $\Lambda \gg \Delta$ the contribution of χ_{3l} in the first equation can be neglected. But then the remaining two equations constitute the Dirac equation for the bispinor $\binom{\varphi_c}{\chi_1}$ in the chosen special coordinate frame. In the coordinate representation and in an arbitrary coordinate system they assume the usual form

$$[\varepsilon - \Delta - V(\mathbf{r})]\,\varphi_c(\mathbf{r}) + i\hbar s\,\sigma\nabla\chi_1(\mathbf{r}) = 0,$$
$$[\varepsilon + \Delta - V(\mathbf{r})]\,\chi_1(\mathbf{r}) + i\hbar s\,\sigma\nabla\varphi_c(\mathbf{r}) = 0, \tag{20}$$

in which the constant $s = p/\sqrt{3}\,m_0$ plays the role of the velocity of light, and the effective mass is given by

$$m = \frac{\Delta}{s^2} = m_0\frac{3m_0\Delta}{p^2}. \tag{21}$$

Elimination of one of the functions from the system (20) leads directly to Eq. (4), except that the commutator of the potential into kinetic energy is replaced by the term

$$(\sigma\,\mathrm{grad}\ln[\varepsilon \pm \Delta - V(\mathbf{r})])(\sigma\,\mathrm{grad})$$
$$= \frac{\partial\ln[\varepsilon \pm \Delta - V(\mathbf{r})]}{\partial r}\left(\frac{\mathbf{r}}{r}\nabla + \frac{\sigma\mathbf{L}}{r}\right), \tag{22}$$

where $\mathbf{L}$ is the momentum operator, and the signs + and − appear in the equations for χ_1 and φ_c, respectively. The first part of (22) is written already for a spherically symmetrical potential.

Let us consider by way of an example the known[8] solutions of the Dirac equation in a Coulomb field $V(r) = Ze^2/r$ (Ze is the charge of the center, divided by the effective dielectric constant, which generally speaking differs from its value at infinity). We note only the results of greatest interest to us. The energy levels are determined by the two quantum numbers n and γ:

$$\varepsilon_{n\gamma} = \Delta/\sqrt{1 + \alpha^2/(n+\gamma)^2}, \qquad \alpha = Ze^2/\hbar s,$$
$$n = 0, 1, 2, 3, \ldots, \qquad \gamma^2 = (j+\tfrac{1}{2})^2 - \alpha^2, \tag{23}$$

and j is the total momentum. The quantity γ^2 determines the effective centrifugal potential. At sufficiently large Z it becomes negative, and the corresponding stationary levels, as can be seen from (23), vanish, i.e., there is falling into the center and a resultant decrease in Z. All the electronic levels lie in the upper half of the forbidden band ($\epsilon > 0$ for $Z > 0$). This rule is violated, however, if the deviations of the field from a Coulomb one are taken into account. Finally, we note also that the wave function of the continuous spectrum behaves like $r^{\gamma-1}$ as $r \to 0$, for both attractive and repulsive centers. Thus, for $j = 1/2$ the probability of finding an electron near the center increases like $\exp(-2\alpha^2 \ln r)$, which can noticeably increase the cross section for the capture of slow electrons at large Z.

We now turn to a case of greater practical interest, when the spin-orbit splitting is small compared with the width of the forbidden band. Such a model is closer to the real situation occurring in Ge, and particularly in crystals of the GeAs type, where the minima of both bands lie at $\mathbf{k} = 0$. Neglecting the spin-orbit coupling, we can disregard in general the presence of electron spin. Then we have one function φ_c and three functions that transform like the components of the vector φ_v. The vector $\mathbf{p}_{cv}$ is then parallel to φ_v. But then, as can be readily verified, only the band of light holes, defined by the condition $\varphi_{\Lambda v}(\mathbf{k}) \parallel \mathbf{k}$, is directly coupled with the conduction band, and this explains the smallness of the effective mass in this band. The effective masses of the two other hole bands are entirely connected with the values of $D^{vv}_{\alpha\beta}$, which must therefore be left in the equation. We make, however, one more simplification, assuming that the matrices $D_{\alpha\beta}$ have not cubic but spherical symmetry. We then get from (13) the following equations:

$$[\varepsilon - \Delta - V(\mathbf{r})]\,\varphi_c(\mathbf{r}) + i\hbar s\,\mathrm{div}\,\varphi_v(\mathbf{r}) = 0,$$
$$\left[-\frac{\hbar^2}{2M}\,\mathrm{rot}\,\mathrm{rot} - \frac{\hbar^2}{2M'}\,\mathrm{grad}\,\mathrm{div} + \varepsilon + \Delta - V(\mathbf{r})\right]\varphi_v(\mathbf{r})$$
$$+ i\hbar s\,\mathrm{grad}\,\varphi_c(\mathbf{r}) = 0, \tag{24*}$$

where M is the mass of the heavy hole, M′ is connected with the mass of the light holes M_l by the relation $M_l^{-1} = M'^{-1} + m^{-1}$ (here $M_l \approx m$, since $M' \gg m$), and the parameters s and m are determined, as in the preceding case, by (21). The second term M'^{-1} in the second equation can be omitted, since it yields only a small correction to the mass of the light holes. From the same consideration, the curl curl operator in the first term can be replaced by $-\nabla^2$.

Let us investigate now the possible types of solutions of the system (24) for a spherically symmetrical potential $V(\mathbf{r})$.

First, there are solutions satisfying the supplementary condition $\varphi_v = 0$, meaning that there is no contribution of the light holes to the wave function. Then φ_c also vanishes, and the remaining equation

$$[(\hbar^2/2M)\nabla^2 + \varepsilon + \Delta - V(r)]\,\varphi_v(r) = 0 \tag{25}$$

describes the motion of the heavy hole in the poten-

*rot = curl.

tial V(r). In order to satisfy automatically the supplementary condition, we must seek φ_V in the form

$$\varphi_v(r) = \varphi(r)\,\mathbf{Y}_{lm}^{(0)}(\mathbf{n}), \quad \mathbf{Y}_{lm}^{(0)}(\mathbf{n}) = -i/[\sqrt{l(l+1)}]\,[\mathbf{r}\nabla]\,Y_{lm}(\mathbf{n}), \quad \mathbf{n} = \mathbf{r}/r. \tag{26}$$

Then the radial function $\varphi(r)$ satisfies the relation

$$\frac{\hbar^2}{2M}\frac{1}{r^2}\frac{d}{dr}\left(r^2\frac{d\varphi}{dr}\right) + \left[\varepsilon + \Delta - V(r) - \frac{\hbar^2}{2M}\frac{l(l+1)}{r^2}\right]\varphi = 0. \tag{27}$$

Among these states, however, there are no spherically symmetrical ones ($l = 0$), as can be verified directly from (26).

States of this type form a second group of solutions, in which there is no contribution of the heavy holes (curl $\varphi_V = 0$). For these we have $\varphi_c(\mathbf{r}) = \varphi_c(r)$, $\varphi_V(\mathbf{r}) = \mathbf{n}\varphi_V(r)$, and the equation for the radial functions

$$[\varepsilon - \Delta - V(r)]\,\varphi_c(r) + i\hbar s\frac{1}{r^2}\frac{d(r^2\varphi_v)}{dr} = 0,$$

$$[\varepsilon + \Delta - V(r)]\,\varphi_v(r) + i\hbar s\frac{d\varphi_c}{dr} = 0 \tag{28}$$

coincides with the Dirac equation for $j = 1/2$.

The greatest interest attaches to the third group of solutions, in which the potential V(r) tangles together the light and heavy holes, so that none of the supplementary conditions used above is satisfied, i.e., div $\varphi_V \not\equiv 0$, curl $\varphi_V \not\equiv 0$, and $l \neq 0$. These solutions are analyzed in detail in the Appendix. Their main feature is that the behavior of $\varphi(\mathbf{r})$ at large distances from the center is determined by the mass of the electron or of the light hole m, while near the center there is a narrow region with width of order $\hbar/\sqrt{M\Delta}$, where $|\varphi|^2$ increases sharply because of the presence of heavy holes (we have in mind a center that attracts holes, i.e., $V(r) > 0$). At sufficiently large Z, when $\alpha^2 \gtrsim m/M$, such levels appear above the bottom of the conduction band. These are essentially additional electronic levels, but they can be arbitrarily called virtual heavy-hole levels, since they fall into the region of the continuous spectrum but have a wave function of symmetry different than the neighboring continuous-spectrum functions. The capture of a hole by such a center, i.e., the departure of a randomly captured electron from such a center, can therefore occur sufficiently slowly, i.e., these will be quasistationary levels of relatively low width. The cross section for the capture on them should have a resonant character of the Breit-Wigner type.

If the lifetime of such a level is sufficiently long, then there is a noticeable probability of transition from this level to levels of the same center in the forbidden band, which are therefore stationary in the full meaning of the word. This is how an electron is captured by a negatively charged center. In real multiply charged centers, an important role can be played in such capture by the presence of other electrons in the center. Indeed, for example, the ground-state level of the three-electron system Cu^{---} in Ge lies at 0.26 eV, under the bottom of the conduction band. It is quite probable, however, that this system also has excited levels, some of which can fall in the conduction band. This makes possible the already described resonant capture of electrons at these excited levels with subsequent cascade[9] transition into the ground state.

A detailed quantitative analysis of the recombination process calls for calculation of the polarization corrections to the potential and for a more accurate account of the real band structure. All these are beyond the scope of the present paper, which is aimed at showing the possibility in principle of describing deep levels in semiconductors within the framework of an approximation of the effective-mass type, and explain at the same time such qualitative features of these levels as the ability of capturing carriers of either polarity, for example.

In conclusion I wish to express my deep gratitude to G. E. Pikus, whose remarks played an important role in the performance of this work, and particularly in its formulation.

APPENDIX

The system of equations (24) leads, after elimination of the component $\varphi_c(\mathbf{r})$, to the following equation for the vector $\varphi_V(\mathbf{r})$:

$$\left\{\frac{\hbar^2}{2M}\operatorname{rot}\operatorname{rot}\varphi_v(\mathbf{r}) - \frac{\hbar^2\Delta}{m}\operatorname{grad}\frac{\operatorname{div}\varphi_v(\mathbf{r})}{\varepsilon - \Delta - V(\mathbf{r})} - [\varepsilon + \Delta - V(\mathbf{r})]\,\varphi_v(\mathbf{r})\right\} = 0. \tag{A.1}$$

For the case of interest to us, that of an acceptor impurity, $V(\mathbf{r}) > 0$ and if $\epsilon < \Delta$ the determination of the lowest level reduces to the variational problem

$$\min\Phi\{\varphi_v\} = \min\int\left\{\frac{\hbar^2}{2M}|\operatorname{rot}\varphi_v|^2 + \frac{\hbar^2}{m}\frac{\Delta}{\Delta - \varepsilon + V(\mathbf{r})}|\operatorname{div}\varphi_v|^2 - V(\mathbf{r})|\varphi_v|^2\right\}d\mathbf{r} = -\varepsilon - \Delta \tag{A.2}$$

with the additional normalization condition

$$\int\left\{\frac{\hbar^2}{m\Delta}\frac{\Delta^2}{[\Delta - \varepsilon + V(\mathbf{r})]^2}|\operatorname{div}\varphi_v|^2 + |\varphi_v|^2\right\}d\mathbf{r} = 1. \tag{A.3}$$

Equation (A.2) calls, however, for additional explanations, since the integrand in it depends on ϵ as a parameter. Therefore, strictly speaking, we have in mind the following procedure: We introduce under the integral sign in lieu of ϵ a new parameter $\eta(\epsilon \to \eta)$, carry out the variation for a specified $\eta < \Delta$, determine the extremal $\Phi_{min}(\eta)$ and $\varphi_V(\eta)$, and then solve the transcendental equation

$$\Phi_{min}(\varepsilon) = -\varepsilon - \Delta, \qquad (A.4)$$

which determines the sought-for level. We shall see, however, that the real problem is much simpler.

Indeed, by virtue of the condition $m \ll M$ the second term in the integrand of (A.2) is very large and positive in all cases when div φ_V differs noticeably from zero. Therefore the requirement that the functional be a minimum leads to the condition div $\varphi_V \approx 0$, which is satisfied, as can be readily verified, with such an accuracy that the second term becomes much smaller than the first in (A.2). Then the terms containing div φ_V in (A.2) and (A.3) can be omitted, and the problem is reduced to the usual form, but with the limitation that the variation is carried out on the class of functions satisfying the condition div $\varphi_V = 0$. One possibility of satisfying this condition was indicated above, namely formula (26). Expansion of φ_V in spherical vectors shows that in addition to (26) there are solutions of the type

$$\varphi_v(\mathbf{r}) = \varphi_{\parallel}(r)\, \mathbf{Y}_{lm}^{(-1)}(\mathbf{n}) + \varphi_{\perp}(r)\, \mathbf{Y}_{lm}^{(+1)}(\mathbf{n}), \qquad (A.5)$$

which form with (26) a complete system. Here

$$\mathbf{Y}_{lm}^{(-1)}(\mathbf{n}) = \mathbf{n} Y_{lm}(\mathbf{n}), \quad \mathbf{Y}_{lm}^{(+1)}(\mathbf{n}) = \frac{r}{\sqrt{l(l+1)}} \nabla Y_{lm}(\mathbf{n}). \qquad (A.6)$$

A particular case of (A.6) are the solutions (28) considered above. For the functions (A.5) we have

$$\operatorname{div} \varphi_v(r) = \left\{ \frac{1}{r^2} \frac{d}{dr}(r^2 \varphi_{\parallel}) - \frac{\sqrt{l(l+1)}}{r} \varphi_{\perp} \right\} Y_{lm}(\mathbf{n}),$$

so that when $l \neq 0$ the condition div φ_V can be satisfied by putting

$$\varphi_{\perp}(r) = \frac{1}{r\sqrt{l(l+1)}} \frac{d}{dr}(r^2 \varphi_{\parallel}), \qquad (A.7)$$

after which one independent radial function $\varphi_{\parallel}(r)$ is left for variation, and Eqs. (A.2) and (A.3) reduce to

$$\Phi_l\{\varphi_{\parallel}\} = \int \left\{ \frac{\hbar^2}{2Mr^2} \left[\frac{d^2}{dr^2} \frac{r^2 \varphi_{\parallel}}{\sqrt{l(l+1)}} + \sqrt{l(l+1)}\, \varphi_{\parallel} \right]^2 - V(r) \left[\varphi_{\parallel}^2 + \frac{1}{r^2} \left(\frac{d}{dr} \frac{r^2 \varphi_{\parallel}}{\sqrt{l(l+1)}} \right)^2 \right] \right\} r^2 dr, \qquad (A.8)$$

$$\int \left\{ \frac{1}{r^2} \left(\frac{d}{dr} \frac{r^2 \varphi_{\parallel}}{\sqrt{l(l+1)}} \right)^2 + \varphi_{\parallel}^2(r) \right\} r^2\, dr = 1. \qquad (A.9)$$

In a potential of the Coulomb type the lowest level is six-fold degenerate (with account of the spin) with $l = 1$, as expected for a vector particle. Account of the spin-orbit interaction in the band structure would split it into four-fold and doubly degenerate levels with $j = 3/2$ and $j = 1/2$, as is observed at the shallow levels in Ge[3].

Using the two-parameter family of functions $\varphi_{\parallel}(r) = A(\alpha, \gamma)(1 + 2\gamma\alpha r) \exp(-\alpha r)$, we can obtain for the energy of the ground state in a Coulomb field

$$\varepsilon_0 \approx +1.4 \frac{Z^2 e^4 M}{2\hbar^2} - \Delta, \quad \alpha = 1.7 \frac{Me^2}{\hbar^2}, \quad \gamma = \frac{1}{4}. \qquad (A.10)$$

The solution obtained from (A.8), as in (26), describes essentially the states of heavy holes, as demonstrated at least by the absence of the mass m from these equations. In other words, φ_V, like any other vector, can be resolved into solenoidal (heavy holes) and potential (light holes) components:

$$\varphi_v(\mathbf{r}) = \operatorname{rot} \mathbf{B}(\mathbf{r}) + \operatorname{grad} P(\mathbf{r}), \qquad (A.11)$$

and the condition div $\varphi_V = 0$ excludes the possible presence of heavy holes, while the contribution of the states of the conduction band also vanishes by virtue of the first equation of (24). However, unlike (26), the relation (A.7) is approximate, so that generally speaking it is not compatible with the exact equations for $\varphi_{\parallel}(r)$ and $\varphi_{\perp}(r)$, i.e., div φ_V is not rigorously equal to zero in this case and is a small quantity of order higher than $\sqrt{m/M}$.

Using the solution obtained as a zeroth approximation, we obtain the component $\varphi_c(\mathbf{r})$. To this end, multiplying the second equation of (24) by $-i\hbar\sqrt{\Delta/m}$ div, we get

$$\frac{\hbar^2}{2m} \nabla^2 \varphi_c(\mathbf{r}) + \frac{[\varepsilon - V(\mathbf{r})]^2 - \Delta^2}{2\Delta} \varphi_c(\mathbf{r}) = -\frac{i\hbar}{2\sqrt{m\Delta}} \varphi_v(\mathbf{r}) \operatorname{grad} V(\mathbf{r}). \qquad (A.12)$$

For a spherically symmetrical potential, the right half reduces by virtue of (A.5) and (A.6) to $\varphi_{\parallel}(r)[dV(r)/dr]\, Y_{lm}(\mathbf{n})$, and therefore $\varphi_c(\mathbf{r}) = \varphi_c(r)\, Y_{lm}(\mathbf{n})$. The radial function $\varphi_c(r)$ can be expressed in the usual fashion in terms of the solutions of the homogeneous radial equation

$$\frac{\hbar^2}{2m} \left\{ \frac{1}{r^2} \frac{d}{dr} \left(r^2 \frac{df}{dr} \right) - \frac{l(l+1)}{r^2} f \right\} + \frac{[\varepsilon - V(r)]^2 - \Delta^2}{2\Delta} f = 0, \qquad (A.13)$$

one of which, $f^{(0)}(r)$, is regular at zero, and the other, $f^{(\infty)}(r)$, is regular at infinity:

$$\varphi_c(r) = \frac{f^{(0)}(r)}{\hbar s} \int_r^{\infty} \varphi_{\parallel}(r') \frac{dV(r')}{dr'} f^{(\infty)}(r') r'^2 dr' + \frac{f^{(\infty)}(r)}{\hbar s} \int_0^r \varphi_{\parallel}(r') \frac{dV(r')}{dr'} f^{(0)}(r')\, r'^2\, dr'. \qquad (A.14)$$

Compared with the rapidly decreasing $\varphi_{\parallel}(r)$ and $\varphi_{\perp}(r)$, the functions $f^{(0)}(r)$ and $f^{(\infty)}(r)$ are slowly varying. The corresponding characteristic lengths are of the order $\hbar\sqrt{\Delta/m(\Delta^2-\epsilon^2)}$, i.e., $\sim\sqrt{M/m}\,\alpha^{-1}$ Therefore at large distances from the center ($\gg\alpha^{-1}$) the main contribution to the total wave function is made by the second term of (A.14) which in this region is of the form

$$\varphi_c(r)\approx Cf^{(\infty)}(r),\quad C=\frac{1}{\hbar s}\int_0^\infty \varphi_{\parallel}(r)\frac{dV(r)}{dr}f^{(0)}(r)\,r^2\,dr. \tag{A.15}$$

The smallness of C is ensured by the fact that in the integral of (A.15) the function $f^{(0)}(r)$, which is normalized to a volume on the order of $(M/m\alpha^2)^{3/2}$, is integrated over a much smaller volume $\sim\alpha^{-3}$, determined by the decrease of $\varphi_{\parallel}(r)$. For the Coulomb case under consideration we have for $l=1$

$$\varphi_c(r)\approx\frac{Ze^2}{\hbar s}\frac{\gamma+3}{2}\frac{\Gamma(\gamma+1)\Gamma(1/2-\mu+\gamma)}{\Gamma(2\gamma+1)}\left(\frac{2\varkappa\hbar^2}{1.7\,Ze^2M}\right)^{\gamma+1}W_{\mu,\gamma}(2\varkappa r)$$

$$=\frac{Ze^2}{\hbar s}\frac{\gamma+3}{2}\frac{\Gamma(\gamma+1)\Gamma(1/2-\mu+\gamma)}{\Gamma(2\gamma+1)}$$

$$\times\left[2\frac{m}{M}\frac{\Delta^2-\varepsilon^2}{\Delta(\varepsilon_0+\Delta)}\right]^{(\gamma+1)/2}W_{\mu,\gamma}(2\varkappa r);$$

$$\mu=-\alpha\sqrt{\frac{\varepsilon^2}{\Delta^2-\varepsilon^2}},\quad \gamma^2=\left(l+\frac{1}{2}\right)^2-\alpha^2,$$

$$\varkappa=\sqrt{\frac{m(\Delta^2-\varepsilon^2)}{\hbar^2\Delta}}. \tag{A.16}$$

Here $W_{\mu,\gamma}(2\kappa r)$ is the Whittaker function. The relative amplitude of this component is of order m/M when $\epsilon_0\sim\Delta$, i.e., it is sufficiently small. However, it decreases much more slowly than $\varphi_{\parallel}(r)$ at large distances.

Let us consider, finally, the case when V(r) is so large that the energy of the hole ground state ϵ_0 is larger than Δ, i.e., it falls in the conduction band. At first glance the variational procedure employed is not applicable at all in this case, since the coefficient preceding $|\mathrm{div}\,\varphi_V|^2$ in (A.3) is not positive definite. This is obviously due to the fact that there are arbitrarily high conduction band states (in the scheme under consideration), so that the sought hole state can certainly not correspond to the absolute minimum of the functional. But we are seeking the lowest hole state. Consequently, it should be orthogonal in the zeroth approximation to all the conduction-band states, which is equivalent to requiring that $\varphi_c(r)=0$, i.e., $\mathrm{div}\,\varphi_V=0$. Thus, the problem reduces again to variation of the approximate functional (A.8) and its solution is given by (A.10). The small contribution of the conduction-band states is also expressed by (A.14)—(A.16), in which, however, $f^{(0)}(r)$, $f^{(\infty)}(r)$, and $W_{\mu,\gamma}$ are already functions of the continuous spectrum. The virtual levels thus obtained should play an important role in the capture of carriers by the center or in photoconductivity, and also give the resonance scattering of the conduction electrons.

[1] W. Kohn and J. M. Luttinger, Phys. Rev. **97**, 869 (1955) and **98**, 915 (1955).
[2] W. Kohn and D. Schechter, Phys. Rev. **99**, 1903 (1955).
[3] W. Kohn, Sol. State Phys. **5**, 257 (1957).
[4] E. O. Kane, J. Phys. Chem. Sol. **1**, 249 (1957).
[5] L. V. Keldysh, JETP **33**, 994 (1957), Soviet Phys. JETP **6**, 763 (1958).
[6] L. D. Landau and E. M. Lifshitz, Quantum Mechanics, Pergamon, 1958.
[7] Roth, Lax, and Zwerdling, Phys. Rev. **114**, 90 (1959).
[8] A. I. Akhiezer and V. B. Berestetskiĭ, Kvantovaya elektrodinamika (Quantum Electrodynamics), 1953, Sec. 12.
[9] M. Lax, Phys. Rev. **119**, 1502 (1960).

Translated by J. G. Adashko
62

SOVIET PHYSICS–SOLID STATE VOL. 6, NO. 9 MARCH, 1965

POSSIBLE INSTABILITY OF THE SEMIMETALLIC STATE TOWARD COULOMB INTERACTION

L. V. Keldysh and Yu. V. Kopaev

P. N. Lebedev Physics Institute, Academy of Sciences, USSR, Moscow
Translated from Fizika Tverdogo Tela, Vol. 6, No. 9, pp. 2791-2798, September, 1964
Original article submitted April 18, 1964

It is shown that in a semimetal having the same number of electrons and holes, it is possible for the ground state to be unstable to pairing of the electrons in one band with the holes in another as a result of Coulomb interaction between them. This forms a dielectric which, however, has the energy spectrum of a superconductor, and thus behaves like a superconductor in its thermodynamic characteristics. In particular, there is some critical temperature at which it will be converted into a semimetal by a second-order phase transition. The possibility is noted of applying this model to the problem of conversion of semiconductors into semimetals at high temperatures.

Interaction of electrons with one another undoubtedly exerts a substantial effect on the energy spectra of solids. The clearest example of this occurs in superconductivity: even weak attraction of electrons by phonons results in this case in rearrangement of the electron spectrum at the Fermi surface, due to the formation of copper pairs, and in a complete change of all the electronic properties of the material [1-4].

On the other hand, Arkhipov [5] has advanced some qualitative ideas which show that a Coulomb interaction results in the electrons sticking to the holes, thus converting the metal into a dielectric when the conditions are such that the Coulomb energy is greater than the Fermi energy.

In the present paper we consider a simple model of a semimetal containing $\underline{n}$ electrons in one band and $\underline{n}$ holes in another. It will be shown that Coulomb interaction makes such a state unstable to pairing of the electrons of the one zone with the holes in the other. As a result, the system is a dielectric at low temperatures. However, just as in a superconductor, the gap in the excitation spectrum will decrease with increase in temperature, and there will be some critical temperature T_c at which second order phase transition occurs: the gap closes up, and the system is converted into a semimetal.* A dielectric of this type is similar to a superconductor in its thermodynamic characteristics in the low-temperature range $T < T_c$. It is interesting to note in this connection that the gaps in the spectra of the majority of semiconductors decrease with increase in temperature, i.e., they all have a tendency to change to the metallic state at high temperatures [6]. It was shown by Godefroy and Aigrain that the decrease in gap width may also be accounted for by electron-phonon interaction.

We shall assume that the energy spectrum of the original semimetal consists of two bands, an electron band $\varepsilon_1(\mathbf{p})$, and a hole band $\varepsilon_2(\mathbf{p})$, as shown in Fig. 1. Here ε is the energy of the current carrier, and $\mathbf{p}$ is the quasi-momentum of the carriers.

Assume for simplicity that the extrema of $\varepsilon_1(\mathbf{p})$ and $\varepsilon_2(\mathbf{p})$ occur at $p = 0$, and that both functions are isotropic, i.e., they depend only on $p = |\mathbf{p}|$. Then it follows from the electrical neutrality condition that the Fermi momenta p_0 as given by the relations

$$\varepsilon_1(p_0) = \mu \text{ and } \varepsilon_2(p_0) = \mu \quad (1)$$

will be the same in both bands (μ is the chemical

*Similar results, so we have heard, were obtained by V. M. Galitskii.

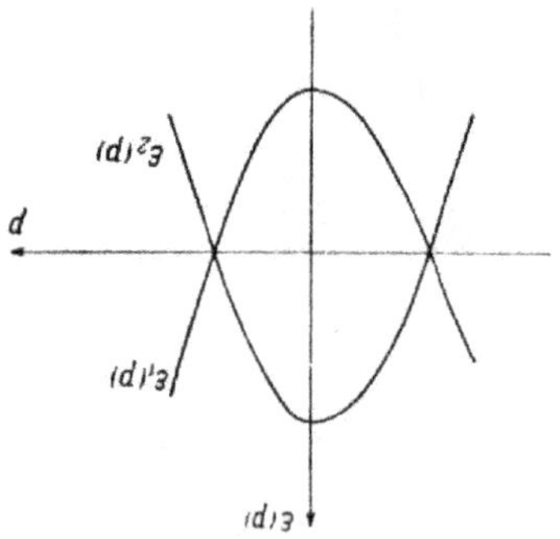

Fig. 1. Energy spectrum of the original semimetal.

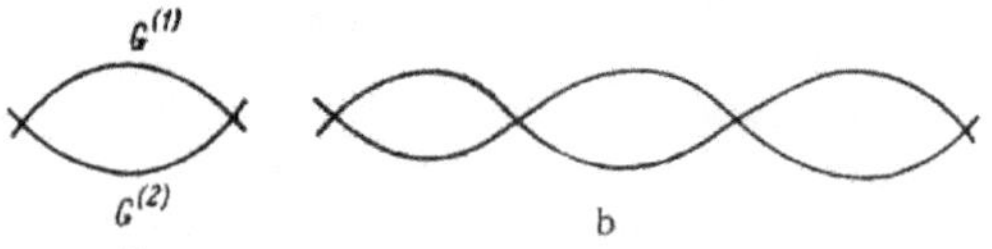

Fig. 2. Vertex describing scattering of an electron by a hole (a), and the corrections to it (b).

potential of the electrons). The concentrations of the electrons and holes are related to p_0 in the usual way

$$n = \frac{8\pi}{3}\left(\frac{p_0}{2\pi\hbar}\right)^3. \tag{2}$$

In second quantization representation, the Hamiltonian of such a system including Coulomb interaction is of the following form:

$$H = \int \psi_1^+(\mathbf{r})\varepsilon_1(\hat{p})\psi_1(\mathbf{r})\,d\mathbf{r} + \int \psi_2^+(\mathbf{r})\varepsilon_2(\hat{p})\psi_2(\mathbf{r})\,d\mathbf{r}$$
$$+ \sum_{\alpha=1}^{2}\iint \psi_\alpha^+(\mathbf{r}_1)\psi_\alpha^+(\mathbf{r}_2)V(\mathbf{r}_1-\mathbf{r}_2)\psi_\alpha(\mathbf{r}_2)\psi_\alpha(\mathbf{r}_1)\,d\mathbf{r}_1 d\mathbf{r}_2$$
$$+ \iint \psi_1^+(\mathbf{r}_1)\psi_2^+(\mathbf{r}_2)V(\mathbf{r}_1-\mathbf{r}_2)\psi_2(\mathbf{r}_2)\psi_1(\mathbf{r}_1)\,d\mathbf{r}_1 d\mathbf{r}_2, \tag{3}$$

where $V(\mathbf{r}) = -\frac{e^2}{r}e^{-\varkappa r_D}$ is the screened Coulomb potential, $\varkappa_D = r_D^{-1}$ is the reciprocal of the Debye screening radius, e is the effective charge reduced by the dielectric polarizability of the medium, and ψ_1 and ψ_2 are the electron operators in the electron and hole bands, which satisfy the usual commutation conditions

$$[\psi_\alpha^+(\mathbf{r}), \psi_\beta(\mathbf{r}')]_+ = \delta_{\alpha\beta}\delta(\mathbf{r}-\mathbf{r}');$$
$$[\psi_\alpha(\mathbf{r}), \psi_\beta(\mathbf{r}')]_+ = [\psi_\alpha^+(\mathbf{r}), \psi_\beta^+(\mathbf{r}')]_+ = 0, \quad \alpha, \beta = 1,2. \tag{4}$$

We shall now show that the ground state of the system with no interaction becomes unstable to pairing of the electrons of band 1 with the holes of band 2 when interaction is introduced, the nature of the instability being exactly the same as in the case of superconductivity [7]. For this purpose, consider the first correction to the vertex describing scattering of an electron of band 1 by a hole of band 2. The graph is shown in Fig. 2a, and the expression corresponding to the graph is of the form

$$\Gamma^{(1)}(\omega, \mathbf{k}) = -\frac{2ie^2}{(2\pi)^3}$$
$$\times \int \frac{G^{(1)}\left(\varepsilon' + \frac{\omega}{2}; \mathbf{p}' + \frac{\mathbf{k}}{2}\right)G^{(2)}\left(\varepsilon' - \frac{\omega}{2}; \mathbf{p}' - \frac{\mathbf{k}}{2}\right)}{(\mathbf{p}-\mathbf{p}')^2 + \varkappa_D^2}\,d\varepsilon' d\mathbf{p}', \tag{5}$$

where $G^{(1)}$ and $G^{(2)}$ are the Green's functions of the electrons in bands 1 and 2:

$$G^\alpha(\mathbf{r}-\mathbf{r}', t-t') = i\langle T\psi_\alpha^+(\mathbf{r}, t)\psi_\alpha(\mathbf{r}', t')\rangle. \tag{6}$$

In all the formulas Planck's constant $\hbar$ is set equal to unity.

In Fourier representation and zero order in the interaction

$$\left.\begin{aligned} G_0^{(1)}(\varepsilon, \mathbf{p}) &= \frac{1}{\varepsilon - \varepsilon_1(\mathbf{p}) + i\delta\,\mathrm{sign}\,\varepsilon_1(\mathbf{p})};\\ G_0^{(2)}(\varepsilon, \mathbf{p}) &= \frac{1}{\varepsilon - \varepsilon_2(\mathbf{p}) + i\delta\,\mathrm{sign}\,\varepsilon_2(\mathbf{p})}.\end{aligned}\right\} \tag{7}$$

It is not difficult to see that $\Gamma^{(1)}(\omega, \mathbf{k})$ has a logarithmic singularity as $\omega \to 0$ and $\mathbf{k} \to 0$, since in this case the poles of the two G-functions in the integrand of (5) coincide for $p' = p_0$. For example, in the case of quadratic dispersion laws

$$\varepsilon_\alpha(\mathbf{p}) = (-1)^\alpha\left(-\frac{p^2}{2m_\alpha} + \frac{p_0^2}{2m_\alpha}\right) \tag{8}$$

(here, as the reference point of the energies we use the position of the Fermi level, i.e., $\mu = 0$), we have

$$\left.\begin{aligned}\Gamma^{(1)}(\omega, \mathbf{k}=0) &= -\frac{me^2}{2\pi p_0}\ln\frac{\varkappa_D^2}{2p_0^2}\left(\frac{1}{2}\ln\frac{\omega}{-2\omega_0} + \frac{1}{2}\ln\frac{\omega}{2\omega_0}\right);\\ \omega_0 &\approx \varkappa_D v_F.\end{aligned}\right\}$$

Accordingly, this correction is not small for small values of ω, and we must sum up all the diagrams obtained from it by the interactions shown in Fig. 2b. Evaluating this sum is simply a matter of solving the integral equation

$$\Gamma(\omega, \mathbf{k}) = \Gamma^{(0)}(\omega, \mathbf{k}) - \frac{2ie^2}{(2\pi)^3}$$

$$\times \int \frac{G_0^{(1)}\left(\varepsilon' + \frac{\omega}{2};\ \mathbf{p}' + \frac{\mathbf{k}}{2}\right)\Gamma(\varepsilon', \mathbf{p}')G_0^{(2)}\left(\varepsilon' - \frac{\omega}{2};\ \mathbf{p}' - \frac{\mathbf{k}}{2}\right)}{(\mathbf{p}-\mathbf{p}')^2 + \varkappa_D^2} d\varepsilon' d\mathbf{p}'. \quad (10)$$

The solution is of the form [7]

$$\Gamma(\omega, \mathbf{k}=0) = \frac{\Gamma^{(0)}(\omega, \mathbf{k}=0)}{1 + \frac{me^2}{2\pi p_0}\ln\frac{\varkappa_D^2}{2p_0^2}\left(\ln\left|\frac{\omega}{2\omega_0}\right| - \frac{i\pi}{2}\right)}. \quad (11)$$

Continuing $\Gamma(\omega)$ analytically in the upper half-plane ω, we have

$$\Gamma(\omega) = \frac{\Gamma^{(0)}(\omega)}{1 + \frac{me^2}{2\pi p_0}\ln\frac{\varkappa_D^2}{2p_0^2}\left(\ln\left|\frac{\omega}{2\omega_0}\right| - \frac{i\pi}{2} + i\varphi\right)}, \quad (12)$$

i.e., there is a pole at the point

$$\Omega = 2\omega_0 \exp\left(\frac{2\pi v_F}{e^2 \ln\frac{\varkappa_D^2}{2p_0^2}}\right). \quad (13)$$

Since the poles of the vertex part and of the two-particle Green's function coincide, this imaginary pole shows that the ground state is unstable to formation of electron-hole pairs from different bands located at the Fermi surface.

The value of Ω gives the binding energy in the pair, i.e., the magnitude of the energy gap for elementary excitations in the system.

It may be seen from (5) that the poles of the Green's functions $G^{(1)}$ and $G^{(2)}$ in the integrand remain in coincidence (which is what causes the instability in the ground state) for any value of the variable of integration $\mathbf{p}'$ only for the case of an isotropic dispersion law; i.e., the instability does not occur for strong anisotropy.

It is easily seen that a graph of the type shown in Fig. 2a, but with the electron and the hole from the same band, has no singularity as $\mathbf{k} \to 0$:

$$\Gamma^{(1)}(\omega, \mathbf{k}) = -\frac{2ie^2}{(2\pi)^3}$$

$$\times \int \frac{G_0^{(1)}\left(\varepsilon' + \frac{\omega}{2},\ \mathbf{p}' + \frac{\mathbf{k}}{2}\right)G_0^{(1)}\left(\varepsilon' - \frac{\omega}{2};\ \mathbf{p}' - \frac{\mathbf{k}}{2}\right)}{(\mathbf{p}-\mathbf{p}')^2 + \varkappa_D^2} d\varepsilon' d\mathbf{p}'. \quad (14)$$

It is not difficult to show that in this case we get

$$\Gamma^{(1)}(\omega, \mathbf{k}) = -\frac{e^2|\mathbf{k}|}{2\pi\omega}\ln\frac{\varkappa_D^2}{2p_0^2}. \quad (15)$$

We pass now to a derivation of the equations for the complete Green's functions (6), including the interaction which produces the pairing [4]. Here, the only term that we leave in the Hamiltonian of the interaction is the one describing the interaction between carriers from different bands:

$$H_i = \int\int \psi_1^+(\mathbf{r}_1)\psi_2^+(\mathbf{r}_2)V(\mathbf{r}_1 - \mathbf{r}_2)\psi_2(\mathbf{r}_2)\psi_1(\mathbf{r}_1)d\mathbf{r}_1 d\mathbf{r}_2, \quad (16)$$

since we now know that this is the principal interaction.

Starting with the Hamiltonian (3) and using (14), we set up a system of equations for $\widetilde{\Psi}_1$ and $\widetilde{\Psi}_2$ in the Heisenberg representation:

$$\left(i\frac{\partial}{\partial t} - \varepsilon_1(\hat{\mathbf{p}})\right)\tilde{\psi}_1(\mathbf{r}, t) - \int \tilde{\psi}_2^+(\mathbf{r}_1, t)V(\mathbf{r}-\mathbf{r}_1)\tilde{\psi}_2(\mathbf{r}_1, t)\tilde{\psi}_1(\mathbf{r}, t)d\mathbf{r}_1 = 0; \quad (17)$$

$$\left(i\frac{\partial}{\partial t} + \varepsilon_1(\hat{\mathbf{p}})\right)\tilde{\psi}_1^+(\mathbf{r}, t) + \int \tilde{\psi}_1^+(\mathbf{r}, t)\tilde{\psi}_2^+(\mathbf{r}_1, t)V(\mathbf{r}-\mathbf{r}_1)\tilde{\psi}_2(\mathbf{r}_1, t)d\mathbf{r}_1 = 0; \quad (18)$$

$$\left(i\frac{\partial}{\partial t} - \varepsilon_2(\hat{\mathbf{p}})\right)\tilde{\psi}_2(\mathbf{r}, t) - \int \tilde{\psi}_1^+(\mathbf{r}_1, t)V(\mathbf{r}-\mathbf{r}_1)\tilde{\psi}_2(\mathbf{r}, t)\tilde{\psi}_1(\mathbf{r}_1, t)d\mathbf{r}_1 = 0; \quad (19)$$

$$\left(i\frac{\partial}{\partial t} + \varepsilon_2(\hat{\mathbf{p}})\right)\tilde{\psi}_2^+(\mathbf{r}, t) + \int \tilde{\psi}_1^+(\mathbf{r}_1, t)\tilde{\psi}_2^+(\mathbf{r}, t)V(\mathbf{r}-\mathbf{r}_1)\tilde{\psi}_1(\mathbf{r}_1, t)d\mathbf{r}_1 = 0. \quad (20)$$

We introduce, in addition to the Green's functions (6), the following:

$$\left.\begin{aligned} F(\mathbf{r}_1, t_1, \mathbf{r}_2, t_2) &= -i\langle T\tilde{\psi}_2(\mathbf{r}_2, t_2)\tilde{\psi}_1^+(\mathbf{r}_1, t_1)\rangle; \\ F^+(\mathbf{r}_1, t_1, \mathbf{r}_2, t_2) &= -i\langle T\tilde{\psi}_1(\mathbf{r}_1, t_1)\tilde{\psi}_2^+(\mathbf{r}_2, t_2)\rangle. \end{aligned}\right\} \quad (21)$$

We have seen from an analysis of the diagrams drawn above that a condensate will occur in the system of the pairs described by the functions F and F^+ in a way similar to what occurs in the problem of superconductivity [4]. This condensate makes the principal contribution to the interaction at the Fermi surface (not including renormalization of the chemical potential μ, which we leave out), for example:

$$\langle T\tilde{\psi}_1^+(\mathbf{r}_1, t_1)\tilde{\psi}_2^+(\mathbf{r}_2, t_2)\tilde{\psi}_2(\mathbf{r}_2, t_2)\tilde{\psi}_1(\mathbf{r}_1, t_1)\rangle = F(\mathbf{r}_1, t_1, \mathbf{r}_2, t_2)F^+(\mathbf{r}_1, t_1, \mathbf{r}_2, t_2).$$

Then from Eqs. (17) and (19) we obtain, after Fourier transformation,

$$\left.\begin{aligned} (\omega - \varepsilon_1(\mathbf{p}))G^{(1)}(\omega, \mathbf{p}) + \tilde{F}_\omega^+(\mathbf{p})F(\omega, \mathbf{p}) &= 1; \\ (\omega - \varepsilon_2(\mathbf{p}))F(\omega, \mathbf{p}) - \tilde{F}_\omega(\mathbf{p})G^{(1)}(\omega, \mathbf{p}) &= 0, \end{aligned}\right\} \quad (22)$$

where

$$\left.\begin{aligned}\tilde{F}_\omega(\mathbf{p})&=\frac{i}{(2\pi)^4}\int F(\varepsilon,\mathbf{k})V(\mathbf{p}-\mathbf{k})\,d\varepsilon d\mathbf{k};\\ \tilde{F}^+_\omega(\mathbf{p})&=\frac{i}{(2\pi)^4}\int F^+(\varepsilon,\mathbf{k})V(\mathbf{p}-\mathbf{k})\,d\varepsilon d\mathbf{k}.\end{aligned}\right\}\tag{23}$$

From Eqs. (18) and (20) we obtain a system of equations for $G^{(2)}(\omega, \mathbf{p})$ and $F^+(\omega, \mathbf{p})$ similar to (22). Solving (22) for $G^{(1)}(\omega, \mathbf{p})$ and $F(\omega, \mathbf{p})$ we obtain

$$G^{(1)}(\omega,\mathbf{p})=\frac{\omega-\varepsilon_2(\mathbf{p})}{\omega^2-(\varepsilon_1(\mathbf{p})+\varepsilon_2(\mathbf{p}))\omega+\varepsilon_1(\mathbf{p})\varepsilon_2(\mathbf{p})-|\Delta_1(\mathbf{p})|^2};\tag{24}$$

$$F(\omega,\mathbf{p})=-\frac{\Delta_1(\mathbf{p})}{\omega^2-(\varepsilon_1(\mathbf{p})+\varepsilon_2(\mathbf{p}))\omega+\varepsilon_1(\mathbf{p})\varepsilon_2(\mathbf{p})-|\Delta_1(\mathbf{p})|^2},\tag{25}$$

where $\Delta_1(\mathbf{p})=\tilde{F}_\omega(\mathbf{p})$.

The positive pole of the Green's function determines the spectrum of the one-particle excitations of the system:

$$\omega_1=\frac{\varepsilon_1(\mathbf{p})+\varepsilon_2(\mathbf{p})}{2}+\sqrt{\frac{(\varepsilon_1(\mathbf{p})-\varepsilon_2(\mathbf{p}))^2}{4}+|\Delta_1(\mathbf{p})|^2}.\tag{26}$$

Similarly, we obtain the second branch of the excitations from $G^{(2)}(\omega, \mathbf{p})$:

$$\omega_2=-\frac{\varepsilon_1(\mathbf{p})+\varepsilon_2(\mathbf{p})}{2}+\sqrt{\frac{(\varepsilon_1(\mathbf{p})-\varepsilon_2(\mathbf{p}))^2}{4}+|\Delta_2(\mathbf{p})|^2},\tag{27}$$

where

$$\Delta_2(\mathbf{p})=\tilde{F}^+_\omega(\mathbf{p});\quad |\Delta_1(\mathbf{p})|^2=|\Delta_2(\mathbf{p})|^2.\tag{28}$$

It may be seen from (26) and (27) that a gap occurs in the excitation spectrum, and, unlike superconductivity, for arbitrary values of $\varepsilon_1(\mathbf{p})$ and $\varepsilon_2(\mathbf{p})$ the minimum excitation energy occurs, generally speaking, for momenta different from p_0.

For example, for $\varepsilon_1(\mathbf{p})=v_{1F}(|\mathbf{p}|-p_0)$ and $\varepsilon_2(\mathbf{p})=v_{2F}(p_0-|\mathbf{p}|)$, where v_{1F} and v_{2F} are the velocities at the Fermi surface in the 1st and 2nd bands,

$$\left.\begin{aligned}\omega_1&=\frac{(v_{1F}-v_{2F})(|\mathbf{p}|-p_0)}{2}+\sqrt{\frac{(v_{1F}+v_{2F})^2(|\mathbf{p}|-p_0)^2}{4}+|\Delta_1(\mathbf{p})|^2};\\ \omega_2&=\frac{(v_{2F}-v_{1F})(|\mathbf{p}|-p_0)}{2}+\sqrt{\frac{(v_{1F}+v_{2F})^2(|\mathbf{p}|-p_0)^2}{4}+|\Delta_2(\mathbf{p})|^2}.\end{aligned}\right\}\tag{29}$$

Including the boundary conditions [4], we obtain the following expressions for $G^{(1)}$ and F:

$$G^{(1)}(\omega,\mathbf{p})=\frac{u^2_{p-}}{\omega-\frac{\varepsilon_1(\mathbf{p})+\varepsilon_2(\mathbf{p})}{2}-\sqrt{\frac{(\varepsilon_1(\mathbf{p})-\varepsilon_2(\mathbf{p}))^2}{4}+|\Delta_1(\mathbf{p})|^2}+i\delta}+\frac{u^2_{p+}}{\omega-\frac{\varepsilon_1(\mathbf{p})+\varepsilon_2(\mathbf{p})}{2}+\sqrt{\frac{(\varepsilon_1(\mathbf{p})-\varepsilon_2(\mathbf{p}))^2}{4}+|\Delta_1(\mathbf{p})|^2}-i\delta};\tag{30}$$

$$F(\omega,\mathbf{p})=\frac{-\Delta_1(\mathbf{p})}{\left(\omega-\frac{\varepsilon_1(\mathbf{p})+\varepsilon_2(\mathbf{p})}{2}-\sqrt{\frac{(\varepsilon_1(\mathbf{p})-\varepsilon_2(\mathbf{p}))^2}{4}+|\Delta_1(\mathbf{p})|^2}+i\delta\right)\times\left(\omega-\frac{\varepsilon_1(\mathbf{p})+\varepsilon_2(\mathbf{p})}{2}+\sqrt{\frac{(\varepsilon_1(\mathbf{p})-\varepsilon_2(\mathbf{p}))^2}{4}+|\Delta_1(\mathbf{p})|^2}-i\delta\right)};$$

$$u^2_{p\mp}=\frac{1}{2}\left(1\pm\frac{1}{2}\,\frac{\varepsilon_2(\mathbf{p})}{\varepsilon_1(\mathbf{p})+\varepsilon_2(\mathbf{p})+\sqrt{(\varepsilon_1(\mathbf{p})-\varepsilon_2(\mathbf{p}))^2+4|\Delta_1(\mathbf{p})|^2}}\right).\tag{31}$$

Similar expressions are obtained for $G^{(2)}$ and F^+. Using the expression (23) for $\tilde{F}_\omega(\mathbf{p})$, and the expression (31) for $F(\omega, \mathbf{p})$, we obtain the equation for finding $\tilde{F}_\omega(\mathbf{p}) = \Delta_1(\mathbf{p})$:

$$\Delta_1(\mathbf{p})=\frac{i}{(2\pi)^4}\int F(\omega',\mathbf{k})V(\mathbf{p}-\mathbf{k})\,d\omega' d\mathbf{k}=-\frac{1\cdot}{(2\pi)^3}\int\frac{\Delta_1(\mathbf{k})V(\mathbf{p}-\mathbf{k})\,d\mathbf{k}}{\sqrt{(\varepsilon_1(\mathbf{k})-\varepsilon_2(\mathbf{k}))^2+4|\Delta_1(\mathbf{k})|^2}}.\tag{32}$$

To convince ourselves that the uncoupling of the two-particle function in Eq. (22) has been done correctly, we can get all these results by means of the method of Bogolyubov [2] and of Tolmachev and Tyablikov [8].

Instead of the field operators ψ_1 and ψ_2, we introduce the particle annihilation and creation operators $a^{(1)}_{\mathbf{k},s}$, $a^{(2)}_{\mathbf{k},s}$, $a^{+(1)}_{\mathbf{k},s}$, $a^{+(2)}_{\mathbf{k},s}$ in the state with momentum $\mathbf{k}$ and spin s in bands 1 and 2. We rewrite the Hamiltonian (3) in the form

$$\left.\begin{array}{l} H=\sum\limits_{\mathbf{k},s}\varepsilon_1(\mathbf{k})a^{+(1)}_{\mathbf{k},s}a^{(1)}_{\mathbf{k},s}+\varepsilon_2(\mathbf{k})a^{+(2)}_{\mathbf{k},s}a^{(2)}_{\mathbf{k},s}+\frac{1}{2V} \\ \times\sum\limits_{\substack{s_1,s_2,\mathbf{k}_1,\mathbf{k}_2,\mathbf{k}_1',\mathbf{k}_2' \\ \mathbf{k}_1+\mathbf{k}_2=\mathbf{k}_1'+\mathbf{k}_2' \\ \mathbf{k}_1\neq\mathbf{k}_1'}} V(\mathbf{k}_1',\mathbf{k}_2',\mathbf{k}_1,\mathbf{k}_2)a^{+(2)}_{\mathbf{k}_2}a^{+(1)}_{\mathbf{k}_1',s_1}a^{(2)}_{\mathbf{k}_2',s_2}a^{(1)}_{\mathbf{k}_1',s_1}; \\ s=\pm\frac{1}{2}, \end{array}\right\} \tag{33}$$

V is the volume of the system.

We make a transformation from the operators $a^{(1)}_{\mathbf{k},s}$, $a^{(2)}_{\mathbf{k},s}$ to the new operators $\alpha^{(1)}$, $\alpha^{(2)}$:

$$a^{(1)}_{\mathbf{k},+}=u^{(1)}_{\mathbf{k}}\alpha^{(1)}_{\mathbf{k},0}+v^{(1)}_{\mathbf{k}}\alpha^{(2)}_{\mathbf{k},1}; \tag{34}$$

$$a^{(2)}_{-\mathbf{k},-}=v^{(1)}_{\mathbf{k}}\alpha^{+(1)}_{\mathbf{k},0}-u^{(1)}_{\mathbf{k}}\alpha^{+(2)}_{\mathbf{k},1}; \tag{35}$$

$$a^{(2)}_{\mathbf{k},+}=u^{(2)}_{\mathbf{k}}\alpha^{(2)}_{\mathbf{k},0}+v^{(2)}_{\mathbf{k}}\alpha^{(1)}_{\mathbf{k},1}; \tag{36}$$

$$a^{(1)}_{-\mathbf{k},-}=v^{(2)}_{\mathbf{k}}\alpha^{+(2)}_{\mathbf{k},0}-u^{(2)}_{\mathbf{k}}\alpha^{+(1)}_{\mathbf{k},1}. \tag{37}$$

It is easily seen that a transformation of this sort is canonical; i.e., the anticommutation relations are satisfied for the condition

$$u^{(1)^2}_{\mathbf{k}}+v^{(1)^2}_{\mathbf{k}}=1,\quad u^{(2)^2}_{\mathbf{k}}+v^{(2)^2}_{\mathbf{k}}=1. \tag{38}$$

The matrix elements corresponding to creation of a particle in band 1 and of a hole in zone 2 from the vacuum contain energy denominators, which, if the particle and the hole have opposite momenta, lead to divergences when integrating over the momenta We note that in the new representation the "dangerous" combinations, i.e., those leading to instability, are the following:

$$\alpha^{(1)}_{\mathbf{k},0}\alpha^{+(2)}_{\mathbf{k},1},\quad \alpha^{(1)}_{\mathbf{k},1}\alpha^{+(2)}_{\mathbf{k},0}. \tag{39}$$

Following the method of [8], we find that to balance out the above terms, in addition to the normalizing condition (38) the coefficients of the transformation (34)-(37) must have imposed on them the additional condition

$$(\varepsilon_1(\mathbf{p})-\varepsilon_2(\mathbf{p}))u^{(1)}_{\mathbf{p}}v^{(1)}_{\mathbf{p}}$$

$$=\frac{1}{V}\left(u^{(1)^2}_{\mathbf{p}}-v^{(1)^2}_{\mathbf{p}}\right)\sum_{\mathbf{p}'}V(\mathbf{p}-\mathbf{p}')u^{(1)}_{\mathbf{p}'}v^{(1)}_{\mathbf{p}'} \tag{40}$$

and similarly for $u^{(2)}_{\mathbf{p}}$, $v^{(2)}_{\mathbf{p}}$. Using the normalizing condition, we find

$$u^{(1)^2}_{\mathbf{p}}=\frac{1}{2}\left(1+\frac{\varepsilon_1(\mathbf{p})-\varepsilon_2(\mathbf{p})}{\sqrt{(\varepsilon_1(\mathbf{p})-\varepsilon_2(\mathbf{p}))^2+4\Delta_1^2(\mathbf{p})}}\right); \tag{41}$$

$$v^{(1)^2}_{\mathbf{p}}=\frac{1}{2}\left(1-\frac{\varepsilon_1(\mathbf{p})-\varepsilon_2(\mathbf{p})}{\sqrt{(\varepsilon_1(\mathbf{p})-\varepsilon_2(\mathbf{p}))^2+4\Delta_1^2(\mathbf{p})}}\right), \tag{42}$$

where

$$\Delta_1(\mathbf{p})=\frac{1}{V}\sum_{\mathbf{p}'}V(\mathbf{p}-\mathbf{p}')u^{(1)}_{\mathbf{p}'}v^{(1)}_{\mathbf{p}'}. \tag{43}$$

Then (40) gives an integral equation for finding Δ_1:

$$\Delta_1(\mathbf{p})=\frac{1}{(2\pi)^3}\int\frac{V(\mathbf{p}-\mathbf{p}')\Delta_1(\mathbf{p}')d\mathbf{p}'}{\sqrt{(\varepsilon_1(\mathbf{p}')-\varepsilon_2(\mathbf{p}'))^2+4\Delta_1^2(\mathbf{p}')}}. \tag{44}$$

A similar equation is found for $\Delta_2(\mathbf{p})$.

These equations are in essence the same as (32), but in deriving them we also included the spin of the electrons and showed that the pairing occurs in the singlet state.

The potential of $V(\mathbf{p}-\mathbf{p}')$ is of the form

$$V(\mathbf{p}-\mathbf{p}')=\frac{4\pi e^2}{(\mathbf{p}-\mathbf{p}')^2+\varkappa_D^2}; \tag{45}$$

and, for the condition $\varkappa_D^2 \ll p_0^2$, this interaction may be regarded as weak, as we had already assumed above in decoupling the two-particle function and balancing out the dangerous diagrams. But then $\Delta_1(\mathbf{p})$ does not change appreciably unless there are momentum changes of the order of magnitude $\varkappa_D$. But in the narrow energy region that we are interested in at the Fermi surface, Δ_1 may be assumed independent of p, and the solution of Eq. (32) or (44) is of the form

$$\Delta=2\omega_0\exp\left(-\frac{2\pi v_F}{e^2\ln\frac{\varkappa_D^2}{2p_0^2}}\right), \tag{45}$$

where $\omega\approx v_F\varkappa_D$.

We have thus found an energy spectrum similar to the spectrum of superconductors, but with the difference that the width of the gap drops with increase in the initial electron concentration n. At low temperatures, $kT < \Delta$, we will have a dielectric (semiconductor), while for $kT > \Delta$, we will have a metal. The weak Coulomb interaction ideas that we have used correspond to the assumption that the mean correlation energy e^2/r_D is much less than the kinetic energy ε_F. This condition may be satisfied even for a relatively small electron concentration, let us say of $n \sim 10^{18}$-10^{19}, if the effective mass of the electrons is small, while the dielectric constant is large. But this is precisely the situation that is characteristic of the majority of semimetals and degenerate semiconductors: the effective masses are of the order of 10^{-1}-10^{-2} times the mass of a free electron, and the dielectric constants are of the order 10-100. On the other hand, if we try to apply these ideas to the problem mentioned above of converting ordinary semiconductors into semimetals by heating, the situation may be the other way around; i.e., the gap with Δ is greater than the initial Fermi energy, so that the interaction is strong. It is to be expected that the qualitative results in this case will be similar to those obtained above, but a completely different approach is required in making a quantitative treatment.

LITERATURE CITED

1. L. N. Cooper, Phys. Rev., 104, 1189 (1957).
2. I. Bardeen, L. N. Cooper, and I. R. Schrieffer, Phys. Rev., 108, 1175 (1957).
3. N. N. Bogolyubov, ZhÉTF, 34, 58 [Soviet Physics – JETP, Vol. 7, p. 41]; 73 (1958) [Soviet Physics – JETP, Vol. 7, p. 51].
4. L. P. Gor'kov, ZhÉTF, 34, 735 (1958) [Soviet Physics – JETP, Vol. 7, p. 505].
5. R. G. Arkhipov, ZhÉTF, 43, 349 (1962) [Soviet Physics – JETP, Vol. 16, p. 251].
6. L. R. Godefroy and P. Aigrain, Proc. Int. Conference on the Physics of Semiconductors, Exeter (1962), p. 234.
7. A. A. Abrikosov, L. P. Gor'kov, and I. E. Dzyaloshinskii, Quantum Field Theory Methods in Statistical Physics [in Russian] (Fizmatgiz, 1962).
8. V. V. Tolmachev and S. V. Tyablikov, ZhÉTF, 34, 66 (1958) [Soviet Physics – JETP, Vol. 7, p. 46].

SOVIET PHYSICS JETP VOLUME 20, NUMBER 4 APRIL, 1965

DIAGRAM TECHNIQUE FOR NONEQUILIBRIUM PROCESSES

L. V. KELDYSH

P. N. Lebedev Physics Institute, Academy of Sciences, U.S.S.R.

Submitted to JETP editor April 23, 1964

J. Exptl. Theoret. Phys. (U.S.S.R.) 47, 1515-1527 (October, 1964)

A graph technique analogous to the usual Feynman technique in field theory is developed for calculating Green's functions for particles in a statistical system which under the action of an external field deviates to any arbitrary extent from the state of thermodynamic equilibrium. It is found that in order to describe such a system it is necessary to introduce two Green's functions for each type of particles. The equation for one of these functions is a generalization of the usual Boltzmann kinetic equation.

THE use of quantum field theory methods has turned out to be very effective for the formulation and for the solution of problems in statistical physics. A number of prescriptions has been given for the description of thermodynamic characteristics of equilibrium systems [1-6] and of the ground state of a many-body system [7-9]. In the papers of Konstantinov and Perel' [10], Dzyaloshinskiĭ [11] and Mills [12] a technique has been developed for the calculation of time dependent quantities (kinetic coefficients) in systems in thermodynamic equilibrium. In carrying out this program the deviation of the system from the Gibbs equilibrium distribution was assumed to be small, i.e., the external field was taken to be sufficiently small. The present paper is devoted to the application of analogous methods to the description of essentially nonequilibrium systems, for example, such systems as the system of electrons in a strong electric field. In the simplest cases in order to describe such systems Boltzmann's kinetic equation or the system of coupled equations proposed by Bogolyubov [13] are used.

We shall obtain below a system of equations for the Green's functions of the particles in the system, and we shall show that one of these equations is essentially analogous to the kinetic equation. The appearance of such an equation—an additional one in comparison with systems in thermodynamic equilibrium—in the problem under consideration is quite natural, since we must have an equation which describes the distribution of particles in the system. In the equilibrium case this is given by the Gibbs canonical distribution. Our diagram technique will be close to Mills' technique for equilibrium systems. This technique has the advantage that it is completely analogous to the usual Feynman technique in field theory, with the only difference that the number of Green's functions appearing in it is increased. In the course of the discussion there appears automatically a function which describes the distribution of the particles in the system, and the equation for this function plays the role of the kinetic equation.

Following the method of Konstantinov and Perel' we shall define the density matrix ρ for the system by the usual equation

$$i\partial\rho / \partial t = [H_i, \rho]_- \equiv H_i(t)\rho(t) - \rho(t)H_i(t) \quad (1)$$

with the boundary condition

$$\rho(t = -\infty) = \rho_0 = \exp\{\Psi_0 - H_0(-\infty)) / kT\}, \quad (2)$$

where $H = H_0 + H_i$ is the Hamiltonian of the system

$$H_0(t) = \int \psi_0^+(\mathbf{r}t)\left\{\varepsilon\left(-i\hbar\nabla - \frac{e}{c}\mathbf{A}\right) + e\Phi\right\} \times \psi_0(\mathbf{r}t)\,d\mathbf{r} + H_T. \quad (3)$$

The first term in H_0 is the energy of noninteracting electrons in an external electromagnetic field $\{\mathbf{A}(\mathbf{r}t), \Phi(\mathbf{r}t)\}$, the second is the Hamiltonian of the heat bath in contact with the system under consideration, $\epsilon(\mathbf{p})$ is the dispersion law for the electrons, Ψ_0 is the initial free energy.

The field operators ψ_0 and ψ_0^+ are defined in the following manner: let $\varphi_{\mathbf{p}}(\mathbf{r}t)$ be the complete system of functions determined by the equation

$$\left\{\varepsilon\left(-i\nabla - \frac{e}{c}\mathbf{A}(\mathbf{r}t)\right) + e\Phi(\mathbf{r}t) - i\frac{\partial}{\partial t}\right\}\varphi_{\mathbf{p}}(\mathbf{r}t) = 0 \quad (4)$$

and by the boundary condition at $t \to -\infty$

$$\varphi_{\mathbf{p}}(\mathbf{r}t) \to \exp\{i(\mathbf{pr} - \varepsilon_{\mathbf{p}}t)\}, \quad (5)$$

where we consider the external field to be switched off at $t \to -\infty$, i.e., $(\mathbf{A}, \Phi) \to 0$. Then the field operator $\psi_0(\mathbf{r}t)$ is defined by the relation

$$\psi_0(\mathbf{r}t) = \sum_{\mathbf{p}} a_{\mathbf{p}} \varphi_{\mathbf{p}}(\mathbf{r}t), \tag{6}$$

where a_p are the usual Fermi operators which satisfy the commutation relations

$$[a_{\mathbf{p}}^+ a_{\mathbf{p}'}]_+ \equiv a_{\mathbf{p}}^+ a_{\mathbf{p}'} + a_{\mathbf{p}'} a_{\mathbf{p}}^+ = \delta_{\mathbf{p}\mathbf{p}'}. \tag{7}$$

Utilizing the fact that the functions $\varphi_p(\mathbf{r}t)$ form at any arbitrary instant of time a complete orthogonal system, and formulas (6) and (7), it can be easily shown that for coincident times

$$[\psi_0^+(\mathbf{r}t),\ \psi_0(\mathbf{r}'t)]_+ = \delta(\mathbf{r}-\mathbf{r}') \tag{8}$$

and in virtue of this the operators $\psi_0(\mathbf{r}t)$ satisfy the free equations of motion

$$i\partial\psi_0 / \partial t = [\psi_0,\ H_0]_-, \tag{9}$$

i.e., we have from the outset defined the field operators in the interaction representation.

We note that the operator $H_0(t)$ differs in the Schrödinger representation and in the interaction representation in contrast to the usual case, when all the external fields are independent of the time. Having in mind such a definition of the field operators we have from the outset written equation (1) in the interaction representation leaving in it only the operator for the interaction energy H_i which, for the sake of definiteness, we shall in future write in the form

$$H_i(t) = g \int \psi_0^+(\mathbf{r}t)\psi_0(\mathbf{r}t)\varphi_0(\mathbf{r}t)\,d\mathbf{r}, \tag{10}$$

where g is a dimensionless coupling constant. Such an expression describes the interaction between electrons and phonons in solids, and also can be utilized to describe the Coulomb interaction of charged particles if we write a separate equation for the Coulomb field.

Equation (1) for the density matrix $\rho(t)$ can be formally solved with the aid of the S-matrix

$$S(t, -\infty) = T \exp\left\{-\frac{i}{\hbar}\int_{-\infty}^{t} H_i(\tau)\,d\tau\right\}, \tag{11}$$

which satisfies the equation

$$i\partial S / \partial t = H_i(t)S(t). \tag{12}$$

The symbol T in (11) denotes a time-ordered product defined in the usual manner. Then we have

$$\rho(t) = S(t, -\infty)\rho_0 S^+(t, -\infty)$$

$$= S(t, -\infty)\rho_0 S(-\infty, t). \tag{13}$$

The density matrix defined in this manner depends explicitly on the time. The average value of an arbitrary operator $L_0(t)$ at time t has the form

$$\langle L_0(t)\rangle = \mathrm{Tr}\{\rho(t)L_0(t)\}, \tag{14}$$

where the subscript zero on the operator L_0 shows that this operator is taken in the interaction representation, i.e., its time dependence is determined by the free equation of motion in the external field

$$i\partial L_0 / \partial t = [L_0(t), H_0(t)]_-, \tag{15}$$

since the density matrix $\rho(t)$ itself was defined by us in the interaction representation. In future, however, we shall have occasion to deal with correlation functions of several field operators taken at different instants of time. In these cases it will be more convenient for us to transfer the entire time dependence to these operators, and to regard the density matrix as being independent of the time, i.e., to go over to the Heisenberg representation. In this case the field operators satisfy the complete equations of motion

$$i\partial\psi / \partial t = [\psi,\ H]_-. \tag{16}$$

For the time-independent density matrix we can take the value of the matrix determined by expression (13) at a certain fixed instant of time, for example, t = 0, having thus included in it all the changes which the distribution ρ_0 (2) had undergone when the external field and the interaction in the system were switched on. In order to convince ourselves that such a procedure is correct we shall, for example, show that in the case of a system in thermodynamic equilibrium it leads to the Gibbs canonical distribution, as it ought. In order to do this we utilize the well-known relation[14,15]

$$S(0, -\infty)H_0 S^+(0, -\infty)$$

$$= H - i\delta g \frac{dS(0, -\infty)}{dg} S^+(0, -\infty), \tag{17}$$

where δ is a parameter describing the adiabatic switching-on of the interaction, i.e., the interaction constant g in (10) is considered to be time dependent, in accordance with $g(t) = ge^{\delta t}$. After going to the limit $\delta \to 0$ the second term in (17) turns into a c-number equal to the sum of all the vacuum loops and describing the correction to the free energy of the system[1-3]. Therefore, for the given case of an equilibrium system

$$\rho = \rho(0) = S(0, -\infty)\rho_0 S^+(0, -\infty) = e^{(\Psi - H)/kT}, \tag{18}$$

where H is the total Hamiltonian, while Ψ is the total free energy.

Having defined the density matrix as $\rho(0)$ we must utilize such Heisenberg field operators $\psi(\mathbf{r}, t)$, which at t = 0 go over into the free $\psi_0(\mathbf{r}t)$. In other words,

$$\psi(\mathbf{r}t) = S(0, t)\psi_0(\mathbf{r}t)S(t, 0), \quad (19)$$

where, by definition,

$$S(t', t) = T\exp\left\{-i\int_t^{t'} H_i(\tau)d\tau\right\} \quad (20)$$

$$= S(t', -\infty)S^+(t, -\infty).$$

Utilizing these relations and the definition of the Heisenberg density matrix ρ, the average value of the T-product of arbitrary Heisenberg operators L(t), M(t') ... can be brought to the form

$$\langle TL(t)M(t')\ldots\rangle = \mathrm{Tr}\{\rho TL(t)M(t')\ldots\}$$

$$= \mathrm{Tr}\{S(0, -\infty)\rho_0 S(-\infty, 0)TL(t)M(t')\ldots\}$$

or, commuting in the argument of Tr the operator S(0, $-\infty$) and going over to operators in the interaction representation,

$$\langle TL(t)M(t')\ldots\rangle$$

$$= \mathrm{Tr}\{\rho_0 S(-\infty, 0)[TS(0, t)L_0(t)S(t, t')M_0(t')\ldots]$$

$$\times S(0, -\infty)\} = \mathrm{Tr}\{\rho_0 T_c[S_c L_0(t)M(t')\ldots]\}, \quad (21)$$

where T_C indicates ordering along the contour c, going from $-\infty$, passing the points t, t' ... and then returning back to $-\infty$; S_C is the complete S-matrix defined along the whole contour c. Ordering along c means that the points on the return branch of this contour correspond to later times than the points on the original direct branch, and of any two points on the return path the later one is that which is closer to $-\infty$.

However, evaluation of integrals along the contour c is inconvenient. Therefore, we perform in (21) the following identity transformation: we insert in the argument of T_C the factor $S(t_0, +\infty)\times S(+\infty, t_0)$ which is identically equal to unity, taking the point t_0 to lie on the contour c later than that one of the points t, t' ... which lies on the extreme right. This is equivalent to cutting the contour C. to the right of the point lying on the extreme right and inserting into it a piece which goes to $+\infty$ and returns. The resultant contour C traverses the whole time axis from $-\infty$ to $+\infty$ and then back again from $+\infty$ to $-\infty$. Formula (21) is thus brought to the final form

$$\langle TL(t)M(t')\ldots\rangle = \mathrm{Tr}\{\rho_0 T_C(S_C L_0(t)M_0(t')\ldots]\}, \quad (22)$$

with all the arguments t, t' ... considered as lying on the positive ($-\infty$, $+\infty$) branch of the contour C.

But use of the contour C enables us to write in the form of a T-product also the usual correlation function of two operators L(t) and M(t'). Indeed, we shall assign to points lying on the positive (proceeding from $-\infty$ to $+\infty$) branch of the contour the subscript +, and to points lying on the negative branch the subscript $-$. Then evidently we have

$$\langle L(t)M(t')\rangle = \mathrm{Tr}\{\rho_0 T_C[S_C L_0(t_-)M_0(t_+')]\}, \quad (23)$$

since in accordance with the definition of the T_C-product the point t_+' always precedes the point t_-. This possibility leads to the fact that the diagram technique obtained below enables us to evaluate directly not only the Green's functions for the particles, but also the usual average values of operators.

The averages of T-products of the type (22) and (23) can be decomposed in the usual manner[2] into a sum of Feynman graphs evaluated in accordance with the following rules.

1. An n-th order graph contains 2n points along the real time axis. Moreover, each point can lie either on the positive or on the negative branch of the contour C, and this is denoted respectively by the subscripts + or $-$. An integration with respect to the coordinates of all the internal points is carried out over the whole volume and over time from $-\infty$ to $+\infty$. Each internal point is put in correspondence with a factor g, if its subscript is +, and with the factor $-g$, if its subscript is $-$.

2. One phonon and two electron lines emerge from (or enter) each point. Each of these lines is set in correspondence with one of four functions depending on the signs of the points which this line joins:

a) if the line goes from the point $-$ to the point +, it is set in correspondence with the function

$$G_0^+(\mathbf{r}t_+, \mathbf{r}'t_-') = -i\mathrm{Tr}\{\rho_0 T_C[\psi_0(\mathbf{r}t_+), \psi_0^+(\mathbf{r}'t_-')]\}$$

$$= i\langle\psi_0^+(\mathbf{r}'t_+')\psi_0(\mathbf{r}t_-)\rangle_0. \quad (24)$$

the symbol $\langle\ldots\rangle_0$ denotes averaging using the matrix ρ_0;

b) if the line goes from the point + to the point $-$, it is set in correspondence with the function

$$G_0^-(\mathbf{r}t_-, \mathbf{r}'t_+') = -i\mathrm{Tr}\{\rho_0 T_C[\psi_0(\mathbf{r}t_-)\psi_0^+(\mathbf{r}'t_+')]\}$$

$$= -i\langle\psi_0(\mathbf{r}t_-)\psi_0^+(\mathbf{r}'t_+')\rangle_0; \quad (25)$$

c) if the line goes from the point + to the point +, then it is set in correspondence with the usual causal Green's function

$$G_0^c(\mathbf{r}t_+, \mathbf{r}'t_+') = -i\mathrm{Tr}\{\rho_0 T_C[\psi_0(\mathbf{r}t_+)\psi_0^+(\mathbf{r}'t_+')]\}$$

$$= -i\langle T\psi_0(\mathbf{r}t_+)\psi_0^+(\mathbf{r}'t_+')\rangle_0; \quad (26)$$

d) and, finally, the line connecting the point $-$ with the point $-$ is set in correspondence with the function

$$\tilde{G}_0^c(\mathbf{r}t_-, \mathbf{r}'t_-') = -i\mathrm{Tr}\{\rho_0 T_C[\psi_0(\mathbf{r}t_-)\psi_0^+(\mathbf{r}'t_-')]\}$$

$$= -i\langle\tilde{T}\psi_0(\mathbf{r}t_-)\psi_0^+(\mathbf{r}'t_-')\rangle_0. \quad (27)$$

In the last formula $\tilde{T}$ denotes the product ordered with respect to the reversed time, i.e.,

$$\tilde{T}\psi_0(t)\psi_0^+(t') = \begin{cases} \psi_0(t)\psi_0^+(t'), & \text{if } t < t', \\ -\psi_0^+(t')\psi_0(t), & \text{if } t > t'. \end{cases}$$

We shall henceforth omit everywhere the indices + and − in the time arguments of the Green's functions, since the index ascribed to the Green's function itself uniquely determines the positions of its final and initial points on the branches of the contour C. The functions G_0^C and $\tilde{G}_0^C$ are related to $G_0^\pm$ by the obvious relations

$$G_0^c(\mathbf{r}t, \mathbf{r}'t') = \theta(t-t')G_0^-(\mathbf{r}t, \mathbf{r}'t')$$
$$+ \theta(t'-t)G_0^+(\mathbf{r}t, \mathbf{r}'t'), \quad (28)$$

$$\tilde{G}_0^c(\mathbf{r}t, \mathbf{r}'t') = \theta(t-t')G_0^+(\mathbf{r}t, \mathbf{r}'t')$$
$$+ \theta(t'-t)G_0^-(\mathbf{r}t, \mathbf{r}'t'); \quad (29)$$

$$\theta(x) = {}^1/_2(1 + \mathrm{sign}\, x).$$

The Green's functions for Bose particles are defined in an analogous manner.

The further rules for calculation are completely analogous to the usual Feynman rules, and, therefore, we shall not state them in detail. We only note that a diagram with 2n points contains the factor i^{n+2F}, where F is the number of closed loops in it, and that the unconnected diagrams must be omitted since they cancel with the trace of the density matrix which we have for this reason omitted in advance in the denominators of all the formulas (21)−(29).

It should also be noted that the applicability of Wick's theorem to the expansion of the T-products (22), (23) is not entirely obvious. Strictly speaking, when four or more field operators having the same momentum are averaged over the distribution ρ_0, the T-product does not decompose into a sum of products of T-products taken a pair at a time. But for systems of Fermi-particles and of phonons (phonons correspond to a real field) considered by us Wick's theorem is completely rigorous. In other cases the contribution of two or more pairs of operators whose momenta are the same is generally of order 1/N, where N is the number of degrees of freedom, since integration is carried out over all the internal momenta. An exception can occur in the case of systems in which phenomena of the type of Bose-condensation are observed. The technique under consideration is not directly applicable to these systems.

We now define the complete Green's functions:

$$G^c(\mathbf{r}t, \mathbf{r}'t') = -i\mathrm{Tr}\{\rho_0 T_C[\psi_0(\mathbf{r}t_+)\psi_0^+(\mathbf{r}'t_+')S_C]\}, \quad (30)$$

$$\tilde{G}^c(\mathbf{r}t, \mathbf{r}'t') = -i\mathrm{Tr}\{\rho_0 T_C[\psi_0(\mathbf{r}t_-)\psi_0^+(\mathbf{r}'t_-')S_C]\}, \quad (31)$$

$$G^\pm(\mathbf{r}t, \mathbf{r}'t') = -i\mathrm{Tr}\{\rho_0 T_C[\psi_0(\mathbf{r}t_\pm)\psi_0^+(\mathbf{r}'t_\mp')S_C]\}, \quad (32)$$

which are sums of all the connected diagrams with one entering and one emerging line which differ from one another by the location of the initial and the final points on different branches of the contour C. The following relations involving them are preserved

$$G^c(\mathbf{r}t, \mathbf{r}'t') = \theta(t-t')G^-(\mathbf{r}t, \mathbf{r}'t') + \theta(t'-t)G^+(\mathbf{r}t, \mathbf{r}'t'), \quad (33)$$

$$\tilde{G}^c(\mathbf{r}t, \mathbf{r}'t') = \theta(t-t')G^+(\mathbf{r}t, \mathbf{r}'t') + \theta(t'-t)G^-(\mathbf{r}t, \mathbf{r}'t'), \quad (34)$$

i.e., only the two functions G^+ and G^- are independent, and even they are related by the obvious expression

$$\lim_{t'\to t}\{G^+(\mathbf{r}t, \mathbf{r}'t') - G^-(\mathbf{r}t, \mathbf{r}'t')\} = -i\delta(\mathbf{r}-\mathbf{r}'). \quad (35)$$

The subsequent technique of calculating the various quantities and the form of the equations are both considerably simplified if, following Mills[12], we go over to the matrix form of writing the equations. We define a two-rowed matrix G in the following manner:

$$G = \begin{pmatrix} G^c & G^- \\ G^+ & \tilde{G}^c \end{pmatrix}. \quad (36)$$

Then, as can be easily seen, summation over the subscript ± at each point reduces to matrix multiplication of matrices corresponding to lines emerging from that point. But since there are three such lines—two electron lines and one phonon line—it is necessary to connect them by means of a vertex matrix of the third rank

$$\gamma_{ij}{}^k = \delta_{ij}(\sigma_z)_{jk}, \quad (37)$$

where the subscripts refer to the electron lines, and the superscript to the phonon line; σ_z is the third Pauli matrix; the fact that for k = 2 it is negative takes into account the factor −1 for points on the return branch of the contour C (in matrix notation the points + correspond to subscripts 1, and the points − correspond to subscripts 2).

An arbitrary matrix element can now be written in the form of a product of blocks of the type

$$G_{li}\gamma_{ij}{}^k G_{jm} D_{kn},$$

in which summation is assumed to be carried out over repeated subscripts. D is the matrix for the Green's functions for Bose-particles defined in analogy with (36). With this notation we no longer need to distinguish the points denoted by + and −. The technique for the evaluation of any arbitrary graph now becomes a purely Feynman one with the only difference that to each line we now set in correspondence the matrix (36), and to each point the vertex matrix (37).

Summation of the diagrams for the complete Green's function leads to an equation of the type of Dyson's equation:

$$G(\mathbf{r}t, \mathbf{r}'t') = G_0(\mathbf{r}t, \mathbf{r}'t')$$

$$+ ig^2 \int G_0(\mathbf{r}t, \mathbf{r}_1t_1)\Sigma(\mathbf{r}_1t_1, \mathbf{r}_2t_2)G(\mathbf{r}_2t_2, \mathbf{r}'t')\,d\mathbf{r}_1d\mathbf{r}_2dt_1dt_2. \tag{38}$$

The self energy matrix Σ is defined by relation (39) in which the complete vertex matrix $\Gamma^k_{ij}(x, x'; y)$ is the sum of all the irreducible diagrams with one incoming (at the point x) and one outgoing (at the point x′) electron line and one external phonon line at the point y:

$$\Sigma_{ij}(x, x')$$

$$= \int \gamma_{ii'}{}^{k} G_{i'j'}(x, x_1)\Gamma_{j'j}{}^{k'}(x_1, x'\ y)D_{k'k}(y, x)\,d^4x_1d^4y,$$

$$x = (\mathbf{r}, t). \tag{39}$$

Dyson's equation in integral form (38) appears to be unsatisfactory since it contains the function G_0 which explicitly depends on the initial distribution ρ_0. At the same time from physical considerations it follows that if the system is in contact with a heat bath, then after sufficient time has elapsed it must reach a state which is, generally speaking, not stationary, since the external field depends on the time, but which is independent of the initial conditions. This inconsistency can be easily avoided by going over to the differential form of Dyson's equation, i.e., by multiplying (38) by the operator matrix

$$\hat{G}_0^{-1} = G_0^{-1}(x)\sigma_z, \tag{40}$$

where the differential operator G_0^{-1} is defined by

$$G_0^{-1}(x) = i\frac{\partial}{\partial t} - \varepsilon\Big(-i\nabla - \frac{e}{c}\mathbf{A}(x)\Big) - e\Phi(x), \tag{41}$$

$x = (\mathbf{r}t)$. Then Eq. (38) is reduced to the form

$$\hat{G}_0^{-1}(x)G(x, x')$$

$$= ig^2 \int \Sigma(x, x_1)G(x_1, x')\,dx_1 + \delta(x - x'). \tag{42}$$

In deriving this formula we have utilized the obvious relations

$$G_0^{-1}(x)G_0^{\pm}(x, x') = 0, \qquad G_0^{-1}(x)G_0^{c}(x, x') = \delta(x - x'),$$

$$G_0^{-1}(x)\tilde{G}_0^{c}(x, x') = -\delta(x - x').$$

Equation (42) contains only complete Green's functions which do not explicitly depend on ρ_0. At the same time its solution is determined only up to an arbitrary solution of the homogeneous equation

$$\hat{G}_0^{-1}(x)h(x, x') = 0. \tag{43}$$

We shall discuss the problem of the unique choice of the solution and of the role of initial conditions later.

Equation (42) represents a system of four integro-differential equations. However, between the four unknown functions G_{ij} there exist two linear relations (33) and (34) which, naturally, are consistent with the equations. Therefore, by means of a certain linear canonical transformation of the matrices G, γ and D we can reduce the number of independent equations to two. Such a transformation is given by

$$G \to \frac{1 - i\sigma_y}{\sqrt{2}}\begin{pmatrix} G^c & G^- \\ G^+ & \tilde{G}^c \end{pmatrix}\frac{1 + i\sigma_y}{\sqrt{2}} = \begin{pmatrix} 0 & G^a \\ G^r & F \end{pmatrix}, \tag{44}$$

where three new Green's functions have been introduced: the advanced one G^a, the retarded one G^r, and the correlation function F which, as we shall see later, is essentially a single-particle density matrix. The matrix σ_y is the second of the three Pauli matrices:

$$\sigma_x = \begin{pmatrix} 0 & 1 \\ 1 & 0 \end{pmatrix}, \quad \sigma_y = \begin{pmatrix} 0 & -i \\ i & 0 \end{pmatrix}, \quad \sigma_z = \begin{pmatrix} 1 & 0 \\ 0 & -1 \end{pmatrix}.$$

The functions G^a, G^r, and F are defined as follows in terms of average values of Heisenberg operators using the Heisenberg density matrix:

$$G^a(x, x') = -i\theta(t - t')\langle[\psi(x), \psi^+(x')]_+\rangle, \tag{45}$$

$$G^r(x, x') = i\theta(t' - t)\langle[\psi(x), \psi^+(x')]_+\rangle, \tag{46}$$

$$F(x, x') = -i\langle[\psi(x), \psi^+(x')]_-\rangle. \tag{47}$$

They are related to the G-functions used above by means of the relations

$$G^a = G^c - G^+ = -\tilde{G}^c + G^-$$

$$= -\theta(t - t')(G^+ - G^-), \tag{48}$$

$$G^r = G^c - G^- = -\tilde{G}^c + G^+$$

$$= \theta(t' - t)(G^+ - G^-), \tag{49}$$

$$F = G^c + \tilde{G}^c = G^+ + G^-. \tag{50}$$

The D-matrix transforms in a similar fashion. The vertex matrix γ^k_{ij} and the operator $\hat{G}_0^{-1}$ are also altered:

$$\gamma_{ij}{}^1 = 2^{-1/2}\delta_{ij}, \quad \gamma_{ij}{}^2 = 2^{-1/2}(\sigma_x)_{ij}, \tag{51}$$

$$\hat{G}_0^{-1} = \sigma_x G_0^{-1}(x). \tag{52}$$

The self energy matrix Σ is brought to the form

$$\Sigma = \begin{pmatrix} \Omega & \Sigma^r \\ \Sigma^a & 0 \end{pmatrix}; \qquad (53)$$

$$\Sigma^r = \Sigma^c + \Sigma^- = -(\tilde{\Sigma}^c + \Sigma^+),$$
$$\Sigma^a = \Sigma^c + \Sigma^+ = -(\tilde{\Sigma}^c + \Sigma^-),$$
$$\Omega = \Sigma^c + \tilde{\Sigma}^c = -(\Sigma^+ + \Sigma^-). \qquad (54)$$

The difference between (53), (54) and (44), (48)—(50) is related to the fact that the quantities $\Sigma^\pm$ in contrast to Σ^c and $\tilde{\Sigma}^c$ contain the factor -1, which appears in the vertex (37) for k = 2, an odd number of times.

The general equations (39) and (42) are obviously invariant under the transformation (44); we simply go over to a new representation for the matrices G and D. But the number of independent equations is now obviously equal to two since the functions G^a and G^r are Hermitian conjugates and the equations for them follow from each other. The physical meaning of the functions G^r and F is quite different. Roughly speaking, G^r and G^a characterize the dynamic properties of the particles in the system under consideration, while F characterizes their static distribution. This can already be seen from the definition of these functions in the absence both of an external field and of an interaction, i.e., for $t \to -\infty$. Then, going over to Fourier transforms with respect to the differences $\mathbf{r}-\mathbf{r}'$ and $t-t'$ we obtain

$$G_0(\mathbf{p}\varepsilon) = \int G_0(\mathbf{r}-\mathbf{r}', t-t') \exp\{-i[\mathbf{p}(\mathbf{r}-\mathbf{r}') - \varepsilon(t-t')]\}\, d(\mathbf{r}-\mathbf{r}')\, d(t-t') \qquad (55)$$

and in the limiting case mentioned above

$$G_0^r(\mathbf{p}\varepsilon) = G_0^{a*}(\mathbf{p}\varepsilon) = 1/(\varepsilon - \varepsilon_\mathbf{p} - i\delta), \quad \delta \to +0, \qquad (56)$$

$$G_0^+(\mathbf{p}\varepsilon) = 2\pi i n_\mathbf{p} \delta(\varepsilon - \varepsilon_\mathbf{p}),$$
$$G_0^-(\mathbf{p}\varepsilon) = -2\pi i(1-n_\mathbf{p})\delta(\varepsilon - \varepsilon_\mathbf{p}), \qquad (57)$$
$$F_0(\mathbf{p}\varepsilon) = 2\pi i(2n_\mathbf{p} - 1)\delta(\varepsilon - \varepsilon_\mathbf{p}), \qquad (58)$$

where

$$n_\mathbf{p} = \left[\exp\frac{\varepsilon_\mathbf{p} - \mu}{kT} + 1\right]^{-1}$$

is the Fermi distribution function, μ is the chemical potential in the equilibrium state. Thus, the functions F and $G^\pm$ essentially represent the distribution function for the particles written in different forms.

When the interaction is taken into account the functions G^r and G^a also indirectly depend on the distribution of the particles, but the fact that $G^\pm(\mathbf{r}t, \mathbf{r}'t)$ and $F(\mathbf{r}t, \mathbf{r}'t)$ taken at coincident times $t' = t$ directly determine the distributions of electrons and of holes follows directly from their definition (32). We shall now show that the equation for G^+ (or the equation for F equivalent to it) contains within itself the usual Boltzmann kinetic equation under the assumptions under which the latter exists, i.e., when the external field is quasiclassical and the interaction is weak in the sense

$$\hbar/\tau \ll \bar{\varepsilon}, \qquad (59)$$

where τ is the time taken to traverse a mean free path, and $\bar{\epsilon}$ is the average energy of the particles.

The equation for G^+ has the form

$$G_0^{-1}(x)G^+(x,x') = ig^2 \int \{\Sigma^+(x,x'')G^c(x'',x') + \tilde{\Sigma}^c(x,x'')G^+(x'',x')\}\, dx''. \qquad (60)$$

The corresponding Hermitian conjugate equation is

$$G_0^{-1*}(x')G^+(x,x') = ig^2 \int \{\tilde{G}^c(x,x'')\Sigma^+(x'',x') + G^+(x,x'')\Sigma^c(x'',x')\}\, dx''. \qquad (61)$$

We add these two equations, then go over from the variables x, x′ to the new variables $(\mathbf{r}_s t_s) = x_s = (x+x')/2$ and $(\mathbf{r}_a t_a) = x_a = (x-x')/2$, and then take a Fourier transform of the type (55) with respect to the difference variable x_a. As a result terms of the type

$$ig^2 \int \Sigma(x_s + x_a', p')G(x_s - x_a + x_a', p'') \exp\{i[(p'-p)x_a + (p''-p')x_a']\}\, dx_a dx_a' dp' dp'', \qquad p = (\mathbf{p}\varepsilon),$$

will appear on the right hand side of the equation.

The quasiclassical nature of the external field means that quantities of the type $p\nabla_s$ are small. At the same time the right hand side of the equation is proportional to the small interaction constant g^2. Therefore, neglecting small terms of the form $g^2(p\nabla_s)$ and introducing a further change of variables $\mathbf{p} \to \mathbf{p} - e\mathbf{A}(x_s)/c$, $\epsilon \to \epsilon - e\Phi(x_s)$, we obtain

$$\left\{\frac{\partial}{\partial t_s} + \mathbf{v}_\mathbf{p}\nabla_s + \left(e\mathbf{E} + \left[\frac{e\mathbf{v}_\mathbf{p}}{c}\mathbf{H}\right]\right)\nabla_\mathbf{p}\right\} G^+(x_s p)$$
$$= g^2\{\Sigma^+(x_s p)[G^c(x_s p) + \tilde{G}^c(x_s p)] + [\Sigma^c(x_s p) + \tilde{\Sigma}^c(x_s p)]G^+(x_s p)\}, \qquad (62)^*$$

where $\mathbf{v}_\mathbf{p} = \nabla_\mathbf{p}\epsilon_\mathbf{p}$ is the velocity of the particles, E and H are the electric and the magnetic fields:

*$[\mathbf{v}_\mathbf{p}\mathbf{H}] = \mathbf{v}_\mathbf{p} \times \mathbf{H}$.

$$\mathbf{E} = -\nabla\Phi - c^{-1}\partial\mathbf{A}/\partial t; \qquad \mathbf{H} = \mathrm{rot}\,\mathbf{A}. \qquad (63)^*$$

In deriving (62) we have expanded in the left hand side of the equation the operator $G_0^{-1}(x) - G_0^{-1*}(x')$ in series in terms of the small difference $x_a = (x - x')/2$ up to the first order on the basis of the same quasiclassical condition. The right hand side of (62) can be further simplified by utilizing (50) and (54):

$$\left\{\frac{\partial}{\partial t} + \mathbf{v}\nabla + e\left(\mathbf{E} + \left[\frac{\mathbf{v}}{c}\,\mathbf{H}\right]\right)\nabla_{\mathbf{p}}\right\} G^{+}(xp)$$

$$= g^2\{\Sigma^{+}(xp)G^{-}(xp) - \Sigma^{-}(xp)G^{+}(xp)\}. \qquad (64)$$

The right hand side of (64) describes the effect of collisions on the distribution of the particles, with the first term being related to particles entering the state of momentum **p** from all the other states with momenta **p′**, while the second term is related to the particles leaving the state. In order to bring this equation to the usual form of the Boltzmann equation we introduce the distribution function f(**prt**) which, as has been noted previously, coincides with G^{+} for coincident times t′ = t:

$$f(\mathbf{p}\mathbf{r}t) = \frac{1}{2\pi i}\int_{-\infty}^{\infty} G^{+}(\mathbf{r}t;\,\mathbf{p}\varepsilon)\,d\varepsilon. \qquad (65)$$

Integration of equation (64) over ϵ then yields

$$\left\{\frac{\partial}{\partial t} + \mathbf{v}\nabla + e\left(\mathscr{E} + \left[\frac{\mathbf{v}}{c}\mathbf{H}\right]\right)\nabla_{\mathbf{p}}\right\} f(\mathbf{p}\mathbf{r}t)$$

$$= \frac{g^2}{2\pi i}\int_{-\infty}^{\infty}\{\Sigma^{+}(\mathbf{r}t;\,\mathbf{p}\varepsilon)G^{-}(\mathbf{r}t;\,\mathbf{p}\varepsilon)$$

$$-\Sigma^{-}(\mathbf{r}t;\,\mathbf{p}\varepsilon)G^{+}(\mathbf{r}t;\,\mathbf{p}\varepsilon)\}\,d\varepsilon. \qquad (66)$$

In accordance with the assumption that the interaction is small we must evaluate the right hand side of (66) up to the first order in g^2, i.e., we must substitute into (66) and (39) the functions $G_0^{\pm}$, satisfying the free equations

$$G_0^{-1}G_0^{\pm}(\mathbf{r}t;\,\mathbf{p}\varepsilon) = 0, \qquad (67)$$

but normalized with respect to an as yet unknown distribution function f by equations (65) and (35). Symbolically they can be written in the form

$$G_0^{+}(\mathbf{r}t;\,\mathbf{p}\varepsilon) = 2\pi i\delta(\varepsilon - \varepsilon_{\mathbf{p}} + i\hat{B})f(\mathbf{p}\mathbf{r}t), \qquad (68)$$

$$G_0^{-}(\mathbf{r}t;\,\mathbf{p}\varepsilon) = -2\pi i\delta(\varepsilon - \varepsilon_{\mathbf{p}} + i\hat{B})(1 - f(\mathbf{p}\mathbf{r}t)), \qquad (69)$$

where $\hat{B}$ is the operator in the left hand side of (66). The quantity $\delta(G_0^{-1})$ in formulas (68), (69), is an integral operator with a kernel satisfying equation (67) and equal to $\delta(\mathbf{r} - \mathbf{r}')$ for t′ = t. It can be easily expressed in terms of the functions $\varphi_{\mathbf{p}}$ (4).

*rot = curl.

In order to evaluate $\Sigma^{\pm}$ we must also define $D_0^{\pm}$. In the majority of problems of solid state physics the phonons can be regarded as being in thermodynamic equilibrium. Then

$$D_0^{\pm}(\mathbf{k}\omega) = -i\pi|m_{\mathbf{k}}|^2\{(1 + N_{\mathbf{k}})\delta(\omega \pm \omega_{\mathbf{k}})$$

$$+ N_{\mathbf{k}}\delta(\omega \mp \omega_{\mathbf{k}})\}, \qquad (70)$$

where **k** and $\omega_{\mathbf{k}}$ are the propagation vector and the frequency of the phonon, $m_{\mathbf{k}}$ is the matrix element for the interaction with the electron, and

$$N_{\mathbf{k}} = \left[\exp\frac{\omega_{\mathbf{k}}}{kT} - 1\right]^{-1} \qquad (71)$$

is the Planck distribution function.

Substituting (68)–(70) into (66) we obtain

$$\hat{B}f(\mathbf{p}\mathbf{r}t) = 2\pi^2 g^2\int|m_{\mathbf{k}}|^2\{[(1 + N_{\mathbf{k}})\delta(\varepsilon_{\mathbf{p}} - \varepsilon_{\mathbf{p}+\mathbf{k}} + \omega_{\mathbf{k}})$$

$$+ N_{\mathbf{k}}\delta(\varepsilon_{\mathbf{p}} - \varepsilon_{\mathbf{p}+\mathbf{k}} - \omega_{\mathbf{k}})]f(\mathbf{p}+\mathbf{k},\,\mathbf{r}t)(1 - f(\mathbf{p}\mathbf{r}t))$$

$$- [(1 + N_{\mathbf{k}})\delta(\varepsilon_{\mathbf{p}} - \varepsilon_{\mathbf{p}+\mathbf{k}} - \omega_{\mathbf{k}}) + N_{\mathbf{k}}\delta(\varepsilon_{\mathbf{p}} - \varepsilon_{\mathbf{p}+\mathbf{k}} + \omega_{\mathbf{k}})]$$

$$\times(1 - f(\mathbf{p}+\mathbf{k},\,\mathbf{r}t))f(\mathbf{p}\mathbf{r}t)\}\,d\mathbf{k}/(2\pi)^3, \qquad (72)$$

i.e., the usual form of Boltzmann's equation. Thus, in deriving it we have utilized two conditions: the quasiclassical nature of the external field

$$\frac{d}{dx}(e\mathbf{A},\,e\Phi) \ll \bar{\varepsilon}\bar{p} \qquad (73)$$

and the smallness of the interaction

$$\bar{\varepsilon} \gg g^2\Sigma^{\pm} \sim \hbar/\tau. \qquad (74)$$

The quasiclassical condition is not obligatory. Proceeding in exactly the same manner, but without expanding (60) and (61) in terms of the external field, we would have obtained the so-called quantum kinetic equation[16,17]. But the condition of smallness of the interaction (74) is necessary in this case also. If it is violated, then the functions $G^{\pm}$ do not have a δ-like character (68), (69) as functions of ϵ, and the integration over ϵ in equation (66) cannot be effectively carried out. In such a case we shall not obtain, generally speaking, a closed equation for the distribution function f(**prt**), i.e., for the G^{+}-function for coincident times t′ = t. The equation exists only for the complete G-function considered as a function of all its arguments. In the most general case this corresponds to Dyson's equations (39) and (42). If the interaction is not weak, they can be solved if the system has some other parameters, for example density, which enable us to select a set of principal diagrams. However, in this case there will arise the previously mentioned problem of the nonuniqueness of the solution of (42) up to an arbitrary solution of the free equation. We shall now show that such arbitrariness does not in fact exist, and that there is a unique prescription for the choice of the correct solution.

In order to do this we compare Eqs. (42) and (38) in the representation (44), by writing them separately for each of the functions G^a, G^r and F. In differential form

$$G_0^{-1}G^a = 1 + \Sigma^a G^a, \quad (75)$$

$$G_0^{-1}G^r = 1 + \Sigma^r G^r, \quad (76)$$

$$G_0^{-1}F = \Omega G^a + \Sigma^r F; \quad (77)$$

in integral form, corresponding to equation (38),

$$G^a = G_0^a(1 + \Sigma^a G^a), \quad (75a)$$

$$G^r = G_0^r(1 + \Sigma^r G^r), \quad (76a)$$

$$F = F_0(1 + \Sigma^a G^a) + G_0^r(\Omega G^a + \Sigma^r F). \quad (77a)$$

Comparison of (75)—(77) with (75a)—(77a) shows that equations (75), (76) do not contain any arbitrariness: they are solved by the advanced operator G_0^a or the retarded operator G_0^r respectively. For $t' \to t$ their solutions must go over into $\delta(r-r')$. A possible lack of uniqueness, i.e., dependence on initial conditions, is associated only with the term proportional to F_0 in equation (77a), since F_0 depends on the initial distribution, while in the differential form of the equation this term is completely absent. However, it can be easily seen that this term is identically equal to zero. Indeed, we substitute (75) into (77a). Then we have

$$F = F_0 G_0^{-1} G^a + G_0^r(\Omega G^a + \Sigma^r F) = G_0^r(\Omega G^a + \Sigma^r F), \quad (78)$$

since $F_0 G_0^{-1} = 0$. Thus, (77) is solved by means of the function G_0^r operating from the left. Equation (77) and its solution only appear to be nonsymmetric with respect to retarded and advanced functions. In place of (77) we can consider the Hermitian conjugate equation equivalent to it

$$G_0^{-1}F = G^r\Omega + F\Sigma^a, \quad (79)$$

which is solved by the function G_0^a operating from the right. Thus, solutions of Eqs. (75)—(77) are determined uniquely and independently of the initial distribution at $t = -\infty$.

We should now make clear to what systems the results obtained above are applicable. Evidently, we are dealing with well established solutions, i.e., such solutions which are determined by the effect of the external field at the present and at earlier instants of time, by the interaction with the heat bath, but not by the initial conditions. Generally speaking, these are not stationary solutions if the external field depends on the time. The lack of dependence on the initial conditions means physically that the initial conditions, even if they had been imposed, were specified in the sufficiently remote past, formally at $t \to -\infty$, while in fact we are dealing with times large in comparison to the relaxation time of the system. However, in discussions of kinetic phenomena, another way of posing the problem is possible: at time t_0 an arbitrary density matrix $\rho(t_0)$ is given and we are dealing with the relaxation of this state towards some well established regime. Such a problem, apparently, can not in principle be described by any diagram technique, since the latter is always based on the concept of the possibility of expressing two-, three-, etc. particle Green's functions in terms of the one-particle Green's function at least in the form of an infinite series. But an arbitrary specification of the density matrix is equivalent to an arbitrary and independent specification of all these quantities.

The existence of well established solutions independent of initial conditions is evident in the case of systems in contact with a heat bath. In the case of an adiabatic system the problem is less clear and the technique under discussion is applicable with certainty only in the case when the system had been actually in a state of thermodynamic equilibrium and was then taken out of that state by a given external field. The problem of approach to equilibrium from an arbitrary initial state has been discussed in papers by Van Hove[18], Prigogine and Resibois[19], Fujita[20] and others.

In conclusion I wish to express my gratitude to V. L. Gurevich, I. E. Dzyaloshinskiĭ, D. A. Kirzhnits, and E. S. Fradkin for discussions of a number of points dealt with in the present article.

[1] T. Matsubara, Progr. Theoret. Phys. (Kyoto) **14**, 351 (1955).

[2] Abrikosov, Gor'kov and Dzyaloshinskii, JETP **36**, 900 (1959), Soviet Phys. JETP **9**, 636 (1959); Metody kvantovoĭ teorii polya v statisticheskoĭ fizike (Methods of quantum field theory in statistical physics), Fizmatgiz, 1962.

[3] E. S. Fradkin, JETP **36**, 1286 (1959), Soviet Phys. JETP **9**, 912 (1959); Nuclear Phys. **12**, 465 (1959).

[4] J. M. Luttinger and J. C. Ward, Phys. Rev. **118**, 1417 (1960).

[5] P. C. Martin and J. Schwinger, Phys. Rev. **115**, 1342 (1959).

[6] V. L. Bonch-Bruevich and Sch. M. Kogan, Ann. Phys. **9**, 125 (1960), V. L. Bonch-Bruevich and S. V. Tyablikov, Metod funktsiĭ Grina v statisticheskoĭ fizike (The method of Green's functions in statistical physics), Fizmatgiz, 1961.

[7] A. Salam, Progr. Theoret. Phys. (Kyoto) **9**, 550 (1953).

[8] V. L. Bonch-Bruevich, JETP **28**, 121 (1955); **30**, 343 (1956), Soviet Phys. JETP **1**, 169 (1955); **3**, 278 (1956).

[9] V. M. Galitskiĭ and A. B. Migdal JETP **34**, 139 (1957), Soviet Phys. JETP **7**, 96 (1958).

[10] O. V. Konstantinov and V. I. Perel', JETP **39**, 197 (1960), Soviet Phys. JETP **12**, 142 (1961).

[11] I. E. Dzyaloshinskiĭ, JETP **42**, 1126 (1962), Soviet Phys. JETP **15**, 778 (1962).

[12] R. Mills, Preprint, 1962.

[13] N. N. Bogolyubov, Lektsii po kvantovoĭ statistike (Lectures on Quantum Statistics), Kiev, Publishing House "Sovetskaya Shkola" (Soviet School), 1949.

[14] M. Gell-Mann and F. Low, Phys. Rev. **84**, 350 (1951).

[15] D. A. Kirzhnits, Polevye metody teorii mnogikh chastits (Field Methods in Many Body Theory), Atomizdat, 1963.

[16] J. E. Moyal, Proc. Cambridge Phil. Soc. **45**, 99 (1949).

[17] V. P. Silin and A. A. Rukhadze, Élektromagnitnye svoĭstva plazmy i plazmopodobnykh sred (Electromagnetic Properties of Plasma and Plasma-like Media), Atomizdat, 1961.

[18] L. Van Hove, Physica **23**, 441 (1957).

[19] J. Prigogine and P. Resibois, Physica **27**, 629 (1961). P. Resibois, Physica, **29**, 721 (1963).

[20] S. Fujita, Physica **28**, 281 (1962).

Translated by G. Volkoff
211

SOVIET PHYSICS JETP VOLUME 20, NUMBER 5 MAY, 1965

IONIZATION IN THE FIELD OF A STRONG ELECTROMAGNETIC WAVE

L. V. KELDYSH

P. N. Lebedev Physics Institute, Academy of Sciences, U.S.S.R.

Submitted to JETP editor May 23, 1964

J. Exptl. Theoret. Phys. (U.S.S.R.) 47, 1945-1957 (November, 1964)

Expressions are obtained for the probability of ionization of atoms and solid bodies in the field of a strong electromagnetic wave whose frequency is lower than the ionization potential. In the limiting case of low frequencies these expressions change into the well known formulas for the probability of tunnel auto-ionization; at high frequencies they describe processes in which several photons are absorbed simultaneously. The ionization probability has a number of resonance maxima due to intermediate transition of the atom to an excited state. In the vicinity of such a maximum the ionization cross section increases by several orders of magnitude. The positions and widths of the resonances depend on the field strength in the wave. It is shown that for optical frequencies the mechanism under consideration, of direct ionization by the wave field, may be significant in the case of electric breakdown in gases, and especially in condensed media.

AN essential feature of the tunnel effect, a feature of importance in practical applications, is the practical absence of time lag. In other words, the probability of tunneling remains constant up to the highest frequencies of the radio band. The reason for this is that the tunneling time is determined essentially by the mean free time of the electron passing through a barrier of width

$$l = I / eF,$$

where I—ionization potential and F—electric field intensity. The average electron velocity is of the order of $(I/m)^{1/2}$ (m—electron mass). Therefore, up to frequencies on the order of

$$\omega_t = eF / \sqrt{2mI}$$

the tunnel effect is determined simply by the instantaneous value of the field intensity.

At higher frequencies there should appear a frequency dependence of the tunneling probability, since the electron does not have time to jump through the barrier within one cycle. Values typical of the tunnel effect in semiconductors are $I \sim 1$ eV, $m \sim 10^{-28}$ g, $F \sim 10^5$ V/cm, yielding $\omega_t \sim 10^{13}$ sec^{-1}, i.e., the dispersion can be noticeable at infrared and optical frequencies. A similar estimate is obtained for atoms, where $I \sim 10$ eV, $m = 10^{-27}$ g, and $F \sim 10^7$ V/cm.

With the appearance of lasers, the question of the tunnel effect at such frequencies became timely, since this is apparently the most effective mechanism for the absorption of high-power radiation in the transparency region. By transparency region we understand here the region of frequencies $\hbar\omega < I$, in which the substance is transparent to low-power radiation. Recent papers report gas breakdown in the focus of a laser beam [1,2]. This group of questions was considered theoretically by Bunkin and Prokhorov [3].

At first glance there exists at such high frequencies still another absorption mechanism, which competes with the tunnel effect. We have in mind the multi-photon absorption, in which the transition of the electron into a free state is accompanied by simultaneous absorption of several quanta. We shall show, however, that the nature of these two effects is essentially the same. We shall obtain a common formula which goes over into the usual formula for the tunnel effect [4-6] at low frequencies and very strong fields, when $\omega \ll \omega_t$, and describes multi-photon absorption when $\omega \gg \omega_t$.

In the simplest case of ionization of atoms, the general formula for the ionization probability is

$$w = A\omega\left(\frac{I_0}{\hbar\omega}\right)^{3/2}\left(\frac{\gamma}{(1+\gamma^2)^{1/2}}\right)^{5/2} S\left(\gamma, \frac{\tilde{I}_0}{\hbar\omega}\right)$$
$$\times \exp\left\{-\frac{2\tilde{I}_0}{\hbar\omega}\left[\sinh^{-1}\gamma - \gamma\frac{(1+\gamma^2)^{1/2}}{1+2\gamma^2}\right]\right\},$$
$$\gamma = \omega / \omega_t = \omega(2mI_0)^{1/2} / eF, \tag{1}$$

where the effective ionization potential is defined by

$$\tilde{I}_0 = I_0 + e^2F^2 / 4m\omega^2 = I_0(1 + 1 / 2\gamma^2); \tag{2}$$

$S(\gamma, \tilde{I}_0/\hbar\omega)$ is a relatively slowly varying function of the frequency and of the field, defined by formula (18) below, and A is a numerical coefficient of the order of unity.

Formula (1) describes the indirect transition of an electron from the ground state of the atom to the free state. However, in the case when one of the higher harmonics of the incident monochromatic wave is close to resonance with the electronic transition, in which the atom goes over into the s-th excited state, an appreciable role can be assumed by a two-step process consisting of the excitation of the atom followed by ionization of the excited state. The probability of such a resonant process is described by the formula

$$w^{(r)} = \frac{1}{4}\sum_s \left| J_{n_s+1}\left(\frac{eF\sigma_s}{\hbar\omega}\right) + J_{n_s-1}\left(\frac{eF\sigma_s}{\hbar\omega}\right)\right|^2 \times \frac{|V_{0s}|^2 w_s}{(I_0 - I_s + {}^1/_2 e^2F^2\alpha_s - n_s\hbar\omega)^2 + \hbar^2(\omega_s + \gamma_s)^2}, \quad (3)$$

where J_n—Bessel functions; w_s—probability of ionization of the s-th excited state, described by formula (1) in which, however, I_0 must be replaced by I_s—ionization potential of the s-th level; σ_s and α_s—coefficients that describe the Stark shift of the s-th level in a static electric field:

$$I_s(F) = I_s - eF\sigma_s - {}^1/_2 e^2F^2\alpha_s; \quad (4)$$

V_{0s}—matrix element of the transition from the ground state to the s-th state in a homogeneous electric field:

$$V_{0s} = \int \psi_s^*(\mathbf{r})\, e\mathbf{F}\mathbf{r}\psi_0(\mathbf{r})\, d^3r;$$

n_s—integer closest to $(I_0 - I_s)/\hbar\omega$; γ_s—radiation width of the level.

We present now a detailed derivation and analysis of formulas (1) and (3). On the basis of the results we shall then derive briefly analogous expressions for the ionization probability in a solid.

The electric field of the wave exerts the strongest action on the states of the continuous spectrum and on the degenerate levels of the excited bound states. Therefore, as the first step, we take into account these effects. The wave function of the free electron in an electric field

$$\mathbf{F}(t) = \mathbf{F}\cos\omega t \quad (5)$$

is of the form

$$\psi_{\mathbf{p}}(\mathbf{r}, t) = \exp\left\{\frac{i}{\hbar}\left[\left(\mathbf{p} + \frac{e\mathbf{F}}{\omega}\sin\omega t\right)\mathbf{r} - \int_0^t \frac{1}{2m}\left(\mathbf{p} + \frac{e\mathbf{F}}{\omega}\sin\omega\tau\right)^2 d\tau\right]\right\}, \quad (6)$$

and for the s-th bound state

$$\psi_s(\mathbf{r}, t) = \psi_s(\mathbf{r})\exp\left\{\frac{i}{\hbar}\left(I_s t - \frac{eF\sigma_s}{\omega}\sin\omega t\right)\right\}. \quad (7)$$

We have in mind here the hydrogen atom, in which the Stark effect is nonlinear in the field. In the wave function of the final state (6), we have neglected for the time being the influence of the Coulomb field of the ionized atom. The role of this interaction, as well as of the quadratic Stark effect, will be discussed below.

Let us calculate now the probability of the transition from the ground state to a free-electron state of the type (6). The difference between our procedure and the usual perturbation theory lies thus only in the fact that we calculate the probability of transition not to a stationary final state, but to a state (6) that already takes exact account of the main effect of the electric field—the acceleration of the free electron. The matrix elements of the transition between the bound states, on the other hand, are taken into account only in the lower orders of perturbation theory, since they are proportional to eFa_0, and as will be shown below, the transition matrix elements which we take into account in the continuous spectrum are proportional to $eFa_0\sqrt{I_0/\hbar\omega}$. The ratio $\hbar\omega/I_0$ will be assumed to be sufficiently small. For the lasers presently in existence it is of the order of 0.1.

The probability of direct transition from the ground state to the continuous spectrum is of the form

$$w_0 = \frac{1}{\hbar^2}\lim_{T\to\infty} \mathrm{Re}\int \frac{d^3p}{(2\pi\hbar)^3}\int_0^T dt\cos\omega T\cos\omega t \times V_0\left(\mathbf{p} + \frac{e\mathbf{F}}{\omega}\sin\omega T\right) V_0\left(\mathbf{p} + \frac{e\mathbf{F}}{\omega}\sin\omega t\right) \times \exp\left\{\frac{i}{\hbar}\int_T^t \left[I_0 + \frac{1}{2m}\left(\mathbf{p} + \frac{e\mathbf{F}}{\omega}\sin\omega\tau\right)^2\right] d\tau\right\}, \quad (8)$$

$$V_0(\mathbf{p}) = \int e^{-i\mathbf{p}\mathbf{r}/\hbar}\, e\mathbf{F}\mathbf{r}e^{-r/a_0}\frac{d^3r}{(\pi a_0^3)^{1/2}} = 8(\pi a_0^3)^{1/2} e\hbar\mathbf{F}\nabla_{\mathbf{p}}(1 + p^2a_0^2/\hbar^2)^{-2}. \quad (9)$$

Let us expand the expression

$$L(\mathbf{p}, t) = V_0\left(\mathbf{p} + \frac{e\mathbf{F}}{\omega}\sin\omega t\right) \times \exp\left\{\frac{i}{\hbar}\int_0^t \left[I_0 + \frac{1}{2m}\left(\mathbf{p} + \frac{e\mathbf{F}}{\omega}\sin\omega\tau\right)^2\right] d\tau\right\} \quad (10)$$

in a Fourier series in t:

$$L(\mathbf{p}, t) = \sum_{n=-\infty}^{\infty} \exp\left\{\frac{i}{\hbar}\left(I_0 + \frac{p^2}{2m} + \frac{e^2F^2}{4m\omega^2} - n\hbar\omega\right)t\right\} L_n(\mathbf{p}), \quad (11)$$

$$L_n(\mathbf{p}) = \frac{1}{2\pi}\int_{-\pi}^{\pi} V_0\left(\mathbf{p} + \frac{e\mathbf{E}}{\omega}\sin x\right) \times \exp\left\{\frac{i}{\hbar\omega}\left(n\hbar\omega x - \frac{e\mathbf{F}\mathbf{p}}{2m\omega}\cos x - \frac{e^2F^2}{8m\omega^2}\sin 2x\right)\right\} dx. \tag{12}$$

Substituting this expansion in (8), we obtain after the usual transformations

$$w_0 = \frac{2\pi}{\hbar}\int \frac{d^3p}{(2\pi\hbar)^3}\sum_{n=-\infty}^{\infty}\frac{1}{4}\left|L_{n+1}(\mathbf{p}) + L_{n-1}(\mathbf{p})\right|^2 \times \delta\left(I_0 + \frac{p^2}{2m} + \frac{e^2F^2}{4m\omega^2} - n\hbar\omega\right), \tag{13}$$

or, using the presence of δ-functions

$$w_0 = \frac{2\pi}{\hbar}\int \frac{d^3p}{(2\pi\hbar)^3}\left|L(\mathbf{p})\right|^2 \times \sum_{n=-\infty}^{\infty}\delta\left(I_0 + \frac{p^2}{2m} + \frac{e^2F^2}{4m\omega^2} - n\hbar\omega\right), \tag{14}$$

$$L(\mathbf{p}) = \frac{1}{2\pi}\oint V_0\left(\mathbf{p} + \frac{e\mathbf{F}}{\omega}u\right)\exp\left\{\frac{i}{\hbar\omega}\int_0^u \left[I_0 + \frac{1}{2m}\left(\mathbf{p} + \frac{e\mathbf{F}}{\omega}v\right)^2\right]\frac{dv}{(1-v^2)^{1/2}}\right\} du. \tag{15}$$

The integral with respect to u in (15) is taken along a closed contour which encloses the segment (−1, 1).

Formula (14) has the explicit form of the sum of multi-photon processes. The exponential in (15) is rapidly oscillating, so that the integral can be calculated by the saddle-point method. The saddle points are determined by the condition

$$I_0 + \frac{1}{2m}\left(\mathbf{p} + \frac{e\mathbf{F}}{\omega}u_s\right)^2 = 0.$$

It is easy to verify, however, that for the hydrogen atom the matrix element $V_0(\mathbf{p} + e\mathbf{F}u/\omega)$ has a pole at the same points, by virtue of the relation $2mI_0 = \hbar^2/a_0^2$. The presence of this pole is neither accidental nor typical of the hydrogen atom, but reflects the universally known fact that the scattering amplitude has poles at complex momentum values corresponding to bound states[5]. Taking these singularities into account, we can easily verify that the contribution of each saddle point to (15) is equal to

$$2\sqrt{\pi a_0^3}\,\frac{I_0}{eFa_0}\,\frac{\hbar\omega}{(1-u_s^2)^{1/2}} \times \exp\left\{\frac{i}{\hbar\omega}\int_0^{u_s}\left[I_0 + \frac{1}{2m}\left(\mathbf{p} + \frac{e\mathbf{F}}{\omega}v\right)^2\right]\frac{dv}{(1-v^2)^{1/2}}\right\}.$$

The positions of the saddle points depend on p. However, contributions to the total probability of ionization (14) are made only by small p, satisfying the condition $p^2 \ll 2mI_0$. Consequently we can put $\mathbf{p} = 0$ in the pre-exponential factor, and we can expand in the exponential in powers of p up to second order inclusive. Substituting then the resultant expression in (14) and integrating with respect to p, we transform (14) to

$$w_0 = \sqrt{\frac{2I_0}{\hbar}}\,\omega\left(\frac{\gamma}{\sqrt{1+\gamma^2}}\right)^{3/2} S\left(\gamma, \frac{\tilde{I}_0}{\hbar\omega}\right) \times \exp\left\{-\frac{2\tilde{I}_0}{\hbar\omega}\left[\sinh^{-1}\gamma - \gamma\frac{\sqrt{1+\gamma^2}}{1+2\gamma^2}\right]\right\}; \tag{16}$$

Here

$$\gamma = \omega\sqrt{2mI_0}/eF, \tag{17}$$

and the function $S(\gamma, \tilde{I}_0/\hbar\omega)$, which varies slowly compared with an exponential function, is of the form

$$S(\gamma, x) = \sum_{n=0}^{\infty}\exp\left\{-2[\langle x+1\rangle - x + n] \times \left(\sinh^{-1}\gamma - \frac{\gamma}{\sqrt{1+\gamma^2}}\right)\right\} \times \Phi\left\{\left[\frac{2\gamma}{\sqrt{1+\gamma^2}}\left(\langle x+1\rangle - x + n\right)\right]^{1/2}\right\}. \tag{18}$$

The symbol $\langle x\rangle$ denotes the integer part of the number x, and the function $\Phi(z)$, defined by

$$\Phi(z) = \int_0 e^{y^2 - z^2}dy, \tag{19}$$

is expressed in terms of the well known probability integral.

The function $S(\gamma, \tilde{I}_0/\hbar\omega)$ describes the spectrum structure connected with the discreteness of the number of absorbed photons. It obviously has characteristic threshold singularities of the type $(\tilde{I}_0 - n\hbar\omega)^{1/2}$. The effective threshold energy of absorption $\tilde{I}_0$, defined by formula (2), exceeds the ionization potential I_0 by the value of the average oscillation energy of the electron in the field of the wave. This is precisely the quantity which enters into the δ-function that expresses the law of energy conservation in the general formula (14).

For the case of low frequencies and very strong fields, when $\gamma \ll 1$, the main contribution in expression (18) for S is made by large $n \sim \gamma^{-3}$. Consequently, going over from summation over n to integration, we obtain

$$S(\gamma, \tilde{I}_0/\hbar\omega) \approx \sqrt{3\pi}/4\gamma^2,$$

as a result of which the ionization probability is described by the formula

$$w_0 = \frac{\sqrt{6\pi}}{4}\frac{I_0}{\hbar}\left(\frac{eF\hbar}{m^{1/2}I_0^{3/2}}\right)^{1/2} \times \exp\left\{-\frac{4}{3}\frac{\sqrt{2m}\,I_0^{3/2}}{e\hbar F}\left(1-\frac{m\omega^2 I_0}{5e^2F^2}\right)\right\}. \qquad (20)$$

As $\omega \to 0$, the exponential in (20) coincides with the known expression for the tunnel auto-ionization of atoms in an electric field[4,5], whereas the exponential factor is different from the correct one. This is connected with the fact that we have used for the wave functions of the final state the functions of the free electron (6), i.e., we have neglected the Coulomb interaction in the final state, which, as is well known, changes the power of F in the pre-exponential expression, without changing the exponential itself. For the same reason, the pre-exponential factor in (16) must also be corrected, but this is of less significance, since all the dependences are determined essentially by the exponential function. A crude quasiclassical analysis shows that, apart from a numerical coefficient, the inclusion of this interaction results in a correction factor $I_0\gamma/\hbar\omega\,(1+\gamma^2)^{1/2}$, which we have taken into account in (1).

In the opposite limiting case of high frequencies and not very strong fields $\gamma \gg 1$, the fundamental role in (18) is played by the zeroth term, and formula (1) describes in this case the probability of simultaneous absorption of several photons:

$$w_0 = A\omega\left(\frac{I_0}{\hbar\omega}\right)^{3/2}\exp\left\{2\left\langle\frac{\tilde I_0}{\hbar\omega}+1\right\rangle - \frac{\tilde I_0}{\hbar\omega}\left(1+\frac{e^2F^2}{2m\omega^2 I_0}\right)\right\} \times\left(\frac{e^2F^2}{8m\omega^2 I_0}\right)^{\langle \tilde I_0/\hbar\omega+1\rangle}\Phi\left[\left(2\left\langle\frac{\tilde I_0}{\hbar\omega}+1\right\rangle - \frac{2\tilde I_0}{\hbar\omega}\right)^{1/2}\right]. \qquad (21)$$

After introducing the correction for the Coulomb interaction in the final state, formulas (1) and (21) should be applicable, accurate to a numerical factor of the order of unity, to a description of the ionization of any atom, not only hydrogen. Indeed, as can be seen from the preceding deduction, this probability is determined essentially by the action of the field on the final state of the free electron, and not on the ground state of the atom. Therefore the use of another wave function for the ground state changes only the matrix element $V_0(p)$, i.e., the pre-exponential factor.

So far we have considered a transition from the ground state directly to the continuous-spectrum state. Let us consider now, in the next higher order of perturbation theory, a process in which the electron first goes over into an excited state (7), and then into the continuous spectrum. The probability of such a transition is equal to

$$w^{(r)} = \frac{1}{\hbar^2}\sum_s \lim_{T\to\infty} \operatorname{Re}\int\frac{d^3p}{(2\pi\hbar)^3}L_s^{(r)}(\mathbf{p}, T)\cos\omega T \times \int_0^T L_s^{(r)*}(\mathbf{p}, t)\cos\omega t\,dt, \qquad (22)$$

$$L_s^{(r)}(\mathbf{p}, t) = \frac{1}{i\hbar}V_s\left(\mathbf{p}+\frac{e\mathbf{F}}{\omega}\sin\omega t\right)\exp\left\{\frac{i}{\hbar}\int_0^t\left[I_s + \frac{1}{2m}\left(\mathbf{p}+\frac{e\mathbf{F}}{\omega}\sin\omega\tau\right)^2 - eF\sigma_s\cos\omega\tau\right]d\tau\right\} \times \int_0^t V_{0s}\cos\omega t'\exp\left\{\frac{i}{\hbar}\int_0^{t'}[I_0 - I_s + eF\sigma_s\cos\omega\tau]\,d\tau\right\}dt'. \qquad (23)$$

After transformations analogous to (11)—(13), this expression reduces to the form

$$w^{(r)} = \frac{1}{4}\sum_{sn}\left|J_{n-1}\left(\frac{eF\sigma_s}{\hbar\omega}\right)+J_{n+1}\left(\frac{eF\sigma_s}{\hbar\omega}\right)\right|^2 \times\frac{|V_{0s}|^2}{(I_0 - I_s - n\hbar\omega)^2}\frac{2\pi}{\hbar}\int\frac{d^3p}{(2\pi\hbar)^3} \times\sum_{n'}\frac{1}{4}\left|L_{n'-1}(\mathbf{p})+L_{n'+1}(\mathbf{p})\right|^2 \times\left[\delta\left(I_s+\frac{p^2}{2m}+\frac{e^2F^2}{4m\omega^2}-n'\hbar\omega\right) + \delta\left(I_0+\frac{p^2}{2m}+\frac{e^2F^2}{4m\omega^2}-(n'+n)\hbar\omega\right)\right], \qquad (24)$$

where J_n—Bessel function of order n.

Carrying out summation over n′ and integration over p, and retaining in the sum over n only the term n_s closest to resonance $|I_0 - I_s - n_s\hbar\omega| \ll \hbar\omega$, we obtain formula (3), but with a denominator in which we take account also of the shift of the atomic levels in the second order of the Stark effect, and of the attenuation of the level I_s due to its ionization w_s and its spontaneous emission γ_s. Formula (3) describes also the radiation of atoms at frequencies $\hbar\omega_s = I_0 - I_s$, if we replace w_s by γ_s in its numerator.

The quantity $eF\sigma_s/\hbar\omega$, which enters in (3), is of the order of $1/\gamma$, and therefore formally when $\gamma \gg 1$ the probabilities $w^{(r)}$ and w_0 can be of the same order of magnitude. Numerically, however, $w^{(r)}$ is in the mean very small compared with w_0, owing to the factor $1/n_s!$ in the definition of J_n. Directly near resonance, however, $w^{(r)}$ becomes many orders larger than w_0:

$$\left(\frac{w^{(r)}}{w_0}\right)_{max} \sim \left(\frac{1}{n_s!}\right)^2\left(\frac{m\hbar\omega^3}{e^2F^2}\right)^{2I_s/\hbar\omega}$$

This resonance is of course very narrow and

therefore difficult to observe. It is interesting to note that if the frequency of the incident radiation lies in a relatively broad region, such that $|I_0 - I_s - n_s\hbar\omega| \lesssim e^2F^2\alpha_s/2$, then we can observe just as sharp a resonant change in the ionization probability when the field intensity in the wave is changed, for the resonant frequency itself is displaced in this case.

We note, finally, that in the case of the purely quadratic Stark effect ($\sigma_s = 0$) the situation does not change qualitatively, although quantitatively $w^{(r)}$ becomes somewhat smaller. The Bessel-function arguments in (3) then contain in lieu of $eF\sigma_s$ the quantity $e^2F^2\alpha_s/4$, and their index $n_s - 1$ is replaced by $(n_s - 1)/2$ for odd n_s and $n_s/2 - 1$ for even n_s. In the case of even n_s, V_{0s} is replaced by $V_{0s}^{(1)}$—the correction to V_{0s} in first-order perturbation theory.

The formulas presented above pertained to the ionization of individual atoms, i.e., to gases. We now turn to ionization by a strong electromagnetic wave in a crystal. In this case the ionization process reduces to the transfer of the electron from the valence band into the conduction band, in other words, to the creation of an electron-hole pair. Therefore the energy of the final state is not simply the energy of the free electron, but is equal to the sum of the energies of the electron and of the hole:

$$\varepsilon(\mathbf{p}) = \varepsilon_c(\mathbf{p}) - \varepsilon_v(\mathbf{p}). \tag{25}$$

Here the indices c and v denote the conduction and valence bands, and $\epsilon_{c,v}(\mathbf{p})$ is the dependence of the energy on the quasimomentum in these bands. The quasimomenta of the initial and final states, should naturally be the same, since the homogeneous electric field cannot change the momentum of a system that is neutral on the whole.

The Bloch wave functions of an electron, accelerated by the field inside each of the bands, have a form analogous to (6):

$$\psi_{\mathbf{p}}^{c,v}(\mathbf{r}, t) = u_{\mathbf{p}(t)}^{c,v}(\mathbf{r}) \exp\left\{\frac{i}{\hbar}\left[\mathbf{p}(t)\mathbf{r} - \int_0^t \varepsilon_{c,v}(\mathbf{p}(\tau))\,d\tau\right]\right\},$$

$$\mathbf{p}(t) = \mathbf{p} + (e\mathbf{F}/\omega)\sin\omega t, \tag{26}$$

where $u_{\mathbf{p}}^{c,v}(\mathbf{r})$ are periodic functions that have the translational symmetry of the lattice. Calculations perfectly similar to (8)—(15) lead to a general formula for the ionization probability

$$w = \frac{2\pi}{\hbar}\int \frac{d^3p}{(2\pi\hbar)^3}\,|L_{cv}(\mathbf{p})|^2 \sum_n \delta(\overline{\varepsilon(\mathbf{p})} - n\hbar\omega), \tag{27}$$

where, however,

$$\overline{\varepsilon(\mathbf{p})} = \frac{1}{2\pi}\int_{-\pi}^{\pi} \varepsilon\left(\mathbf{p} + \frac{e\mathbf{F}}{\omega}\sin x\right)dx, \tag{28}$$

$$L_{cv}(\mathbf{p}) = \frac{1}{2\pi}\oint V_{cv}\left(\mathbf{p} + \frac{e\mathbf{F}}{\omega}u\right) \times \exp\left\{\frac{i}{\hbar\omega}\int_0^u \varepsilon\left(\mathbf{p} + \frac{e\mathbf{F}}{\omega}v\right)\frac{dv}{(1-v^2)^{1/2}}\right\}du. \tag{29}$$

The matrix element of the optical transition from the valence to the conduction band is determined in the following fashion:

$$V_{cv}(\mathbf{p}) = i\hbar \int u_{\mathbf{p}}^{c*}(\mathbf{r})\, e\mathbf{F}\nabla_{\mathbf{p}} u_{\mathbf{p}}^{v}(\mathbf{r})\, d^3\mathbf{r}. \tag{30}$$

The saddle points of the integral (29) are now determined by the relation

$$\varepsilon(\mathbf{p} + e\mathbf{F}u_s/\omega) = 0. \tag{31}$$

The matrix element $V_{cv}(\mathbf{p} + e\mathbf{F}u/\omega)$ has at these points, as for isolated atoms, a pole with a universal residue value[6,7]

$$\operatorname{res} V_{cv}(\mathbf{p} + e\mathbf{F}u_s/\omega) = \pm i\hbar\omega/4. \tag{32}$$

The essential difference from the case of atoms is, however, the fact that the function $\epsilon(\mathbf{p} + e\mathbf{F}u/\omega)$ has at the point u_s not a simple zero, but a branch point of the root type. Calculating $L_{cv}(\mathbf{p})$ with allowance for these singularities, we find that the contribution to (29) made by each of the saddle points is equal to

$$\frac{\hbar\omega}{3}\exp\left\{\frac{i}{\hbar\omega}\int_0^{u_s}\varepsilon\left(\mathbf{p} + \frac{e\mathbf{F}}{\omega}u\right)\frac{du}{(1-u^2)^{1/2}}\right\}.$$

Further calculations are perfectly analogous to those that lead to formula (16), but call for the use of a concrete type of dispersion law $\epsilon(\mathbf{p})$. For typical semiconductors[6,7] we have

$$\varepsilon(\mathbf{p}) = \Delta(1 + p^2/m\Delta)^{1/2}, \tag{33}$$

where Δ—width of the forbidden band separating the valence band from the conduction band and m—reduced mass of the electron and the hole,

$$1/m = 1/m_e + 1/m_h.$$

In this case, having in mind small $\mathbf{p}$ (but not $\mathbf{p} + e\mathbf{F}u/\omega$) and introducing dimensionless variables x and y defined by

$$x^2 = p_{\parallel}^2/m\Delta; \quad y^2 = p_{\perp}^2/m\Delta, \tag{34}$$

where $p_{\parallel}$ and $p_{\perp}$ are the quasimomentum components parallel and perpendicular to the direction of the field $\mathbf{F}$, we obtain

$$\bar{\varepsilon}(\mathbf{p})=\frac{2}{\pi}\Delta\Big\{\frac{\sqrt{1+\gamma^2}}{\gamma}E\Big(\frac{1}{\sqrt{1+\gamma^2}}\Big)+\frac{\gamma}{\sqrt{1+\gamma^2}}E\Big(\frac{1}{\sqrt{1+\gamma^2}}\Big)\frac{x^2}{2}+\frac{\gamma}{\sqrt{1+\gamma^2}}K\Big(\frac{1}{\sqrt{1+\gamma^2}}\Big)\frac{y^2}{2}\Big\} \tag{35}$$

and for the argument of the exponential in (2.9) we get

$$\frac{i}{\hbar\omega}\int_0^{u_0}\varepsilon\Big(\mathbf{p}+\frac{e\mathbf{F}}{\omega}u\Big)\frac{du}{\sqrt{1-u^2}}$$
$$=-\frac{\Delta}{\hbar\omega}\Big\{\frac{\sqrt{1+\gamma^2}}{\gamma}\Big[K\Big(\frac{\gamma}{\sqrt{1+\gamma^2}}\Big)-E\Big(\frac{\gamma}{\sqrt{1+\gamma^2}}\Big)\Big]$$
$$+\frac{x^2}{2}\frac{\gamma}{\sqrt{1+\gamma^2}}\Big[K\Big(\frac{\gamma}{\sqrt{1+\gamma^2}}\Big)-E\Big(\frac{\gamma}{\sqrt{1+\gamma^2}}\Big)\Big]$$
$$+\frac{y^2}{2}\frac{\gamma}{\sqrt{1+\gamma^2}}K\Big(\frac{\gamma}{\sqrt{1+\gamma^2}}\Big)$$
$$\pm i\gamma x\Big[1-2\sqrt{1+\gamma^2}E\Big(\frac{\gamma}{\sqrt{1+\gamma^2}}\Big)$$
$$+\frac{2}{\sqrt{1+\gamma^2}}K\Big(\frac{\gamma}{\sqrt{1+\gamma^2}}\Big)\Big]\Big\}. \tag{36}$$

The functions K and E in (35) and (36) are complete elliptic integrals of the first and second kind. The term in (36), which is linear in x, will henceforth be left out, for when account is taken of both saddle points it gives rise in L(p) to a rapidly oscillating factor of the type 2cos x, which reduces after squaring and integrating with respect to x to a factor 2, which we can take into account directly in the final answer. The quantity γ in (35)—(42) differs from (17) by a factor $1/\sqrt{2}$:

$$\gamma=\omega\sqrt{m\Delta}/eF.$$

The ionization probability is given by the expression

$$w=\frac{2\omega}{9\pi}\Big(\frac{\sqrt{1+\gamma^2}}{\gamma}\frac{m\omega}{\hbar}\Big)^{3/2}Q\Big(\gamma,\frac{\tilde{\Delta}}{\hbar\omega}\Big)\exp\Big\{-\pi\Big\langle\frac{\tilde{\Delta}}{\hbar\omega}+1\Big\rangle$$
$$\times\Big[K\Big(\frac{\gamma}{\sqrt{1+\gamma^2}}\Big)-E\Big(\frac{\gamma}{\sqrt{1+\gamma^2}}\Big)\Big]\Big/E\Big(\frac{1}{\sqrt{1+\gamma^2}}\Big)\Big\}, \tag{37}$$

where $\tilde{\Delta}$—effective ionization potential:

$$\tilde{\Delta}=\frac{2}{\pi}\Delta\frac{\sqrt{1+\gamma^2}}{\gamma}E\Big(\frac{1}{\sqrt{1+\gamma^2}}\Big). \tag{38}$$

The symbol $\langle x\rangle$ again denotes the integer part of the number x, and $Q(\gamma, \tilde{\Delta}/\hbar\omega)$ is a function analogous to S in (16):

$$Q(\gamma,x)=\Big[\pi/2K\Big(\frac{1}{\sqrt{1+\gamma^2}}\Big)\Big]^{1/2}$$
$$\times\sum_{n=0}^{\infty}\exp\Big\{-\pi\Big[K\Big(\frac{\gamma}{\sqrt{1+\gamma^2}}\Big)$$
$$-E\Big(\frac{\gamma}{\sqrt{1+\gamma^2}}\Big)\Big]n\Big/E\Big(\frac{1}{\sqrt{1+\gamma^2}}\Big)\Big\}$$
$$\times\Phi\Big\{\Big[\pi^2(2\langle x+1\rangle-2x+n)\Big/2K\Big(\frac{1}{\sqrt{1+\gamma^2}}\Big)$$
$$\times E\Big(\frac{1}{\sqrt{1+\gamma^2}}\Big)\Big]^{1/2}\Big\}. \tag{39}$$

In the derivation of (37) we have made use of an identity known from the theory of elliptic integrals:

$$K(x)E(\sqrt{1-x^2})+K(\sqrt{1-x^2})E(x)$$
$$-K(x)K(\sqrt{1-x^2})=\pi/2.$$

Asymptotically, the behavior of the ionization probability in a solid (37) is similar to the variation of the ionization probability in gases (16), both for low and for high frequencies. In the case of low frequencies and strong fields $\gamma\ll 1$, the ionization probability reduces to the formula for the tunnel effect[6,8]:

$$w=\frac{2}{9\pi^2}\frac{\Delta}{\hbar}\Big(\frac{m\Delta}{\hbar^2}\Big)^{3/2}\Big(\frac{e\hbar F}{m^{1/2}\Delta^{3/2}}\Big)^{5/2}$$
$$\times\exp\Big\{-\frac{\pi}{2}\frac{m^{1/2}\Delta^{3/2}}{e\hbar F}\Big(1-\frac{1}{8}\frac{m\omega^2\Delta}{e^2F^2}\Big)\Big\}. \tag{40}$$

At first glance (40) differs from the corresponding formula (19) in[6] by a factor of the order $(e\hbar F/m^{1/2}\Delta^{3/2})^{1/2}$ in front of the exponential. This is connected with the fact that (40), like the general formula (27), describes the average ionization probability over a time which is much larger than the period of the external field $2\pi/\omega$, while formula (19) of[6] yields the instantaneous tunneling probability in a slowly varying field. If we introduce in this last formula a field of the type $F(t) = F\cos\omega t$ and average it over the time, then the result agrees with (40).

In the opposite limiting case $\gamma\gg 1$, we again obtain a formula that describes the probability of multi-quantum absorption:

$$w=\frac{2}{9\pi}\omega\Big(\frac{m\omega}{\hbar}\Big)^{3/2}\Phi\Big[\Big(2\Big\langle\frac{\tilde{\Delta}}{\hbar\omega}+1\Big\rangle-\frac{2\tilde{\Delta}}{\hbar\omega}\Big)^{1/2}\Big]$$
$$\times\exp\Big\{2\Big\langle\frac{\tilde{\Delta}}{\hbar\omega}+1\Big\rangle\Big(1-\frac{e^2F^2}{4m\omega^2\Delta}\Big)\Big\}\Big(\frac{e^2F^2}{16m\omega^2\Delta}\Big)^{\langle\tilde{\Delta}/\hbar\omega+1\rangle}, \tag{41}$$

$$\tilde{\Delta}=\Delta+e^2F^2/4m\omega^2. \tag{42}$$

Ionization in solids can proceed also via intermediate states, similar to the process described by formula (3). The role of the excited states can then be played by the excitons, if their binding energy is commensurate with or larger than the energy of the quantum of incident radiation $\hbar\omega$, or else by the local level of the impurities and lattice defects, if their concentration is sufficiently high. The corresponding formula is perfectly analogous to (3) and will not be written out here.

Owing to the presence of the dielectric probability, the influence of the Coulomb interaction of the free electron and hole on the ionization probability is negligibly small[6].

We now present some quantitative estimates. Having in mind the hydrogen atom (I_0 = 13.6 eV) and emission in the red region of the spectrum at frequency $\omega \approx 3 \times 10^{15}$ sec^{-1}, we obtain $\langle I_0/\hbar\omega + 1\rangle = 8$. The field intensity is best expressed in terms of the power of the incident wave W and the transverse cross section of the beam[1)] R^2:

$$\frac{1}{\gamma^2}=\frac{e^2F^2}{2m\omega^2I_0}=\frac{2\pi e^2}{mc\omega^2I_0}\frac{W}{R^2}\approx 2.5\cdot 10^{-22}\frac{W}{R^2}.$$

After substituting in (21) we obtain

$$\omega_0 \approx 10^{17}(2\cdot 10^{-9}\,W/R^2)^8\,\mathrm{sec}^{-1}.$$

Thus, for example, for W = 50 MW and $R^2 = 10^{-6}$ cm^2 we get $\omega_0 \approx 10^9$ sec^{-1}, i.e., within a time on the order of 0.001 μsec the gas is completely ionized. However, at lower values of the power or in the case of poor focusing, the effect decreases very rapidly. Thus, for W = 50 MW but $R^2 = 10^{-4}$ cm, the frequency is $\omega_0 \approx 10^{-7}$ sec^{-1}.

Analogously, for ionization of the first excited level of hydrogen, $I_1 = 0.25\, I_0$, we get

$$\omega_1 \approx 10^{16}(8\cdot 10^{-9}\,W/R^2)^2\,\mathrm{sec}^{-1}.$$

The Stark splitting for this level is $3eF\hbar^2/me^2$. Substituting these expressions in (3) we obtain for the resonant part of the ionization probability the following estimate:

$$w^{(r)}\approx\frac{w_1}{[(n_1-1)!]^2}\left(\frac{eF\sigma_1}{2\hbar\omega}\right)^{2n_1}\frac{(\hbar\omega)^2}{(I_0-I_1-n_1\hbar\omega)^2}\sim$$

$$\sim 10^{12}\left(6\cdot 10^{-9}\frac{W}{R^2}\right)^8\left(\frac{\hbar\omega}{I_0-I_1-6\hbar\omega}\right)^2.$$

If, for example, the distance from resonance in the denominator of this formula is of the order of 2×10^{-2} eV, then $w^{(r)}$ exceeds w_0 by more than four orders of magnitude.

In the case of a molecular gas, the condition of resonance can be assumed to be satisfied almost always, since the energy which is lacking for resonance can be obtained from (or given up to) the vibrational degrees of freedom, which are strongly excited in some fields. The total ionization probability is in this case of the order of

$$w\sim\omega\left(\frac{eF\sigma_s}{\hbar\omega}\right)^{2n_s}\frac{w_s}{w_s+\gamma_s}$$

The main contribution to the ionization is made by the excited states with the highest ionization potentials.

In a real situation, the resonant ionization is decisive in a much wider frequency range, since the power of the incident radiation varies with time, and the resonant frequencies in (3) are shifted by an amount on the order of

$$\Delta(\hbar\omega_r)=\frac{1}{2}e^2F^2\alpha_s=2\pi\frac{e^2}{c}\alpha_sW.$$

If $|I_0 - I_S - n_S\hbar\omega| < \Delta(\hbar\omega_r)$, then at some instant of time the resonance condition is satisfied exactly, and this instant makes the main contribution to the total ionization. The total ionization probability is determined by the time integral of formula (3):

$$\int w^{(r)}\,dt=\frac{2\pi}{e^2\alpha_s}\frac{|V_{0s}|^2}{\hbar}\left(\frac{dF^2}{dt}\right)^{-1}\frac{1}{4}\left|J_{n_s-1}\left(\frac{eF\sigma_s}{\hbar\omega}\right)\right|^2. \quad (43)$$

In the right side of this formula there should be substituted the resonant value of the field.

If we introduce a dimensionless coefficient of the order of unity

$$a_s=|V_{0s}|^2/e^2F^2\alpha_sI_0,$$

then (43) can be rewritten in a more illustrative form

$$\int w^{(r)}\,dt=2\pi a_s\frac{I_0\tau}{4\hbar}\left|J_{n_s-1}\left(\frac{eF\sigma_s}{\hbar\omega}\right)\right|^2, \quad (44)$$

where τ—characteristic time of the order of the duration of the radiation pulse, determined by the relation

$$\frac{1}{\tau}=\frac{1}{F^2}\frac{dF^2}{dt}. \quad (45)$$

Under the same assumptions used to obtain the previous estimate, we obtain (n_S = 6)

$$\int w^{(r)}\,dt\sim 10^{12}\tau(6\cdot 10^{-9}\,W/R^2)^5.$$

Thus, for a pulse duration $\tau \sim 100\,\mu$sec, radiation of 50 MW power, focused in a region with dimensions $R \sim 3\times 10^{-3}$ cm produces practically full ionization.

In breakdown of gases under real conditions, the role of the processes considered in the present

[1)]In all the formulas that follow W is in MW and R in cm.

paper is determined by their competition with the cascade multiplication of the free electrons. The field intensity at which the cascade breakdown occurs increases rapidly with decreasing dimensions of the region where the field is effective, or with decrease in the pulse duration. Therefore for a sufficiently focused beam $R \lesssim 10^{-3}$ cm, or for sufficiently short pulses, the breakdown should apparently be determined by the direct ionization of the atoms in the field of the wave.

In the case of condensed media, the situation is much more favorable for the mechanism in question, for owing to the rapid energy dissipation by the free electrons, the breakdown fields are much higher there, and the ionization potentials lower.

In conclusion I take the opportunity to thank G. A. Askar'yan, V. L. Ginzburg, and M. A. Krivoglaz for valuable remarks made during the discussion of the present paper.

[1] E. K. Dammon and R. G. Tomlinson, Appl. Opt. **2**, 546 (1963).

[2] R. G. Meyerhand and A. F. Haught, Phys. Rev. Lett. **11**, 401 (1963).

[3] F. V. Bunkin and A. M. Prokhorov, JETP **46**, 1090 (1964), Soviet Phys. JETP **19**, 739 (1964).

[4] J. R. Oppenheimer, Phys. Rev. **31**, 66 (1928).

[5] L. D. Landau and E. M. Lifshitz, Kvantovaya mekhanika (Quantum Mechanics), Fizmatgiz, 1963.

[6] L. V. Keldysh, JETP **33**, 994 (1957), Soviet Phys. JETP **6**, 763 (1958).

[7] E. O. Kane, J. Phys. Chem. Solids **1**, 249 (1957).

[8] E. O. Kane, J. Appl. Phys. **12**, 181 (1960).

Translated by J. G. Adashko
280

SOVIET PHYSICS JETP VOLUME 21, NUMBER 6 DECEMBER, 1965

CONCERNING THE THEORY OF IMPACT IONIZATION IN SEMICONDUCTORS

L. V. KELDYSH

P. N. Lebedev Physics Institute, Academy of Sciences, U.S.S.R.

Submitted to JETP editor January 9, 1965

J. Exptl. Theoret. Phys. (U.S.S.R.) **48**, 1692-1707 (June, 1965)

The energy distribution of the electrons in a covalent semiconductor in the presence of a strong electric field is determined. It is shown that the number of ionizing electrons increases with increasing field **E**, first approximately like $\exp(-\mathrm{const}\cdot E^{-1})$, and in extremely strong fields like $\exp(-\mathrm{const}\cdot E^{-2})$.

THE average probability of impact ionization for electrons in a strong electric field is determined essentially by the probability that the electron will acquire in the field an energy equal to the ionization threshold energy ϵ_i. The increase in the electron energy depends on the relation between two factors: acceleration in the external field and energy dissipation by collision with phonons. From this point of view, we can visualize two extreme possibilities, as a result of which the electron would acquire an energy ϵ_i. First, it can receive this energy from the field without experiencing accidentally even a single collision; second, the same energy can be attained gradually, after many collisions, such, however, that in each collision the electron loses on the average less energy than it receives from the field during the time between two collisions.

The first of these possibilities was pointed out already in the first papers of Townsend on discharge in gases, and was investigated theoretically by Shockley for the case of ionization in semiconductors.[1] In this case the probability of impact ionization w_i is proportional to the probability that the electron will cover without collision a path $L = \epsilon_i/eE$, i.e.,

$$w_i \sim \exp(-\varepsilon_i / eEl), \tag{1}$$

where l is the mean free path. A formula of the type (1) was first used by Chynoweth[2] to describe the experimental data on impact ionization in semiconductors.

The second possibility was investigated by Druyvesteyn,[3] Davydov,[4] and many other authors and was applied to an investigation of the breakdown in covalent semiconductors by Wolff.[5] In this case the increase of the electron energy has the character of energy diffusion. The stationary distribution of the electrons with respect to the energy is described by the distribution function

$$f_0(\varepsilon) = \mathrm{const}\cdot\exp\left(-\frac{\varepsilon^2}{e^2E^2l^2}\delta\right). \tag{2}$$

where δ is a small quantity on the order of the ratio of the energy lost by the electron in one collision to the total energy. For elastic collisions in gases, δ is of the order of the ratio of the mass of the electron to the mass of the ion,[3] while for collisions with acoustic phonons in semiconductors[4] $\delta \sim mc^2/kT$, where c is the speed of sound and T the temperature. Finally, for the case of interaction between electrons and optical phonons in semiconductors[5] or with molecular vibrations in gases, we have $\delta \sim \hbar\omega/\epsilon$, where $\hbar\omega$ is the energy of the emitted vibrational quantum. In the latter case, the velocity distribution of the electrons (2) becomes Maxwellian with effective temperature $kT_e \sim (eEl)^2/\hbar\omega$.[5]

The essential result contained in (2) is the fact that the number of the ionizing electrons, which is proportional to $f_0(\epsilon_i)$, increases with increasing field like

$$w_i \sim \exp(-\mathrm{const}/E^2). \tag{3}$$

Thus, the two possibilities considered lead to essentially different dependences of the probabilities of the impact ionization on the field, (1) and (3). At first glance the result (3), derived from the kinetic equation, appears to be better founded. However, the experimental data[2,6,7] confirm the existence of a relation of the type (1) over a wide range of field intensities **E**. This contradiction was clarified by Baraff,[8] who showed that in the region of relatively weak fields, when $eEl \lesssim \hbar\omega$ (ω is the average frequency of the emitted phonons), the kinetic equation also leads to a dependence of the type (1), while the diffusion approximation cannot be employed for its solution. If we assume for the mean free path in semiconductors a value $l \sim 10^{-6}$ cm, and for the phonon energies $\hbar\omega \sim 5 \times 10^{-2}$ eV, then the field in which the diffusion approximation and its corollary (3) are valid should be much stronger than 5×10^4 V/cm. However, the breakdown fields in germanium and silicon are of the order of 10^5 V/cm, and consequently, even if the dependence (3) could be observed, this would happen only in fields directly preceding breakdown.

Baraff's results[8] were obtained by numerically integrating the kinetic equation for a series of chosen values of the parameters and, in addition, pertained to temperatures that were sufficiently close to zero. The purpose of the present paper is to solve the problem of impact ionization in semiconductors in analytic form, and for arbitrary values of the field E and of the temperature T. We shall show below that in the region of large energies $\epsilon \gg \hbar\omega$, which are the only ones of interest for the impact-ionization problem, the distribution function of the electrons interacting with the acoustic and optical phonons in a covalent crystal is of the form

$$f_0(\varepsilon) = \text{const}\cdot\varepsilon^{\nu}\exp\left(-\frac{\varepsilon}{eFl}s_0(E, T)\right), \tag{4}$$

where ϵ is the electron energy, the parameter $s_0(E, T)$ is determined as the positive root of the transcendental equation

$$(1+\lambda)\operatorname{ch}\frac{\hbar\omega}{2kT}\Big/\left[\lambda\operatorname{ch}\frac{\hbar\omega}{2kT}+\operatorname{ch}\left(\frac{\hbar\omega}{2kT}-s_0\frac{\hbar\omega}{eFl}\right)\right] + \frac{1}{2s_0}\ln\frac{1-s_0}{1+s_0} = 0, \quad s_0 > 0, \tag{5*}$$

and the exponent ν is expressed in terms of s_0:

$$\nu = -\frac{1}{2}\left\{1 - \frac{\hbar\omega}{eFl}\operatorname{th}\left(\frac{\hbar\omega}{2kT}\right)\frac{s_0}{1-s_0^2} \times\left[\frac{1+\lambda}{1-s_0^2} + \frac{1+\lambda}{2s_0}\ln\frac{1-s_0}{1+s_0} - s_0\frac{\hbar\omega}{eFl}\left(\frac{1}{2s_0}\ln\frac{1-s_0}{1+s_0}\right)^2 \times\operatorname{sh}\left(\frac{\hbar\omega}{2kT}-s_0\frac{\hbar\omega}{eFl}\right)\Big/\operatorname{ch}\frac{\hbar\omega}{2kT}\right]^{-1}\right\}. \tag{6†}$$

From these formulas F is the effective electric field intensity, which coincides with its true value E in the case of a semiconductor with a scalar effective carrier mass (independent of the direction). In the presence of anisotropy

$$F = (m/m_{\parallel})^{1/2}E, \tag{7}$$

where m is the effective mass, determined from the state density,

$$m = (m_x m_y m_z)^{1/3},$$

and $m_{\parallel}$ is the effective mass for motion along the field direction

$$\frac{1}{m_{\parallel}} = \frac{\cos^2\gamma_x}{m_x} + \frac{\cos^2\gamma_y}{m_y} + \frac{\cos^2\gamma_z}{m_z};$$

m_x^{-1}, m_y^{-1}, and m_z^{-1} are the principal values of the effective-mass tensor, and γ_x, γ_y, and γ_z are the angles between the field E and the principal axes of this tensor. Thus, formula (7) contains, in particular, the anisotropy of the impact ionization and of the breakdown fields. The dimensionless parameter λ determines the relative contribution of the scattering by acoustic and optical phonons to the total mean free path

$$\lambda = \frac{l_{op}}{l_{ac}}, \quad \frac{1}{l} = \frac{1}{l_{op}} + \frac{1}{l_{ac}} = \frac{1+\lambda}{l_{op}}.$$

We now consider qualitatively the behavior of the parameter s_0 as the field intensity is varied. In extremely strong fields, $eFl \gg \hbar\omega$, we have

$$s_0 \approx \frac{3}{1+\lambda}\frac{\hbar\omega}{eFl}\operatorname{th}\frac{\hbar\omega}{2kT}.$$

In this case the distribution function (4) is close to that investigated by Wolff,[5] and consequently the diffusion approximation describes the real situation quite well. We assume now that $2kT \ll \hbar\omega$, and investigate the region of fields such that $kT \lesssim eFl \ll \hbar\omega$. In this region s_0 is close to unity, but it never reaches this value, remaining smaller:

$$s_0 \approx 1 - 2\exp\left\{-2(1+\lambda)\operatorname{ch}\frac{\hbar\omega}{2kT}\Big/\left[\lambda\operatorname{ch}\frac{\hbar\omega}{2kT} + \operatorname{ch}\left(\frac{\hbar\omega}{2kT} - \frac{\hbar\omega}{eFl}\right)\right]\right\}.$$

When $eFl \approx 2kT$ the value of s_0 reaches its maximum

$$s_{0m} \approx 1 - 2\exp\left\{-\frac{2(1+\lambda)}{\lambda + \operatorname{sech}(\hbar\omega/2kT)}\right\}, \tag{8}$$

and in still smaller fields it begins to decrease. In this entire range of fields, when $eFl \ll \hbar\omega$, the distribution function (4) is proportional to $\exp(-\epsilon/eFl)$, i.e., it describes electrons accelerated by the field to an energy ϵ without colli-

*ch = cosh.

†th = tanh; sh = sinh.

sions, in other words, it corresponds to the picture considered by Shockley.[1] At the smallest fields $eFl \ll kT$ the parameter s_0 tends to eFl/kT, and the distribution function goes over into a Maxwellian distribution with a temperature equal to the lattice temperature.

If the effective-mass approximation, i.e., the assumption that the electron energy is quadratically dependent on the momentum, cannot be valid at energies on the order of ϵ_i, then the mean free path can no longer be regarded as energy independent, i.e., $l = l(\epsilon)$. In this case, as will be shown below, the distribution function $f_0(\epsilon)$ becomes not Maxwellian, but of the form

$$f_0(\varepsilon) \sim \exp\left\{-\int\limits_{\varepsilon_0}^{\varepsilon} \frac{d\varepsilon'}{eFl(\varepsilon')} s_0(\varepsilon')\right\}, \tag{4a}$$

where ϵ_0 is the bottom of the band and $s_0(\epsilon)$ is determined by the same transcendental equation (5), the only difference being that the constant l in it should be replaced by the function $l(\epsilon)$. The qualitative analysis presented above of the dependence s_0 on the field intensity $\mathbf{E}$ is consequently applicable also to this case, but for a quantitative comparison of the theory with experiment the difference between (4) and (4a) may turn out to be significant.

We present also a formula for the impact-ionization coefficient $\kappa(\mathbf{E}, T)$—the average number of ionization collisions of the electron per unit length of path. In the case described by (4) we have

$$\varkappa(\mathbf{E}, T) = a\left(\frac{m_{\|}}{m}\right)^{1/2} \frac{eF}{\varepsilon_i}\left(\frac{\varepsilon_i s_0}{eFl + \hbar\omega e^{(s_0-1)/s_0}}\right)^{\nu+2}$$

$$\times\, \varphi_k\left\{\frac{\varepsilon_i s_0}{eFl}\left(\frac{s_0^2}{12p}\right)^{1/k}\right\} \exp\left(-\frac{\varepsilon_i}{eFl} s_0\right), \tag{9}$$

where a is a numerical coefficient of the order of unity, $\varphi_k(z)$ a function defined by formula (75) below, and p and k are constants characterizing the energy dependence of the impact-ionization cross section near threshold, in accordance with formula (72). The asymptotic behavior of $\varphi_k(z)$ at both small and large values of the argument has a power-law character:

$$\varphi_k(z) \approx k!z^{-k}, \qquad z \gg 1,$$

$$\varphi_k(z) \approx 2\Gamma\left(2\frac{k+1}{k+2}\right)\left(\frac{z}{k+2}\right)^{-k/(k+2)}, \qquad z \ll 1.$$

The following explanation is in order with respect to formula (9). The most important factors which are contained in it—the exponential function and φ_k—follow from formulas (4) and (4a), and therefore are not subject to any doubt, in contradistinction from the factor preceding them, which is essentially connected with the normalization constant in $f_0(\epsilon)$. The latter cannot be determined accurately when $eFl \ll \hbar\omega$, for then the greater part of the electrons have energies $\epsilon \lesssim \hbar\omega$, for which our solutions (4) and (4a) are generally inapplicable. Therefore the factor in the parentheses in (9) was obtained by interpolation between the limiting cases $eFl \ll kT$, $kT \ll eFl \ll \hbar\omega$, and $\hbar\omega \ll eFl$. The specific form of the interpolating formula is of no importance, since the role of this entire term in (9) is very small compared with the two factors following it.

We now proceed to derive formulas (4)–(7). The electron momentum distribution function $f(\mathbf{p})$ is determined by the usual kinetic equation

$$e\mathbf{E}\nabla_{\mathbf{p}} f(\mathbf{p}) + S_{\mathbf{p}}^- f(\mathbf{p}) = S_{\mathbf{p}}^+\{f\}, \tag{10}$$

where $\mathbf{p}$ is the electron momentum, $S_{\mathbf{p}}^-$ the probability of collision with the phonon:

$$S_{\mathbf{p}}^- = \frac{2\pi}{\hbar}\int |m_{\mathbf{k}}|^2 \{(1 + N_{\omega_{\mathbf{k}}})\,\delta(\varepsilon_{\mathbf{p}} - \varepsilon_{\mathbf{p}+\mathbf{k}} - \hbar\omega_{\mathbf{k}})$$

$$+ N_{\omega_{\mathbf{k}}}\delta(\varepsilon_{\mathbf{p}} - \varepsilon_{\mathbf{p}+\mathbf{k}} + \hbar\omega_{\mathbf{k}})\}\frac{d^3k}{(2\pi\hbar)^3}, \tag{11}$$

$m_{\mathbf{k}}$—the matrix element of the interaction between the electron and the phonon with wave vector $\mathbf{k}$, $\omega_{\mathbf{k}}$ the phonon frequency, and $N_{\omega_{\mathbf{k}}}$ the equilibrium number of such phonons:

$$N_{\omega_{\mathbf{k}}} = \left(\exp\frac{\hbar\omega_{\mathbf{k}}}{kT} - 1\right)^{-1}, \tag{12}$$

$\epsilon_{\mathbf{p}}$ is the energy of an electron with momentum $\mathbf{p}$; $S_{\mathbf{p}}^+\{f\}$—the number of electrons arriving per unit time in the state with momentum $\mathbf{p}$ from all other states as a result of emission or absorption of phonons:

$$S_{\mathbf{p}}^+\{f\} = \frac{2\pi}{\hbar}\int |m_{\mathbf{k}}|^2 \{(1 + N_{\omega_{\mathbf{k}}})\,\delta(\varepsilon_{\mathbf{p}} - \varepsilon_{\mathbf{p}+\mathbf{k}} + \hbar\omega_{\mathbf{k}})$$

$$+ N_{\omega_{\mathbf{k}}}\delta(\varepsilon_{\mathbf{p}} - \varepsilon_{\mathbf{p}+\mathbf{k}} - \hbar\omega_{\mathbf{k}})\}\, f(\mathbf{p}+\mathbf{k})\frac{d^3k}{(2\pi\hbar)^3}. \tag{13}$$

We shall assume that there are two branches of phonons—acoustic and optical. The energy losses are essentially connected with the emission of optical phonons, for which we assume that

$$m_{\mathbf{k}} = m_0 = \text{const}, \qquad \omega_{\mathbf{k}} = \omega = \text{const}.$$

The energy lost to emission of acoustical phonons can be neglected in the energy region of interest to us $\epsilon_{\mathbf{p}} \gg \hbar\omega$, i.e., we can leave out the $\hbar\omega_{\mathbf{k}}$ of the acoustical phonons in the arguments of the δ functions of (11) and (13). Using also the approximations customarily made for acoustical phonons

$$N_{\omega_{\mathbf{k}}} \approx 1 + N_{\omega_{\mathbf{k}}} \approx kT/\hbar\omega_{\mathbf{k}} \sim |\mathbf{k}|^{-1}, \quad |m_{\mathbf{k}}|^2 \sim |\mathbf{k}|,$$

we verify that only $f(\mathbf{p} + \mathbf{k})$ and $\epsilon(\mathbf{p} + \mathbf{k})$ in the integrand of (13) depend on $\mathbf{k}$. Going over therefore to a new integration variable $\mathbf{p}' = \mathbf{p} + \mathbf{k}$ and integrating over the surface $\epsilon_{\mathbf{p}'} = \text{const}$, we obtain

$$S_{\mathbf{p}}^{-} = \frac{1}{\tau_{ac}(\mathbf{p})} + \frac{1}{\tau_{op}(\mathbf{p})}\left\{\frac{e^{\beta}}{2\,\mathrm{ch}\,\beta}\frac{\Omega(\varepsilon_{\mathbf{p}} - \hbar\omega)}{\Omega(\varepsilon_{\mathbf{p}})} + \frac{e^{-\beta}}{2\,\mathrm{ch}\,\beta}\frac{\Omega(\varepsilon_{\mathbf{p}} + \hbar\omega)}{\Omega(\varepsilon_{\mathbf{p}})}\right\}, \tag{14}$$

$$S_{\mathbf{p}}^{+}\{f\} = \frac{f_0(\varepsilon_{\mathbf{p}})}{\tau_{ac}(\mathbf{p})} + \frac{1}{\tau_{op}(\mathbf{p})}\left\{\frac{e^{\beta}\Omega(\varepsilon_{\mathbf{p}} + \hbar\omega)}{2\Omega(\varepsilon_{\mathbf{p}})\,\mathrm{ch}\beta} f_0(\varepsilon_{\mathbf{p}} + \hbar\omega) + \frac{e^{-\beta}\,\Omega(\varepsilon_{\mathbf{p}} - \hbar\omega)}{2\Omega(\varepsilon_{\mathbf{p}})\,\mathrm{ch}\,\beta} f_0(\varepsilon_{\mathbf{p}} - \hbar\omega)\right\}. \tag{15}$$

We have introduced here the following notation: $\beta = \hbar\omega/2kT$; $f_0(\epsilon_{\mathbf{p}})$—symmetrical part of the distribution function:

$$f_0(\varepsilon) = \frac{1}{\Omega(\varepsilon)}\int f(\mathbf{p})\,\delta(\varepsilon - \varepsilon_{\mathbf{p}})\frac{d^3p}{(2\pi\hbar)^3}, \tag{16}$$

$$\Omega(\varepsilon) = \int \delta(\varepsilon - \varepsilon_{\mathbf{p}})\frac{d^3p}{(2\pi\hbar)^3}; \tag{17}$$

$\tau_{op}^{-1}(\mathbf{p})$—frequency of collisions with the optical phonons:

$$\tau_{op}^{-1}(\mathbf{p}) = \frac{2\pi}{\hbar}|m_0|^2\frac{\Omega(\varepsilon_{\mathbf{p}})}{\mathrm{th}\,\beta}; \tag{18}$$

$\tau_{ac}^{-1}(\mathbf{p})$—frequency of collisions with acoustical phonons, determined in analogous fashion.

We now note that the energy dependence of τ_{ac}^{-1} and τ_{op}^{-1} is determined by the same factor $\Omega(\epsilon)$, and therefore their ratio does not depend on the energy:

$$\lambda = \frac{\tau_{op}(\mathbf{p})}{\tau_{ac}(\mathbf{p})} = \text{const.} \tag{19}$$

Using this circumstance and expanding (14) and (15) in powers of $\hbar\omega/\epsilon_{\mathbf{p}}$ up to first order, we obtain

$$S_{\mathbf{p}}^{-} \approx \frac{1}{\tau(\mathbf{p})}\left\{1 - \frac{\hbar\omega}{1+\lambda}\frac{d\ln\Omega(\varepsilon_{\mathbf{p}})}{d\varepsilon_{\mathbf{p}}}\,\mathrm{th}\,\beta\right\}, \tag{20}$$

$$S_{\mathbf{p}}^{+}\{f\} \approx \frac{1}{(1+\lambda)\,\tau(\mathbf{p})}\left\{\lambda + \frac{\mathrm{ch}(\beta + \hbar\omega d/d\varepsilon_{\mathbf{p}})}{\mathrm{ch}\,\beta} + \hbar\omega\frac{d\ln\Omega(\varepsilon_{\mathbf{p}})}{d\varepsilon_{\mathbf{p}}}\frac{\mathrm{sh}(\beta + \hbar\omega d/d\varepsilon_{\mathbf{p}})}{\mathrm{ch}\,\beta}\right\} f_0(\varepsilon_{\mathbf{p}}). \tag{21}$$

In the last formula we have introduced, for convenience in notation, the operators $\cosh(\beta + \hbar\omega d/d\epsilon_{\mathbf{p}})$ and $\sinh(\beta + \hbar\omega d/d\epsilon_{\mathbf{p}})$, whose action is defined by the well-known relations

$$\exp(\pm\hbar\omega d/d\varepsilon_{\mathbf{p}})\,f_0(\varepsilon_{\mathbf{p}}) \equiv f_0(\varepsilon_{\mathbf{p}} \pm \hbar\omega).$$

The mean free time $\tau(\mathbf{p})$ is connected with τ_{ac} and τ_{op} in the following fashion:

$$\tau^{-1}(\mathbf{p}) = \tau_{ac}^{-1}(\mathbf{p}) + \tau_{op}^{-1}(\mathbf{p}) = (1+\lambda)\,\tau_{op}^{-1}(\mathbf{p}). \tag{22}$$

We now solve (10) with respect to the distribution function $f(\mathbf{p})$, regarding $S_{\mathbf{p}}^{+}\{f\}$ as the free term of the equation

$$f(\mathbf{p}) = \int_{-\infty}^{0} S_{\mathbf{p}+e\mathbf{E}t}^{+}\{f\}\exp\left\{\int_0^t S_{\mathbf{p}+e\mathbf{E}t'}^{-}\,dt'\right\}dt. \tag{23}$$

Substituting (20) and (21) in (23) and integrating the first two terms in $S_{\mathbf{p}+e\mathbf{E}t}^{+}$ by parts, we obtain

$$f(\mathbf{p}) = \left[\lambda + \frac{\mathrm{ch}(\beta + \hbar\omega d/d\varepsilon_{\mathbf{p}})}{\mathrm{ch}\beta}\right]\frac{f_0(\varepsilon_{\mathbf{p}})}{1+\lambda} - \int_{-\infty}^{0}\exp\left\{\int_0^t\left[1 - \hbar\omega\frac{d\ln\Omega(\varepsilon_{\mathbf{p}+e\mathbf{E}t'})}{d\varepsilon_{\mathbf{p}+e\mathbf{E}t'}}\frac{\mathrm{th}\,\beta}{1+\lambda}\right]\right.$$
$$\left.\times\frac{dt'}{\tau(\mathbf{p}+e\mathbf{E}t')}\right\}\left\{\frac{d\varepsilon_{\mathbf{p}+e\mathbf{E}t}}{dt}\left[\lambda + \frac{\mathrm{ch}(\beta + \hbar\omega d/d\varepsilon_{\mathbf{p}+e\mathbf{E}t})}{\mathrm{ch}\,\beta}\right]\right.$$
$$\times\frac{f_0'(\varepsilon_{\mathbf{p}+e\mathbf{E}t})}{1+\lambda} - \frac{\hbar\omega}{\tau(\mathbf{p}+e\mathbf{E}t)}\frac{d\ln\Omega(\varepsilon_{\mathbf{p}+e\mathbf{E}t})}{d\varepsilon_{\mathbf{p}+e\mathbf{E}t}}$$
$$\times\left[\frac{\mathrm{th}\,\beta}{1+\lambda}\left(\lambda + \frac{\mathrm{ch}(\beta + \hbar\omega d/d\varepsilon_{\mathbf{p}+e\mathbf{E}t})}{\mathrm{ch}\,\beta}\right)\right.$$
$$\left.\left. + \frac{\mathrm{sh}(\beta + \hbar\omega d/d\varepsilon_{\mathbf{p}+e\mathbf{E}t})}{\mathrm{ch}\,\beta}\right]\frac{f_0(\varepsilon_{\mathbf{p}+e\mathbf{E}t})}{1+\lambda}\right\}dt. \tag{24}$$

Here f_0' is the derivative of the function f_0.

Formula (24) is a generalization of the usual relation

$$f(\mathbf{p}) = f_0(\varepsilon_{\mathbf{p}}) - e\mathbf{E}\tau(\mathbf{p})\frac{d\varepsilon_{\mathbf{p}}}{d\mathbf{p}}\frac{df_0(\varepsilon_{\mathbf{p}})}{d\varepsilon_{\mathbf{p}}}, \tag{25}$$

into which it goes over if we put $e\mathbf{E}t \sim e\mathbf{E}\tau_{\mathbf{p}} \ll \mathbf{p}$ and $\hbar\omega/\epsilon_{\mathbf{p}} \to 0$. In our problem, however, in accordance with the statements made above, an important role may be played by electrons whose range accidentally is anomalously large, i.e., $t \gg \tau_{\mathbf{p}}$. The velocity distribution of these electrons is certainly strongly anisotropic, and the approximation (25) is not suitable for their description, since it is based on the assumption that the asymmetrical part of the distribution function is much smaller than the symmetrical part.

Formula (24) reduces the problem of determining the total distribution function $f(\mathbf{p})$ to the finding of its symmetrical part $f_0(\epsilon_{\mathbf{p}})$. In order to obtain an equation for f_0, we should average Eq. (24) over the constant-energy surface in accordance with the definition (16).

We present the subsequent calculations,

assuming the electron dispersion equation to be anisotropic but quadratic:

$$\varepsilon_{\mathbf{p}} = \frac{p_x^2}{2m_x} + \frac{p_y^2}{2m_y} + \frac{p_z^2}{2m_z}. \tag{26}$$

In this case our problem can be reduced to the problem with quadratic and isotropic dispersion

$$\varepsilon_{\mathbf{q}} = q^2/2m \tag{27}$$

by changing the scales along the x, y, and z axes:

$$q_i = (m/m_i)^{1/2} p_i \tag{28}$$

(i = x, y, z) and introducing the effective electric field **F**, whose components are connected with the components of the true field **E** by relations analogous to (28):

$$F_i = (m/m_i)^{1/2} E_i, \tag{29}$$

$$F = (m/m_{\parallel}) E. \tag{30}$$

For the average mass m we choose the effective mass determined from the energy-state density

$$m = (m_x m_y m_z)^{1/3}. \tag{31}$$

We now introduce for convenience in subsequent manipulations a set of new symbols. The effective mean free path l we define by the relation

$$l = \left(\frac{2\varepsilon_{\mathbf{p}}}{m}\right)^{1/2} \tau_{\mathbf{p}} = \left(\frac{2\varepsilon_{\mathbf{q}}}{m}\right)^{1/2} \tau_{\mathbf{q}}. \tag{32}$$

It is obvious that it is independent of the energy. All the energies are measured in units of eFl, i.e.,

$$\zeta_{\mathbf{q}} = \frac{q^2}{2meFl}. \tag{33}$$

The optical-phonon energy expressed in these units will be denoted by α:

$$\alpha = \hbar\omega / eFl. \tag{34}$$

Finally, u and v denote the cosines of the angles between the field **F** and the vectors **q** and **q** + e**F**t respectively:

$$u = \cos\{\mathbf{F}, \mathbf{q}\}, \quad v = \cos\{\mathbf{F}, \mathbf{q} + e\mathbf{F}t\}. \tag{35}$$

The quantities $\zeta = \zeta_{\mathbf{q}}$, $\zeta' = \zeta_{\mathbf{q}+e\mathbf{F}t}$, u, and v are connected by the obvious relation

$$\zeta(1-u^2) = \zeta'(1-v^2). \tag{36}$$

It follows from (17) and (33) that

$$\frac{d\ln\Omega(\zeta)}{d\zeta} = \frac{1}{2\zeta} \tag{37}$$

We now multiply (24) by $\Omega^{-1}(\epsilon)\delta(\epsilon - \epsilon_{\mathbf{p}})$ and integrate it over all the momenta **p**. Using then formulas (26)—(37), we obtain the following equation for the function $f_0(\zeta)$:

$$\begin{aligned}
&\left\{1 - \frac{1}{1+\lambda}\left[\lambda + \frac{\operatorname{ch}(\alpha + \beta d/d\zeta)}{\operatorname{ch}\beta}\right]\right\} f_0(\zeta) + \frac{1}{2}\int_{-1}^{1} du \int_{-1}^{u} \frac{dv}{1+\lambda} \\
&\quad \times \exp\left\{-\int_v^u \left[\frac{2\zeta(1-u^2)}{(1-\eta^2)^2} - \frac{\alpha \operatorname{th}\beta}{(1+\lambda)(1-\eta^2)}\right] d\eta\right\} \\
&\quad \times \left\{\frac{2\zeta(1-u^2)}{(1-v^2)^2} v \left[\lambda + \frac{\operatorname{ch}(\beta + \alpha d/d\zeta')}{\operatorname{ch}\beta}\right] f_0' \right. \\
&\quad \times \left(\zeta\frac{1-u^2}{1-v^2}\right) - \frac{\alpha}{1-v^2}\left[\frac{\operatorname{th}\beta}{1+\lambda}\left(\lambda + \frac{\operatorname{ch}(\beta + \alpha d/d\zeta')}{\operatorname{ch}\beta}\right)\right. \\
&\quad \left.\left. + \frac{\operatorname{sh}(\beta + \alpha d/d\zeta')}{\operatorname{ch}\beta}\right] f_0\left(\zeta\frac{1-u^2}{1-v^2}\right)\right\} = 0.
\end{aligned} \tag{38}$$

The derivative with respect to ζ' in (38) denotes differentiation with respect to the total argument $\zeta' = \zeta(1+u^2)/(1-v^2)$.

Equation (38) was obtained by expanding with respect to $\hbar\omega/\epsilon$, and is consequently valid only in the region $\epsilon \gg \hbar\omega$, i.e., $\zeta \gg \alpha$, and consequently it is meaningful to seek its solution only in this region. This is most conveniently done by taking the Laplace transform, i.e., going over to the function

$$\tilde{f}(s) = \int_0^\infty e^{-s\zeta} f_0(\zeta)\, d\zeta. \tag{39}$$

Then

$$f_0(\zeta) = \frac{1}{2\pi i}\int_{c-i\infty}^{c+i\infty} e^{s\zeta}\tilde{f}(s)\, ds, \tag{40}$$

where c > Re s_0, and s_0 is the singular point of $\tilde{f}(x)$ farthest to the right. At large ζ, the asymptotic value of $f_0(\zeta)$ is determined only by the position and the character of the extreme right singular point s_0. Therefore, going over from (38) to the expression for $\tilde{f}(s)$, we should investigate only the singular points of this equation.

Here, however, it will be more convenient for us to write the equation not for $\tilde{f}(s)$, but for the related function $\Phi(s)$:

$$\begin{aligned}
\Phi(s) &= \int_0^\infty \left\{\lambda + \frac{\operatorname{ch}(\beta + \alpha d/d\zeta)}{\operatorname{ch}\beta}\right\}\frac{f_0(\zeta)}{1+\lambda} e^{-s\zeta} d\zeta \\
&= \left(\lambda + \frac{\operatorname{ch}(\beta + \alpha s)}{\operatorname{ch}\beta}\right)\frac{\tilde{f}(s)}{1+\lambda} \\
&\quad - \frac{e^{\beta + \alpha s}}{2(1+\lambda)\operatorname{ch}\beta}\int_0^\alpha e^{-s\zeta} f_0(\zeta)\, d\zeta.
\end{aligned} \tag{41}$$

We multiply (38) by $e^{-s\zeta}$ and integrate with respect to ζ from zero to infinity, representing

$f_0(\zeta')$ in the integrand in terms of $\widetilde{f}(s)$ with the aid of (40). After several elementary transformations we obtain for $\Phi(s)$ the following equation:

$$\left[\frac{(1+\lambda)\,\mathrm{ch}\,\beta}{\lambda\,\mathrm{ch}\,\beta+\mathrm{ch}(\beta+\alpha s)}-\frac{1}{2s}\ln\frac{1+s}{1-s}\right]\Phi(s)$$

$$-\frac{1}{2\pi i}\int_{c-i\infty}^{c+i\infty}[K(s,s')\Phi(s')-g(s,s')]\,ds'=h(s), \quad (42)$$

where

$$K(s,s')=\frac{1}{2}\int_{-1}^{1}\frac{du}{1-u^2}\int_{-1}^{u}\frac{dv}{1-v^2}\left(\frac{1+u}{1-u}\frac{1-v}{1+v}\right)^{\alpha\,\mathrm{th}\,\beta/2(1+\lambda)}$$

$$\times\left(\frac{s+u}{1-u^2}-\frac{s'+v}{1-v^2}+\frac{1}{2}\ln\frac{1+u}{1-u}\frac{1-v}{1+v}\right)^{-1}$$

$$\times\left[\frac{s'(1-v^2)}{(1+s'v)^2}+\frac{\alpha\,\mathrm{th}\,\beta}{(1+\lambda)(1+s'v)}\right.$$

$$\left.+\frac{\alpha\,\mathrm{sh}(\beta+\alpha s')}{\lambda\,\mathrm{ch}\,\beta+\mathrm{ch}(\beta+\alpha s')}\right],$$

$$g(s,s')=\frac{\alpha}{2}\int_{-1}^{1}\frac{du}{1-u^2}\int_{-1}^{u}\frac{dv}{1-v^2}\left(\frac{1+u}{1-u}\frac{1-v}{1+v}\right)^{\alpha\,\mathrm{th}\,\beta/2(1+\lambda)}$$

$$\times\left(\frac{s+u}{1-u^2}-\frac{s'+v}{1-v^2}+\frac{1}{2}\ln\frac{1+u}{1-u}\frac{1-v}{1+v}\right)^{-1}$$

$$\times\frac{e^{\beta+\alpha s'}}{2(1+\lambda)\,\mathrm{ch}\,\beta}\left[1-\frac{\mathrm{sh}(\beta+\alpha s')}{\lambda\,\mathrm{ch}\,\beta+\mathrm{ch}(\beta+\alpha s')}\right]$$

$$\times\int_0^{\alpha}f_0(\zeta)e^{-s'\zeta}d\zeta, \quad (44)$$

$$h(s)=\frac{1}{2}\int_1^{\infty}\frac{\Phi(s')\,ds'}{s'(s-s')}-\frac{e^{\beta+\alpha s}}{2[\lambda\,\mathrm{ch}\,\beta+\mathrm{ch}(\beta+\alpha s)]}$$

$$\times\int_0^{\alpha}f_0(\zeta)e^{-s\zeta}d\zeta. \quad (45)$$

In the derivation of (42) we assume that $\mathrm{Re}\,s_0 < c < \mathrm{Re}\,s$. We now note that the kernels $K(s,s')$ and $g(s,s')$ have as functions of s' a logarithmic cut along the real semi-axis $(s,+\infty)$ in the half-plane $\mathrm{Re}\,s' > c$, and also an additional branch point $s' = 1$. The points determined by the zeros of the expression

$$\lambda\,\mathrm{ch}\,\beta+\mathrm{ch}\,(\beta+\alpha s),$$

are not singular for $\Phi(s)$, as can be readily verified by multiplying all the equations in (42) by this expression. We then see directly that at these points

$$(1+\lambda)\,\mathrm{ch}\,\beta\,\Phi(s)+\frac{1}{2}e^{\beta+\alpha s}\int_0^{\alpha}f_0(\zeta)e^{-s\zeta}d\zeta=0,$$

and consequently the singularities corresponding to them in $K(s,s')\Phi(s')$ and $g(s,s')$ cancel out.

We consider further the values of s in the strip $-1 < \mathrm{Re}\,s < 1$, since $s = 1$ is likewise not a singular point of $\Phi(s)$. The singularities of the different terms of the equation at these points cancel each other. Thus, for example, the logarithm in the first term in the left side of (42) cancels a similar divergence of the integral in the first term of (45). Shifting the contour of integration in (42) to the right, we reduce this equation to the form

$$\left[\frac{(1+\lambda)\,\mathrm{ch}\,\beta}{\lambda\,\mathrm{ch}\,\beta+\mathrm{ch}(\beta+\alpha s)}-\frac{1}{2s}\ln\frac{1+s}{1-s}\right]\Phi(s)$$

$$-\int_s^{\infty}\varkappa(s,s')\Phi(s')\,ds'=h(s)-\int_s^{\infty}\gamma(s,s')\,ds', \quad (46)$$

where $\kappa(s,s')$ and $\gamma(s,s')$ are the jumps in the functions $K(s,s')$ and $g(s,s')$ on the cut $s \le s' < \infty$.

On the segment $s \le s' < 1$ we have

$$\varkappa(s,s')=\frac{1}{2\pi i}[K(s,s'+i\delta)-K(s,s'-i\delta)]$$

$$=\frac{1}{2}\int_{-1}^{1}\frac{du}{1-u^2}\int_{-1}^{u}\frac{dv}{1-v^2}\left(\frac{1+u}{1-u}\frac{1-v}{1+v}\right)^{\alpha\,\mathrm{th}\,\beta/2(1+\lambda)}$$

$$\times\left[\frac{s'(1-v^2)}{(1+s'v)^2}+\frac{\alpha\,\mathrm{th}\,\beta}{(1+\lambda)(1+s'v)}\right.$$

$$\left.+\frac{\alpha\,\mathrm{sh}(\beta+\alpha s')}{\lambda\,\mathrm{ch}\,\beta+\mathrm{ch}(\beta+\alpha s')}\right]$$

$$\times\delta\left(\frac{s+u}{1-u^2}-\frac{s'+v}{1-v^2}+\frac{1}{2}\ln\frac{1+u}{1-u}\frac{1-v}{1+v}\right). \quad (47)$$

The terms in the left side of (46) have singularities at those points where the function $\Phi(s)$ has singularities, while the terms in the right side have no singularities at these points.

The only singular point of $\Phi(s)$ in the half-plane $\mathrm{Re}\,s > -1$ is determined by the zero of the coefficient of $\Phi(s)$ in the left side of (46):

$$\frac{(1+\lambda)\,\mathrm{ch}\,\beta}{\lambda\,\mathrm{ch}\,\beta+\mathrm{ch}(\beta+\alpha s_0)}-\frac{1}{2s_0}\ln\frac{1+s_0}{1-s_0}=0. \quad (48)$$

The point s_0 is also the farthest singularity of $\Phi(s)$ on the right. In order to investigate the behavior of $\Phi(s)$ in the vicinity of s_0, we expand all the coefficients in (46) in powers of $(s - s_0)$ near this point, retaining the first nonvanishing term. This equation then takes the form

$$(s-s_0)\Phi(s)+\nu\int^{s}\Phi(s')\,ds'=\mathrm{const}, \quad (49)$$

where

$$\nu = \varkappa(s_0, s_0)\left\{\frac{d}{ds_0}\left[\frac{(1+\lambda)\operatorname{ch}\beta}{\lambda\operatorname{ch}\beta+\operatorname{ch}(\beta+\alpha s_0)} - \frac{1}{2s_0}\ln\frac{1+s_0}{1-s_0}\right]\right\}^{-1}. \quad (50)$$

The solution of (49) is of the form const$\cdot(s - s_0)^{-\nu-1}$. More strictly speaking this denotes that $\Phi(s)$ is of the form

$$\Phi(s) = (s - s_0)^{-\nu-1}\varphi_1(s) + \varphi_2(s), \quad (51)$$

where $\varphi_1(s)$ and $\varphi_2(s)$ are functions that are regular at the point s_0. They cannot be obtained without a complete solution of Eq. (46). For our purposes, however, this is not necessary.

Equation (48), which determines s_0, has two roots. One is $s_0 = 0$ and the other $s_0 < 0$. However, the singular point corresponds only to the second, negative root, since $\nu = -1$ when $s_0 = 0$, and consequently the function (51) is regular at this point. Indeed,

$$\varkappa(s, s) = \frac{1}{2s}\left[\frac{1}{1-s^2} - \frac{1}{2s}\ln\frac{1+s}{1-s}\right] + \frac{\alpha\operatorname{th}\beta}{2(1+\lambda)}\,\frac{1}{1-s^2} + \frac{\alpha}{4s}\ln\frac{1+s}{1-s}\,\frac{\operatorname{sh}(\beta+\alpha s)}{\lambda\operatorname{ch}\beta+\operatorname{ch}(\beta+\alpha s)}. \quad (52)$$

Substituting this result in (50), we obtain

$$\nu = -\frac{1}{2}\left\{1 + \alpha\frac{s_0\operatorname{th}\beta}{1-s_0^2}\left[\frac{1+\lambda}{1-s_0^2} - \frac{1+\lambda}{2s_0}\ln\frac{1+s_0}{1-s_0} + \alpha s_0\frac{\operatorname{sh}(\beta+\alpha s_0)}{\operatorname{ch}\beta}\left(\frac{1}{2s_0}\ln\frac{1+s_0}{1-s_0}\right)^2\right]^{-1}\right\}. \quad (53)$$

The point s_0 is also the extreme right singularity for the function $\tilde{f}(s)$. Substituting (51) and (41) in (40) and shifting the contour of integration in (40) to the left, we obtain the following asymptotic representation for $f_0(\zeta)$ at large ζ:

$$f_0(\zeta) = C\zeta^\nu e^{s_0\zeta}. \quad (54)$$

The constant C is proportional to $\varphi_1(s_0)$ and therefore cannot be calculated in the approximation employed by us. However, its influence on the probability of impact ionization is negligibly small, and we confine ourselves to an estimate of its order of magnitude. This is easily done on the basis of the following considerations. Physically, C is connected with the normalization constant in the distribution function. In the case of large fields $\alpha \ll 1$, when most electrons have an energy on the order of $\alpha^{-1}eFl \gg eFl \gg \hbar\omega$ and are consequently in the region of applicability of the solution (54), C simply coincides with the normalization constant

$$C = n\frac{(2\pi^3)^{1/2}\hbar^3}{m^{3/2}}\left[\frac{3}{1+\lambda}\,\frac{\hbar\omega}{e^2F^2l^2}\operatorname{th}\frac{\hbar\omega}{2kT}\right]^{3/2} \sim \left(\frac{\hbar}{m\omega}\,\frac{3\alpha^2}{1+\lambda}\operatorname{th}\beta\right)^{3/2} n, \quad (55)$$

where n is the total number of electrons.

If $\alpha \gg 1$, then the bulk of the electrons has energies smaller than $\hbar\omega$ and is not described by the function (54). In this case C is the number of electrons in the tail of the distribution function, and can be estimated in the following fashion. In this case the electrons fill practically uniformly the region of energies $\epsilon_p < \hbar\omega$, and $n(\epsilon)$, the number of electrons reaching an energy ϵ, is proportional to the probability of passing without collision through the energy range from $\epsilon_p < \hbar\omega$ to ϵ. This probability is in turn proportional to

$$\exp\left(-\int_0^t S^-_{\mathbf{p}+e\mathbf{E}t'}dt'\right),$$

where t is determined from the condition $\epsilon_{\mathbf{p}+e\mathbf{E}t} = \epsilon$. An elementary calculation of this type yields under the condition $1 \ll \alpha \lesssim \beta$

$$n(\zeta) \sim n\alpha^{-1}\exp\left\{-\zeta + \frac{\alpha\operatorname{th}\beta}{2(1+\lambda)}\ln\zeta - \frac{\alpha\operatorname{th}\beta}{2(1+\lambda)}\left(\ln\frac{\alpha}{4} - 1\right)\right\}. \quad (56)$$

On the other hand, the quantity $n(\zeta)$ can be determined also as

$$n(\zeta) = \frac{\sqrt{2}(meFl)^{3/2}}{\pi^2\hbar^3}\int_\zeta f_0(\zeta)\sqrt{\zeta}\,d\zeta.$$

Substituting here (54) and using the fact that $s_0 \to -1$ when $\alpha \gg 1$, and that the exponent

$$\nu \approx -\frac{1}{2} + \frac{\alpha}{2(1+\lambda)}\operatorname{th}\beta,$$

we obtain

$$n(\zeta) \approx \frac{\sqrt{2}}{\pi^2}\left(\frac{m\omega}{\hbar\alpha}\right)^{3/2} C\exp\left\{-\zeta + \frac{\alpha\operatorname{th}\beta}{2(1+\lambda)}\ln\zeta\right\}. \quad (57)$$

Comparing (56) and (57) we obtain, leaving out numerical factors,

$$C \sim n\left(\frac{\hbar}{m\omega}\right)^{3/2}\sqrt{\alpha}\exp\left\{-\frac{\alpha\operatorname{th}\beta}{2(1+\lambda)}\left(\ln\frac{\alpha}{4} - 1\right)\right\}. \quad (58)$$

It is now easy to construct an interpolation formula for C, which when $\alpha \ll 1$ coincides with (55), and when $\alpha \gg 1$ with (58); in the intermediate region $\alpha \sim 1$ the formula gives a result of the correct order of magnitude. For example,

$$C(\alpha,\beta)\sim -n\frac{s_0}{\alpha}\left(-\frac{\hbar}{m\omega}\alpha s_0\right)^{3/2}$$

$$\times\exp\left\{-\frac{\alpha\,\mathrm{th}\,\beta}{2(1+\lambda)}\left(\ln\frac{\alpha}{4}-1\right)\right\}. \quad (59)$$

We emphasize once more that a more accurate calculation of $C(\alpha,\beta)$ is of no interest in the problem of impact ionization, for in the worst case $C(\alpha,\beta)$ results in corrections to the exponential in (54), the relative magnitude of which is of the order of $\hbar\omega/\epsilon_i \sim 10^{-2}$.

More important are the other simplifications made in the derivation of (54), particularly the assumption that ϵ_p is a quadratic function of the momentum up to an energy on the order of ϵ_i, and that the influence of the impact ionization process itself on the function $f_0(\epsilon)$ in the region $\epsilon \le \epsilon_i$ is small. The first of these limitations can be eliminated. We now present a derivation which is somewhat less rigorous than the preceding one, but which leads to the same results and makes it possible to take into account the non-parabolic nature of the function ϵ_p. For concreteness we confine ourselves here to the most interesting case, when the dispersion law is not parabolic but isotropic:

$$\varepsilon_p=\Delta\left(1+\frac{p^2}{m\Delta}\right)^{1/2}, \quad (60)$$

where Δ is the half-width of the forbidden band, from the middle of which the energies are reckoned in the given case. This type of dependence of ϵ_p is precisely the one which is typical, as is well known, for the majority of semiconductors of the $A^{III}B^{V}$ group and for a number of other semiconductors. In this case

$$\Omega(\varepsilon)=\frac{m^{3/2}\Delta^{1/2}}{2\pi^2\hbar^3}\frac{\varepsilon}{\Delta}\left(\frac{\varepsilon^2}{\Delta^2}-1\right)^{1/2}$$

and the mean free path $l(\epsilon)$ depends on ϵ like

$$l(\varepsilon)=\frac{\varepsilon^2}{\Delta^2}l, \quad (61)$$

where l is the mean free path in the region of energies $\hbar\omega \ll (\epsilon^2-\Delta^2)^{1/2} \ll \Delta$.

Since we are interested only in calculating the exponential in the distribution function $f_0(\epsilon)$, we now omit in (24) all the terms of order $\hbar\omega/\epsilon$. We change over again to the variables

$$\varepsilon=\varepsilon_p,\quad \varepsilon'=\varepsilon_{p+eEt};$$

$$u=\cos\{\mathbf{E},\mathbf{p}\},\quad v=\cos\{\mathbf{E},\mathbf{p}+e\mathbf{E}t\}$$

$$=\left[1-\frac{\varepsilon^2-\Delta^2}{\varepsilon'^2-\Delta^2}(1-u^2)\right]^{1/2}. \quad (62)$$

After averaging (24) over the angles we obtain

$$f_0(\varepsilon)=\left[\lambda+\frac{\mathrm{ch}(\beta+\hbar\omega d/d\varepsilon)}{\mathrm{ch}\,\beta}\right]\frac{f_0(\varepsilon)}{1+\lambda}$$

$$-\frac{1}{2}\int_{-1}^{1}du\int_{\Gamma}^{\varepsilon}\exp\left\{-\int_{\varepsilon'}\left[1-\frac{\varepsilon^2-\Delta^2}{\varepsilon''^2-\Delta^2}(1-u^2)\right]^{-1/2}\right.$$

$$\times\left.\frac{d\varepsilon''}{eEl(\varepsilon'')}\right\}\left[\lambda+\frac{\mathrm{ch}(\beta+\hbar\omega d/d\varepsilon')}{\mathrm{ch}\,\beta}\right]\frac{f_0'(\varepsilon')}{1+\lambda}d\varepsilon' \quad (63)$$

The integral with respect to ϵ' in (62) is taken along the contour Γ which goes from $+\infty$ along the real axis, but in the lower half-plane, to the branch point $\epsilon_B'^2-\Delta^2=(\epsilon^2-\Delta^2)(1-u^2)$, turning around this point and then proceeding to $\epsilon'=\epsilon$. The point ϵ itself is assumed to lie on the upper branch of the contour Γ, if $u>0$ and on the lower if $u<0$. When ϵ' varies along the contour Γ, the quantity v (62) increases monotonically from -1 to u. The branch point corresponds to $v=0$. The integral with respect to ϵ'' is taken along a segment of the same contour Γ.

We now seek a solution of (63) in the form

$$f_0(\varepsilon)=\mathrm{const}\cdot\exp\left\{-\int_{\Delta}^{\varepsilon}s(\varepsilon')\frac{d\varepsilon'}{eEl(\varepsilon')}\right\}. \quad (64)$$

Substituting (64) in (63), we shall assume that

$$\frac{d^n}{d\varepsilon^n}f_0(\varepsilon)\approx\left[-\frac{s(\varepsilon)}{eEl(\varepsilon)}\right]^n f_0(\varepsilon),$$

since the other terms which appear upon differentiation will be on the order of eEl/ϵ or smaller, and we neglect terms of this order, because we seek only the asymptotic value of $f_0(\epsilon)$ when $\epsilon \gg eEl$. As a result we obtain from (63) the following equation for $s(\epsilon)$:

$$\frac{\mathrm{ch}\,\beta-\mathrm{ch}[\beta-\alpha(\varepsilon)s(\varepsilon)]}{(1+\lambda)\mathrm{ch}\,\beta}-\frac{1}{2}\int_{-1}^{1}du\int_{\Gamma}^{\varepsilon}\frac{d\varepsilon'}{eEl(\varepsilon')}$$

$$\times\exp\left\{-\int_{\varepsilon'}^{\varepsilon}\left(\left[1-\frac{\varepsilon^2-\Delta^2}{\varepsilon''^2-\Delta^2}(1-u^2)\right]^{-1/2}\right.\right.$$

$$\left.\left.-s(\varepsilon'')\right)\frac{d\varepsilon''}{eEl(\varepsilon'')}\right\}\left[\lambda+\frac{\mathrm{ch}[\beta-\alpha(\varepsilon')s(\varepsilon')]}{\mathrm{ch}\,\beta}\right]\frac{s(\varepsilon')}{1+\lambda}=0. \quad (65)$$

The value of $s(\epsilon)$, as we shall show below, is always less than unity. Therefore

$$\frac{1}{v}-s(\varepsilon'')=\left[1-\frac{\varepsilon^2-\Delta^2}{\varepsilon''^2-\Delta^2}(1-u^2)\right]^{-1/2}-s(\varepsilon'')>0$$

for $v>0$, i.e., on the upper edge of the cut. If $v<0$, i.e., on the lower edge of the cut, this difference is negative.

Thus, the exponential in (65) contains a negative quantity, the absolute value of which increases monotonically with increasing distance between the points ϵ' and ϵ, and consequently the main contribution to the integral with respect to ϵ' is made in the asymptotic region $\epsilon \gg eEl$ by the region $\epsilon' \sim \epsilon$. Taking this circumstance into account, we expand the exponential in powers of $(\epsilon - \epsilon')$ up to first order inclusive:

$$\int_{\varepsilon'}^{\varepsilon}\left[\left(1-\frac{\varepsilon^2-\Delta^2}{\varepsilon''^2-\Delta^2}(1-u^2)\right)^{-1/2}-s(\varepsilon'')\right]\frac{d\varepsilon''}{eEl(\varepsilon'')}$$

$$\approx\left(\frac{1}{u}-s(\varepsilon)\right)\frac{\varepsilon-\varepsilon'}{eEl(\varepsilon)},$$

and replace the remaining quantities that depend on ϵ' by their values at $\epsilon' = \epsilon$. We note furthermore that from the definition of the contour Γ, the difference $\epsilon' - \epsilon$ is positive when $u < 0$ and negative when $u > 0$ in the region making the main contribution to the integral. Carrying out now all the integrations, we reduce (65) to the form

$$\frac{(1+\lambda)\operatorname{ch}\beta}{\lambda\operatorname{ch}\beta+\operatorname{ch}[\beta-\alpha(\varepsilon)s(\varepsilon)]}-\frac{1}{2s(\varepsilon)}\ln\frac{1+s(\varepsilon)}{1-s(\varepsilon)}=0. \quad (66)$$

Thus, an account of the non-parabolic nature of the dispersion law introduces nothing essentially new to the problem of impact ionization, but makes the distribution function (64) non-Maxwellian.

It is easy to include in the scheme under consideration also the impact ionization process. To this end we must add to S_p (20) a term in the form $\tau_i^{-1}(p)$, describing the ionization collision probability of an electron with momentum $\mathbf{p}$. The subsequent derivations are perfectly analogous to (63)—(66), so that we present immediately their final result. The distribution function can again be represented in the form (64), but with $l(\epsilon)$, the total mean free path, determined by the scattering by phonons and by the ionization collisions:

$$l(\varepsilon)=\left(\frac{2\varepsilon}{m}\right)^{1/2}\tau(\varepsilon),\quad \tau^{-1}(\varepsilon)=\tau_{ac}^{-1}(\varepsilon)+\tau_{op}^{-1}(\varepsilon)+\tau_i^{-1}(\varepsilon), \quad (67)$$

while the parameter $s(\epsilon)$ is determined by an equation which differs somewhat from (66):

$$\frac{1+\lambda}{1-\mu(\varepsilon)}\frac{\operatorname{ch}\beta}{\lambda\operatorname{ch}\beta+\operatorname{ch}[\beta-\alpha(\varepsilon)s(\varepsilon)]}$$

$$-\frac{1}{2s(\varepsilon)}\ln\frac{1+s(\varepsilon)}{1-s(\varepsilon)}=0; \quad (68)$$

$\mu(\epsilon)$ is the relative contribution of the ionization collisions

$$\mu(\varepsilon)=\tau(\varepsilon)/\tau_i(\varepsilon). \quad (69)$$

At energies $\epsilon < \epsilon_i$ this ratio is equal to zero, and then (68) coincides with (66).

An experimentally determined characteristic of the impact ionization process is usually the coefficient of impact ionization $\kappa(E, T)$, defined as the ratio of the average probability of impact ionization to the average electron drift velocity v_d:

$$\varkappa(\mathbf{E},T)=\frac{1}{nv_d}\int_0^\infty\tau_i^{-1}(\varepsilon)f_0(\varepsilon)\Omega(\varepsilon)\,d\varepsilon\equiv\frac{\sigma(\mathbf{E},T)}{nv_d}f_0(\varepsilon_i). \quad (70)$$

For a distribution function in the form (64) $\sigma(E, T)$ is of the form

$$\sigma(\mathbf{E},T)=\int_{\varepsilon_i}^\infty\tau_i^{-1}(\varepsilon)\exp\left\{-\int_{\varepsilon_i}^{\varepsilon}s(\varepsilon')\frac{d\varepsilon'}{eEl(\varepsilon')}\right\}\Omega(\varepsilon)\,d\varepsilon. \quad (71)$$

Compared with $f_0(\epsilon_i)$, this quantity, like v_d, is a slowly varying function of the field.

The main contribution to (71) is made by the region of energies close to threshold: $\epsilon - \epsilon_i \lesssim eEl(\epsilon_i)s^{-1} \ll \epsilon_i$. In this region the probability of impact ionization can be represented in the form

$$\tau_i^{-1}(\varepsilon)=\tau^{-1}(\varepsilon_i)p\left(\frac{\varepsilon-\varepsilon_i}{\varepsilon_i}\right)^k, \quad (72)$$

where p is a dimensionless constant which, generally speaking, is much larger than unity,[9] and the exponent k can assume values equal to 1, 2, and 3 depending on whether the crystal is isotropic or not[10] and depending on how large its dielectric constant is.[9]

In order to calculate (71) in explicit form, we now note the following: in a narrow region of energies close to a threshold, which makes the main contribution to (71), the condition $\mu(\epsilon) \ll 1$ is apparently always satisfied. Therefore, in the calculation of $\sigma(E, T)$, we can put $l(\epsilon) \approx l(\epsilon_i)$ and

$$s(\varepsilon)=\frac{1}{2}[s(\varepsilon_i)+(s^2(\varepsilon_i)+12\mu(\varepsilon))^{1/2}]. \quad (73)$$

Indeed, expression (73) follows directly from (68), if

$$\alpha(\varepsilon)\approx\alpha(\varepsilon_i)\ll 1$$

and $\mu \ll 1$. On the other hand, if

$$\alpha(\varepsilon)\approx\alpha(\varepsilon_i)\gg\sqrt{3\mu(\varepsilon)}$$

the quantity μ in (68) plays no role at all and $s(\epsilon) \approx s(\epsilon_i)$. By virtue of the condition $\mu \ll 1$, these regions overlap and consequently (73) holds true everywhere. Using (72) and (73), we can easily transform (71) into

$$\sigma(\mathbf{E},T)=\frac{\Omega(\varepsilon_i)}{\tau(\varepsilon_i)}eEl(\varepsilon_i)\varphi_k\left\{\frac{\varepsilon_i s(\varepsilon_i)}{eEl(\varepsilon_i)}\left(\frac{s^2(\varepsilon_i)}{12p}\right)^{1/k}\right\}, \quad (74)$$

$$\varphi_k(z) = z\int_0^\infty \exp\left\{-\frac{z}{2}\int_0^x [1+(1+y^k)^{1/2}]\,dy\right\} x^k\,dx. \quad (75)$$

The calculation of the normalization constant in $f_0(\epsilon)$ and of the drift velocity v_d in (70) is less definite. Both terms, however, vary very little as functions of E, compared with the exponential and with φ_k, so that it is sufficient to be able to write down at least their order of magnitude. This can be done by writing down their correct values in two limiting cases: $eFl \ll kT$ and $eFl \gg \hbar\omega$, when the propagation function is almost isotropic and the function (4) gives a solution which describes correctly the majority of the electrons, after which one can use (59) to interpolate in the intermediate region $kT \ll eFl \ll \hbar\omega$. This method yielded the factor

$$\left(\frac{\varepsilon_i s_0}{eFl + \hbar\omega e^{(s_0-1)/s_0}}\right)^{\nu+2}$$

in (9). As a rule, impact ionization is observed in such fields that $\alpha \lesssim 1$. Then, apart from factors of the order of unity, this term can be calculated from the formulas that pertain to the case $\alpha \ll 1$. In formula (9) this reduces simply to the fact that it is necessary to put $\hbar\omega \exp[(s_0 - 1)/s_0] = 0$.

Formulas (4)—(9) thus yield the solution of the problem of impact ionization in covalent semiconductors, for which as is well known, $m_k = \text{const}$, at arbitrary values of the field and of the temperature, and also under rather general assumptions concerning the character of the band structure and the interaction between the electrons and the phonons. There is, however, one limitation, which has been used in the derivation: we have assumed that the probability of scattering by an optical phonon will not depend on the scattering angle, i.e., $m_k = \text{const}$. This is precisely why we can write a closed equation such as (38) for the function $f_0(\epsilon)$, dependent only on the energy. Qualitative deductions concerning the variation of the law of the growth of the number of ionizing electrons with the field in fields such that $eEl \sim \hbar\omega$ apparently remain also in force when $m_k \neq \text{const}$, but the quantitative estimates may differ. In particular, some caution is necessary when the results of the present work are applied to semiconductors with a noticeable fraction of ionic bonding, for in this case m_k contains a term proportional to $|k|^{-1}$ for sufficiently small phonon momenta, i.e., small scattering angles.

[1] W. Shockley, Czech. J. Phys. **B11**, 81 (1961).
[2] A. G. Chynoweth, Phys. Rev. **109**, 1537 (1959).
[3] M. J. Druyvesteyn and F. M. Penning, Revs. Modern Phys. **12**, 87 (1940).
[4] B. I. Davydov, JETP **7**, 1069 (1937).
[5] P. A. Wolff, Phys. Rev. **95**, 1415 (1954).
[6] A. G. Chynoweth, J. Appl. Phys. **31**, 1161 (1960). Logan, Chynoweth, and Cohen, Phys. Rev. **128**, 2518 (1962).
[7] J. L. Moll and R. Vanoverstraeten, Electronics **6**, 147 (1963).
[8] G. A. Baraff, Phys. Rev. **128**, 2507 (1962).
[9] L. V. Keldysh, JETP **37**, 713 (1959), Soviet Phys. JETP **10**, 509 (1960).
[10] W. Franz, Encycl. Phys. **17**, 155 (1956).

Translated by J. G. Adashko
241

SOVIET PHYSICS USPEKHI VOLUME 8, NUMBER 3 NOVEMBER-DECEMBER 1965
537.312.6

SUPERCONDUCTIVITY IN NONMETALLIC SYSTEMS

L. V. KELDYSH

Usp. Fiz. Nauk 86, 327-333 (June, 1965)

THERE has been greatly renewed interest in the superconductivity phenomenon in recent years. This is explained to a considerable degree by the fact that after Bardeen, Cooper, and Schrieffer (BCS),[1] Bogolyubov,[2], Gor'kov[3] and others constructed in 1957 the microscopic theory of superconductivity, it became possible, at least in principle, to aim at finding superconductors with various specified properties.

Until recently, superconductivity was observed only in metals and at very low temperatures, not higher than 20°K. In addition, the superconducting state is usually destroyed in even relatively weak magnetic fields (on the order of 10^2–10^3 G), so that the currents flowing through such a superconductor can likewise not be large. All these circumstances greatly limit the possibilities of utilizing superconductivity, and it is therefore natural that one of the central problems of the theory of superconductivity at present is the question of whether one can hope to obtain a substance in which the superconducting state would exist at considerably higher temperatures and higher magnetic fields.

The main result of the BCS theory is that in a system of Fermi particles at a sufficiently low temperature any arbitrarily small attraction between the particles leads to a radical realignment of the state. Particles (we shall henceforth have in mind electrons) stick together to form pairs, and this is furthermore done in such a way that all pairs are in the same state, i.e., they have the same total momentum. In the equilibrium state, i.e., in the absence of current, the momenta of all the pairs are equal to zero. Consequently, the electrons sticking together are those having equal but opposite momenta. In other words, the electron pairs, which have the properties of Bose particles, form a Bose condensate, which behaves like a charged superfluid liquid. It is essential in this case that the pairing is a collective effect: the bound state of two electrons arises only when a large number of other pairs are in the same state, and the binding energy of each pair increases with increasing number of pairs in this state. Therefore any change in the momentum of the pair, connected with its removal from the condensate, should be accompanied by a breaking up of the pair and consequently calls for the expenditure of appreciable energy. It is the latter circumstance which leads to the stability of the superconducting current. The binding energy Δ in each pair at zero temperature, when all the electrons near the Fermi surface are paired, is determined by the width of that electron energy region ϵ_0, in which there is an effective attracting interaction by the density of the electronic states over an energy interval near the Fermi surface N and by the average value of the energy of attraction of two electrons V:

$$\Delta_0 = 2\varepsilon_0 e^{-\frac{1}{NV}}. \qquad (1)$$

With increasing temperature T, the thermal motion breaks some of the pairs, so that their number in the condensate decreases, and at the same time, as noted above, $\Delta(T)$, the binding energy in each pair, also decreases. At some critical temperature T_c, the order of magnitude of which is close to Δ_0/k (k is Boltzmann's constant), $\Delta(T_c)$ vanishes, i.e., the superconducting state disappears.

The binding energy Δ determines also the critical value of the magnetic field H_c in which the superconducting state is destroyed. The point is that in the presence of a magnetic field a superconducting current is produced in the superconductor, the magnetic field of which cancels completely the external magnetic field inside the volume of the superconductor. This effect of forcing out the magnetic field from the superconductor (the so-called Meissner effect) is one of the most characteristic features of the superconductivity phenomenon. The energy of the compensating magnetic field per unit volume of the superconductor is $H^2/8\pi$, where H is the intensity of the external field. It is obvious that in a sufficiently strong field, when this energy becomes larger than the decrease of energy connected with the pairing on going from the normal state to the superconducting state, the existence of the superconducting phase will be thermodynamically unfavorable and this phase will become destroyed. The energy released upon formation of one pair is equal to 2Δ, and the electrons participating in pair production are for the most part in a narrow region of energies near the Fermi surface, with width on the order of Δ. The number of such electrons is of the order of $N\Delta$, and the total decrease in the energy on going to the superconducting state is of the order of $N\Delta^2$. Equating this energy to the energy of the critical magnetic field $H_c^2/8\pi$, we get

$$H_c \sim \sqrt{N}\Delta. \qquad (2)$$

The density of states near the Fermi surface N can hardly be much larger in any substances than in typical superconducting metals. Therefore, an increase in the critical field H_c can be attained in substances with large binding energy Δ, i.e., with a high critical temperature, if such substances can be found.

However, the problem of increasing the critical magnetic field in which superconductivity can still exist has found a somewhat unexpected solution in 1961, when it was observed that certain alloys (Nb_3Sn, NdZr) remain superconducting in magnetic fields up to 10^5 G.[4] This phenomenon was predicted by Abrikosov [5] on the basis of a semiphenomenological theory of superconductivity, proposed by Ginzburg and Landau. [6] He showed that in some alloys, in fields larger than H_c, there can occur the so-called mixed state, characterized by the fact that the magnetic field penetrates into a superconductor in the form of thin filaments which thread through the sample. Between these filaments, the sample remains superconducting, and it is naturally sufficient even for a small fraction of the volume of the sample to be superconducting in order that the resistance of the entire sample be equal to zero. On the other hand, the thermodynamically unfavorable situation for the superconducting state improves radically in this case, since the magnetic field is forced out of the superconductor only partially. By now alloys have been obtained which retain their superconducting properties in fields close to 200 kG. This discovery was of extreme importance for the use of the phenomenon of superconductivity in physics research, and in engineering, especially in the production of strong magnetic fields. Magnets based on superconducting solenoids are already extensively used.

The prospects of further utilization of superconductors in engineering would seem to be unlimited, were it not for the lamentable circumstance that in all the cases known to date the superconducting state is realized at very low, and therefore difficult to attain, temperatures. Thus the question of the future of superconductivity is connected primarily with the problem of increasing the critical temperatures, if this is at all possible.

The magnitude of the critical temperature T_c, which is connected with the energy of pair production Δ, is determined essentially, as can be seen from (1), by the attraction force V which leads to pairing of the electrons. In metals this attraction, according to an idea by Froehlich, is a result of interaction between the electrons and vibrations of the crystal lattice—phonons. Exchange of phonons between two electrons leads to their attraction to each other. This interaction is effective for electrons having energies in a narrow region near the Fermi surface, of width $\epsilon_0 \sim kT_D$, where T_D is the so-called Debye temperature, the order of magnitude of which is equal to the maximum energy of the phonons. The electron-phonon interaction in metals is itself quite weak, i.e., $NV \ll 1$ and there are theoretical indications that it cannot be strong in principle, [7] for when the interaction is strong the crystal lattice becomes unstable, i.e., it is rearranged into some other modification. Taking into account the fact that the Debye temperatures for typical metals are low, on the order of 100–200°K, we see from (1) that the critical temperatures under such conditions should be very small, much smaller than 100°K, as is indeed observed in reality. This raises the question of whether superconductivity cannot be produced as a result of some other stronger interaction, and whether one should not seek such a possibility in other substances, which do not belong to the customary well-investigated class of metals, or else in metals but under unusual conditions.

Recently, there have been published in this direction several theoretical papers, of which the greatest interest was evoked by the paper of Little [8] concerning the possibility of obtaining a superconducting state in long organic molecules. The attraction between the electrons, proposed in this paper, is of the Coulomb type and is therefore connected with energies on the order of 1 eV, corresponding to temperatures on the order of thousands of degrees, two orders of magnitude larger than the energies characteristic of the interaction connected with phonons. The schematic model considered by Little consists of a long one-dimensional chain of atoms, along which electrons move freely, i.e., there is metallic conductivity, and of side chains. With respect to the latter it is assumed that they are strongly polarizable. From the quantum mechanical point of view this means, that the molecules comprising the side chains have at least one low excited level (with excitation energy 1–2 eV) possessing a large dipole moment. In such a model the free electron of the central chain, moving near one of the side chains, polarizes the latter in such a way that a considerable positive charge is induced on the near end of the chain. This charge attracts other electrons from the central chain, as a result of which an effective attraction occurs of these electrons to the electron which has caused the initial polarization of the side chain. According to Little's estimates, the attraction can turn out to be larger than the direct Coulomb repulsion acting between the electrons in the chain, as a result of which the total interaction between them will have the character of attraction with an average neighboring-electron interaction energy ≈ 1.5–2 eV. Examining further the system so obtained with the aid of the methods of the BCS theory, the author reaches the conclusion that a superconducting state should occur in it, and that the role of ϵ_0 in the formula for the effective region of the energies (1), in which the pairing interaction is effective, is played by the excitation energy of that level of the molecule—the side chain, which possesses a large dipole moment. This energy, as indicated above, is assumed to be of the order of 2 eV, which leads to a critical temperature close to 2000°K for the superconducting transition. This result, if correct, is of phenomenal interest, and it is therefore not surprising that Little's paper has attracted universal attention. Even if we disregard the possible technical applications, the presence of a superconducting state in organic molecules, together with its unique high degree of ordering, can, as indicated by the author, be of cardinal significance for many biological processes.

However, Little's paper has raised many objections, both of fundamental character and with respect to the conclusiveness of its deductions. The initial model itself is reasonable, since metallic conductivity in some organic molecules does seem to exist[10] and examples of molecules which have large polarizabilities and could play the role of the side chains are directly indicated in the article.[8] Moreover, an interesting supplementary result of this article is the statement that the transition into the superconducting state can occur even in the case when the molecule in the initial state does not have metallic conductivity, under the condition that the energy of pair production Δ exceeds the initial binding energy of the electrons in the central chain.

The main objection is connected, however, with the fact that Little's result contradicts the known theorem that in a one-dimensional system there can be no phase transition into an ordered state.[11] From the physical point of view this is connected with the fact that in a one-dimensional chain of atoms, each atom is connected with the others only through its nearest neighbors. Therefore an accidental sufficiently large fluctuation displacement in one point leads immediately to a loss of correlation between the atoms situated on the right and on the left of this point. In two- and three-dimensional cases this obviously is not the case, since the correlation connection between two given atoms which are far from each other is realized not only through the atoms lying on the line joining the two given atoms, but also through a very large number of other atoms, which lie on the side of this shortest path. Therefore for a complete loss of correlation between two remote atoms one would have to have an appreciable fluctuation over an entire plane between these two atoms, an event of very low probability. Formally this circumstance is expressed by the fact that the fluctuations in a one-dimensional system are so large, that they break up the ordered state.

Strictly speaking, this reasoning does not pertain directly to Little's model, since the latter is based on the assumption that the forces have a short-range character, and couple only the nearest neighbors. However, there is a widespread opinion that this result is of a more general character, and that phase transitions to an ordered phase in a one-dimensional chain are generally impossible. Ferrel[12] consider this question especially as applied to Little's model. He showed that the oscillations of the electron density, which take place in a three-dimensional superconductor, but play a minor role there, lead in the one-dimensional case to a destruction of the superconducting state even at arbitrarily low temperatures. In other words, even zero-order oscillations of the electronic density completely destroy the electron ordering characterizing the superconducting state. This result, to be sure, is also not absolutely convincing, since the character of the spectrum or even the very existence of electron-density oscillations in the Little model have not been reliably established. In addition, as indicated by the authors of [13], Ferrel's deduction pertains only to an infinite linear chain, whereas in a macromolecule, which consists of a large but finite number of links, the oscillations of the electron density can lead only to an appreciable decrease in the critical temperature. Assuming that the central chain in Little's model consists of 10^5 atoms, they found that the oscillations of the electron density decrease the pair binding energy Δ by approximately one order of magnitude.

In connection with the question of the possibility of a superconducting transition in a one-dimensional system, interest attaches also to the results of Lattinger,[14] who considered a very schematic example of a one-dimensional electron system, greatly differing from the model used by Little. He showed that an examination of this model by the methods of the BCS theory leads to the deduction that superconductivity exists in it. At the same time, this model admits of an exact solution, showing that the ground state of the system is not superconducting.

Objections of another kind are connected with the insufficiently correct analysis made by Little of the polarization interaction between electrons itself, an interaction which plays the main role in this theory. The estimates made by Little are based on perturbation theory, whereas the energy of interaction between an electron and the polarized side chain is much larger than the excitation energy of this level, which is connected with the occurrence of the dipole moment. It is quite clear that this circumstance should lead to a considerable shift of these levels and to a cardinal realignment of the side chain, and perhaps of the entire molecule as a whole. Therefore Little's calculations show that the molecular model proposed by him is apparently unstable and should rearrange itself spontaneously into some other state. It is not at all obvious here that this new state will be superconducting and not dielectric. From a somewhat different point of view, we can state, that the attraction used by Little is the result of the fact that his model leads to the occurrence of a negative dielectric constant. The stability of such a system, in which like charges (not only electrons in the central chain) attract each other, calls for a careful analysis.

Thus, Little's results can apparently not be regarded as proved with any degree of reliability. However, the ideas on which it is based, and the questions which arise in its analysis, are of exceeding interest, and in this probably lies the main value of this paper. In particular, one cannot exclude the possibility of the appearance of superconductivity as a result of a strong Coulomb attraction in systems which are not one-dimensional but have negative dielectric constants. Systems in which the dielectric is negative at least in some region of the frequency, are known and widespread, and can also be made synthetically. An interesting possibility for the occurrence of additional attraction between electrons was indicated by Vonsovskiĭ and Svirskiĭ.[15]

The main objection against the possibility of existence of superconductivity in Little's model, as indicated above, is connected with the one-dimensional nature of this model. Therefore particular interest attaches to the surface conductivity proposed somewhat earlier by Ginzburg and Kirzhnits,[16] i.e., superconductivity of a solid surface. The point is that on the surface of a crystal, as first shown by Tamm,[17] additional electronic states can arise which attenuate rapidly on going inside the crystal. The electrons situated at such levels can, however, move along the surface. The authors of [16] have shown that in the presence of attraction between the electrons there can arise in such a system also a superconducting state. Formally in such a two-dimensional model the fluctuations are also infinite. However, their divergence is very weak—logarithmic—and therefore for any body with finite dimensions it is quite inessential. It is interesting that superconductivity on surface levels could exist in principle also in the case when by its volume properties the substance is a dielectric.

In this case the interaction of the electrons can also be very specific. It can arise, for example, as a result of interaction between the electrons and the Rayleigh surface waves.

Ginzburg has also indicated that to intensify the attraction between the electrons one can use Little's mechanism, by introducing into the crystal highly polarizable impurity atoms, which in this model can lead to a sharp increase in the critical temperature.

Numerous possibilities are also afforded by an investigation of superconductivity in semiconductors, where the concentrations of the electrons, and also the character of their interaction with phonons and with one another can vary over a wide range. Unlike the preceding cases considered above and so far predicted purely theoretically, superconductivity in many semiconductors has already been observed experimentally. It was first considered theoretically by Gurevich, Larkin, and Firsov[18] and later by Cohen.[19] The first experimental results were obtained with the compound GeTe[20] and strontium titanate, $SrTiO_3$.[21] The latter case is of particular interest, and we shall stop to discuss it in more detail. The point is that strontium titanate is similar in many respects to barium titanate—a typical ferroelectric.[22] Although in strontium titanate itself the ferroelectric transition does not take place, it is very "close" to such a transition. Its dielectric constant at low temperatures reaches tremendous values ($\sim 10^3$–10^4). Therefore the Coulomb repulsion of the electrons from one another—the main factor preventing the appearance of superconductivity—is practically missing in this substance. From the microscopic point of view, the ferroelectric transition, according to present day notions, arises as a result of the fact that the frequency of one of the so-called optical vibrations of the lattice tends to zero. This means that the quasi-elastic force preventing the corresponding type of deformation tends to zero, and the crystal lattice becomes unstable, i.e., rearranged. But on the other hand, from the theory of dispersion of electromagnetic waves in crystals it is known that the frequencies of optical vibrations correspond to absorption lines in crystals, and some region of frequencies, higher than this characteristic frequency, is the region of anomalous dispersion, i.e., in this region the dielectric constant is negative. Therefore, if this region of frequencies makes an appreciable contribution to the interaction between the electrons, then the interaction will be attractive. In essence such an attraction mechanism is analogous to that considered by Little, but in his model the negative dielectric constant is due to the electronic polarizability, whereas here it is due to the polarizability of the ionic lattice. It is therefore essential that in strontium titanate at low temperature one of the frequencies of the optical vibrations turns out to be very low, i.e., the region of anomalous dispersion lies low. It is possible that this is precisely the explanation of the fact that, as reported by the authors of [21], superconductivity appears in $SrTiO_3$ even at very low electron concentrations, $\sim 10^{17}$.

In conclusion we emphasize once more that the presently gained understanding of the nature of the superconductivity phenomenon has made it possible to expand greatly the number of substances in which this phenomenon can exist, and has raised the hope that superconductors which differ greatly in their properties from ordinary metals can be obtained. The number of such possibilities, as we have already seen, is very large, and it is difficult to imagine that all will turn out to be fruitless. However, the main question, whether a superconductor can be obtained with sufficiently high critical temperature (at least on the order of 100°K), still remains open.

[1] Bardeen, Cooper, and Schrieffer, Phys. Rev. **108**, 1175, (1957).

[2] N. N. Bogolyubov, JETP **34**, 58 and 73 (1958), Soviet Phys. JETP **7**, 41 and 51 (1958).

[3] L. P. Gor'kov, JETP **34**, 735 (1958), Soviet Phys. JETP **7**, 505 (1958).

[4] Kunzler, Buchler, Hsu, and Wernick, Phys. Rev. Letts. **6**, 89 (1961); J. E. Kunzler, Rev. Mod. Phys. **33**, 501 (1961).

[5] A. A. Abrikosov, JETP **32**, 1442 (1957), Soviet Phys. JETP **5**, 1174 (1957).

[6] V. L. Ginzburg and L. D. Landau, JETP **20**, 1064 (1950).
[7] A. B. Migdal, JETP **34**, 1438 (1958), Soviet Phys. JETP **7**, 996 (1958).
[8] W. A. Little, Phys. Rev. **A134**, 1416 (1964).
[9] W. A. Little, Scientific American **212**(2), 21 (1965).
[10] K. G. Kepler, J. Chem. Phys. **39**, 3528 (1963).
[11] L. D. Landau and E. M. Lifshitz, Statisticheskaya fizika (Statistical Physics), Nauka, 1964.
[12] R. A. Ferrel, Phys. Rev. Letts. **13**, 330 (1964).
[13] Dewamec, Lehman, and Wolfram, Phys. Rev. Letts. **13**, 749 (1964).
[14] J. M. Luttinger, Journ. Math. Phys. **4** (1963).
[15] S. V. Vonsovskiĭ and M. S. Svirskiĭ, JETP **47**, 1354 (1964), Soviet Phys. JETP **20**, 914 (1965).
[16] V. L. Ginzburg and D. A. Kirzhnits, JETP **46**, 397 (1964), Soviet Phys. JETP **19**, 269 (1964).
[17] I. E. Tamm, Physik. Z. Sowjetunion **1**, 733 (1932).
[18] Gurevich, Larkin, and Firsov, FTT **4**, 185 (1962), Soviet Phys. Solid State **4**, 131 (1962).
[19] M. L. Cohen, Revs. Modern Phys. **36**, 240 (1964).
[20] Hein, Gibson, Maselsky, Miller, and Hulm, Phys. Rev. Letts. **12**, 320 (1964).
[21] Schooley, Hosler, and Cohen, Phys. Rev. Letts. **12**, 474 (1964).
[22] B. M. Vul, DAN SSSR **4**, 139 (1945).

Translated by J. G. Adashko

SOVIET PHYSICS JETP VOLUME 27, NUMBER 3 SEPTEMBER, 1968

COLLECTIVE PROPERTIES OF EXCITONS IN SEMICONDUCTORS

L. V. KELDYSH and A. N. KOZLOV

Submitted October 10, 1967

Zh. Eksp. Teor. Fiz. 54, 978–993 (March, 1968)

The problem of exciton interaction in semiconductors is considered in its multi-electron formulation. Expressions are obtained, in the approximation linear in the concentration, for the ground-state energy and for the law of dispersion of elementary excitations. Conditions for the Bose condensation of excitons are investigated and it is shown that low-density system of excitons behaves like a weakly nonideal Bose-gas. Furthermore, all quantities (the chemical potential, the rate of collective excitations) that depend on the two-particle scattering amplitude in the nonideal Bose-gas case are expressed in our analysis by the same formulas through the four-fermion interaction amplitude (two electrons and two holes) which includes, apart from the two-exciton scattering amplitude, the scattering amplitudes of two and three fermions as well as the terms connected with the Pauli statistics for the electrons and holes, and resulting from the fact that excitons are compound particles. These terms yield an essential positive contribution to the exciton scattering amplitude and may in principle ensure the stability of the ground Bose-condensed state even if there is a weak attraction between the excitons.

IN recent years a number of authors [1-4] indicated that excitons in crystals can reveal properties characteristic of Bose-particle systems, particularly a tendency to Bose condensation. This circumstance is quite interesting, at least because the small effective mass of the excitons can make the condensation temperature for them sufficiently high even at relatively low concentrations. Indeed, for an ideal Bose gas, as is well known [5],

$$kT_c = 3.31\hbar^2 M^{-1} N^{2/3},$$

where N is the concentration and M the mass of the particle. For large-radius excitons, and only these will be discussed here, we have $M \sim 10^{-27} - 10^{-28}$ g and the condensation temperature at $N \sim 10^{18}$ cm^{-3} is $T_c \sim 100$°K. Exciton concentrations of $10^{17} - 10^{18}$ cm^{-3} are presently perfectly realistic, since various methods of excitation of semiconductor lasers give apparently electron and hole concentrations of the same order of magnitude. At such densities, the interaction between the excitons (e.g., the Van der Waals interaction) becomes noticeable, i.e., they form an ideal Bose gas. The theory of a weakly-nonideal Bose gas was developed in sufficient detail [6,7]. However, the possibility of regarding the system of excitons as a weakly-nonideal gas is not obvious. The point is that excitons in semiconductors constitute a rather loosely-coupled state of two Fermi particles - an electron and a hole. The binding energy ϵ_0 and the exciton radius a_0 are determined in the simplest case by the well known Bohr formulas for the hydrogen atom:

$$\varepsilon_0 = \frac{1}{2}\frac{e^4 m}{\varkappa^2\hbar^2} \sim 10^{-2}\ \text{ev}, \quad a_0 = \frac{\varkappa\hbar^2}{me^2} \sim 10^{-6}\ \text{cm}, \qquad (1)$$

where e is the electron charge, κ the dielectric constant ($\kappa \sim 10$), and m the reduced effective mass of the electron and hole:

$$m = m_e m_h / (m_e + m_h) \sim 10^{-28}\ \text{g}.$$

At the concentrations considered above, $N \sim 10^{17} - 10^{18}$ cm^{-3}, the average distance between excitons $N^{-1/3}$ is of the same order as their radius a_0. Under such conditions, an important role is assumed by the internal structure of the exciton and by the fact that the Fermi particles of which the excitons are made up obey the Pauli principle. Two electrons (or two holes) contained in different excitons cannot come close to each other if their spins are parallel. Consequently, at $N^{-1/3} \sim a_0$ the excitons greatly deform each other even if no account is taken of the direct dynamic interaction, merely by virtue of the Pauli principle for the electrons and the holes, and the excitons can therefore not be regarded as structureless Bose particles.

In order to clarify this problem in somewhat greater detail, we introduce the operator Q_P^+ for the creation of an exciton with momentum P, and express this operator in terms of the operators for the creation of an electron, $a^+_{P/2+p}$ and hole $b^+_{P/2-p}$ (p - momentum of relative motion):

$$Q_P^+ = \sum_p \varphi(p)\, a^+_{P/2+p} b^+_{P/2-p}, \qquad (2)$$

where

$$\varphi(p) = \frac{8\sqrt{\pi}\, a_0^{3/2}}{[1 + (pa_0/\hbar)^2]^2} \qquad (3)$$

is the normalized wave function of the ground state of the hydrogenlike exciton.

Using the definition (2) and the usual Fermi commutation relations for the operators a_p and b_p, we can easily obtain the following commutation relations for the exciton creation and annihilation operators:

$$[Q_P, Q^+_{P'}] = \delta_{P,P'} - \sum_p \varphi\left(p + \frac{P}{2}\right)\varphi\left(p + \frac{P'}{2}\right)(a^+_{p+P'} a_{p+P} + b^+_{p+P'} b_{p+P}). \qquad (4)$$

The second term in the right side of (4) is an operator whose matrix elements, as can be readily shown, are of the order of Na_0^3, where N is the concentration of the electrons and holes. Thus, (4) corresponds to the commutation relations for Bose particles only accurate to terms of the order Na_0^3. The fact that a bound complex of two fermions is, strictly speaking, not a Bose particle was already indicated earlier in [8]. We note also the following circumstance, which will be of importance later. Effects connected with the deviation of the exciton statistics from Bose statistics come into play in the same order of magnitude as the effects connected with the nonideal nature of the Bose gas. Indeed, the chemical potential of a weakly-nonideal Bose gas is

$$\mu = \frac{4\pi\hbar^2}{Ma^2} Na^3, \tag{5}$$

where a is the scattering length. But in our problem, involving a system of particles interacting by Coulomb's law, the only parameter with the dimension of length is a_0, and therefore the exciton-exciton scattering length should be of the order of a_0. If we take further into account the fact that the masses of the electrons and of the holes are of the same order of magnitude and therefore $M = m_e + m_h \sim m = m_e m_h/M$, then (5) is reduced to the form $\mu \sim \epsilon_0(Na_0^3)$. But corrections of exactly this order should arise, as we have already seen, as a result of the fact that strictly speaking the excitons do not obey the Bose statistics. Therefore the problem of an interacting system of excitons, even in the lower orders in the concentration, cannot be equated to the problem of a weakly-nonideal Bose gas. It is the aim of the present paper to examine this question consistently.

We shall show below that a system of excitons actually does have many properties similar to the properties of a weakly-nonideal Bose gas. In particular, at sufficiently low temperatures, the excitons becomes condensed in a state with momentum $P = 0$; the correction to the energy of the ground state E_0 is quadratic in the concentration, and the correction to the chemical potential μ is linear in the exciton concentration:

$$E_0 / V = -N\varepsilon_0(1 - {}^1\!/_2 fNa_0^3), \tag{6}$$

$$\mu = -\varepsilon_0 + f\varepsilon_0 Na_0^3. \tag{7}$$

In these formulas, V is the volume of the system and f is a dimensionless parameter of the order of unity, an expression for which will be given below (formula (35)). The dependence of the energy of the moving exciton on its momentum has the usual form for a Bose gas

$$\omega(\mathbf{P}) = \sqrt{s^2\mathbf{P}^2 + (\mathbf{P}^2/2M)^2}, \tag{8}$$

i.e., it satisfies the Landau criterion for superfluidity, and the speed of "sound" s is connected with the correction to the energy by the usual hydrodynamic relation

$$Ms^2 = (\mu + \varepsilon_0) = f\varepsilon_0 Na_0^3. \tag{9}$$

The essential difference, however, between formulas (6)–(9) and the corresponding formulas for a weakly-nonideal Bose gas is the fact that the coefficient f (the sign of which is chosen opposite to that customarily used for the scattering amplitude in accordance with [9]) is not expressed directly in terms of the amplitude for the scattering of two free excitons by each other. Roughly speaking, the definition of f includes scattering amplitudes of three different types: exciton-exciton, electron-exciton, or hole-exciton, and the corrections for the amplitudes of scattering of electrons and holes by one another, connected with the presence of the exciton condensate. The latter are connected with the fact that the presence of the excitons leads to a change in the parameters of the electrons and holes (e.g., their effective masses) and of the effective interaction between them, in the same linear order in the concentration, and this in turn gives rise to a change in the internal energy of the exciton and its binding energy, and makes a contribution to all the quantities described by formulas (6)–(9).

This circumstance is very significant, since the theory of a weakly-nonideal Bose gas shows, as is well known, that such a gas can exist at low temperatures only when the forces between the particles are on the average repulsive, or, more accurately speaking, when the scattering amplitude is positive. Otherwise the gas state - state with low density - is unstable. In our case a similar criterion holds, but not for the scattering amplitude but for the quantity f, and the latter will be shown subsequently to differ from the amplitude of scattering of two excitons in the presence of an essentially positive and rather large term. Therefore the exciton gas can exist also in the presence of weak interaction between the excitons, provided this attraction does not lead to the formation of bound molecule-like states. This is all the more important, since at large distances between the excitons a Van der Waals attraction is certainly present. The system is stabilized in this case by the Fermi statistics of the electrons.

Similar results for the energy and for the chemical potential were already obtained by Popov [10], but for a system that differs essentially from that considered by us. Popov considered a system of Fermi particles of one kind, and to ensure its stability he proposed that the interaction forces between the particles depend essentially on the spins: attraction for one mutual orientation of the spins and repulsion for the other.

In an experimental investigation of semiconductors with large exciton density, the results presented above should become manifest in the fact that, at large concentrations, the exciton line in the optical spectrum should shift towards larger energies by an amount $\delta\mu = f\epsilon_0 Na_0^3$. The exciton binding energy decreases by an amount of the same order, i.e., the threshold of the interband transitions approaches the exciton line.

At the same time, an additional band appears in the luminescence spectrum, the upper edge of which is shifted away from the main line into the region of low frequencies by an amount equal to the binding energy of the exciton μ. This band is a result of exciton collisions, in which one of the excitons recombines and the other breaks up into an electron and a hole. With further increase of the concentration, the intensity of the additional band increases, and its upper edge approaches the main line. At concentrations $Na_0^3 \sim 1$ the binding energy of the exciton tends to zero, i.e., the excitons disintegrate into a Fermi gas of electrons and holes, and the additional band merges with the region of the continuous spectrum. Strictly speaking, our analysis is not valid at such high concentrations, but the conclusion that the excitons vanish is confirmed by an analysis of the opposite limiting case in [11,12], where it is shown that when $Na_0^3 \gg 1$ the gap in the electron spectrum, i.e., the effective energy of their binding with the holes, tends exponentially to zero at a perfectly isotropic dispersion law, and that it vanishes in the presence of anisotropy even at zero temperature. The presence of a condensed state and superfluidity should apparently become manifest also in an anomalously large exciton diffusion.

We proceed now to a quantitative investigation of our problem. The Hamiltonian of the system of electrons and holes interacting in accordance with Coulomb's law has in the second-quantization representation the form

$$\hat{\mathscr{H}} = \sum_{\mathbf{p}} [(\varepsilon_{\mathbf{p}}^{e} - \mu_e) a_{\mathbf{p}}^{+} a_{\mathbf{p}} + (\varepsilon_{\mathbf{p}}^{h} - \mu_h) b_{\mathbf{p}}^{+} b_{\mathbf{p}}] + \frac{1}{2} \sum_{\mathbf{pp'k}} V_{\mathbf{k}} \{a_{\mathbf{p}}^{+} a_{\mathbf{p'}}^{+} a_{\mathbf{p'+k}} a_{\mathbf{p-k}} + b_{\mathbf{p}}^{+} b_{\mathbf{p'}}^{+} b_{\mathbf{p'+k}} b_{\mathbf{p-k}} - 2a_{\mathbf{p}}^{+} b_{\mathbf{p'}}^{+} b_{\mathbf{p'+k}} a_{\mathbf{p-k}}\}, \quad V_{\mathbf{k}} = \frac{4\pi e^2 \hbar^2}{\varkappa \mathbf{k}^2}, \tag{10}$$

where a_p and b_p are the Fermi operators for electron and hole annihilation, ϵ_p^e and ϵ_p^h are the dependences of their energy on the momentum p, and μ_e and μ_h are the chemical potentials, determined by the conditions

$$\sum_{\mathbf{p}} \langle a_{\mathbf{p}}^{+} a_{\mathbf{p}} \rangle = \sum_{\mathbf{p}} \langle b_{\mathbf{p}}^{+} b_{\mathbf{p}} \rangle = NV. \tag{11}$$

The symbol $\langle \ldots \rangle$ denotes averaging over the ground state, and κ is the dielectric constant of the semiconductor. Although it is easy to continue the analysis in rather general form, we shall confine ourselves, in order not to make the subsequent formulas too cumbersome, to the case of the simplest dispersion law

$$\varepsilon_{\mathbf{p}}^{e} = \mathbf{p}^2 / 2m_e, \qquad \varepsilon_{\mathbf{p}}^{h} = \mathbf{p}^2 / 2m_h \tag{12}$$

and, moreover, we put for the time being $m_e = m_h = 2m$ (m = reduced mass). In addition, we disregard the spins of the electrons and the holes. The final result will be presented for the more general case $m_e \neq m_h$, with allowance for the spin structure in all the formulas.

We note, finally, one more assumption which has already been made by choosing the Hamiltonian in the form (10). Regarding the electrons and the holes as two independent types of particles, we neglect the possibility of the transition of the electron from one band to the other, and in particular we omit from the Hamiltonian the corresponding matrix elements of the potential V. This assumption however, is fully justified, for owing to the orthogonality of the wave functions of the different bands these matrix elements are small compared with those retained in (10) (their relative order is $\epsilon_0/\Delta \sim 10^{-2}$, where Δ is the width of the forbidden band). Because of this we can independently reckon the energies of the electrons and holes from the edge of the corresponding band, as was done in (12); the exciton energy is reckoned in this case from the width of the forbidden band. In exactly the same manner, the momenta of the electron and of the hole are reckoned from their values at the bottom of each of the bands.

Taking the foregoing assumptions under consideration, we introduce now Coulomb measurement units, i.e., we put $m = \hbar = e^2/\kappa = 1$. Then

$$a_0 = 1, \quad \varepsilon_0 = {}^1/_2, \quad V_{\mathbf{k}} = 4\pi / \mathbf{k}^2, \qquad \varepsilon_{\mathbf{p}}^{e} = \varepsilon_{\mathbf{p}}^{h} = {}^1/_2 \varepsilon_{\mathbf{p}} = \mathbf{p}^2/4, \quad \mu_e = \mu_h = \mu/2. \tag{13}$$

Here $\mu = \mu_e + \mu_h$ is the chemical potential of the excitons. Finally, we introduce the dimensionless exciton concentration

$$n = Na_0^3. \tag{14}$$

The quantity n is thus the only parameter of the problem, since μ should be expressed in terms of n with the aid of relations (11), which now take the form

$$\sum_{\mathbf{p}} \langle a_{\mathbf{p}}^{+} a_{\mathbf{p}} \rangle = \sum_{\mathbf{p}} \langle b_{\mathbf{p}}^{+} b_{\mathbf{p}} \rangle = n. \tag{15}$$

Formula (15) and all the succeeding ones are referred to a unit volume of the system.

We have already stated above that our problem corresponds to the "gas" situation, i.e., $n \ll 1$. In this sense, it is the opposite of the problem investigated in[11], which was formally analogous to the problem of superconductivity, i.e., it corresponded to weak attraction of the electrons and the holes. In the case $n \ll 1$, the interaction energy is much larger than the kinetic energy of the ideal Fermi gas of the electrons and holes, and therefore the latter can under no consideration be used as the initial approximation for solving our problem. It is clear from physical considerations that the ground state of the system is made up of excitons, i.e., of bound electron-hole pairs. It is therefore natural to start with the Bogolyubov canonical transformation[13], which is known from the theory of superconductivity and is described by the unitary operator

$$S = \exp\Big\{ \sum_{\mathbf{p}} \varphi_{\mathbf{p}} (a_{\mathbf{p}}^{+} b_{-\mathbf{p}}^{+} - b_{-\mathbf{p}} a_{\mathbf{p}}) \Big\}, \tag{16}$$

$$S a_{\mathbf{p}} S^{+} = u_{\mathbf{p}} a_{\mathbf{p}} + v_{\mathbf{p}} b_{-\mathbf{p}}^{+}, \qquad S b_{\mathbf{p}} S^{+} = u_{\mathbf{p}} b_{\mathbf{p}} - v_{\mathbf{p}} a_{-\mathbf{p}}^{+}, \tag{17}$$

where

$$u_{\mathbf{p}} = \cos\varphi_{\mathbf{p}}, \qquad v_{\mathbf{p}} = \sin\varphi_{\mathbf{p}}, \qquad u_{\mathbf{p}}^2 + v_{\mathbf{p}}^2 = 1. \tag{18}$$

The function φ_p should be determined in this case from the condition of minimum energy and of stability of the ground vacuum state of the system.

The Hamiltonian (10) is transformed as follows:

$$S\hat{\mathscr{H}}S^{+} = U\{\varphi_{\mathbf{p}}\} + \hat{\mathscr{H}}_0 + \hat{\mathscr{H}}_i, \tag{19}$$

where $U\{\varphi_p\}$ is a numerical (not operator) functional of φ_p, which is separated after reducing the transformed Hamiltonian to the normal form

$$U\{\varphi_{\mathbf{p}}\} = \sum_{\mathbf{p}} (\varepsilon_{\mathbf{p}} - \mu) v_{\mathbf{p}}^2 - \sum_{\mathbf{p},\mathbf{p'}} V_{\mathbf{p}-\mathbf{p'}} (u_{\mathbf{p}} v_{\mathbf{p}} u_{\mathbf{p'}} v_{\mathbf{p'}} + v_{\mathbf{p}}^2 v_{\mathbf{p'}}^2). \tag{20}$$

The operator $\hat{\mathscr{H}}_0$ includes terms that are bilinear in the Fermi operators:

$$\hat{\mathscr{H}}_0 = \sum_{\mathbf{p}} \Big[(u_{\mathbf{p}}^2 - v_{\mathbf{p}}^2) \Big(\frac{\varepsilon_{\mathbf{p}} - \mu}{2} - \sum_{\mathbf{p'}} V_{\mathbf{p}-\mathbf{p'}} v_{\mathbf{p'}}^2 \Big) + 2u_{\mathbf{p}} v_{\mathbf{p}} \sum_{\mathbf{p'}} V_{\mathbf{p}-\mathbf{p'}} u_{\mathbf{p'}} v_{\mathbf{p'}} \Big] (a_{\mathbf{p}}^{+} a_{\mathbf{p}} + b_{\mathbf{p}}^{+} b_{\mathbf{p}}) + \sum_{\mathbf{p}} \Big[\Big(\varepsilon_{\mathbf{p}} - \mu - 2 \sum_{\mathbf{p'}} V_{\mathbf{p}-\mathbf{p'}} v_{\mathbf{p'}}^2 \Big) u_{\mathbf{p}} v_{\mathbf{p}} - (u_{\mathbf{p}}^2 - v_{\mathbf{p}}^2) \sum_{\mathbf{p'}} V_{\mathbf{p}-\mathbf{p'}} u_{\mathbf{p'}} v_{\mathbf{p'}} \Big] (a_{\mathbf{p}}^{+} b_{-\mathbf{p}}^{+} + b_{-\mathbf{p}} a_{\mathbf{p}}). \tag{21}$$

The operator $\hat{\mathscr{H}}_i$ contains fourfold combinations of the Fermi operators:

$$\hat{\mathcal{H}}_i = \sum_{pp'k} V_k \{ {}^1/_2 \gamma_{p,\,p-k} \gamma_{p',\,p'+k} (a_p^+ a_{p'}^+ a_{p'+k} a_{p-k} + b_p^+ b_{p'}^+ b_{p'+k} b_{p-k} - 2a_p^+ b_{p'}^+ b_{p'+k} a_{p-k}) + \gamma_{p,\,p-k} \tilde{\gamma}_{p',\,p'+k} (a_p^+ a_{p'}^+ b_{-p'-k}^+ a_{p-k} - a_{p'}^+ b_{-p'-k}^+ b_p b_{p-k} + \text{h.c.}) + {}^1/_2 \tilde{\gamma}_{p,\,p-k} \tilde{\gamma}_{p',\,p'+k} \times (a_p^+ a_{p'}^+ b_{-p'-k}^+ b_{-p+k}^+ + a_p^+ b_{-p+k}^+ b_{p'+k} a_{-p'} + \text{h.c.}) \}, \tag{22}$$

where

$$\gamma_{p,\,p'} = u_p u_{p'} + v_p v_{p'} = \cos(\varphi_p - \varphi_{p'}),$$
$$\tilde{\gamma}_{p,\,p'} = u_p v_{p'} - v_p u_{p'} = \sin(\varphi_{p'} - \varphi_p), \tag{23}$$

and obviously the following relation is satisfied.

$$\gamma_{pp'}^2 + \tilde{\gamma}_{pp'}^2 = 1. \tag{24}$$

The first term in the interaction Hamiltonian $\hat{\mathcal{H}}_i$ describes electron and hole scattering processes similar to those included in the initial Hamiltonian $\hat{\mathcal{H}}$, except that in each vertex, where the momentum of the former particle changes from p to q, there appears an additional factor γ_{pq}. A graphic representation of these matrix elements is given in Fig. 1a.

The matrix elements of the second term in $\hat{\mathcal{H}}_i$ is shown in Fig. 1b. They correspond to processes in which the Fermi particle is scattered and an electron-hole pair from the vacuum is created (or annihilated). The vertex at which the scattering takes place corresponds in the matrix element to the same factor $\gamma_{p,p-k}$, and the vertex at which creation (or annihilation) of an electron with momentum p and a hole with momentum $-p' - k$ is produced corresponds to the factor $\tilde{\gamma}_{p,p'+k}$. We note also that the matrix elements $\hat{\mathcal{H}}_i$ corresponding to creation of a pair by an electron or a hole have opposite signs.

In order not to write out the indices e and h on the diagrams, we propose henceforth that if creation (or annihilation) of an electron-hole pair occurs at any one vertex, the upper of the lines drawn from this vertex corresponds to the electron and the lower to the hole.

Finally, the last term in $\hat{\mathcal{H}}_i$ corresponds to processes in which two pairs are produced (or annihilated), or else one pair is produced from vacuum and the other is annihilated. A graphic representation of these processes is shown in Figs. 1c, d. To each vertex on these diagrams there corresponds a factor $\tilde{\gamma}$.

Thus, following the transformation (19), our problem becomes in some respects similar to the problem of a weakly-nonideal Bose gas in the Belyaev analysis[7]: the perturbation-theory diagrams include, besides the processes describing the particle scattering, also vertices in which creation of particles from the vacuum (condensate) takes place, or else their annihilation

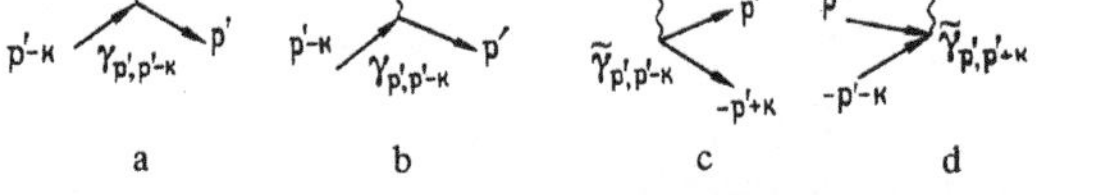

a b c d

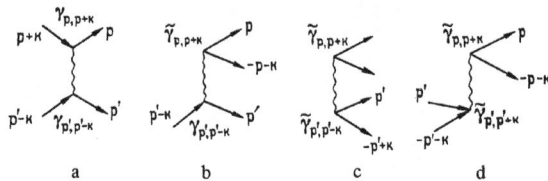

FIG. 1.

(falling into the condensate). This analogy becomes even closer if account is taken of the fact that $\tilde{\gamma} \sim \sqrt{n}$, as will be shown below; consequently, $\tilde{\gamma}$ in our case plays the same role as the operators of creation and annihilation of condensate particles in Belyaev's technique. To verify this, we perform the transformation (16) also in the normalization condition (15). Then, adding both equations of (15), we get

$$\sum_p \{ v_p^2 + {}^1/_2 (u_p^2 - v_p^2) \langle a_p^+ a_p + b_p^+ b_p \rangle + u_p v_p \langle a_p^+ b_{-p}^+ + b_{-p} a_p \rangle \} = n. \tag{25}$$

However, as is clear from physical considerations (and will be confirmed by the subsequent analysis), the mean values $\langle a_p^+ a_p \rangle$ and $\langle b_p^+ b_p \rangle$ should vanish. Indeed, all the levels of the single-particle Fermi excitations should lie at energies close to zero (bottom of the band) and higher energies, and the chemical potentials of the electrons and holes are essentially negative ($\mu_e = \mu_H = \mu/2 - \approx 1/4$), inasmuch as the chemical potential of the excitons μ should obviously be somewhere near the level of the free exciton. But then all the levels of the single-particle Fermi excitations will be empty, as stated above.

To avoid misunderstanding, we emphasize that although we use as before the terms electron and hole, in fact, following the transformation (17), a_p and b_p, are operators of certain new Fermi quasiparticles corresponding to elementary excitations in the system under consideration, and going over into ordinary electrons and holes only when $n \to 0$.

The last term in (25) must also be set equal to zero, i.e.,

$$\langle a_p^+ b_{-p}^+ \rangle = \langle b_{-p} a_p \rangle = 0. \tag{26}$$

Condition (26) is not satisfied, of course, automatically, but we can use the leeway we still possess in the choice of the function φ_p in (16), in order to ensure satisfaction of (26). Moreover, we shall verify below that condition (26) is necessary to ensure stability of the ground (vacuum) state of the system chosen by us. In other words, (26) should be regarded as an equation defining φ_p, with the normalization obtained simultaneously from (25):

$$\sum_p v_p^2 = \sum_p \sin^2 \varphi_p = n. \tag{27}$$

It follows directly from (27) that $v_p \sim \sqrt{n}$, and then from (18) we get $u_p = 1 - O(n)$. Using these estimates and the definitions (23), we get $\tilde{\gamma} \sim \sqrt{n}$ and $\gamma \sim 1 - O(n)$. Thus, the scattering of the quasiparticles by one another is renormalized by the transformation (16) only in order n; on the other hand, the appearing new processes of creation and annihilation of particle pairs contain factors of order $\sqrt{n}$, as do the processes connected with the emergence of the particles from the condensate in Belyaev's technique.

We now proceed to obtain the explicit form of (26), and to prove that it is a necessary condition for the stability of the ground state. To this end, we note first that the transformed Hamiltonian $\hat{\mathcal{H}}$ admits of creation of single electron-hole pairs with a total momentum equal to zero from vacuum. The corresponding matrix elements are contained in $\hat{\mathcal{H}}_0$, and can also be obtained

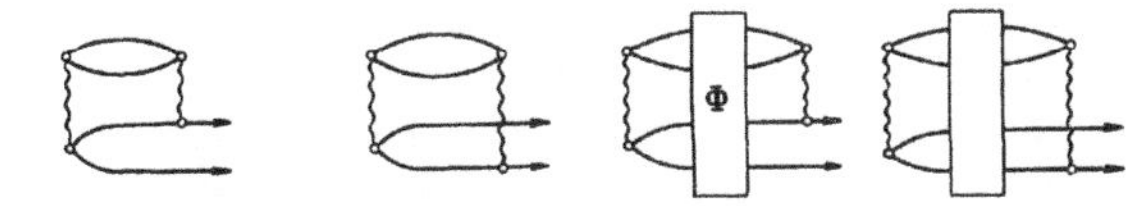

a b c d

FIG. 2.

in higher orders of perturbation theory from $\mathscr{H}_i$. For example, the diagrams shown in Figs. 2a, b, which describe the creation of one pair, can be constructed from the matrix elements of Figs. 1b, c. The somewhat more complicated diagram of Fig. 2c can be obtained by making Fig. 2a more complicated by introducing in it a block Φ, that describes all possible processes of scattering of two electrons and two holes by one another. It is easy to see that the diagrams of Figs. 2a and 2c are of the same order, since the quasiparticle scattering processes are of zero order in the concentration. Both these diagrams are of order $n^{3/2}$, since they contain each three vertices $\tilde{\gamma}$. If we confine ourselves to this order, as we shall do in what follows, then the block Φ must be replaced by Φ_0, which describes the interaction of two electrons and two holes in the absence of other particles, and which is obtained from Φ by replacing all the scattering vertices γ by unity.

We now call attention to the fact that addition, to the diagrams of the perturbation-theory series, of parts connected with the remaining part of the diagram by only one pair of lines - electron and hole - with zero total momentum, leads immediately to divergences. Indeed, taking into account the interaction, such a pair of lines should be replaced by a complete two-particle propagation function G_2 of the electron and hole, as shown for the diagram of Fig. 2c in Fig. 3a. But the function G_2 describes also the bound states of the electron and the hole, i.e., excitons, with the exciton levels corresponding to poles of G_2 relative to the total pair energy. Since the pair in question was created from vacuum, its total momentum is equal to zero, and the total energy is $\mu = \mu_e + \mu_h$. But the energy of an exciton with zero momentum, by definition, should equal μ. Therefore an electron-hole pair with total momentum equal to zero and with energy μ should correspond to a pole of G_2, and consequently the diagram of Fig. 3a becomes infinite, as well as all other diagrams containing single electron-hole pairs created from vacuum. The only possibility of eliminating this divergence is to stipulate mutual cancellation of all the diagrams that lead to creation of one pair from vacuum. This condition, as is well known, should indeed define the function φ_p. We carry out this cancellation with accuracy to terms of order $n^{3/2}$ inclusive. The divergences can be eliminated by selecting the function φ_p.

In order to analyze this process in somewhat greater detail, we introduce the function

$$F(\mathbf{p}, \varepsilon) = -i \int \langle T b_{-\mathbf{p}}(t+\tau) a_{\mathbf{p}}(t) \rangle e^{i\varepsilon\tau} d\tau, \tag{28}$$

which is analogous to the well known pair function introduced by Gor'kov in superconductivity theory. Here $a_p(t)$ and $b_{-p}(t)$ are the Heisenberg operators of annihilation of an electron and a hole. This function can be represented symbolically by the diagram of Fig. 3b, where the block Σ_{eh} is the sum of all diagrams with one incoming pair of lines - one electron and one hole line - and with lines irreducible with respect to such a pair inside. The simplest elements of Σ_{eh} are the coefficients of the operator $a_p^+ b_{-p}^+$ in $\mathscr{H}_0$, and the diagrams of Figs. 2c, d. All the remaining diagrams contained in Σ_{eh} are of higher order in the concentration. Strictly speaking, it is impossible to satisfy the condition $\Sigma_{eh} \equiv 0$, since we have at our disposal the function φ_p, that depends only on the momentum p, and Σ_{eh} depends on the 4-component quantity $p = \{\mathbf{p}, \epsilon\}$, where ϵ is the relative frequency of the electron and the hole. However, to cancel out the divergences, it suffices, as we shall presently show, to satisfy the weaker condition

$$\int \Sigma_{eh}(\mathbf{p}, \varepsilon) G_e(\mathbf{p}, \varepsilon) G_h(-\mathbf{p}, -\varepsilon) \frac{d\varepsilon}{2\pi} = 0. \tag{29}$$

The point is that the pole term in G_2, which is the only one that needs to be cancelled out, depends on ϵ only via the entering single-particle Green's functions $G_e(p)$ and $G_h(-p)$. Therefore the integration with respect to ϵ in the pole term of the diagram in Fig. 3b reduces to the integral in formula (29) and, if φ_p is chosen such as to satisfy (29), then the diagram on Fig. 3b becomes convergent. Taking (21) into account, we rewrite, accurate to terms of order $n^{3/2}$ inclusive, the condition (29) in the form

$$\Big(\varepsilon_p - \mu - 2 \sum_{\mathbf{p}'} V_{\mathbf{p}-\mathbf{p}'} v_{\mathbf{p}'}^2\Big) u_{\mathbf{p}} v_{\mathbf{p}} - (u_{\mathbf{p}}^2 - v_{\mathbf{p}}^2) \sum_{\mathbf{p}'} V_{\mathbf{p}-\mathbf{p}'} u_{\mathbf{p}'} v_{\mathbf{p}'}$$
$$+ (\mu - \varepsilon_{\mathbf{p}}) \int \Lambda(\mathbf{p}\ \varepsilon)\, i d\varepsilon / 2\pi = 0, \tag{30}$$

where $\Lambda(p, \epsilon)$ denotes the sum of the diagrams of Figs. 2c, d. The factor $\mu - \epsilon_p$ in front of the last term in (30) is the result of the fact that Σ_{eh} contains the coefficient of $a_p^+ b_{-p}^+$ in (21). After it is multiplied by $G_e(p, \epsilon) G_h(-p, -\epsilon)$, and following integration with respect to ϵ, a factor $(\mu - \epsilon_p)^{-1}$ arises, and Eq. (30) is obtained from (29) by multiplication by $\mu - \epsilon_p$.

In the lower approximation in v_p, i.e., accurate to terms of order $\sqrt{n}$, Eq. (30) reduces to the ordinary Coulomb equation

$$(\varepsilon_{\mathbf{p}} - \mu_0) v_{\mathbf{p}} - \int V_{\mathbf{p}-\mathbf{p}'} v_{\mathbf{p}'} \frac{d^3 p'}{(2\pi)^3} = 0. \tag{31}$$

Its solution in conjunction with the normalization condition (27) takes the form

$$v_{\mathbf{p}} = \sqrt{n}\, \psi_0(\mathbf{p}), \quad \mu_0 = -\varepsilon_0, \tag{32}$$

where ϵ_0 and $\psi_0(p)$ are the binding energy and the wave function of the ground state of the exciton, determined by formulas (1) and (3). The correction to the chemical potential $\mu - \mu_0$ can now be determined from (30) with the aid of ordinary perturbation theory, provided we substitute in the terms of order $n^{3/2}$, which

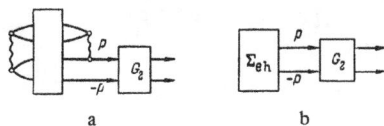

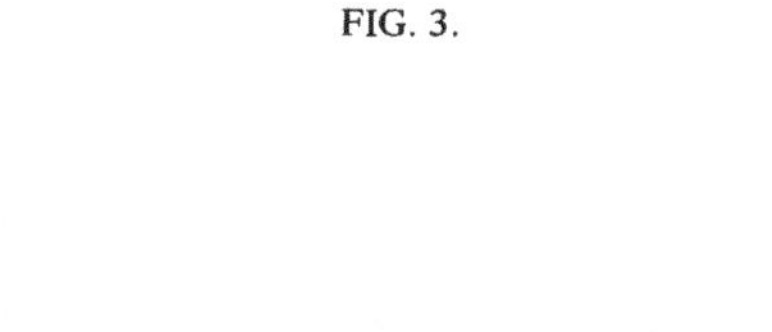

a b

FIG. 3.

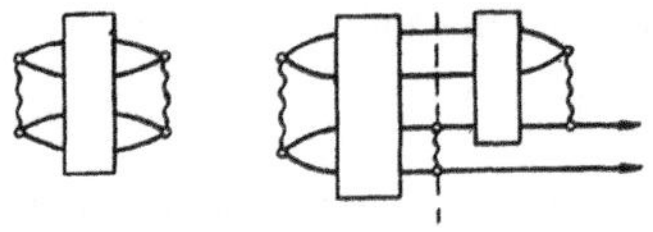

FIG. 4.

were omitted from (31), the zeroth approximation for v_p from (32) and we regard them as a small perturbation. After simple calculations we obtain (in Coulomb units)

$$\mu - \mu_0 = \frac{13\pi}{3} n + \frac{1}{n}\Lambda = \left(\frac{13\pi}{3} + \lambda\right) n, \tag{33}$$

where

$$\Lambda = \int \Lambda(p, \varepsilon) v_p (\mu_0 - \varepsilon_p) \frac{i d^4 p}{(2\pi)^4} = n^2 \lambda. \tag{34}$$

We shall show that the quantity Λ, defined by formula (34), is described by the graphic block shown in Fig. 4a. To this end we draw in greater detail, for example, the diagram of Fig. 2c, drawing in it a vertical section corresponding to the instant of time of the last interaction on the $G_h(-p)$ line. One of the corresponding diagrams is shown in Fig. 4b. Since the interaction V_k does not depend on the transferred frequency, the entire left part of this diagram does not depend on ϵ. In its right side, the functions that depend on ϵ are $G_e(p)$, $G_h(-p)$, and the three-particle propagation function for which ϵ is the summary energy. The singularities of the last two functions with respect to ϵ lie at $\epsilon > 0$ (i.e., in the lower half-plane), and the singularities of $G_h(-p)$ lie at $\epsilon < 0$ (i.e., in the upper half-plane). We note now that the product $(\mu_0 - \epsilon_p) G_e(p) G_h(-p)$ simply equals $G_e(p) + G_h(-p)$. Then, taking into account the forgoing, the only term making a contribution is that containing $G_h(-p)$, but not $G_e(p)$. Therefore when integrating the contribution from diagram 2c in (34), we can leave out the product $(\mu_0 - \epsilon_p) G_e(p)$. Repeating this reasoning for the contribution from diagram 2d, and noting that, with the required accuracy, $v'_p - v_p \approx \tilde{\gamma}_{pp'}$, we can readily verify that Λ is given by the diagram of Fig. 4a, in accordance with the statement made above, and is, as clearly seen from this figure, a quantity of the order of n^2, while λ is a dimensionless constant of the order of unity. From a comparison of formulas (7) and (33) it follows that

$$f = 13\pi/3 + \lambda, \tag{35}$$

where

$$\lambda = \int \left[\psi_0\left(\frac{\mathbf{k}}{2} - \mathbf{p}\right) - \psi_0\left(\frac{\mathbf{k}}{2} + \mathbf{p}\right)\right]\left[\psi_0\left(-\frac{\mathbf{k}}{2} - \mathbf{q}\right) - \psi_0\left(-\frac{\mathbf{k}}{2} + \mathbf{q}\right)\right]$$
$$\times \Phi(k, p, q; k', p', q')\left[\psi_0\left(\frac{\mathbf{k}'}{2} - \mathbf{p}'\right) - \psi_0\left(\frac{\mathbf{k}'}{2} + \mathbf{p}'\right)\right]\left[\psi_0\left(-\frac{\mathbf{k}'}{2} - \mathbf{q}'\right)\right.$$
$$\left. - \psi_0\left(-\frac{\mathbf{k}'}{2} + \mathbf{q}'\right)\right](2\pi)^{-24} d^4k\, d^4p\, d^4q\, d^4k'\, d^4p'\, d^4q', \tag{36}$$

Φ is the complete single-time propagation function of two electrons and two holes with total momentum equal to zero and with total energy equal to $-2\epsilon_0$.

We emphasize now that in obtaining these results we actually used the assumption that two excitons cannot form a bound state. Indeed, if such a state were to exist, then the quantity $\Lambda(p, \epsilon)$, regarded as a function of μ, would have a pole at $\mu < -\epsilon_0$. In the vicinity of this pole, neither the perturbation theory used by us to solve (30), nor this equation itself, which is obtained by choosing the principal diagrams with respect to the powers of the concentration, would be valid. The validity of the assumption that there are no bound states of two excitons will be discussed later.

We now proceed to consider the spectrum of elementary excitations in the exciton system. Obviously, single-particle excitations (band states of electrons and holes) will be separated from the ground state by a gap having a width approximately equal to ϵ_0. The lowest excited states should be the two-particle states corresponding, in the limit as $n \to 0$, to the motion of the individual exciton as a whole. They are determined by the poles of the two-particle Green's function $G_2(P; p, p')$ where $P = \{\mathbf{P}, E\}$ is the summary momentum and frequency of the electron-hole pair, and $p = \{\mathbf{p}, \epsilon\}$ and $p' = \{\mathbf{p}', \epsilon'\}$ are the relative momentum and frequency of the electron-hole pair. The presence of an exciton condensate makes it necessary to introduce into consideration, besides the ordinary pair function G_2, also the function $\tilde{G}_2(P; p, p')$, which is the sum of all the connected diagrams describing creation of two electron-hole pairs from vacuum. A diagram of this function is shown in Fig. 5a. The equivalent diagram shown in Fig. 5b is more convenient for tracing out the diagrams, and therefore will also be used later.

It is easy to verify that the functions G_2 and $\tilde{G}_2$ are determined by the system of equations of Fig. 6, which is analogous in some respect to Belyaev's system of equations[7] for a non-ideal Bose gas. In the equations of Fig. 6 we have grouped together the terms in such a way that G_2 denotes not the two-particle Green's function itself, but only that part corresponding to the connected diagrams. The main difference between these equations and the purely algebraic equations of Belyaev lies in the fact that the equations of Fig. 6 are integral with respect to the momenta of the relative motion of the electron and the hole, and therefore describe both the motion of the exciton and its internal structure. Near the poles corresponding to the exciton-gas density oscillations, the main contribution to the functions G_2 and G_2 is made by diagrams that break up into two parts connected by a single electron line and a single hole line, directed to one side, i.e., homogene-

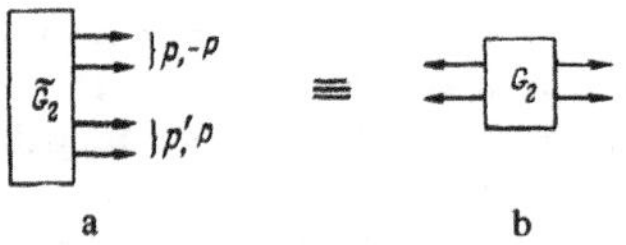

FIG. 5.

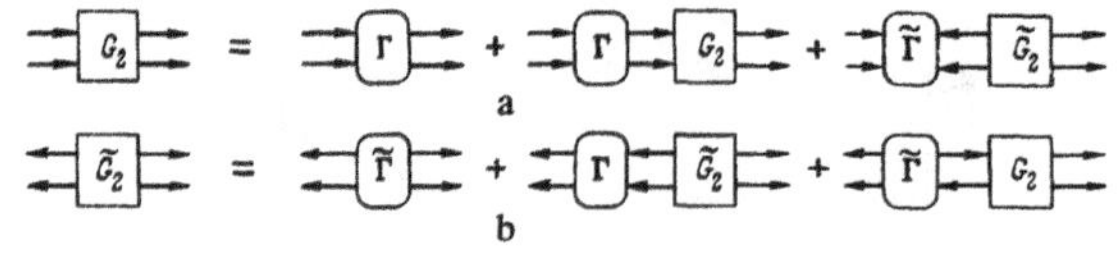

FIG. 6.

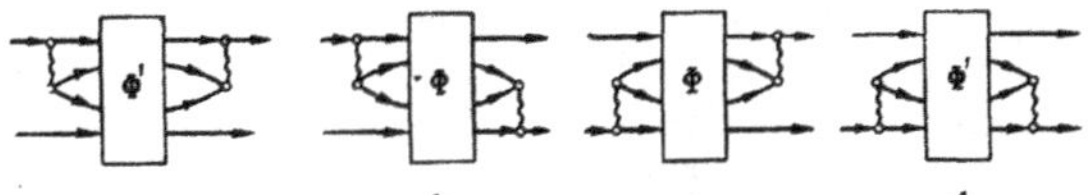

a b c d

FIG. 7.

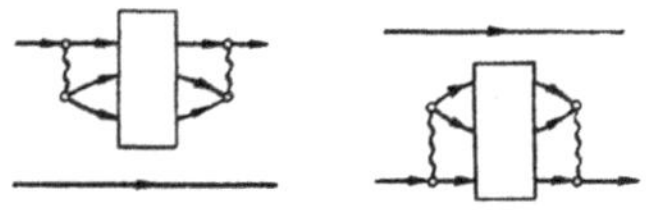
a b

FIG. 8.

ous terms with respect to G_2 and G_2 in the right sides of the equations of Fig. 6. The inhomogeneous terms contain the vertices $\Gamma(P, p, p')$ and $\tilde{\Gamma}(P; p, p')$, which have no pole character and constitute sums of diagrams that are irreducible in the indicated sense. Accurate to terms linear in n we have

$$\Gamma(P;\, p,\, p') = -V_{\mathbf{p}-\mathbf{p}'}\gamma_{\mathbf{p}+\mathbf{P}/2,\, \mathbf{p}'+\mathbf{P}/2}\gamma_{\mathbf{p}-\mathbf{P}/2,\, \mathbf{p}'-\mathbf{P}/2} + v'(P;\, p,\, p'), \tag{37}$$

where the first term corresponds to Fig. 1a, and the second is the sum of the diagrams of Fig. 7a-d. The block Φ in these diagrams, just as in Fig. 2, contains all possible scatterings of two electrons and two holes by one another in the zeroth order in the concentration, while the block Φ′ in the diagrams of Figs. 7a,d differs from Φ only in that it does not contain the diagrams shown in Figs. 8a, b, since they are included in the self-energy corrections to the functions G_e and G_h. If we join these corrections to the diagrams of Fig. 7, then the block Φ′ is completed to form the block Φ.

In order not to complicate the derivations that follow, we shall proceed in this fashion, i.e., we shall assume that the function $v'(P; p, p')$ is determined by the diagrams of Fig. 7, in which Φ′ is replaced by Φ, and the single-particle Green's functions G_e and G_h are taken throughout without the self-energy corrections of second and higher orders in the interaction, i.e., in accordance with (21), also in an approximation that is linear in n

$$G_e(p) = G_h(p) = \left\{\varepsilon - \frac{\varepsilon_{\mathbf{p}} - \mu}{2} - \int V_{\mathbf{p}-\mathbf{p}_1}(v_{\mathbf{p}}v_{\mathbf{p}_1} - v_{\mathbf{p}_1}{}^2)\frac{d\mathbf{p}}{8\pi^3} + i\delta\right\}^{-1}. \tag{38}$$

In analogy with formula (37), the vertex part of $\tilde{\Gamma}$ is the sum of all the diagrams describing the creation of two pairs from vacuum and of internal lines that are irreducible in terms of the pair and go to one side. In the approximation linear in n we have

$$\tilde{\Gamma}(P;\, p,\, p') = V_{\mathbf{p}-\mathbf{p}'}\tilde{\gamma}_{\mathbf{p}+\mathbf{P}/2,\, -\mathbf{p}'+\mathbf{P}/2}\tilde{\gamma}_{\mathbf{p}'-\mathbf{P}/2,\, -\mathbf{p}+\mathbf{P}/2} + \tilde{v}(P;\, p,\, p'). \tag{39}$$

The first of the terms in (39) corresponds to the matrix element of Fig. 1c, and the second to the sum of the diagrams of Figs. 9a-d. The diagrams of Fig. 9 have the general property that the outermost interaction occurs on them between one of the particles of the first electron-hole pair and the particle from the second pair. This property reflects the requirement that the diagrams entering in $\tilde{\Gamma}$ be irreducible with respect to each of the particle pairs.

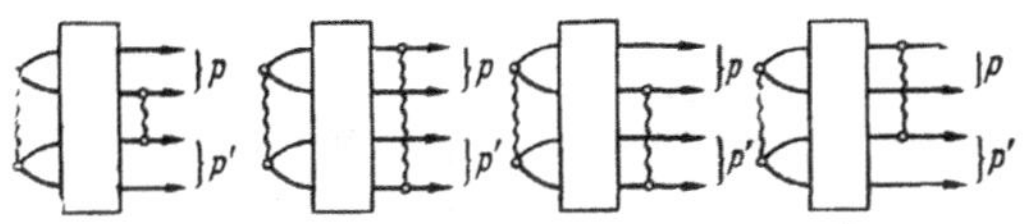

a b c d

FIG. 9.

As seen from (37) and (39), $\Gamma = V_{p-p'} + O(n)$, and $\tilde{\Gamma} \sim n$. Consequently, the system of Fig. 6 can be solved with the aid of perturbation theory, putting in the zeroth order $\Gamma = V_{p-p'}$ and $\tilde{\Gamma} = 0$. We shall assume also that the summary frequency E and the kinetic energy $P^2/2M$ of the exciton are small compared with unity, and employing perturbation theory, we confine ourselves to terms of the first order in all three parameters n, E, and $P^2/2M$. It can be readily seen that accordingly we must put $P = 0$ in (37) and (39).

We now write out, using the assumed approximation, the equations of Fig. 6 in analytic form

$$\begin{aligned} G_2(P;\, p, p') = &-G_e(p + P/2)G_h(-p + P/2)\int \frac{id^4p_1}{(2\pi)^4}V_{\mathbf{p}-\mathbf{p}_1}G_2(P;\, p_1, p') \\ &+ G_e(p)G_h(-p)\int \{V_{\mathbf{p}-\mathbf{p}_1}\gamma^2_{\mathbf{p}\mathbf{p}_1} + v'(p, p_1)\}G_2(P;\, p_1, p')\frac{id^4p_1}{(2\pi)^4} \\ &+ G_e(p)G_h(-p)\int \{V_{\mathbf{p}-\mathbf{p}_1}\tilde{\gamma}^2_{\mathbf{p}\mathbf{p}_1} + \tilde{v}(p, p_1)\}\tilde{G}_2(P;\, p_1, p')\frac{id^4p_1}{(2\pi)^4} \\ &+ G_e(p + P/2)G_h(-p + P/2)\Gamma(p, p')G_e(p' + P/2)G_h(-p' + P/2), \end{aligned} \tag{40}$$

$$\begin{aligned} \tilde{G}_2(P;\, p, p') = &-G_e(p - P/2)G_h(-p - P/2)\int \frac{id^4p_1}{(2\pi)^4}V_{\mathbf{p}-\mathbf{p}_1}\tilde{G}_2(P;\, p_1, p') \\ &+ G_e(p)G_h(-p)\int \{V_{\mathbf{p}-\mathbf{p}_1}\gamma^2_{\mathbf{p}\mathbf{p}_1} + v'(p, p_1)\}\tilde{G}_2(P;\, p_1, p')\frac{id^4p_1}{(2\pi)^4} \\ &+ G_e(p)G_h(-p)\int \{V_{\mathbf{p}-\mathbf{p}_1}\tilde{\gamma}^2_{\mathbf{p}\mathbf{p}_1} + \tilde{v}(p, p_1)\}G_2(P;\, p_1, p')\frac{id^4p_1}{(2\pi)^4} \\ &+ G_e(p)G_h(-p)\tilde{\Gamma}(p, p')G_e(p')G_h(-p'). \end{aligned} \tag{41}$$

In the zeroth order in the perturbation, the pole part of the functions G_2 and $\tilde{G}_2$ is determined by the first terms in the right sides of (40) and (41). We see therefore that

$$\begin{aligned} G_2(P;\, p, p') &= G_e(p)G_h(-p)A(P;\, \mathbf{p}, \mathbf{p}')G_e(p')G_h(-p'), \\ \tilde{G}_2(P;\, p, p') &= G_e(p)G_h(-p)\tilde{A}(P;\, \mathbf{p}, \mathbf{p}')G_e(p')G_h(-p'). \end{aligned} \tag{42}$$

Substituting (42) in those terms of (40) and (41) which pertain to the perturbation, and integrating with respect to ϵ, we obtain a system in which the integration takes place already over the three-dimensional momentum p, and in the zeroth order the homogeneous parts of both equations reduce to a Coulomb equation of the type (31). The solution can be obtained in standard fashion with the aid of the Green's function of the Schrödinger equation with Coulomb potential. This procedure is straightforward but quite laborious, and we therefore confine ourselves to presenting the final results.

The pole terms of the two-particle Green's functions are of the form (42), with

$$\begin{aligned} A(P;\, \mathbf{p}, \mathbf{p}') &= \left(\varepsilon_0 + \frac{\mathbf{p}^2}{2m^*}\right)\psi_0(\mathbf{p})\mathcal{G}(P)\psi_0(\mathbf{p}')\left(\varepsilon_0 + \frac{\mathbf{p}'^2}{2m^*}\right) \\ \tilde{A}(P;\, \mathbf{p}, \mathbf{p}') &= \left(\varepsilon_0 + \frac{\mathbf{p}^2}{2m^*}\right)\psi_0(\mathbf{p})\tilde{\mathcal{G}}(P)\psi_0(\mathbf{p}')\left(\varepsilon_0 + \frac{\mathbf{p}'^2}{2m^*}\right), \end{aligned} \tag{43}$$

where

$$\mathcal{G}(P) = \frac{\mu - \mu_0 + \mathbf{P}^2/2M + E}{E^2 - E^2(\mathbf{P})} = \frac{1 + N_\mathbf{P}}{E - E(\mathbf{P})} - \frac{N_\mathbf{P}}{E + E(\mathbf{P})}, \tag{44}$$

$$\tilde{\mathcal{G}}(P) = -\frac{\mu - \mu_0}{E^2 - E^2(\mathbf{P})}, \tag{45}$$

$$E(\mathbf{P}) = \left[\frac{\mu - \mu_0}{M}\mathbf{P}^2 + \left(\frac{\mathbf{P}^2}{2M}\right)^2\right]^{1/2} \tag{46}$$

$$N_\mathbf{P} = \frac{1}{2}\left\{\frac{\mu - \mu_0 + \mathbf{P}^2/2M}{E(\mathbf{P})} - 1\right\}. \tag{47}$$

According to (43)—(47), the factors $\mathcal{G}$(P(and $\tilde{\mathcal{G}}$(P) of the two-particle Green's functions G_2 and $\tilde{G}_2$, which depend on the summary 4-momentum of the electron-hole pair P, coincide in form with the single-particle Green's function for a weakly non-ideal Bose gas [7], and the energy of the elementary excitations E(P) and the occupation numbers of the "supercondensate" excitons N_P are connected in the usual manner with the correction to the chemical potential $\mu - \mu_0$. In formulas (43) m* denotes the reduced mass of the electron of the hole (without the previous limitation $m_e = m_h$). It can be shown, in addition, that allowance for the spin variables does not change any of the final formulas.

Let us make one more remark explaining formulas (44)—(47). The correction to the chemical potential $\mu - \mu_0$, which enters in these formulas, is due to such terms of the Eqs. (40) and (41), which contain the quantities v′(P; p, p′) and $\tilde{v}$(P; p, p′). During the course of the solution, as can be readily seen, operations are performed on these functions, corresponding the closing of the diagrams of Figs. 7—9, as a result of which they reduce to the already known blocks of type shown in Fig. 4a.

We now discuss the results from the point of view of the possibility of their experimental observation. We have already mentioned that our analysis, strictly speaking, is not valid if the excitons form a bound state of the hydrogen-molecule type. Such a state apparently arises unavoidably in those cases when the mass of one of the particles (usually a hole) is much larger than that of the other, i.e., for example, in the majority of semiconductors of the type $A^{III}B^{V}$, where $m_e + m_h \sim 0.1$, inasmuch as in this case the problem of the interaction of two excitons does not differ in principle from the problem of interaction of two hydrogen atoms. The situation changes radically, however, if the masses of the electron and hole are of the same order. In this case the relative contribution made by the kinetic energy of the exciton motion to the total energy of the system increases strongly (the analog of zero-point oscillations of the atoms in the hydrogen molecule). Indeed, if a bound state with radius a_1 is produced, then the average kinetic energy of relative motion of the excitons should be, by virtue of the uncertainty principle, $\gtrsim \hbar^2/(m_e + m_h)a_1^2$. When $m_e/m_h \ll 1$, this quantity is small compared with the interaction energy, the order of which is $\epsilon_0 \sim \hbar^2/m_e a_0^2$. But when $m_e/m_h \sim 1$, these two energies are of the same order, if $a_1 \sim a_0$, which should prevent formation of a bound state or, at any rate, should decrease noticeably its binding energy ϵ_1 and increase the radius a_1.

The limiting case $m_e = m_h$ was considered by Hylleras and Ore [14] (bound state of two positrons). The binding energy obtained in this case $\epsilon_1 \approx 10^{-2}\epsilon_0$ corresponds in our problem to $\epsilon_1 \approx 10^{-4}$ eV and can be disregarded, since at temperatures $T \gtrsim 1°K$ this bound state no longer exists. It vanishes apparently also at zero temperature, if the exciton density is such that $na_1^3 \gg 1$, where $a_1 \sim \hbar/\sqrt{m\epsilon_1} \gg a_0$, i.e., if the average distance between excitons is smaller than the radius a_1, but still much larger than a_0. In this case the results obtained by us for T = 0 are valid in the region of intermediate concentrations $a_1^{-3} \ll n \ll a_0^{-3}$.

1 S. A. Moskalenko, Fiz. Tverd. Tela 4, 276 (1962) [Sov. Phys.-Solid State 4, 199 (1962)].

2 S. A. Moskalenko, P. I. Khadzhi, and A. I. Bobrysheva, ibid. 5, 1444 (1963) [5, 1051 (1963)].

3 J. M. Blatt, K. W. Boer, and W. Brandt, Phys. Rev. 126, 1691 (1962).

4 R. C. Casella, J. Appl. Phys. 34, 1703, (1963).

5 L. D. Landau and E. M. Lifshitz, Statisticheskaya fizika, Gostekhizdat, 1951 [Statistical Physics, Addison-Wesley, 1958].

6 N. N. Bogolyubov, Izv. AN SSSR ser. fiz. 11, 77 (1947).

7 S. T. Belyaev, Zh. Eksp. Teor. Fiz. 34, 417 (1958) [Sov. Phys.-JETP 7, 289 (1958)].

8 Ya. B. Zel'dovich, ibid. 37, 569 (1960) [10, 403 (1961)].

9 A. A. Abrikosov, L. P. Gor'kov, and I. E. Dzyaloshinskiĭ, Metody kvantovoi teorii polya v statisticheskoy fizike, Fizmatgiz, 1963 [Quantum Field Theoretical Methods in Statistical Physics, Pergamon, 1965].

10 V. N. Popov, Zh. Eksp. Teor. Fiz. 50, 1550 (1966) [Sov. Phys.-JETP 23, 1034 (1966)].

11 L. V. Keldysh and Yu. V. Kopaev, Fiz. Tverd. Tela 6, 2791 (1964) [Sov. Phys.-Solid State 6, 2219 (1965)].

12 A. N. Kozlov and L. A. Maksimov, Zh. Eksp. Teor. Fiz. 48, 1184 (1965); 49, 1284 (1965); 50, 131 (1966) [Sov. Phys. JETP 21, 790 (1965); 22, 889 (1966); 23, 88 (1966)].

13 N. N. Bogolyubov, ibid. 34, 58 and 73 (1958) [7, 41 and 51 (1958)].

14 E. Hylleras and A. Ore, Phys. Rev. 71, 493 (1947).

Translated by J. G. Adashko
112

L. V. Keldysh. Electron-hole Drops in Semiconductors

At sufficiently low temperatures, the non-equilibrium electrons and holes introduced into a pure semiconductor are bound together into excitons—systems similar to positronium, but differing from it in having macroscopically large Bohr radii ($a_0 \sim 10^{-6}$ cm) and very low binding energies ($\epsilon_0 \sim 10^{-2}$ eV).

Such a change in the length and energy scales in a system coupled by Coulomb forces is due to the decrease of the Coulomb interaction as a result of the large dielectric constants of the semiconductors, $\kappa \geq 10$, and the small effective masses of the electrons and holes, $m \sim 0.1\, m_0$ (m_0—mass of free electron). Substitution of these values into the known Bohr formulas for the binding energy and the radius of the hydrogenlike atom

$$\varepsilon_0 = e^4 m/2\varkappa^2\hbar^2, \qquad a_0 = \varkappa\hbar^2/me^2$$

leads to the estimates indicated above. An increase of the length scale by two orders of magnitude and a decrease of the energy scale by three orders of magnitude compared with the length and energy scales in ordinary substances is characteristic also of all the phenomena considered below which occur in a system of electrons and holes in a semiconductor. In particular, the criterion of high exciton density, wherein an important role is assumed by the interaction between them, corresponds obviously to concentrations $n_0 \sim a_0^{-3} \sim 10^{18}\ \mathrm{cm}^{-3}$, and the region of temperatures at which all these phenomena should be observed is $kT \lesssim 0.1\epsilon_0$, i.e., $T \lesssim 10°$ K.

If the electron concentration is large enough, the interaction between them can lead to a "liquefaction" of the exciton gas,[1] i.e., to the formation of a relatively dense electron-hole phase, in which all the particles are coupled by mutual attraction forces and the average distance between them is of the order of a_0, while their concentration is $n_0 \sim a_0^{-3} \sim 10^{17}-10^{18}\ \mathrm{cm}^{-3}$. This phase differs from the usual electron-hole plasma in semiconductors in the same manner as liquid metals (e.g., mercury) differ from an electron-ion plasma: it is contained by internal forces and has a perfectly well defined equilibrium density n_0. It does not diffuse over the entire sample, and occupies only that part of the sample volume which can be uniformly filled with a density n_0 at a specified total number of electrons and holes introduced into the sample. The transition from the gas of free excitons to the electron-hole "liquid" should have many characteristic features of a first-order phase transition. In particular, when the average exciton concentration in the sample reaches a certain value $n_C(T)$ that depends on the temperature T ($n_C(T) \ll n_0$ at sufficiently low temperatures), the system should become laminated into two phases: regions filled with the liquid phase—"drops"—with density n_0, and regions filled with an exciton gas having a much lower density. With further increase of the number of electrons and holes introduced into the sample, the volume of the liquid phase increases, but its density n_0 does not change so long as it does not fill the entire sample. A rigorous theoretical investigation of the properties of the liquid phase entails considerable difficulties, but its main properties can be predicted from general considerations. The absence of heavy ions from the sample makes it impossible to produce in such a phase any spatial ordering such as crystallization at arbitrary temperatures, since the amplitudes of the zero-point oscillations of the particles should be of the order of a_0, i.e., of the average distance between the particles. For the same reason, it is not very likely that such a liquid phase can consist of exciton molecules—biexcitons. The large zero-point oscillations and the low coupling energy of the biexciton should lead to an intense interaction of each particle with all the nearest neighbors, to a strong electron exchange, and as a consequence to a collectivization of all the electrons and holes. Therefore the phase under consideration is more likely to be similar to a liquid metal.

The electron-hole drops in pure semiconductors should have quite high mobility, since the scattering of the electrons and of the holes by the phonons, which is sufficiently small at low temperatures to start with, is suppressed even more by the presence of Fermi degeneracy in the drop, and the density of the effective mass in the drop is very small. Therefore such external actions as inhomogeneous deformations or inhomogeneous magnetic fields can relatively easily accelerate the drops to velocities of the order of the velocity of sound. It is not very likely that the drop can exceed this velocity, owing to the coherent emission of phonons. However, even these velocities suffice to move the drops over a distance on the order of several centimeters during the lifetime of the non-equilibrium electrons and holes.

An electron-hole liquid can acquire properties such as superfluidity or superconductivity. In the special case when the effective masses of the electrons and the holes are almost isotropic, it can be proved[3] that when the temperature is decreased there occurs a collective binding of the electrons with the holes, leading to the transition of the liquid into a dielectric superfluid state. In the presence of noticeable anisotropy, the question of the feasibility of such a transition remains open. In order for superconductivity to occur it is necessary that electrons be attracted to electrons (or holes to holes). Theoretically it is impossible to prove reliably or refute reliably the existence of such an attraction. Qualitatively it is possible as a result of the interaction of two electrons and one hole (or two holes) or with the vibrations of the liquid density ("phonon").

Finally, it should be noted that theoretically one cannot exclude one more possibility of the behavior of the system of excitons with decreasing concentration, an alternative to that described above. If the exchange repulsion of the biexcitons prevails over the Van der Waals attraction and if the coupling energy per particle in the biexciton is larger than in the liquid phase, then a Bose-condensed superfluid gas of biexcitons can exist at low temperatures,[4] and the coupling energy of the biexcitons tends to zero only gradually with increasing concentration.[5] Even if the coupling in the biexcitons is weaker than in a liquid, but at distances larger than a_0 repulsion between them prevails, there can exist a Bose condensate of biexcitons as a certain metastable phase. For "atomic" excitons, such a behavior is impossible, owing to the predominance of the attraction forces.

Until recently, the considerations advanced concerning the "liquefaction" of the excitons[1, 2] remained hypothetical to a considerable degree. A "metallization" of excitons was observed,[6] but the character of this transition remained unclear. During the last year and a half, however, there appeared a large number of experimental papers whose results are interpreted by their authors as the observation of drops of electron-hole liquid in such semiconductors as germanium and silicon.

In [7] they investigated the absorption of light in pure germanium in the exciton lines (in the region of the so-called direct exciton) at helium temperatures, as a function of the number of electron-hole pairs introduced into the sample. Analyzing their results, the authors of [7] reached the conclusion that in the concentration region $\bar{n} \sim 10^{15}-10^{16}$ cm^{-3} (here and below $\bar{n}$ denotes the concentration of the electrons and holes averaged over the volume without allowance for the possible lamination into phases), the sample breaks up into regions in which the exciton absorption remains practically constant, and regions in which the exciton line vanishes completely as a result of the screening of the Coulomb interaction between the electron and the hole, which appear in this region as free charges ("metallic" regions). An estimate of the equilibrium concentration n_0 in the drops of the liquid metallic phase, according to the data of [7], yields $n_0 \sim 2 \times 10^{16}$ cm^{-3}. It is shown in [8] that in the spectrum of the recombination radiation of the electron-hole pairs in germanium at low temperatures $T \lesssim 4.2°$ K, besides the usual emission line of the free excitons, after a certain critical electron and hole concentration is reached there appears a new line, corresponding to lower energies of the photon emitted upon recombination, i.e., to larger coupling energy between the electrons and the holes. With further growth of the concentration or with decreasing temperature, this line becomes rapidly dominating in the emission spectrum. The position and shape of this line, and also its practically complete independence of a great variety of impurities when their concentration is varied from values much smaller than n to values much larger than n, do not make it possible to ascribe this emission to any known impurities. At the same time, all its characteristic features, including the concentration and temperature dependences, as shown in [8], can be satisfactorily explained if this radiation is ascribed to recombination of the electrons and holes in the drops of the liquid phase of the type considered above with an equilibrium concentration $n_0 \approx 2 \times 10^{17}$ cm^{-3}. An analogous radiation was observed in silicon in [9, 10]. Further confirmation of the foregoing point of view was obtained in [11]. It turned out that under conditions of uniaxial compression of the germanium crystal, the above-described emission line behaves in a completely anomalous manner: whereas the lines of the free excitons and the different impurity levels, which fall in this energy interval, shift upon deformation in almost identical fashion (the binding energies do not change), the line ascribed to the emission of the drops is hardly displaced at small deformations along the $\langle 111 \rangle$ axis, so that its distance from the exciton line (the coupling energy) decreases by a factor of two, and only then, at still larger deformations, does it begin to shift together with all the other lines, i.e., no further decrease of the coupling energy takes place. Such a behavior of the coupling energy in the metallic liquid phase is attributed in [11] to the fact that in germanium, by virtue of the known singularities of the structure of its electron spectrum, deformation along the $\langle 111 \rangle$ axis decreases by a factor of four the density of the electronic states near the bottom of the conduction band. At a specified concentration of the electrons n_0, this would lead to a growth of the Fermi energy and of the electron-gas pressure, i.e., in order to maintain the equilibrium, the value of n_0, and with it also the binding energy per particle, should decrease (the drop "expands"). Whenever the deformations were not homogeneous through the sample, a number of new anomalies was observed, particularly a catastrophic drop (by two orders of magnitude) of the emission intensity in the described line. These anomalies can be interpreted as an indirect indication of the acceleration of the drops by the deformation gradients (the energy of the drop depends on the deformations). The drop of intensity of the radiation is in this case explained by the fact that the nuclei of the drops of the liquid phase, having time to grow to the equilibrium value determined by $\bar{n}$ and T, go off from the region where the non-equilibrium electrons and holes are produced and where consequently the drops can grow. Finally, it was shown in [12] that under the same conditions ($\bar{n}$ and T) at which there appears the recombination-radiation line ascribed in [8, 11] to drops of an electron-hole "liquid," pure germanium begins to absorb in the far infrared region, where heretofore it was perfectly transparent. This absorption has, as a function of the wavelength λ, a distinct maximum in the region $\lambda \sim 100\ \mu$, which was interpreted as plasma resonance in the absorption (or scattering) by metallic drops whose linear dimensions are much larger than the wavelength λ. From the position of this resonance it is possible to estimate directly the concentration n_0 of the particles in the drop. It also turned out to be $\approx 2 \times 10^{17}\ cm^{-3}$.

Thus, at the present time there is an entire series of facts that agree satisfactorily with the hypothesis that a condensed electron-hole phase exists in semiconductors. Some of these facts can be explained just as well as being due to the fact that at low temperatures the excitons become bound into "molecules" (biexcitons).[9] However, within the framework of the biexciton picture, there is still no satisfactory explanation of such facts as the absorption in the infrared region, the anomalous behavior of the radiation under uniaxial deformations, and the vanishing of the absorption line of the direct exciton. Therefore the existence of condensed-phase drops seems to be quite likely, but only further experiments can prove it (or refute it) conclusively. Such convincing experiments might be, for example, direct observation of the motion of the drops over a macroscopic distance, or scattering of light by these drops.

[1] L. V. Keldysh, Proc. Internat. Conf. on the Physics of Semiconductors, Moscow, 1968, p. 1307.

[2] L. V. Keldysh and A. A. Rogachev, Paper at Session of the Division of General Physics and Astronomy, USSR Academy of Sciences, September 1968.

[3] L. V. Keldysh and Yu. V. Kopaev, Fiz. Tverd Tela 6, 2791 (1964) [Sov. Phys.-Solid State 6, 2219 (1965)].

[4] S. A. Moskalenko, ibid. 4, 276 (1962) [4, 199 (1962)]; J. M. Blatt, K. W. Bőer, and W. Brandt, Phys. Rev. 126, 1691 (1962).

[5] L. V. Keldysh and A. N. Kozlov, Zh. Eksp. Teor. Fiz. 54, 978 (1968) [Sov. Phys.-JETP 27, 521 (1968)].

[6] V. M. Asnin, A. A. Rogachev, and S. M. Ryvkin, Fiz. Tekh. Poluprov. 1, 1740 (1967) [Sov. Phys.-Semicond. 1, 1445 (1968)]; ZhETF Pis. Red. 7, 464 (1968) [JETP Lett. 7, 360 (1968)].

[7] V. M. Asnin and A. A. Rogachev, ZhETF Pis. Red. 9, 415 (1969) [JETP Lett. 9, 248 (1969)].

[8] Ya. E. Pokrovskiĭ and K. I. Svitstunova, ibid. 9, 435 (1969) [9, 261 (1969)].

[9] I. R. Haynes, Phys. Rev. Lett. 17, 86 (1966).

[10] S. M. Ryvkin and A. A. Yaroshevskiĭ, Fiz. Tekh. Poluprov. No. 8 (1969) [Sov. Phys.-Semicond. No. 2, 1970].

[11] V. S. Bagaev, T. I. Galkina, O. V. Gogolin, and L. V. Keldysh, ZhETF Pis. Red. 10, 309 (1969) [JETP Lett. 10, 195 (1969)].

[12] V. S. Vavilov, V. A. Zayats, and V. N. Murzin, ibid. 10, 304 (1969) [10, 192 (1969)].

Physics – Uspekhi **60** (11) 1180 – 1186 (2017)

FROM THE ARCHIVE

PACS numbers: 03.75.Kk, **67.10.**–**j**, **71.35.**–**y**

IN MEMORY OF LEONID VENIAMINOVICH KELDYSH

Coherent states of excitons

L V Keldysh

DOI: https://doi.org/10.3367/UFNe.2017.10.038227

Abstract. **The concept of a coherent exciton state is formulated. It is shown that for this state, a macroscopic wave function can be introduced such that it satisfies a nonlinear equation of the type familiar in the phenomenological theory of a superfluid liquid. The corresponding nondissipative flux is the flux of energy. For excitons interacting with an electromagnetic field, a coupled system of Maxwell equations and Ginzburg–Pitaevskii-type equations (phenomenological theory of Bose liquid) is obtained.**

Keywords: excitons, coherent state of excitons, macroscopic wave function, phenomenological theory of superfluid liquid, Bose-liquid equations, Ginzburg–Pitaevskii equations

There has been a large number of recent theoretical papers on exciton condensation in crystals [1–14]. There are, in fact, three different problems at issue here. One is the thermodynamically equilibrium rearrangement of the electronic spectrum due to the instability of the original spectrum under the electron–electron interaction [6–8]. Another is the Bose condensation of nonequilibrium excitons (for example, those excited by light) [1–5]. The third is the coalescence of excitons into a dense phase, i.e., condensation in the same sense of the word in which any gas condenses into a liquid [14, 15].

Although the problem of Bose condensation of nonequilibrium excitons was the first to appear in the literature, it remains the subject of the most fundamental difference of opinion among researchers. Early studies [1–3] assumed that excitons, which consist of two Fermi particles (an electron and a hole), are bosons and that theoretical results for Bose gases and Bose liquids (which consist of structureless particles) directly apply to a system of excitons. Subsequent analyses [4, 5] showed that the deviation of the exciton statistics from the Bose statistics must be taken into account simultaneously with introducing the exciton–exciton interaction, and that sufficiently large exciton densities make the very concept of the exciton meaningless. But at low densities, a system of excitons does indeed behave like a weakly nonideal Bose gas and, notably, can exhibit superfluid motion in a crystal. However, it was argued recently in [13] that unlike a system of true bosons, a system of excitons cannot be superfluid in principle. In this paper, we focus on the analysis of this situation and on resolving the problem of what the Bose condensation and excitonic superfluidity mean from the physical (observational) standpoint. We show, in particular, that the conclusions in Ref. [13] are based on a misconception.

L V Keldysh Lebedev Physical Institute,
Russian Academy of Sciences,
Leninskii prosp. 53, 119991 Moscow, Russian Federation

This paper was first published in 1972 in the Igor' Evgen'evich Tamm memorial collection *Problems of Theoretical Physics* [19]
Uspekhi Fizicheskikh Nauk **187** (11) 1273 – 1279 (2017)
DOI: https://doi.org/10.3367/UFNr.2017.10.038227
Translated by E G Strel'chenko; edited by A M Semikhatov

By an exciton, as usual, we mean an itinerant electronic excitation in a crystal not associated with charge and mass transfer. In the simplest molecular crystal or semiconductor models, the exciton is, respectively, an excited single-molecule state transferred resonantly between elementary cells of the crystal (Frenkel exciton) or a hydrogen-like bound electron–hole state (Wannier–Mott exciton). Thus, simply by definition, the motion of an exciton cannot involve a flow of matter or electric charge. Excitons transfer their excitation energy and, possibly, properties such as the angular momentum and the electric and magnetic moments whenever appropriate. Therefore, the superfluidity of nonequilibrium excitons can also well imply the existence of undamped energy flows (with a reservation to be made below) or, for example, the existence of polarization, but it does not imply a superfluid mass or charge transfer, whereas the proof of the impossibility of exciton superfluidity in Ref. [13] totally relies on the analysis of mass transfer.

We also note in passing that the formal proof in Ref. [13] has no relation to nonequilibrium excitons because it assumes that all electrons have the same chemical potential and hence the system is in full thermodynamic equilibrium. In actual fact, however, the condensation of nonequilibrium excitons implies that the electron–hole system is not fully in equilibrium in the sense that although the electrons, holes, and excitons are in equilibrium among themselves and with the lattice, the total number of excitons and electron–hole pairs is determined not by thermodynamic equilibrium but by a certain external excitation source. Such a situation readily occurs in real conditions, because recombination is in most cases much slower than the thermalization of electrons and holes and their binding into excitons. For example, in germanium at liquid helium temperatures, the thermalization time $\lesssim 10^{-9}$ s and the exciton formation time from electrons and holes are of the same order of magnitude for the electron and hole concentrations $n_{\mathrm{e,h}} \gtrsim 10^{12}$ cm^{-3} and the exciton lifetime $\gtrsim 10^{-5}$ s. The exciton lifetime can be much longer if the exciton recombination is spin-forbidden.

We can now be more precise about the concept of the superfluid flow of excitons. Clearly, in contrast to liquid helium and superconductors, the superfluid flow of excitons exists not arbitrarily long but only during the exciton lifetime, and the transition of the system of excitons into a superfluid state means that the flow damping time is determined not by the exciton scattering time but by the exciton lifetime, which is longer by several orders of magnitude.

Excitons are most commonly viewed as certain quasiparticles in a crystal, and from this standpoint their Bose condensation is the accumulation of a macroscopic number of such particles in a single state. The same situation, however, can also be described in other terms: as is known [16, 17], excitons are in fact the quanta of normal vibrations of the electron density in a crystal, similar in many respects to plasmons. Their Bose-condensed state is then a coherent definite-phase electron density wave with a finite amplitude (rather than with an amplitude of the order of V^{-1}, where V is the system volume). As regards the statement on superfluidity, this means that introducing effects that are nonlinear in the amplitude results in the complete suppression of scattering processes for such a wave.

We now turn to a more formal analysis of the problem posed. The secondary-quantized electron Hamiltonian of the crystal has the usual form

$$H = -\frac{\hbar^2}{2m_0}\int \psi_\alpha^+(\mathbf{x})\,\nabla^2\psi_\alpha(\mathbf{x})\,\mathrm{d}^3x - \sum_{\mathbf{n},k} Z_k e^2 \int \frac{\psi_\alpha^+(\mathbf{x})\,\psi_\alpha(\mathbf{x})}{|\mathbf{x}-\mathbf{R}_{\mathbf{n},k}|}\,\mathrm{d}^3x + \frac{e^2}{2}\int \frac{\psi_\alpha^+(\mathbf{x})\,\psi_\beta^+(\mathbf{x}')\,\psi_\beta(\mathbf{x}')\,\psi_\alpha(\mathbf{x})}{|\mathbf{x}-\mathbf{x}'|}\,\mathrm{d}^3x\,\mathrm{d}^3x', \tag{1}$$

where $\psi_\alpha^+(\mathbf{x})$ and $\psi_\alpha(\mathbf{x})$ are fermion operators satisfying the commutation relations $[\psi_\alpha(\mathbf{x}), \psi_\beta^+(\mathbf{x}')]_+ = \delta_{\alpha\beta}\delta(\mathbf{x}-\mathbf{x}')$, the symbol $[...]_+$ denotes the anticommutator, $\hbar$, m_0, and e are the Planck constant, the electron mass, and the electron charge, and Z_k and $\mathbf{R}_{\mathbf{n},k}$ are the atomic number and the radius vector of the nucleus at the kth position in the $\mathbf{n}$th unit cell. For simplicity, we consider the nuclei to be rigidly fixed, and the subsequent treatment is carried out first in the mean-field approximation. Then the operator $\psi_\alpha(\mathbf{x})$ can be decomposed into electron (positive-frequency) and hole (negative-frequency) parts:

$$\psi_\alpha(\mathbf{x}) = \psi_\alpha^{(\mathrm{e})}(\mathbf{x}) + \psi_\alpha^{(\mathrm{h})+}(\mathbf{x}),$$

$$\psi_\alpha^{(\mathrm{e})}(\mathbf{x}) = \sum_{j>j_0} a_j\chi_{j\alpha}(\mathbf{x}), \qquad \psi_\alpha^{(\mathrm{h})+}(\mathbf{x}) = \sum_{j\leqslant j_0} a_j\chi_{j\alpha}(\mathbf{x}), \tag{2}$$

$$[a_j, a_{j'}^+]_+ = \delta_{jj'}, \qquad [a_j, a_{j'}]_+ = 0.$$

Here, $\chi_{j\alpha}(\mathbf{x})$ is the set of Hartree–Fock basis functions, where the indices $j \leqslant j_0$ ($j > j_0$) label the states with filled (empty) electronic bands. These functions must evidently have the Bloch form

$$\chi_{j\alpha}(\mathbf{x}) = \exp\left(\frac{\mathrm{i}}{\hbar}\,\mathbf{p}\mathbf{x}\right)u_{\mathbf{p}l\alpha}(\mathbf{x}),$$

where $\mathbf{p}$ is the quasimomentum and l labels bands. Thus, in Eqns (2), $j = \{\mathbf{p}, l\}$, and because our discussion concerns nonmetal crystals, the summation over $j \leqslant j_0$ means a summation over all $\mathbf{p}$ within the first Brillouin zone and over $l \leqslant l_0$.

The function $\chi_{j\alpha}(\mathbf{x})$ satisfies the Hartree–Fock equations

$$\int h_{\alpha\beta}(\mathbf{x},\mathbf{x}')\,\chi_{j\beta}(\mathbf{x}')\,\mathrm{d}^3x' = \varepsilon_j\chi_{j\alpha}(\mathbf{x}), \tag{3}$$

$$h_{\alpha\beta}(\mathbf{x},\mathbf{x}') = \delta_{\alpha\beta}\delta(\mathbf{x}-\mathbf{x}')\left\{-\frac{\hbar^2}{2m_0}\nabla^2 - \sum_{\mathbf{n},k}\frac{Z_k e^2}{|\mathbf{R}_{\mathbf{n},k}-\mathbf{x}|} + \frac{e^2}{2}\int\frac{g_{\beta\beta}(\mathbf{y},\mathbf{y})}{|\mathbf{x}-\mathbf{y}|}\,\mathrm{d}^3y\right\} - e^2\,\frac{g_{\alpha\beta}(\mathbf{x},\mathbf{x}')}{|\mathbf{x}-\mathbf{x}'|}, \tag{4}$$

where

$$g_{\alpha\beta}(\mathbf{x},\mathbf{x}') = \sum_{j\leqslant j_0}\chi_{j\alpha}(\mathbf{x})\,\chi_{j\beta}^*(\mathbf{x}').$$

The function $g_{\alpha\beta}$ is the operator of projection onto the subspace of filled electron states and coincides with the limit value of the electron Green's function

$$G_{\alpha\beta}^{(0)}(\mathbf{x}t;\mathbf{x}'t') = -\frac{\mathrm{i}}{\hbar}\langle(T\psi_\alpha^{(0)}(\mathbf{x}t)\,\psi_\beta^{+(0)}(\mathbf{x}'t'))\rangle_0$$

as $t' \to t-0$. Here, as usual, $\langle T...\rangle_0$ denotes the ground-state average of the chronologically ordered operator product and $\psi_\alpha^{(0)}(\mathbf{x}t)$ is the electron Fermi field operator in the interaction representation.

The exciton states of interest to us are described by the two-particle two-time Green's function

$$G_{\alpha\beta,\gamma\delta}^{(2)}(\mathbf{x},\mathbf{y},t;\mathbf{x}',\mathbf{y}',t') = -\frac{\mathrm{i}}{\hbar}\langle T\psi_\alpha^+(\mathbf{x}t)\,\psi_\beta(\mathbf{y}t)\,\psi_\gamma^+(\mathbf{x}'t')\,\psi_\delta(\mathbf{y}'t')\rangle_0,$$

where $\psi_\alpha(\mathbf{x}t)$ are Heisenberg operators.

Introducing the full set of excited states $|J\mathbf{P}\rangle$ (where $\mathbf{P}$ is the total quasimomentum and J is the set of all other quantum numbers) and their corresponding energy levels $E_{J\mathbf{P}}$, we can write the function $G^{(2)}$ in the form

$$\mathrm{i}\hbar G_{\alpha\beta,\gamma\delta}^{(2)}(\mathbf{x},\mathbf{y},t;\mathbf{x}',\mathbf{y}',t') = \sum_{\mathbf{P},J}\left\{\varphi_{\alpha\beta}^{\mathbf{P}J}(\mathbf{x},\mathbf{y})\varphi_{\gamma\delta}^{\mathbf{P}J*}(\mathbf{x}',\mathbf{y}') \times \exp\left[\frac{\mathrm{i}}{\hbar}\left(\mathbf{P}\,\frac{\mathbf{x}+\mathbf{y}-\mathbf{x}'-\mathbf{y}'}{2} - E_{J\mathbf{P}}(t-t')\right)\right]\right\} \quad \text{at } t>t',$$

$$\mathrm{i}\hbar G_{\alpha\beta,\gamma\delta}^{(2)}(\mathbf{x},\mathbf{y},t;\mathbf{x}',\mathbf{y}',t') = \sum_{\mathbf{P},J}\left\{\varphi_{\alpha\beta}^{\mathbf{P}J*}(\mathbf{x},\mathbf{y})\varphi_{\gamma\delta}^{\mathbf{P}J}(\mathbf{x}',\mathbf{y}') \times \exp\left[-\frac{\mathrm{i}}{\hbar}\left(\mathbf{P}\,\frac{\mathbf{x}+\mathbf{y}-\mathbf{x}'-\mathbf{y}'}{2} - E_{J\mathbf{P}}(t-t')\right)\right]\right\} \quad \text{at } t<t',$$

where

$$\exp\left(\frac{\mathrm{i}}{2\hbar}\mathbf{P}(\mathbf{x}+\mathbf{y})\right)\varphi_{\alpha\beta}^{\mathbf{P}J}(\mathbf{x},\mathbf{y}) = \langle 0|\psi_\alpha^+(\mathbf{x})\,\psi_\beta(\mathbf{y})|J\mathbf{P}\rangle. \tag{5}$$

Because of the translation symmetry of the problem,

$$\varphi_{\alpha\beta}^{\mathbf{P}J}(\mathbf{x}+\mathbf{R}_\mathbf{n},\mathbf{y}+\mathbf{R}_\mathbf{n}) = \varphi_{\alpha\beta}^{\mathbf{P}J}(\mathbf{x},\mathbf{y}), \tag{6}$$

where $\mathbf{R}_\mathbf{n}$ is an arbitrary lattice vector. Passing to the Fourier time representation of the quasimomentum, we obtain

$$G_{\alpha\beta,\gamma\delta}^{(2)}(\mathbf{x},\mathbf{y};\mathbf{x}',\mathbf{y}';\mathbf{P}E) = \sum_J \frac{2E_{J\mathbf{P}}}{E^2-(E_{J\mathbf{P}}-\mathrm{i}\delta)^2}\,\varphi_{\alpha\beta}^{J\mathbf{P}}(\mathbf{x},\mathbf{y})\varphi_{\gamma\delta}^{J\mathbf{P}*}(\mathbf{x}',\mathbf{y}'), \quad \delta\to+0. \tag{7}$$

The discrete values of $E_{J\mathbf{P}}$ (at fixed $\mathbf{P}$), i.e., the poles of Eqn (7), correspond to excitons, and the corresponding function $\varphi^{\mathbf{P}J}_{\alpha\beta}(\mathbf{x},\mathbf{y})\exp[(\mathrm{i}/2\hbar)\mathbf{P}(\mathbf{x}+\mathbf{y})]$ defined by Eqn (5) can be viewed as the exciton wave function, with the respective variables $(\mathbf{x},\alpha)$ and $(\mathbf{y},\beta)$ referring to the electron and the hole. Hence, the only possible definition for the exciton creation and annihilation operators $J\mathbf{P}$ is apparently

$$B^+_{J\mathbf{P}} = \frac{1}{\sqrt{V}}\int \exp\left(\frac{\mathrm{i}}{2\hbar}\,\mathbf{P}(\mathbf{x}+\mathbf{y})\right) \times \psi^+_\alpha(\mathbf{x})\,\varphi^{J\mathbf{P}}_{\alpha\beta}(\mathbf{x},\mathbf{y})\,\psi_\beta(\mathbf{y})\,\mathrm{d}^3x\,\mathrm{d}^3y\,, \tag{8}$$

$$B_{J\mathbf{P}} = \frac{1}{\sqrt{V}}\int \exp\left(-\frac{\mathrm{i}}{2\hbar}\,\mathbf{P}(\mathbf{x}+\mathbf{y})\right) \times \psi^+_\alpha(\mathbf{x})\,\varphi^{J\mathbf{P}+}_{\alpha\beta}(\mathbf{x},\mathbf{y})\,\psi_\beta(\mathbf{y})\,\mathrm{d}^3x\,\mathrm{d}^3y\,, \tag{9}$$

where $\varphi^{J\mathbf{P}+}_{\alpha\beta}(\mathbf{x},\mathbf{y}) = [\varphi^{J\mathbf{P}}_{\beta\alpha}(\mathbf{y},\mathbf{x})]^*$ and V is the normalization volume. By writing the commutator of these operators,

$$[B_{J\mathbf{P}}, B^+_{J'\mathbf{P}'}] = \frac{1}{V}\int \psi^+_\alpha(\mathbf{x})\left\{\exp\left(-\frac{\mathrm{i}}{2\hbar}\,\mathbf{P}'\mathbf{x}\right)\varphi^{J'\mathbf{P}'+}_{\alpha\gamma}(\mathbf{x},\mathbf{z}) \times \exp\left(-\frac{\mathrm{i}}{2\hbar}(\mathbf{P}-\mathbf{P}')\mathbf{z}\right)\varphi^{J\mathbf{P}}_{\gamma\beta}(\mathbf{z},\mathbf{y})\exp\left(\frac{\mathrm{i}}{2\hbar}\,\mathbf{P}\mathbf{y}\right) - \exp\left(\frac{\mathrm{i}}{2\hbar}\,\mathbf{P}\mathbf{x}\right)\varphi^{J\mathbf{P}}_{\alpha\gamma}(\mathbf{x},\mathbf{z})\exp\left(\frac{\mathrm{i}}{2\hbar}(\mathbf{P}-\mathbf{P}')\mathbf{z}\right) \times \varphi^{J'\mathbf{P}'+}_{\gamma\beta}(\mathbf{z},\mathbf{y})\exp\left(-\frac{\mathrm{i}}{2\hbar}\,\mathbf{P}'\mathbf{y}\right)\right\}\psi_\beta(\mathbf{y})\,\mathrm{d}^3x\,\mathrm{d}^3y\,\mathrm{d}^3z\,, \tag{10}$$

it is easy to see that they are not at all of the Bose type in general. The situation is simplified in the mean-field approximation, however, where a complete orthogonal set of one-electron states exists. In this case, the exciton wave function can be presented as a superposition of the products of electron and hole states,

$$\varphi^{J\mathbf{P}}_{\alpha\beta}(\mathbf{x},\mathbf{y}) = \sum_{l>l_0,\,l'\leqslant l_0,\,\mathbf{p}} u_{\mathbf{p}l\gamma}(\mathbf{x})(\varphi^{J\mathbf{P}}_{\alpha\beta})^{\gamma\delta}_{ll'}u^*_{\mathbf{p}-\mathbf{P}l'\delta}(\mathbf{y})\,. \tag{11}$$

The products $\varphi\varphi^+$ and $\varphi^+\varphi$ are then the projection operators onto the mutually orthogonal subspaces of electron and hole states, and therefore, decomposing the operators ψ in Eqn (10) into the electron and hole parts and using the orthonormalization condition for the $\varphi^{J\mathbf{P}}$,

$$\frac{1}{V}\int \varphi^{J\mathbf{P}+}_{\alpha\beta}(\mathbf{x},\mathbf{y})\,\varphi^{J'\mathbf{P}}_{\beta\alpha}(\mathbf{y},\mathbf{x})\,\mathrm{d}^3x\,\mathrm{d}^3y = \delta_{JJ'}\,, \tag{12}$$

we can transform Eqn (10) to the form

$$[B_{J\mathbf{P}}, B^+_{J'\mathbf{P}'}] = \delta_{JJ'}\delta_{\mathbf{P}\mathbf{P}'} - \frac{1}{V}\int\left\{\psi^{(\mathrm{e})+}_\alpha(\mathbf{x})\exp\left(\frac{\mathrm{i}}{2\hbar}\,\mathbf{P}\mathbf{x}\right) \times \varphi^{J\mathbf{P}}_{\alpha\gamma}(\mathbf{x},\mathbf{z})\exp\left(\frac{\mathrm{i}}{2\hbar}(\mathbf{P}-\mathbf{P}')\mathbf{z}\right)\varphi^{J'\mathbf{P}'+}_{\gamma\beta}(\mathbf{z},\mathbf{y}) \times \exp\left(-\frac{\mathrm{i}}{2}\,\mathbf{P}'\mathbf{y}\right)\psi^{(\mathrm{e})}_\beta(\mathbf{y}) + \psi^{(\mathrm{h})+}_\alpha(\mathbf{x})\exp\left(-\frac{\mathrm{i}}{2\hbar}\,\mathbf{P}'\mathbf{y}\right) \times \varphi^{J'\mathbf{P}'+}_{\beta\gamma}(\mathbf{y},\mathbf{z})\exp\left(\frac{\mathrm{i}}{2\hbar}(\mathbf{P}-\mathbf{P}')\mathbf{z}\right)\varphi^{J\mathbf{P}}_{\gamma\alpha}(\mathbf{z},\mathbf{x}) \times \exp\left(\frac{\mathrm{i}}{2\hbar}\,\mathbf{P}\mathbf{x}\right)\psi^{(\mathrm{h})}_\beta(\mathbf{y})\right\}\mathrm{d}^3x\,\mathrm{d}^3y\,\mathrm{d}^3z\,. \tag{13}$$

It follows from this equation that the operators B are close to the Bose type for weakly excited states of the system, their commutation relations deviating from those of Bose operators by a quantity of the order of $n_e a^3$, where n_e is the density of electron excitations and a is the effective exciton radius, which is determined by the way $\varphi(\mathbf{x},\mathbf{y})$ decays at large $|\mathbf{x}-\mathbf{y}|$: $\varphi(\mathbf{x},\mathbf{y}) \lesssim \mathrm{const}\times\exp[-|\mathbf{x}-\mathbf{y}|/a]$ at $|\mathbf{x}-\mathbf{y}| \gg a$.

If excitons were pure bosons and their interaction could be ignored, the coherent exciton states could be defined in the usual way,

$$|\beta, J\mathbf{P}\rangle = \exp\left[\beta B^+_{J\mathbf{P}}\exp\left(\frac{\mathrm{i}}{\hbar}\,E_{J\mathbf{P}}t\right) - \beta^* B_{J\mathbf{P}}\exp\left(-\frac{\mathrm{i}}{\hbar}\,E_{J\mathbf{P}}t\right)\right]|0\rangle\,. \tag{14}$$

However, including deviations of the exciton statistics from the Bose statistics together with the interaction between excitons leads to the fact that both the form of the operator $B_{J\mathbf{P}}$ and the energy value $E_{J\mathbf{P}}$ change as the exciton wave amplitude β is varied. The coherent exciton states must therefore be defined in a more general way:

$$|\varphi\rangle = \exp\left\{\int\left[\psi^+_\alpha(\mathbf{x})\varphi_{\alpha\beta}(\mathbf{x},\mathbf{y})\psi_\beta(\mathbf{y})\exp\left(\frac{\mathrm{i}}{\hbar}\left(\mathbf{P}\frac{\mathbf{x}+\mathbf{y}}{2}-\mu t\right)\right) - \psi^+_\alpha(\mathbf{x})\varphi^+_{\alpha\beta}(\mathbf{x},\mathbf{y})\psi_\beta(\mathbf{y})\exp\left(-\frac{\mathrm{i}}{\hbar}\left(\mathbf{P}\frac{\mathbf{x}+\mathbf{y}}{2}-\mu t\right)\right)\right]\mathrm{d}^3x\,\mathrm{d}^3y\right\}|0\rangle\,. \tag{15}$$

The function $\varphi_{\alpha\beta}$ entering this definition does not coincide with any of the $\varphi^{J\mathbf{P}}_{\alpha\beta}$ but tends to one of them in the low-density limit, and its exact form (as well as the value of μ) must be determined from the Schrödinger equation

$$\left(\mathrm{i}\hbar\,\frac{\partial}{\partial t} - H\right)|\varphi\rangle = 0\,.$$

Letting D_φ denote the operator in the right-hand side of Eqn (15), we transform it to the form

$$D_\varphi D^+_\varphi\left(\mathrm{i}\hbar\,\frac{\partial}{\partial t} - H\right)D_\varphi|0\rangle = 0\,,$$

which is equivalent to

$$\left(\mathrm{i}\hbar D^+_\varphi\,\frac{\partial D_\varphi}{\partial t} - \tilde{H}\right)|0\rangle = 0\,, \tag{16}$$

where $\tilde{H} = D^+_\varphi H D_\varphi$.

Equation (16) cannot be satisfied rigorously by any choice of the function $\varphi_{\alpha\beta}(\mathbf{x},\mathbf{y})$ because the form of function (15) does not take multiparticle correlation effects into account and corresponds in its meaning to describing the state of the system in the mean-field approximation. This situation is not specific to excitons. For any system of interacting bosons, it is only in the mean-field approximation that Eqn (14) determines the coherent states. All the correlation corrections can be calculated by a diagram technique for strongly non-equilibrium states, as discussed in Ref. [18]. In this paper, we confine ourselves, as already mentioned, to the lowest (mean-field) approximation. The function $\varphi_{\alpha\beta}(\mathbf{x},\mathbf{y})$ then has a structure similar to that in Eqn (11), i.e., its expansion in the set of functions $\chi_{j\alpha}(\mathbf{x})$ contains only electron states for the variable $\mathbf{x}$ and only hole states for $\mathbf{y}$. Therefore, the operator

D_φ can be rewritten in the form

$$D_\varphi = \exp\left\{\iint\left[\psi_\alpha^{(e)+}(\mathbf{x})\varphi_{\alpha\beta}(\mathbf{x},\mathbf{y})\right.\right.$$
$$\times \exp\left(\frac{\mathrm{i}}{\hbar}\left(\mathbf{P}\frac{\mathbf{x}+\mathbf{y}}{2} - \mu t\right)\right)\psi_\beta^{(h)+}(\mathbf{y}) - \psi_\beta^{(h)}(\mathbf{y})\varphi_{\alpha\beta}^*(\mathbf{x},\mathbf{y})$$
$$\left.\left.\times \exp\left(-\frac{\mathrm{i}}{\hbar}\left(\mathbf{P}\frac{\mathbf{x}+\mathbf{y}}{2} - \mu t\right)\right)\psi_\alpha^{(e)}(\mathbf{x})\right]\mathrm{d}^3x\,\mathrm{d}^3y\right\}. \quad (17)$$

As is known, this unitary operator performs a linear transformation of the operators $\psi^{(e)}$ and $\psi^{(h)}$. Physically, this means that the partial redistribution of electrons due to the creation of a large number of excitons results in a change in the concept of a hole (the subspace of filled states) in the system, and state (15) is the vacuum state for the operators redefined in this way. The formulas for passing to the new operators are

$$\psi_\alpha^{(e)}(\mathbf{x}) \to D_\varphi^+\psi_\alpha^{(e)}(\mathbf{x})D_\varphi = \int\left\{C_{\alpha\beta}(\mathbf{x},\mathbf{y})\psi_\alpha^{(e)}(\mathbf{y})\right.$$
$$\left.+\exp\left(\frac{\mathrm{i}}{\hbar}\left(\mathbf{P}\frac{\mathbf{x}+\mathbf{y}}{2} - \mu t\right)\right)S_{\alpha\beta}(\mathbf{x},\mathbf{y})\psi_\beta^{(h)+}(\mathbf{y})\right\}\mathrm{d}^3y\,,$$

$$\psi_\alpha^{(h)}(\mathbf{x}) \to D_\varphi^+\psi_\alpha^{(h)}(\mathbf{x})D_\varphi = \int\left\{\tilde{C}_{\alpha\beta}(\mathbf{x},\mathbf{y})\psi_\alpha^{(h)}(\mathbf{y})\right.$$
$$\left.-\exp\left(\frac{\mathrm{i}}{\hbar}\left(\mathbf{P}\frac{\mathbf{x}+\mathbf{y}}{2} - \mu t\right)\right)\psi_\beta^{(e)+}(\mathbf{y})S_{\beta\alpha}(\mathbf{y},\mathbf{x})\right\}\mathrm{d}^3y\,, \quad (18)$$

where

$$C_{\alpha\beta}(\mathbf{x},\mathbf{y}) = \delta_{\alpha\beta}\delta(\mathbf{x}-\mathbf{y}) + \sum_{n=1}^{\infty}\frac{(-1)^n}{(2n)!}(\varphi\varphi^+)^n\,,$$
$$S_{\alpha\beta}(\mathbf{x},\mathbf{y}) = \varphi_{\alpha\beta}(\mathbf{x},\mathbf{y}) + \sum_{n=1}^{\infty}\frac{(-1)^n}{(2n+1)!}\varphi(\varphi^+\varphi)^n\,. \quad (19)$$

The product of the operators φ and φ^+ in Eqns (19) is understood in the sense of integral convolution. With Eqns (18) and (19), Eqn (16) becomes

$$0 = \iint\left\{\psi_\alpha^{(e)+}(\mathbf{x})\tilde{h}_{\alpha\beta}^{(e)}(\mathbf{x},\mathbf{y})\psi_\beta^{(e)}(\mathbf{y}) + \psi_\alpha^{(h)+}(\mathbf{x})\tilde{h}_{\alpha\beta}^{(h)}(\mathbf{x},\mathbf{y})\psi_\beta^{(h)}(\mathbf{y})\right.$$
$$+\psi_\alpha^{(e)+}(\mathbf{x})Q_{\alpha\beta}(\mathbf{x},\mathbf{y})\psi_\beta^{(h)+}(\mathbf{y})\exp\left(-\frac{\mathrm{i}}{\hbar}\left(\mathbf{P}\frac{\mathbf{x}+\mathbf{y}}{2} - \mu t\right)\right)$$
$$+\psi_\beta^{(h)}(\mathbf{y})Q_{\alpha\beta}^*(\mathbf{x},\mathbf{y})\psi_\alpha^{(e)}(\mathbf{x})\exp\left(\frac{\mathrm{i}}{\hbar}\left(\mathbf{P}\frac{\mathbf{x}+\mathbf{y}}{2} - \mu t\right)\right)$$
$$\left.+\frac{e^2}{2}\frac{N[\psi_\alpha^+(\mathbf{x})\psi_\beta^+(\mathbf{y})\psi_\beta(\mathbf{y})\psi_\alpha(\mathbf{x})]}{|\mathbf{x}-\mathbf{y}|}\right\}\mathrm{d}^3x\,\mathrm{d}^3y\,|0\rangle\,, \quad (20)$$

where $N[\ldots]$ is the normally ordered product of the operators $\psi = \psi^{(e)} + \psi^{(h)+}$. Although the electron–electron interaction term looks formally the same as before the transformation, it is different because the operators $\psi^{(e)}$ and $\psi^{(h)}$ and the function $G_{\alpha\beta}^{(0)}$ are different.

In the mean-field approximation, Eqn (20) takes the form

$$\{\psi^{(e)+}\tilde{h}^{(e)}\psi^{(e)} + \psi^{(h)+}\tilde{h}^{(h)}\psi^{(h)}$$
$$+\psi^{(e)+}Q\psi^{(h)+} + \psi^{(h)}Q^+\psi^{(e)}\}|0\rangle = 0 \quad (21)$$

(we again use a symbolic expression where products are understood as integral convolutions). Here, as in Eqn (20), the following notation is used:

$$\tilde{h}^{(e)} = C(h^{(e)} - v)C - S(h^{(h)} - v^+)S^+$$
$$+ CVS^+ + SV^+C - \mu S^+S\,, \quad (22)$$
$$Q = C(h^{(e)} - v)S + S(h^{(h)} + v^+)\tilde{C}$$
$$- CV\tilde{C} + SV^+S - \mu CS\,. \quad (23)$$

The matrix $\tilde{C}$ differs from C by the permutation $\varphi \rightleftarrows \varphi^+$, and the quantities V and v are defined as

$$V_{\alpha\beta}(\mathbf{x},\mathbf{y}) = \frac{e^2}{|\mathbf{x}-\mathbf{y}|}\int S_{\alpha\gamma}(\mathbf{x},\mathbf{z})\,\tilde{C}_{\gamma\beta}(\mathbf{z},\mathbf{y})\,\mathrm{d}^3z\,, \quad (24)$$

$$v_{\alpha\beta}(\mathbf{x},\mathbf{y}) = \frac{e^2}{|\mathbf{x}-\mathbf{y}|}\int S_{\alpha\gamma}(\mathbf{x},\mathbf{z})S_{\gamma\beta}^+(\mathbf{z},\mathbf{y})\,\mathrm{d}^3z\,. \quad (25)$$

We also note the relations $CC + SS^+ = 1$ and $CS = S\tilde{C}$ and the fact that the quantities

$$n_e(\mathbf{x}) = \int S_{\alpha\beta}(\mathbf{x},\mathbf{y})\,S_{\beta\alpha}^+(\mathbf{y},\mathbf{x})\,\mathrm{d}^3y\,,$$
$$n_h(\mathbf{x}) = \int S_{\alpha\beta}^+(\mathbf{x},\mathbf{y})\,S_{\beta\alpha}(\mathbf{y},\mathbf{x})\,\mathrm{d}^3y \quad (26)$$

determine the densities of excited electrons and holes, which are periodic with the crystal lattice period for the class of states considered so far.

The necessary and sufficient condition for Eqn (21) to hold is clearly $Q \equiv 0$, because the first two terms in the left-hand side give zero when acting on the vacuum. Thus, Eqn (21) reduces to

$$Q_{\alpha\beta}(\mathbf{x},\mathbf{y}) = 0\,, \quad (27)$$

which, by Eqns (19) and (23)–(25), is a nonlinear integro-differential equation for $\varphi_{\alpha\beta}(\mathbf{x},\mathbf{y})$, and the exciton chemical potential μ is defined as an eigenvalue of Eqn (27).

The analysis [2] of solutions of Eqn (27) in the general case is hardly possible, and we therefore confine ourselves to relatively low excitation densities $n_{e,h}a^3 \ll 1$, for which Eqn (27) can be expanded in powers of φ using the relations

$$C \simeq 1 - \frac{1}{2}\varphi\varphi^+\,, \quad \tilde{C} \simeq 1 - \frac{1}{2}\tilde{\varphi}^+\varphi\,, \quad S \simeq \varphi - \frac{1}{6}\varphi\varphi^+\varphi\,.$$

The lowest approximation, linear in φ, yields

$$\int\left\{h_{\alpha\gamma}^{(e)}(\mathbf{x},\mathbf{z})\,\varphi_{\gamma\beta}(\mathbf{z},\mathbf{y}) + h_{\gamma\beta}^{(h)}(\mathbf{z},\mathbf{y})\,\varphi_{\alpha\gamma}(\mathbf{x},\mathbf{z})\right\}\mathrm{d}^3z$$
$$-\left(\frac{e^2}{|\mathbf{x}-\mathbf{y}|} + \mu\right)\varphi_{\alpha\beta}(\mathbf{x},\mathbf{y}) = 0\,, \quad (28)$$

which is the Schrödinger equation for Coulomb-interacting electrons and holes. The way they actually interact in a crystal is much more complicated, but it is easy to show that after the summation of all correlation corrections that are linear in φ, Eqn (28) becomes exactly an equation for the exciton wave functions $\varphi^{J\mathbf{P}}$. In what follows, we replace the index J by 0, having the lowest excitonic branch of the spectrum in mind.

Thus, in the lowest approximation,

$$\varphi = \sqrt{n}\,\varphi^{0\mathbf{P}}\,, \qquad \mu^{(0)} = E_{0\mathbf{P}}\,,$$

where the normalization factor $\sqrt{n}$ is determined by the mean exciton concentration. The solvability condition for the next-approximation equation ($\sim \varphi^3$) yields a level shift $\mu^{(1)}$ proportional to n:

$$\mu^{(1)} = \varkappa n\,,$$

$$\varkappa = e^2 \int \mathrm{d}^3x\,\mathrm{d}^3y \left\{ \int \left(\frac{1}{|\mathbf{x}-\mathbf{y}|} + \frac{1}{|\mathbf{x}-\mathbf{z}|} \right) \right.$$
$$\times\, \varphi_{\alpha\beta}(\mathbf{x},\mathbf{y})\varphi^+_{\beta\gamma}(\mathbf{y},\mathbf{z}')\,\varphi_{\gamma\delta}(\mathbf{z}',\mathbf{z})\,\varphi^+_{\delta\alpha}(\mathbf{z},\mathbf{x})\,\mathrm{d}^3z\,\mathrm{d}^3z'$$
$$- \sum_{\alpha,\beta} \frac{1}{|\mathbf{x}-\mathbf{y}|} \left(\left| \int \varphi_{\alpha\gamma}(\mathbf{x},\mathbf{z})\,\varphi^+_{\gamma\beta}(\mathbf{z},\mathbf{y})\,\mathrm{d}^3z \right|^2 \right.$$
$$\left.\left. + \left| \int \varphi^+_{\alpha\gamma}(\mathbf{x},\mathbf{z})\,\varphi_{\gamma\beta}(\mathbf{z},\mathbf{y})\,\mathrm{d}^3z \right|^2 \right) \right\}. \qquad (29)$$

The constant $\varkappa$ can also be calculated for the pure Coulomb case. Summing all the corrections for the constant is a more difficult problem [4], and it is more convenient to simply regard it as a phenomenological parameter.

Until now, we have been considering stationary coherent exciton states with a fixed total quasimomentum $\mathbf{P}$, i.e., with the same mean excitation density n for all elementary cells in the crystal. In a more general case, wave packets are composed of states of this type. To introduce such packets, we specify a transformation D_Φ of a more general type,

$$D_\Phi = \exp\left\{ \int \left[\psi_\alpha^{(\mathrm{e})+}(\mathbf{x})\,\Phi_{\alpha\beta}(\mathbf{x},\mathbf{y},t)\exp\left(\frac{\mathrm{i}}{\hbar}\mu t\right)\psi_\beta^{(\mathrm{h})+}(\mathbf{y}) \right.\right.$$
$$\left.\left. - \psi_\beta^{(\mathrm{h})}(\mathbf{y})\exp\left(-\frac{\mathrm{i}}{\hbar}\mu t\right)\Phi^+_{\alpha\beta}(\mathbf{x},\mathbf{y},t)\,\psi_\alpha^{(\mathrm{e})}(\mathbf{x}) \right] \mathrm{d}^3x\,\mathrm{d}^3y \right\},$$

where we assume that the dependence of $\Phi_{\alpha\beta}(\mathbf{x},\mathbf{y},t)$ on t is slower than $\exp((\mathrm{i}/\hbar)\mu t)$. Similarly to the discussion above, the equation for Φ then takes the form

$$\mathrm{i}\hbar\,\frac{\partial\Phi}{\partial t} - Q\{\Phi\} = 0\,.$$

For this equation, we seek a solution of the form $\Phi(\mathbf{x},\mathbf{y},t) = \Phi(\mathbf{x}+\mathbf{y},t)\varphi^{0\mathbf{P}}(\mathbf{x},\mathbf{y})$ assuming that $|a\nabla\Phi(\mathbf{X},t)/\Phi| \ll 1$, i.e., considering $\Phi(\mathbf{X},t)$ as a slowly varying amplitude. We set $\mathbf{P}$ equal to $\mathbf{P}_\mathrm{m}$, a value for which $E_{0\mathbf{P}}$ reaches a minimum; in its neighborhood, $E_{0\mathbf{P}} \approx E_0 + (\mathbf{P}-\mathbf{P}_\mathrm{m})^2/2m$. Then, in the lowest approximation, disregarding nonlinear terms in the derivatives of $\Phi(\mathbf{X},t)$, we obtain $\mu = E_0$, and the next approximation yields an equation for $\Phi(\mathbf{X},t)$,

$$\mathrm{i}\hbar\,\frac{\partial\Phi}{\partial t} + \frac{\hbar^2}{2m}\nabla^2\Phi - \varkappa|\Phi|^2\Phi = 0\,, \qquad (30)$$

equivalent, as is known, to the phenomenological hydrodynamics equations for a superfluid liquid (for $\varkappa > 0$).

The discussion above has been concerned with excitons that have virtually no interaction with light. The condensation of dipole-active excitons should automatically create an electromagnetic field accompanying them, also in a coherent state. Writing the full Hamiltonian for the particle–field system, we readily obtain the following system of equations for the coherent-state amplitudes:

$$\mathrm{i}\hbar\,\frac{\partial\Phi}{\partial t} + \frac{\hbar^2}{2m}\nabla^2\Phi + (\hbar\omega - E_0)\Phi - \varkappa|\Phi|^2\Phi = \mathbf{E}\mathbf{d}\,,$$
$$\mathrm{rot}\,(\mathrm{rot}\,\mathbf{E}) + \frac{\tilde{\varepsilon}}{c^2}\left(\frac{\partial}{\partial t} - \mathrm{i}\omega\right)^2\mathbf{E} = -\frac{4\pi}{c^2}\,\mathbf{d}\left(\frac{\partial}{\partial t} - \mathrm{i}\omega\right)^2\Phi\,. \qquad (31)$$

Here, ω is the mean field frequency, $\mathbf{d}$ is the dipole moment matrix element for an exciton transition, $\tilde{\varepsilon}$ is the crystal dielectric constant with the contribution from the exciton state under consideration subtracted, and $\mathbf{E}(x,t)$ is the complex field amplitude in terms of which the real field is expressed as

$$\frac{1}{2}\left(\mathbf{E}(x,t)\exp(-\mathrm{i}\omega t) + \mathbf{E}^*(x,t)\exp(\mathrm{i}\omega t)\right).$$

System of equations (31) incorporates the effects of frequency and spatial dispersion and those of nonlinear polarizability.

In the case of high-symmetry crystals, exciton states can be degenerate, requiring several functions Φ to be introduced for their description. While this complicates system (31), the qualitative results remain unchanged.

Note from the Editors

L V Keldysh's paper "Coherent states of excitons" presented in this memorial issue of *Physics–Uspekhi* was first published in 1972 in the Tamm memorial collection [19] and went relatively unnoticed by the physics community. The paper is of quite fundamental importance, however.

In the latter half of the 1960s and the early 1970s, there was an interesting discussion in the literature as to whether Bose condensation and superfluidity are possible in a system of excitons in a semiconductor. Within quite a short time, the discussion grew quite confused, and it was the objective of Keldysh's work to clarify a number of fundamental issues that arose. The paper emphasizes that there are three fundamental problems that should be recognized in the area. One of these is directly related to the rearrangement of the electron spectrum of a semimetal due to the Bose condensation of electron–hole pairs in the ground (equilibrium) state in the Keldysh–Kopaev model of an exciton dielectric [6]. The second problem is related to the possibility of Bose condensation and superfluidity in a nonequilibrium exciton gas produced by the optical pumping of a semiconductor [1, 4]. Finally, the third problem is about the formation (condensation) of electron–hole droplets in highly excited semiconductors as excitons break up to form a sufficiently dense electron–hole Fermi-liquid phase [20]. Experimentally, most real (multi-valley) semiconductors exhibit precisely this last scenario (it is this scenario which Keldysh predicted in his concluding speech at the Ninth International Conference on the Physics of Semiconductors held in Moscow in 1968 [20]).

At the same time, the direct analogy existing between the excitonic insulator model and BCS superconductivity led to contradictory opinions among researchers, some claiming that this model allows a superfliud electron–hole pair condensate (which manifests itself in superconductivity-type phenomena) [8], and others fully ruling out that superfluidity can occur in a system of excitons [11]. It is to explain some of these contradictions that Ref. [19] is now reproduced in *Physics–Uspekhi*.

Leonid Veniaminovich Keldysh in the late 1960s
(which was exactly the period of his active excitonic studies).
(Courtesy of Galina Nikolaevna Mikhailova, co-author of Ref. [23].)

This study is mainly concerned with whether superfluidity can be exhibited by *nonequilibrium* excitons that are produced in a semiconductor exposed to an external excitation source. It is emphasized that moving excitons cannot produce either a flow of matter or an electric current but can imply the existence of undamped (up to the exciton lifetime) energy or polarization flows, but not a superfluid mass or charge transfer, which immediately invalidates the 'general' proof of Kohn and Sherringtone, who additionally, at the very outset, assumed a thermodynamic equilibrium (i.e., a system of the excitonic insulator type).

Most of the paper focuses on the explicit construction of coherent excitonic states taking their generally nonbosonic nature into account. The result, Eqn (30), is directly analogous to the Gross–Pitaevskii equation for a superfluid liquid, with the dipole interaction with an external electromagnetic field (31) taken into account.

As regards the possibility of superfluidity in an excitonic insulator, this debate was completely resolved in a later paper by Keldysh and Guseinov [21], in which it was shown that allowing interband transitions in this model turns a second-order transition into a first-order one, and therefore the excitonic insulator state has no properties distinguishing it from the usual dielectrics.

The author of these lines was at the time Keldysh's postgraduate student at the FIAN (Lebedev Physical Institute) Theory department, named after I E Tamm shortly before. The publication of the memorial collection [19] was a major event in the life of the department. An impressive feature of the collection was the list of referenced authors, which included prominent Soviet and foreign scientists. Some of the department staff used to come to the seminars with a copy of this volume to ask its authors for an autograph. False modesty prevented me from doing this — and quite regrettably, because the autographs made those copies uniquely valuable.

Incidentally, my postgraduate research (electrons in disordered systems) had no relation to the condensation of excitons, which was then Keldysh's primary concern, and therefore my role here is simply that of an unprejudiced witness. Keldysh's prediction of electron–hole droplets in Ref. [20] was followed by a rather long break in his publications in this area. The first sufficiently detailed account of the theoretical foundations of this concept also appeared in a relatively hard-to-access paper collection [22] in 1971. Over a number of years, Keldysh's interests were centered on the experimental confirmation of this phenomenon, sometimes to the extent of coauthoring experimental studies [23]. As is known, the general picture of the formation of electron–hole droplets he gave in Ref. [20] received striking experimental confirmation, and experimental and theoretical research in this field has intensified worldwide [24].

Paper [19] stands alone in this sense, and I can offer some conjectures as to its origin. All of Keldysh's students knew about his large notebooks into which, when at home, he wrote down his calculations on a wide range of solid-state physics problems and where he described his results in detail, often without later publishing them as journal papers.

For example, E G Maksimov told me that in those years he was actively involved in attempts at constructing a consistent theory of electron–phonon interaction in metals, to extend and improve the traditional Fröhlich Hamiltonian approach by correctly using the adiabatic approximation and introducing multiparticle effects. In Maksimov's words, Keldysh also devoted much attention to these problems and occasionally showed his results to Maksimov, but did not publish anything at all.

In my view, Ref. [19] appeared as a reply to Kohn and Sherrington's paper [13], which made some points Keldysh did not agree with. This led him to perform a number of 'private' calculations, as it were, that remained hidden for a number of years in his notebooks until the opportunity came to publish them in the Tamm memorial collection. I may be wrong, but all of us, his students, recall those notebooks quite often. It would be interesting to find them and examine them for interesting results, which they are almost certain to contain and which Keldysh did not manage to publish.

M V Sadovskii

References

1. Moskalenko S A *Sov. Phys. Solid State* **4** 199 (1962); *Fiz. Tverd. Tela* **4** 276 (1962)
2. Blatt J M, Böer K W, Brandt W *Phys. Rev.* **126** 1691 (1962)
3. Casella R C *J. Appl. Phys.* **34** 1703 (1963)
4. Keldysh L V, Kozlov A N *Sov. Phys. JETP* **27** 521 (1968); *Zh. Eksp. Teor. Fiz.* **54** 978 (1968)
5. Agranovich V M, Toshich B S *Sov. Phys. JETP* **26** 104 (1968); *Zh. Eksp. Teor. Fiz.* **53** 149 (1967)
6. Keldysh L V, Kopaev Yu V *Sov. Phys. Solid State* **6** 2219 (1965); *Fiz. Tverd. Tela* **6** 2791 (1964)
7. Des Cloizeaux J *J. Phys. Chem. Solids* **26** 259 (1965)

8. Kozlov A N, Maksimov L A *Sov. Phys. JETP* **21** 790 (1965); *Zh. Eksp. Teor. Fiz.* **48** 1184 (1965); *Sov. Phys. JETP* **22** 889 (1966); *Zh. Eksp. Teor. Fiz.* **49** 1284 (1965); *Sov. Phys. JETP* **23** 88 (1966); *Zh. Eksp. Teor. Fiz.* **50** 131 (1966)
9. Kopaev Yu V *Sov. Phys. Solid State* **8** 175 (1966); *Fiz. Tverd. Tela* **8** 223 (1966)
10. Jérome D, Rice T M, Kohn W *Phys. Rev.* **158** 462 (1967)
11. Kohn W *Phys. Rev. Lett.* **19** 439 (1967)
12. Halperin B I, Rice T M *Rev. Mod. Phys.* **40** 755 (1968)
13. Kohn W, Sherrington D *Rev. Mod. Phys.* **42** 1 (1970)
14. Chesnut D B *J. Chem. Phys.* **41** 472 (1964)
15. Keldysh L V *Sov. Phys. Usp.* **13** 292 (1970); *Usp. Fiz. Nauk* **100** 514 (1970)
16. Knox R S *Theory of Excitons* (New York: Academic Press, 1963); Translated into Russian: *Teoriya Eksitonov* (Moscow: Mir, 1966)
17. Agranovich V M, Ginzburg V L *Crystal Optics with Spatial Dispersion, and Excitons* (Berlin: Springer-Verlag, 1984); Translated from Russian: *Kristallooptika s Uchetom Prostranstvennoi Dispersii i Teoriya Eksitonov* (Moscow: Nauka, 1965)
18. Keldysh L V *Sov. Phys. JETP* **20** 1018 (1965); *Zh. Eksp. Teor. Fiz.* **47** 1515 (1964)

Additional reading

19. Keldysh L V "Kogerentnye sostoyaniya eksitonov" ("Coherent states of excitons"), in *Problemy Teoreticheskoi Fiziki. Pamyati Igorya Evgen'evicha Tamma* (Problems of Theoretical Physics. In Memory of Igor Evgen'evich Tamm) (Ed. V I Ritus) (Moscow: Nauka, 1972) p. 433
20. Keldysh L V "Concluding remarks", in *Trudy IX Mezhdunarodnoi Konf. po Fizike Poluprovodnikov, Moskva, 23–29 Iyulya, 1968 g.* (Proc. of the IX Intern. Conf. on the Physics of Semiconductors, Moscow, July 23–29, 1968) Vol. 2 (Leningrad, Nauka, 1969) pp. 1303 – 1312
21. Guseinov R R, Keldysh L V *Sov. Phys. JETP* **36** 1193 (1973); *Zh. Eksp. Teor. Fiz.* **63** 2255 (1972)
22. Keldysh L V, in *Eksitony v Poluprovodnikakh* (Excitons in Semiconductors) (Ed. B M Vul) (Moscow: Nauka, 1971) p. 5
23. Keldysh L V, Manenkov A A, Milyaev V A, Mikhailova G N *Sov. Phys. JETP* **39** 1072 (1974); *Zh. Eksp. Teor. Fiz.* **66** 2178 (1974)
24. Tikhodeev S G *Sov. Phys. Usp.* **28** 1 (1985); *Usp. Fiz. Nauk* **145** 3 (1985)

SOVIET PHYSICS JETP VOLUME 36, NUMBER 6 JUNE, 1973

NATURE OF THE PHASE TRANSITION UNDER THE CONDITIONS OF AN "EXCITONIC" INSTABILITY IN THE ELECTRONIC SPECTRUM OF A CRYSTAL

R. R. GUSEĬNOV and L. V. KELDYSH

P. N. Lebedev Physics Institute, USSR Academy of Sciences

Submitted March 7, 1972

Zh. Eksp. Teor. Fiz. 63, 2255–2263 (December, 1972)

The effect of the interband matrix elements of the electrons' Coulomb interaction on the exciton spectrum and the nature of the phase transition into the "excitonic insulator" state are investigated in a model of a semiconductor with the following features: the maximum of the valence band is located at the center of the Brillouin zone, and the conduction band minima occur at the edges of the Brillouin zone; the model is also characterized by a narrow energy gap. It is shown that if the exciton binding energy is close enough to the width of the energy gap, the system becomes thermodynamically unstable and undergoes a first-order phase transition. The two-particle excitation (exciton) spectrum in the new phase is also separated from the ground state by an energy gap.

As has been shown by a number of authors,[1-5] under certain conditions the ground state in a crystal may become unstable with respect to the formation of bound electron-hole pairs of the Mott exciton type.

Such an instability arises when the exciton binding energy becomes greater than the width of the forbidden gap. In this connection the single-particle excitation spectrum becomes restructured, a new branch of elementary excitations appears—the two-particle excitations—and, in fact, a new phase appears, which has been called an "excitonic insulator."[4] It has been shown[3] that the transition into the "excitonic insulator" state is a second order phase transition.

These results are obtained by only taking the Coulomb interaction of the electrons and holes into consideration. In terms of diagrams the Coulomb interaction of the particles simply corresponds to their scattering on each other. However, there are also other matrix elements in the total Hamiltonian describing the Coulomb interaction of the electrons; these correspond to diagrams such as those describing the scattering of a particle with the creation of an electron-hole pair, the creation from the vacuum of two electron-hole pairs with total quasimomentum equal to zero, etc.

The object of the present work is to show that the inclusion of these additional terms, which are inevitably present in the interaction Hamiltonian, leads to substantial changes in the spectrum of the elementary excitations and modifies the nature of the phase transition into the "excitonic insulator" state, namely: a region of absolute thermodynamic instability appears and the phase transition becomes a first-order transition.

Earlier we investigated[6] the simplest model having two identical parabolic bands with extrema at the point $k = 0$. It was demonstrated for this model that, when the exciton binding energy becomes larger than the width of the forbidden gap, the two-particle excitation spectrum which appears has, in general, a non-acoustic nature for small momenta, and the phase transition is of first order.

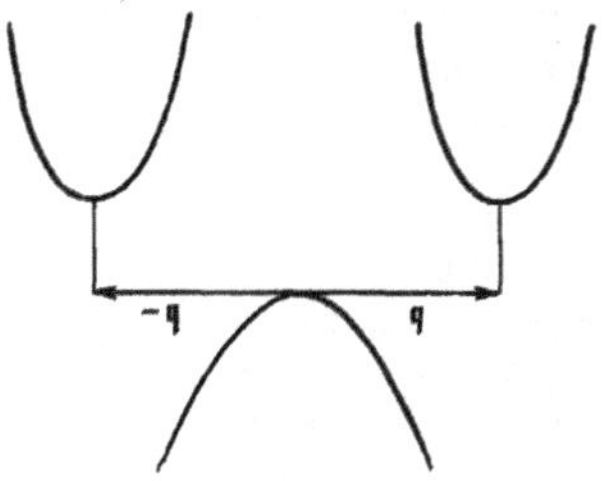

FIG. 1

In the present article we consider a more realistic model with two symmetric minima in the conduction band and a single maximum in the valence band (see Fig. 1). The minima are assumed to be located at opposite boundaries of the Brillouin zone, so that $2\mathbf{q}$ is a reciprocal lattice vector. Since these two minima are physically equivalent, it is sufficient to investigate only one of them. For simplicity the spins of the particles are not taken into consideration, and the whole investigation is carried out for the case of zero temperature.

In the representation of second quantization the Hamiltonian of the system of electrons and holes has the form

$$H=\sum_{\mathbf{p}}[\varepsilon_e(\mathbf{p})a_{\mathbf{p}}^+a_{\mathbf{p}}+\varepsilon_h(\mathbf{p})b_{\mathbf{p}}^+b_{\mathbf{p}}]+\frac{1}{2V}\sum_{\mathbf{p},\mathbf{p}',\mathbf{k}}[V_{\mathbf{k}}(a_{\mathbf{p}}^+a_{\mathbf{p}'}^+a_{\mathbf{p}'+\mathbf{k}}a_{\mathbf{p}-\mathbf{k}}$$

$$+b_{\mathbf{p}}^+b_{\mathbf{p}'}^+b_{\mathbf{p}'+\mathbf{k}}b_{\mathbf{p}-\mathbf{k}}-2a_{\mathbf{p}}^+b_{\mathbf{p}'}^+b_{\mathbf{p}'+\mathbf{k}}a_{\mathbf{p}-\mathbf{k}})+(\tilde{V}_{\mathbf{k}}a_{\mathbf{p}}a_{\mathbf{p}'}b_{\mathbf{k}-\mathbf{p}'}b_{-\mathbf{k}-\mathbf{p}}+\text{h.c.})],$$

where $a_{\mathbf{p}}$ and $b_{\mathbf{p}}$ are fermion operators for the annihilation of electrons and holes; $V_{\mathbf{k}} = 4\pi e^2/\kappa k^2$, κ is the dielectric constant; the dispersion law for the electrons is given by $\epsilon_e(\mathbf{p}) = (1/2)\Delta + \hbar^2(\mathbf{p}+\mathbf{q})^2/2m_e$ and the dispersion law for the holes is given by $\epsilon_h(\mathbf{p}) = (1/2)\Delta + \hbar^2p^2/2m_h$, where Δ denotes the energy gap in the one-particle excitation spectrum. The form of $\tilde{V}_{\mathbf{k}}$ is not specified. It is only assumed that $\tilde{V}_{\mathbf{k}}$ is small and plays the role of a perturbation. In actual fact, we consider the case of a narrow forbidden band, and the excitons which appear are excitons of large radius. Since here all of the characteristic momenta are small

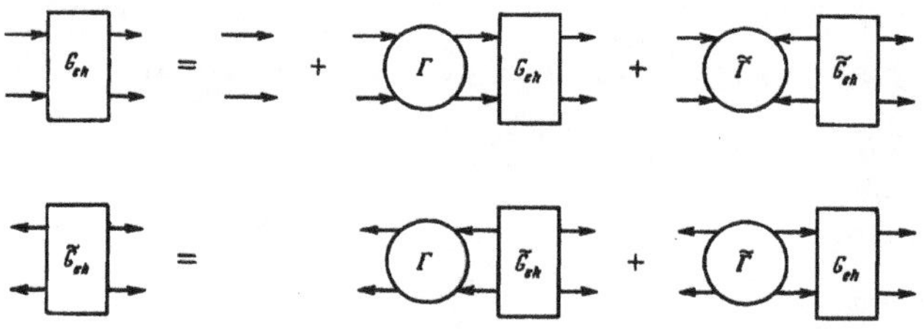

FIG. 2

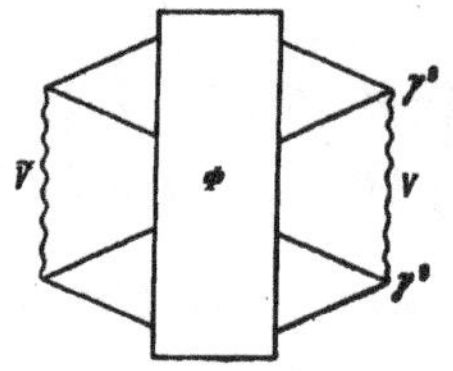

FIG. 3

(in comparison with a reciprocal lattice vector), the quantity $\tilde{V}_k$ turns out to be small due to the orthogonality of the wave functions belonging to different bands.

Not all of the interband matrix elements have been kept in the Hamiltonian (1). However, the inclusion of the remaining terms does not lead to a qualitative change in the results.

The spectrum of the excitons is determined by the poles of the two-particle Green's function $G_{eh}(P, p, p')$ with respect to the variable E, where $P = \{\mathbf{P}, E\}$ denotes the total quasimomentum and energy of the electron-hole pair, and $p = \{\mathbf{p}, \epsilon\}$ and $p' = \{\mathbf{p}', \epsilon'\}$ are the relative quasimomenta and energy of such a pair. The presence in the interaction Hamiltonian of matrix elements describing the simultaneous creation of two electron-hole pairs from the vacuum forces us to also consider the function $\tilde{G}_{eh}(P, p, p')$ in addition to the function $G_{eh}(P, p, p')$; the function $\tilde{G}_{eh}(P, p, p')$ is given by the sum of all connected diagrams describing the creation from the vacuum of two electron-hole pairs with total momentum equal to zero.

The functions $G_{eh}(P, p, p')$ and $\tilde{G}_{eh}(P, p, p')$ are determined by a system of equations[7] similar to the system of Belyaev equations,[1)] which is well-known in the theory of a Bose gas (see Fig. 2). In the present case the vortex Γ is simply the Coulomb interaction iV_k between an electron and a hole, and $\tilde{\Gamma}$ is the sum of all irreducible diagrams of first order in $\tilde{V}_k$, that is, those diagrams which cannot be divided by a vertical cut into two parts, which are joined by single electron and single hole lines running laterally. Thus, the vertex $\tilde{\Gamma}$, being of first order in $\tilde{V}_k$, contains also all possible Coulomb interactions between four particles (two electrons and two holes) to all orders; these interactions are described by diagrams that are irreducible in the sense indicated above. Since the diagrams of first order in V_k are taken into consideration in $\tilde{\Gamma}$, it is clear that the system of equations shown in Fig. 2 is approximate.

The system is solved by using the Green's function for the Coulomb problem:[7]

$$\bar{G}(E,\mathbf{p},\mathbf{p}') = \frac{1}{(2\pi)^3}\sum_{n=1}^{\infty}\frac{\varphi_n^*(\mathbf{p}')\varphi_n(\mathbf{p})}{E-\mathscr{E}_n}, \tag{2}$$

where $\varphi_n(\mathbf{p})$ and $\mathscr{E}_n$ are the eigenfunctions and eigenvalues of the Coulomb equation:

$$\left(E-\frac{\hbar^2\mathbf{p}^2}{2m}\right)\varphi(\mathbf{p})+\frac{1}{(2\pi)^3}\int V_{\mathbf{p}-\mathbf{k}}\,\varphi(\mathbf{k})\,d^3k = 0. \tag{3}$$

The solution of the inhomogeneous equation

$$\left(E-\frac{\hbar^2\mathbf{p}^2}{2m}\right)\varphi(\mathbf{p})+\frac{1}{(2\pi)^3}\int V_{\mathbf{p}-\mathbf{k}}\,\varphi(\mathbf{k})\,d^3k = f(\mathbf{p})$$

is given by

$$\varphi(\mathbf{p}) = \int \bar{G}(E,\mathbf{p},\mathbf{k})\,f(\mathbf{k})\,d^3k. \tag{4}$$

Using (4), from the second equation of the system shown in Fig. 2 one can formally express $\tilde{G}_{eh}(P, p, p')$ linearly in terms of $G_{eh}(P, p, p')$ and substitute this result into the first equation; solving the latter near the lowest pole of $G_{eh}(P, p, p')$ in E, we obtain the following expression for the pole part of the function $G_{eh}(P, p, p')$:

$$G_{eh}(P,p,p') \approx A_{eh}\frac{E+\Delta+\mathscr{E}_1+\hbar^2(\mathbf{P}+\mathbf{q})^2/2M}{E^2-[\Delta+\mathscr{E}_1+\hbar^2(\mathbf{P}+\mathbf{q})^2/2M]^2+|\tilde{v}|^2}, \tag{5}$$

$$A_{eh} = i(2\pi)^4\left(\mathscr{E}_1-\frac{\hbar^2\mathbf{p}^2}{2m}\right)\left(\mathscr{E}_1-\frac{\hbar^2\mathbf{p}'^2}{2m}\right)\varphi_1(\mathbf{p})\varphi_1(\mathbf{p}')$$
$$\times G_e(p-q)G_h(-p)G_e(p'-q)G_h(-p'),$$

where $m = m_e m_h/(m_e + m_h)$, $M = m_e + m_h$, $G_e(p)$ and $G_h(p)$ denote the free Green's functions of the electrons and holes, $q = \{\mathbf{q}, 0\}$,

$$\tilde{v} = \frac{1}{(2\pi)^6}\int \varphi_1(\mathbf{p})\,(\tilde{V}_{\mathbf{q}}^* - \tilde{V}_{\mathbf{p}-\mathbf{p}'-\mathbf{q}}^*)\,\varphi_1(\mathbf{p}')\,d^3p\,d^3p' - \frac{i}{2}\tilde{\Lambda}, \tag{6}$$

and $\tilde{\Lambda}$ is described graphically by the block diagram shown in Fig. 3. In this figure $\gamma^0_{\mathbf{p},\mathbf{p}'} = \varphi_1(\mathbf{p}) - \varphi_1(\mathbf{p}')$, and Φ is the four-particle propagator which contains all possible Coulomb interactions of the two electron-hole pairs and does not contain $\tilde{V}$.

By using Eq. (5) we obtain the following result for the pole part of $\tilde{G}_{eh}(P, p, p')$:

$$\tilde{G}_{eh}(P,p,p') \approx A_{eh}\tilde{v}^*\left[E^2-\left(\Delta+\mathscr{E}_1+\frac{\hbar^2(\mathbf{P}+\mathbf{q})^2}{2M}\right)^2+|\tilde{v}|^2\right]^{-1}. \tag{7}$$

The positive pole of the function $G_{eh}(P, p, p')$ in E determines the exciton spectrum for values of $\mathbf{p}$ close to $-\mathbf{q}$ and for $\Delta + \mathscr{E}_1 \geq |\tilde{v}|$:

$$E(\mathbf{P}) = [(\Delta+\mathscr{E}_1+\hbar^2(\mathbf{P}+\mathbf{q})^2/2M)^2 - |\tilde{v}|^2]^{1/2}. \tag{8}$$

It is clear that the radicand becomes negative for $\Delta + \mathscr{E}_1 < |\tilde{v}|$ and $|\mathbf{p} + \mathbf{q}| \to 0$, and the energy of the excitations becomes imaginary. This indicates that the ground state of the system is unstable with respect to the formation of excitons with quasimomentum $-\mathbf{q}$ for $\Delta + \mathscr{E}_1 < |\tilde{v}|$. The realignment of the electron and hole wave functions which appears in this connection can be described by a canonical transformation of the basis system of the single-particle states,[7] this transformation being formally analogous to the Bogolyubov transformation, which is well-known in the theory of superconductivity. The transformation in question is generated by the unitary operator

$$S = \exp\left\{i\sum_{\mathbf{p}}[\varphi(\mathbf{p})a^+_{\mathbf{p}-\mathbf{q}}b^+_{-\mathbf{p}} - \varphi^*(\mathbf{p})a_{\mathbf{p}-\mathbf{q}}b_{-\mathbf{p}}]\right\}. \tag{9}$$

[1)] This analogy is purely formal. Both of the functions—$G_{eh}(P, p, p')$ and $\tilde{G}_{eh}(P, p, p')$—are the usual two-particle Green's functions of the electrons, but the first is diagonal and the second is nondiagonal with respect to the band indices.

Here $\varphi(\mathbf{p})$ is an arbitrary function which must be chosen from the conditions for stability of the restructured ground state.

It is natural to assume that the realignment is small (that is, the function $\varphi(\mathbf{p})$ is small) when $|\Delta + \mathscr{E}_1| \to |\tilde{V}|$. Therefore, one can find approximate expressions for the coefficients of the Bogolyubov transformation:

$$S^{+}a_{\mathbf{p}}S = u(\mathbf{p}+\mathbf{q})a_{\mathbf{p}} + v(\mathbf{p}+\mathbf{q})b^{+}_{-\mathbf{p}-\mathbf{q}},$$
$$S^{+}b_{\mathbf{p}}S = u(\mathbf{p})b_{\mathbf{p}} - v(\mathbf{p})a^{+}_{-\mathbf{p}-\mathbf{q}}, \tag{10}$$

in terms of the function $\varphi(\mathbf{p})$:

$$u(\mathbf{p}) \approx 1 - {}^{1}/_{2}|\varphi(\mathbf{p})|^2, \quad v(\mathbf{p}) \approx i(1 - {}^{1}/_{6}|\varphi(\mathbf{p})|^2)\varphi(\mathbf{p}). \tag{11}$$

As a result of the transformation the Hamiltonian (1) acquires the following structure:

$$S^{+}HS = U\{\varphi(\mathbf{p})\} + \sum_{\mathbf{p}} [\mathscr{E}_e(\mathbf{p})a^{+}_{\mathbf{p}}a_{\mathbf{p}} + \mathscr{E}_h(\mathbf{p})b^{+}_{\mathbf{p}}b_{\mathbf{p}}] + H_i, \tag{12}$$

where $U\{\varphi(\mathbf{p})\}$ is a numerical functional of $\varphi(\mathbf{p})$,

$$\mathscr{E}_e(\mathbf{p}) = \varepsilon_e(\mathbf{p}) - |\varphi(\mathbf{p}+\mathbf{q})|^2[\varepsilon_h(-\mathbf{p}-\mathbf{q}) + \varepsilon_e(\mathbf{p})]$$
$$-\frac{1}{V}\sum_{\mathbf{p}'} V_{\mathbf{p}-\mathbf{p}'}[|\varphi(\mathbf{p}'+\mathbf{q})|^2 - 2\varphi(\mathbf{p}+\mathbf{q})\varphi^{*}(\mathbf{p}'+\mathbf{q})], \tag{13}$$

$$\mathscr{E}_h(\mathbf{p}) = \varepsilon_h(\mathbf{p}) - |\varphi(\mathbf{p})|^2\left(\frac{\hbar^2\mathbf{p}^2}{2m} + \Delta\right) - \frac{1}{V}\sum_{\mathbf{p}'} V_{\mathbf{p}-\mathbf{p}'}[|\varphi(\mathbf{p}')|^2 - 2\varphi(\mathbf{p})\varphi^{*}(\mathbf{p}')], \tag{14}$$

$$H_i = \sum_{\mathbf{p}} [M_{\mathbf{p}}a^{+}_{\mathbf{p}-\mathbf{q}}b^{+}_{-\mathbf{p}} + M^{*}_{\mathbf{p}}b_{-\mathbf{p}}a_{\mathbf{p}-\mathbf{q}}]$$
$$+\frac{1}{2V}\sum_{\mathbf{p},\mathbf{p}',\mathbf{k}} \{V_{\mathbf{k}}[a^{+}_{\mathbf{p}}a^{+}_{\mathbf{p}'}a_{\mathbf{p}'+\mathbf{k}}a_{\mathbf{p}-\mathbf{k}} + b^{+}_{\mathbf{p}}b^{+}_{\mathbf{p}'}b_{\mathbf{p}'+\mathbf{k}}b_{\mathbf{p}-\mathbf{k}}$$
$$-(2 - |\gamma_{\mathbf{p},\mathbf{p}-\mathbf{k}}|^2 - |\gamma_{\mathbf{p}',\mathbf{p}'+\mathbf{k}}|^2)a^{+}_{\mathbf{p}-\mathbf{q}}b^{+}_{\mathbf{p}'}b_{\mathbf{p}'+\mathbf{k}}a_{\mathbf{p}-\mathbf{q}-\mathbf{k}}]$$
$$+[2iV_{\mathbf{k}}(\gamma^{*}_{\mathbf{p}',\mathbf{k}-\mathbf{p}'}b^{+}_{-\mathbf{p}}a_{\mathbf{p}'-\mathbf{q}}b_{-\mathbf{p}'+\mathbf{k}}b_{-\mathbf{p}-\mathbf{k}} + \gamma^{*}_{-\mathbf{p}'-\mathbf{k},\mathbf{p}'}a^{+}_{-\mathbf{p}}b_{\mathbf{p}'}a_{-\mathbf{p}'-\mathbf{q}-\mathbf{k}}a_{-\mathbf{p}+\mathbf{k}})$$
$$+(\tilde{V}_{\mathbf{k}+\mathbf{q}} - V_{\mathbf{k}}\gamma^{*}_{\mathbf{p}',\mathbf{k}-\mathbf{p}'}\gamma^{*}_{\mathbf{p},-\mathbf{p}-\mathbf{k}})a_{\mathbf{p}-\mathbf{q}}a_{\mathbf{p}'+\mathbf{q}}b_{\mathbf{k}-\mathbf{p}'}b_{-\mathbf{k}-\mathbf{p}} + \text{h.c.}]\}, \tag{15}$$

$$M_{\mathbf{p}} = i\left(\frac{\hbar^2\mathbf{p}^2}{2m} + \Delta\right)\left(1 - \frac{2}{3}|\varphi(\mathbf{p})|^2\right)\varphi(\mathbf{p})$$
$$-\frac{i}{V}\sum_{\mathbf{p}'} V_{\mathbf{p}-\mathbf{p}'}\left\{\left[1 - 2|\varphi(\mathbf{p})|^2 - \frac{2}{3}|\varphi(\mathbf{p}')|^2\right]\varphi(\mathbf{p}') + 2\varphi(\mathbf{p})|\varphi(\mathbf{p}')|^2\right\}$$
$$+\frac{i}{V}\sum_{\mathbf{p}'} (\tilde{V}^{*}_{\mathbf{p}-\mathbf{p}'-\mathbf{q}} - \tilde{V}^{*}_{\mathbf{q}})\varphi^{*}(\mathbf{p}'), \tag{16}$$
$$\gamma_{\mathbf{p},\mathbf{p}'} = \varphi(\mathbf{p}) - \varphi(\mathbf{p}').$$

The transformation of the Hamiltonian has been carried out approximately. Terms giving corrections $\sim\varphi^2$ to the single-particle spectrum, and to Γ and $\tilde{\Gamma}$, have been kept, and also the terms enabling us to obtain the equation of "compensation" (see below) correct to terms of order φ^3 inclusively.

In order to obtain the equation which determines the function $\varphi(\mathbf{p})$, it is necessary to take the following fact into consideration. The Hamiltonian (12) allows the creation from vacuum of electron-hole pairs with total quasimomentum equal to $-\mathbf{q}$. The corresponding matrix elements are contained in the bilinear part of H_i, and can also be obtained in higher orders of perturbation theory from the parts of H_i which are quartic in the operators. For example, in second-order perturbation theory one can construct the matrix element corresponding to the creation (from the vacuum) of an electron-hole pair with total quasimomentum $-\mathbf{q}$ from the terms of H_i which describe the creation (from the vacuum) of two electron-hole pairs and a scattering of the particles involving the annihilation of one electron-hole pair (see Fig. 4).

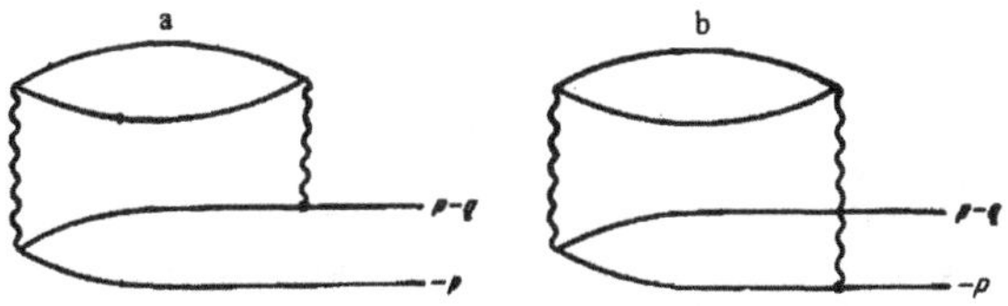

FIG. 4

The appearance in the diagrams of the perturbation-theoretic series of parts, which are connected to the remaining part by single electron and single hole lines moving laterally, with total quasimomentum $-\mathbf{q}$, is an indication that the ground state is unstable with respect to the formation of such pairs in the vicinity of the point $\Delta + \mathscr{E}_1 = |\tilde{V}|$; this has already been mentioned above in connection with the investigation of the exciton spectrum in the restructured state.

However, we may express the arbitrary function $\varphi(\mathbf{p})$ in such a way as to compensate for such "dangerous" diagrams and thereby make the ground state stable.[7] In order to do this we introduce the function

$$F(\mathbf{p}, t_1 - t_2) = -i\langle T\tilde{a}_{\mathbf{p}-\mathbf{q}}(t_1)\tilde{b}_{-\mathbf{p}}(t_2)\rangle. \tag{17}$$

Here $\tilde{a}_{\mathbf{p}}(t)$ and $\tilde{b}_{\mathbf{p}}(t)$ denote the operators in the Heisenberg representation, and the average is taken with respect to the ground state of the Hamiltonian (12).

The compensation of the "dangerous" diagrams lies in the fact that the function $\varphi(\mathbf{p})$ is determined from the condition[7]

$$F(\mathbf{p}, 0) = 0. \tag{18}$$

This condition is simultaneously the condition that the ground state energy be a minimum, and from this point of view it ensures the best choice for the basis of single-particle states.

Taking the smallness of the function $\varphi(\mathbf{p})$ into consideration, let us carry out the compensation to terms of order $\varphi^3(\mathbf{p})$ inclusively, and in this approximation Eq. (18) takes the form

$$M_{\mathbf{p}} - \left(\frac{\hbar^2\mathbf{p}^2}{2m} + \Delta\right)\int_{-\infty}^{\infty} [\tilde{\Lambda}(\mathbf{p}, \varepsilon) - \Lambda(\mathbf{p}, \varepsilon)]\frac{d\varepsilon}{2\pi} = 0. \tag{19}$$

$\tilde{\Lambda}(\mathbf{p}, \epsilon)$ denotes the sum of the diagrams shown in Figs. 5a and 5b, and $\Lambda(\mathbf{p}, \epsilon)$ denotes the sum of the diagrams shown in Figs. 5c and 5d.

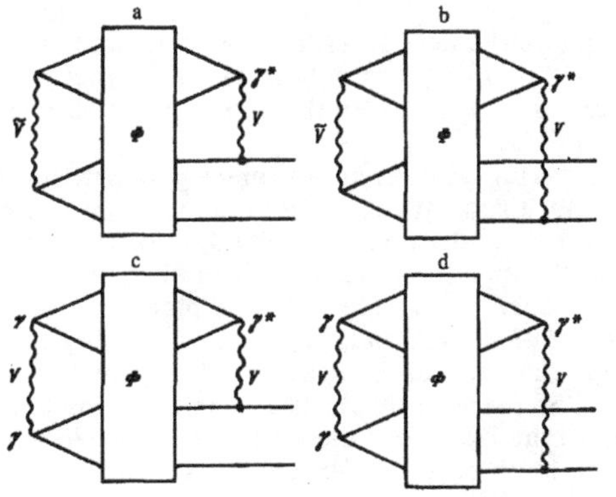

FIG. 5

To the first approximation in $\varphi(\mathbf{p})$ and to zero-order in $\tilde{V}$, Eq. (19) reduces to the Coulomb equation (3) in which Δ plays the role of the energy. Its solution is given by $\varphi(\mathbf{p}) = ce^{i\alpha}\varphi_1(\mathbf{p})$ for $\Delta = -\mathscr{E}_1$. Here c is a real positive coefficient, $\alpha \equiv \arg\varphi(\mathbf{p})$, and $\varphi_1(\mathbf{p})$ and $\mathscr{E}_1$ denote the eigenfunction and the eigenvalue of the hydrogen-like ground state.

Now substituting the value of $\varphi(\mathbf{p})$ into Eq. (19), we obtain an equation for the determination of the coefficient c, and by solving this we find

$$c=\begin{cases} f\sqrt{-\Delta-\mathscr{E}_1+|\tilde{v}|} & \text{for } \Delta+\mathscr{E}_1<|\tilde{v}|, \\ 0 & \text{for } \Delta+\mathscr{E}_1>|\tilde{v}|, \end{cases} \qquad \alpha = {}^1/_2 \arg\tilde{v}.$$

The quantity $\tilde{v}$ was defined earlier by Eq. (6), and f is a real, positive constant:

$$f=\left[\frac{2}{(2\pi)^3}\int\left(\frac{\hbar^2\mathbf{p}^2}{2m}-\mathscr{E}_1\right)\varphi_1^4(\mathbf{p})\,d^3p - \right.$$
$$\left. -\frac{2}{(2\pi)^6}\int\varphi_1^2(\mathbf{p})V_{\mathbf{p}-\mathbf{p}'}\varphi_1^2(\mathbf{p}')\,d^3p\,d^3p'-\frac{i}{2}\Lambda\right]^{-1/2}$$
$$=\left[2\frac{13\pi}{3}a_1^3|\mathscr{E}_1|-\frac{i}{2}\Lambda\right]^{-1/2},$$

where $a_1 = \kappa\hbar^2/me^2$, and Λ is described graphically by a block diagram similar to the one shown in Fig. 3 with, however, the difference that in the present case V_k appears instead of $\tilde{V}_k$ and the vertices on the left part also contain γ^0.

The equations for $G_{eh}(P, p, p')$ and $\tilde{G}_{eh}(P, p, p')$, defined in terms of the new operators, have the same structure as the equations shown in Fig. 2; however, Γ and $\tilde{\Gamma}$ now contain corrections of order φ^2, and $G_e(p)$ and $G_h(p)$ are the free Green's functions of particles whose dispersion laws are given by $\mathscr{E}_e(\mathbf{p})$ and $\mathscr{E}_h(\mathbf{p})$, respectively.

The system of equations is solved in exactly the same way as before the realignment, and as a result we obtain the following expressions for the pole parts of the Green's functions:

$$G_{eh}(P,p,p')\approx A_{eh}\frac{E-\Delta-\mathscr{E}_1+2|\tilde{v}|+\hbar^2(\mathbf{P}+\mathbf{q})^2/2M}{E^2-[-\Delta-\mathscr{E}_1+2|\tilde{v}|+\hbar^2(\mathbf{P}+\mathbf{q})^2/2M]^2+(\Delta+\mathscr{E}_1)^2}, \tag{20}$$

$$\tilde{G}_{eh}(P,p,p')\approx A_{eh}\frac{(\Delta+\mathscr{E}_1)e^{-i2\alpha}}{E^2-[-\Delta-\mathscr{E}_1+2|\tilde{v}|+\hbar^2(\mathbf{P}+\mathbf{q})^2/2M]^2+(\Delta+\mathscr{E}_1)^2} \tag{21}$$

After the realignment $(\Delta + \mathscr{E}_1 < |\tilde{v}|)$ the spectrum of the "excitons" has the following form:

$$E(\mathbf{P}) = [(2|\tilde{v}|-\Delta-\mathscr{E}_1+\hbar^2(\mathbf{P}+\mathbf{q})^2/2M)^2-(\Delta+\mathscr{E}_1)^2]^{1/2}. \tag{22}$$

For $\mathbf{P}$ close to $-\mathbf{q}$ the spectrum has a slotted character with a gap $\delta = 2[|\tilde{v}|(|\tilde{v}| - \Delta - \mathscr{E}_1)]^{1/2}$ which vanishes simultaneously with the realignment at the point $\Delta + \mathscr{E}_1 = |\tilde{v}|$. It is only in this case that the spectrum is acoustic, with the velocity of sound given by $s = (|\tilde{v}|/M)^{1/2}$.

Thus, taking account of the terms proportional to $\tilde{V}$ in the Hamiltonian (1) eliminates the analogy, which has been repeatedly discussed earlier, of an "excitonic insulator" with a superfluid condensate of Bose particles or Cooper pairs. These terms play the role of a "source of excitons" and they remove the degeneracy of the ground state of the system with respect to the phase of the function $\varphi(\mathbf{p})$. The conditions for the validity of the Goldstone theorem are thereby violated, and there is no reason to expect the spectrum of the collective excitations to be acoustic.

In order to clarify the nature of the phase transition it is necessary to investigate the behavior of the ground state energy in the neighborhood of the realignment point. Let us calculate the contribution to the energy coming from the interaction $\tilde{V}$, since the remaining part of the energy does not contain any singularities.

As is well known,[8] the change in the energy of the ground state originating from the interaction $\tilde{H}$ is given by the formula

$$\delta\mathscr{E}=\int_0^1\langle\tilde{H}(g)\rangle\frac{dg}{g}, \tag{23}$$

where g denotes the coupling constant of the interaction contained in $\tilde{H}$. Let us calculate this change of the energy as a function of the volume, first in the absence of the realignment $(\Delta + \mathscr{E}_1 > |\tilde{v}|)$, and then with the realignment $(\Delta + \mathscr{E}_1 < |\tilde{v}|)$ taken into consideration.

1) $\Delta + \mathscr{E}_1 > |\tilde{v}|$.

In this case the average $\langle\tilde{H}\rangle$ is expressed in terms of $\tilde{G}_{eh}(P, p, p')$ in the following manner:

$$\langle\tilde{H}\rangle=\frac{1}{2}\frac{V}{(2\pi)^{16}}\operatorname{Re}\left\{\int(\tilde{V}_{\mathbf{p}}^{*}-\tilde{V}_{\mathbf{p}'-\mathbf{p}-\mathbf{q}}^{*})\tilde{G}_{eh}(P,p,p')\,d^4p\,d^4p'\,d^4P\right\}. \tag{24}$$

This equation is approximate in the same sense that the system of equations shown in Fig. 2 is approximate for the functions $G_{eh}(P, p, p')$ and $\tilde{G}_{eh}(P, p, p')$. Substituting expression (7) for $\tilde{G}_{eh}(P, p, p')$ into (24) and integrating over the four-dimensional momenta p and p′, and also integrating over E, we obtain

$$\langle\tilde{H}\rangle=-\frac{1}{4}\frac{V}{(2\pi)^3}\operatorname{Re}\left\{\tilde{v}^{*}\int\frac{\tilde{V}^{*}(\mathbf{P}+\mathbf{q})}{[(\Delta+\mathscr{E}_1+\hbar^2\mathbf{P}^2/2M)^2-|\tilde{v}|^2]^{1/2}}\,d^3P\right\}, \tag{25}$$

where

$$\tilde{V}(\mathbf{P})=\frac{1}{(2\pi)^6}\int\varphi_1(\mathbf{p})[\tilde{V}_{\mathbf{p}}-\tilde{V}_{\mathbf{p}'-\mathbf{p}-\mathbf{q}}]\varphi_1(\mathbf{p}')\,d^3p\,d^3p'.$$

In order to find $\delta\mathscr{E}$, it is necessary to explicitly introduce the coupling constant g into Eq. (25), substitute $\langle\tilde{H}(g)\rangle$ into Eq. (23) and integrate with respect to g. Thus, we find

$$\delta\mathscr{E}=\frac{1}{4}\frac{V}{(2\pi)^3}\operatorname{Re}\left\{\frac{1}{\tilde{v}}\int\left[\sqrt{\left(\Delta+\mathscr{E}_1+\frac{\hbar^2\mathbf{P}^2}{2M}\right)^2-|\tilde{v}|^2}-\right.\right.$$
$$\left.\left.-\left(\Delta+\mathscr{E}_1+\frac{\hbar^2\mathbf{P}^2}{2M}\right)\right]\tilde{V}^{*}(\mathbf{P}+\mathbf{q})\,d^3P\right\}. \tag{26}$$

It is clear from Eq. (26) that the second derivative of $\delta\mathscr{E}$ with respect to the volume (Δ is assumed to be a monotonic function of the volume), evaluated at the point $\Delta + \mathscr{E}_1 = |\tilde{v}|$, contains an integral which diverges at zero. Isolating the part of $\delta\mathscr{E}$ which gives this singularity, we have

$$\delta\mathscr{E}_{\text{sing}}=\frac{V_0|\tilde{v}|^{1/2}M^{3/2}}{8(2\pi)^2\hbar^3}\operatorname{Re}\left\{\frac{\tilde{V}^{*}(\mathbf{q})}{\tilde{v}}\right\}(\Delta+\mathscr{E}_1-|\tilde{v}|)^2\ln\frac{\Delta+\mathscr{E}_1-|\tilde{v}|}{|\tilde{v}|}, \tag{27}$$

where V_0 denotes the volume at which $\Delta + \mathscr{E}_1 = |\tilde{v}|$.

2) $\Delta + \mathscr{E}_1 = |\tilde{v}|$.

With the realignment taken into consideration, the average $\langle\tilde{H}\rangle$ is expressed in terms of the Green's function $\tilde{G}_{eh}(P, p, p')$ by a formula analogous to Eq. (24); now the Green's function is already defined in terms of the transformed operators. Using expression (21) for $\tilde{G}_{eh}(P, p, p')$ and isolating the part of $\delta\mathscr{E}$ which gives the singularity in the second derivative with respect to the volume (in the same way as before the realignment), we obtain

$$\delta\mathscr{E}_{\rm sing} = -\frac{V_0|\tilde{v}|^{1/2}M^{3/2}}{4(2\pi)^2\hbar^3}\operatorname{Re}\left\{\frac{\tilde{V}^*(q)}{\tilde{v}}\right\}(|\tilde{v}|-\Delta-\mathscr{E}_1)^2\ln\frac{|\tilde{v}|-\Delta-\mathscr{E}_1}{|\tilde{v}|}. \quad (28)$$

From Eqs. (27) and (28) it is seen that near the realignment point a region exists in which the pressure of the system, $P = -\partial\mathscr{E}/\partial V$, is an increasing function of the volume, and furthermore, depending on the sign of the quantity $\operatorname{Re}\{\tilde{V}^*(q)/\tilde{v}\}$, this region will be to the left or to the right of the point $\Delta + \mathscr{E}_1 = |\tilde{v}|$ corresponding to realignment of the spectrum.

The presence of a region of absolute thermodynamic instability leads to the result that, associated with changes of the volume including the point $\Delta + \mathscr{E}_1 = |\tilde{v}|$, the system undergoes a phase transition of the first kind, and moreover this point itself always falls in the region of the thermodynamically unstable states and therefore cannot be reached. The transition under consideration may correspond to a discontinuous change of not only the volume but also of any other parameter which the energy gap significantly depends on (uniaxial deformation, restructuring of the elementary crystal cell, etc.). In addition, the reason for a transition of the first kind may be the interelectron interaction itself,[5] if it leads to the result that attractive forces act between the excitons which are large enough so that the exciton system has a tendency to "liquefy," that is, if its energy for a given total number of excitons were to have a minimum at a certain nonzero value of the exciton density.

Thus, the transition into the "excitonic insulator" state is always a transition of the first kind, and this state itself does not possess any properties which might distinguish it from ordinary dielectrics.

[1] L. V Keldysh and Yu. V. Kopaev, Fiz. Tverd. Tela 6, 2791 (1964) [Sov. Phys.-Solid State 6, 2219 (1965)].

[2] Jacques des Cloizeaux, J. Phys. Chem. Solids 26, 259 (1965).

[3] A. N. Kozlov and L. A. Maksimov, Zh. Eksp. Teor. Fiz. 48, 1184 (1965); 49, 1284 (1965) [Sov. Phys.-JETP 21, 790 (1965); 22, 889 (1966)].

[4] D. Jérome, T. M. Rice, and W. Kohn, Phys. Rev. 158, 462 (1967).

[5] B. I. Halperin and T. M. Rice, Rev. Mod. Phys. 40, 755 (1968).

[6] R. R. Guseĭnov and L. V. Keldysh, Kratkie soobshcheniya po fizike (Brief Communications in Physics) 7, (1971).

[7] L. V. Keldysh and A. N. Kozlov, Zh. Eksp. Teor. Fiz. 54, 978 (1968) [Sov. Phys.-JETP 27, 521 (1968)].

[8] A. A. Abrikosov, L. P. Gor'kov, and I. E. Dzyaloshinskiĭ, Metody kvantovoĭ teorii polya v statisticheskoĭ fizike, Fizmatgiz, 1963 [Quantum Field Theoretical Methods in Statistical Physics, Pergamon, 1965].

Translated by H. H. Nickle
247

Microwave breakdown and exciton condensation in germanium

L. V. Keldysh, A. A. Manenkov, V. A. Milyaev, and G. N. Mikhaĭlova

P. N. Lebedev Physics Institute, USSR Academy of Sciences

(Submitted December 20, 1973)
Zh. Eksp. Teor. Fiz. **66**, 2178–2190 (June 1974)

Exciton breakdown induced in germanium by a pulsed microwave field of frequency $f = 10$ GHz is investigated in detail at helium temperatures. The germanium is optically excited by a pulsed YAG laser ($\lambda = 1.06$ μm). The maximum density of the nonequlibrium carriers is $\sim 10^{15}$ cm^{-3}. The dependences of the threshold breakdown power on the duration and the delay time of the microwave pulse relative to the laser pulse and on the optical excitation level are studied at $T = 1.3$ °K. A theory of the effect is proposed, based on the mechanism of impact ionization of excitons at equilibrium with electron-hole drops. The breakdown effect is also discussed from the viewpoint of the biexciton model. The experimental dependences obtained are in satisfactory agreement with theoretical calculations based on the exciton-drop model.

1. INTRODUCTION

Investigation of excitons in germanium at low temperatures and relatively high concentrations is of considerable interest in connection with the problem of their collective properties. According to the existing theoretical concepts, excitons in germanium form, under certain conditions, bound states regarded by most workers as drops of an electron-hole liquid[1-9] and by others as exciton molecules or biexcitons[10-12]. Consequently, great importance attaches to new experimental approaches that yield additional information on the nature of exciton complexes. Work in this direction, connected with the study of properties of excitons at microwave frequencies, has been recently reported[8, 12-14]. Observation of exciton breakdown in germanium by a microwave field was reported in[15]. The observed characteristics of this breakdown were discussed on the basis of the models of the biexciton gas and of the electron-hole drops, but no reliable conclusion was drawn concerning the nature of the broken-down exciton objects.

We have undertaken a detailed study of the effect of microwave breakdown, for the purpose of obtaining data that would explain the nature of the exciton complexes produced in germanium. Assuming an analogy between the exciton-breakdown mechanism and the known features of laser and microwave breakdown of gases in solids[16-18], we investigated the microwave breakdown in germanium under pulsed conditions, varying the duration of the microwave pulses and their delays relative to the exciting laser pulse. The observed dependences of the threshold breakdown power on the duration and delay time of the microwave pulse, and also on the intensity of the optical excitation, have made it possible to analyze in detail the exciton breakdown process and to propose for this effect a theory based on the mechanism of impact ionization of excitons in the presence of electron-hole drops.

2. EXPERIMENTAL SETUP AND PROCEDURE

We investigated the microwave conductivity of germanium under conditions of pulsed laser excitation. The experiments were performed with a setup whose clock diagram is shown in Fig. 1. The investigated germanium samples 1 were placed in a rectangular microwave resonator 2, in which H_{102} oscillations were excited at f = 10 GHz. The resonator was immersed in a cryostat 3 with liquid helium. The loaded Q of the resonator with the sample was $\sim 10^3$ at T = 1.3°K. The microwave source

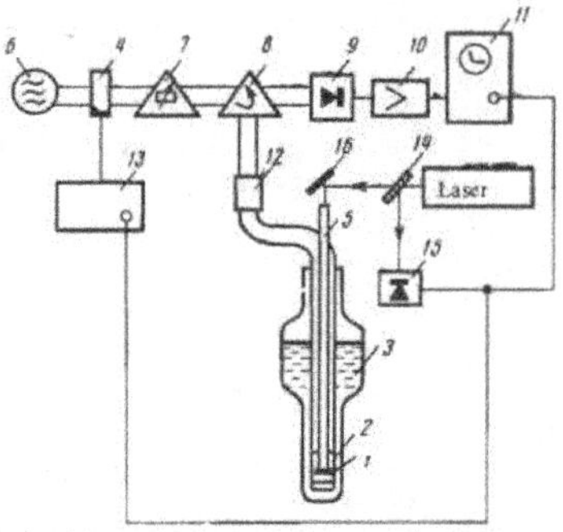

FIG. 1. Block diagram of setup: 1—sample, 2—resonator, 3—helium cryostat, 4—microwave modulator, 5—light pipe, 6—klystron, 7—attenuator, 8—Y circulator, 9—microwave detector, 10—amplifier, 11—oscilloscope, 12—matching section, 13—G5-7A generator, 14—beam splitter, 15—photodiode, 16—mirror.

was a continuously operating klystron with output power ~100 mW. A diode modulator 4 was connected in the waveguide line and made it possible to shape rectangular microwave pulses by applying a voltage from a G5-7A generator. The pulse duration could range from 0.2 to 100 μsec, with rise times ~0.2 μsec. The ratio of the pulse power to the continuous level of the power at the modulator output was ~20dB. The signal reflected from the resonator was observed at a continuous transmitted power level < 0.5 mW. The signal reflected from the resonator was detected with a video receiver and displayed on an oscilloscope screen. The time resolution of the recording apparatus was $\sim 10^{-7}$ sec. When a microwave pulse of ≥ 5 mW power was applied, the microwave conductivity of the sample increased abruptly, thus indicating the appearance of a large number of free carriers in the germanium as a result of the destruction (breakdown) of the excitons.

Optical excitation of the germanium was with a laser based on yttrium-aluminum garnet with neodymium ($\lambda = 1.06$ μ) producing giant pulses of 100 nsec duration, with repetition frequency 100 Hz. The maximum pulse energy was 10^{-4} J. The laser beam passed through a quartz light pipe 5, to the end face of which the sample was secured. The laser-excitation intensity was varied with calibrated neutral filters.

We investigated pure germanium samples with residual impurity density $N_A + N_D$ from 5×10^{12} to 10^{10} cm^{-3}, in the form of plates with typical dimensions 5 × 5 × 0.5 mm, etched in a boiling NaOH + H_2O_2 mixture. Some samples had a smoothly polished surface. The density of the free carriers produced by optical excitation under the conditions of our experiments was estimated from the intensity of the light incident on the sample, and was assumed to be uniform in the entire volume of the sample. The last assumption is based on the following circumstance: Although carriers are generated by light only in a thin surface layer (the coefficient of optical absorption of germanium at $\lambda = 1.06\ \mu$ is $\sim 10^4\ cm^{-1}$), they fill a sample ~1 mm thick within a time $\sim 10^{-8}$ sec, because of their rapid diffusion[19]. The maximum initial carrier density in the samples of our experiments reached $n = 10^{15}\ cm^{-3}$, corresponding to an exciting laser pulse energy 10^{-6} J.

3. EXPERIMENTAL RESULTS

The effect of exciton breakdown in germanium was investigated in the temperature interval 2.5–1.3°K. The principal data presented below were obtained at T = 1.3°K.

At low microwave levels, the waveform of the microwave conductivity signal following the laser pulse is a sum of two exponentials with characteristic times $\tau_1 = 2\ \mu$sec and $\tau_2 = 50\ \mu$sec, which reflect the dynamics of the binding and recombination of the carriers and excitons in the sample after the pulsed laser excitation[1)]. Investigations of the signal waveform following application of a constant magnetic field up to 10 kG have shown that the amplitudes of the indicated exponentials vary in accordance with the cyclotron-resonance lines. We conclude therefore that the observed signal is due to the free carriers, which contribute to the real and imaginary parts of the dielectric constant of the sample. This contribution is determined by the known relations[2]

$$\varepsilon' = -\frac{4\pi n_e e^2}{m(\omega^2+\nu^2)}, \quad \varepsilon'' = \frac{4\pi n_e e^2}{m\omega}\,\frac{\nu}{(\omega^2+\nu^2)}, \qquad (1)$$

where n_e is the density of the free carriers, ω is the frequency of the electromagnetic field, ν is the effective collision frequency, and m is the effective carrier mass.

The observations have shown that the initial section of the signal (exponential with time τ_1) includes both the absorption of the microwave power and the detuning, in accord with (1) at relatively large carrier density (when $\nu \gtrsim \omega$ is possible). This initial section describes the binding of the carriers into excitons and into exciton complexes. The exponential with the time τ_2 is determined mainly by the detuning of the resonator, i.e., by the change of ϵ', in accord with relations (1) at low carriers densities and $\nu \ll \omega$. We assume that the long exponential is connected with residual free carriers that are in equilibrium with excitons and exciton complexes. Measurement of the sign of the frequency detuning of the resonator, corresponding to the signal with the long exponential (τ_2), has shown that the change $\Delta\epsilon'$ is negative. This confirms that the observed signal is due to the residual free carriers[2)].

Application of a microwave pulse that is delayed relative to the exciting laser pulse causes an abrupt burst of sample conductivity, when the pulse power P exceeds a certain threshold value P_d (Fig. 2). The duration of the breakdown spike is ~0.5 μsec. The breakdown has a strongly pronounced threshold character: the spike amplitude increases very rapidly even at a slight excess of power above the threshold $(P-P_d)/P_d \sim 0.01$.

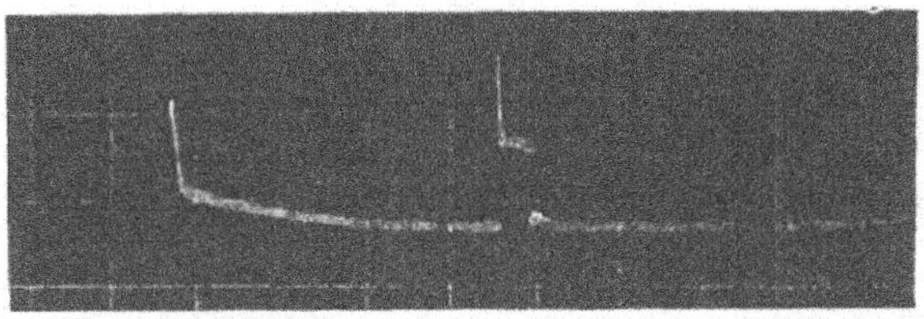

FIG. 2. Waveform of signal reflected from microwave resonator and observed following pulsed laser excitation of germanium. The microwave pulse, on the top of which is seen the breakdown spike, was delayed relatively to the laser pulse by ~80 μsec. The time sweep scale is 20 μsec/div.

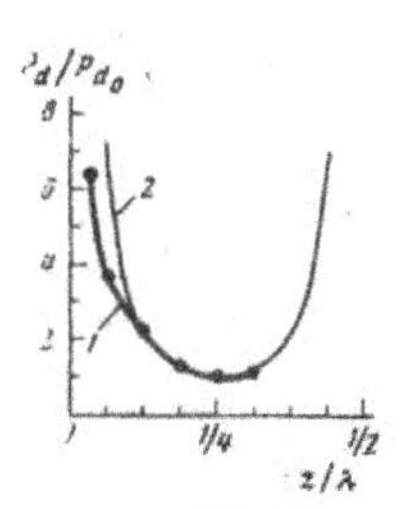

FIG. 3

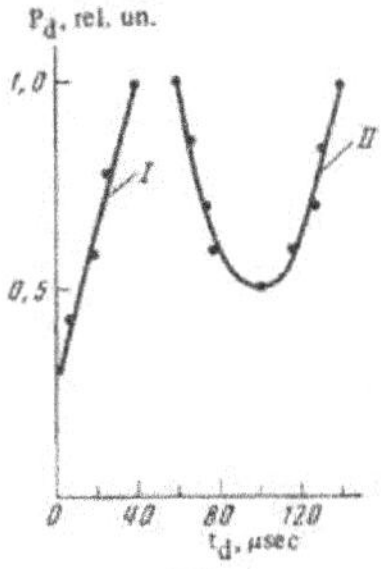

FIG. 4

FIG. 3. Dependence of the exciton breakdown threshold on the sample position along the resonator z axis (curve 1). For comparison, the figure shows the amplitude distribution of the electric microwave field component $(E_0/E_y)^2$ in the resonator (curve 2). P_{d_0} is the value of the breakdown threshold at the maximum of the electric field E_0.

FIG. 4. Dependence of the exciton breakdown threshold P_d on the microwave pulse delay time t_d relative to the exciting laser pulse. The microwave pulse duration is $\tau = 0.2$ sec. The initial density of the electron-hole pairs, is $\bar{n} = 10^{15}\ cm^{-3}$.

This makes it possible to determine very accurately the threshold of the breakdown. Since the threshold is one of the most important characteristics of the breakdown, and is connected with the nature of the broken-down objects, we have measured this quantity under various conditions. By breakdown threshold we mean here and below the minimum resonator microwave input power, at which breakdown is observed.

It was assumed that the breakdown threshold can depend on the relative values of the electric and magnetic components of the microwave field, on the dimensions of the exciton complexes, and on the initial densities of the carriers and excitons. It was therefore of interest to study the dependence of P_d on the position of the samples in the resonator, on the duration τ and on the delay time t_d of the microwave pulse relative to the laser pulse, and on the intensity of the optical excitation. We describe below the experimental results concerning these dependences.

a) Dependence of P_d on the position of the sample in the resonator. In a rectangular microwave resonator, the E and H components of the field are sufficiently well separated, so that the relatively thin samples used by us ($\leq$ 1 mm) could be located either in an electric or in a magnetic microwave field, and we could determine which of these fields is responsible for the breakdown. Figure 3 shows the dependence of the breakdown threshold on the sample position along the z axis of the resonator at fixed durations and delay times of the microwave pulse and of the exciting laser pulse power. We see that the threshold is minimal when the sample is in an electric microwave field, and that there is no breakdown in a magnetic microwave field. The shape of the $P_d(z)$ curve deviates somewhat from the relation $E_y^2(z) = E_0^2 \sin^2 k_z z$ for the y component of the electric field of the normal H_{102} mode in the empty resonator. This deviation may be due to the change in the spatial distribution of the field in the resonator due to the introduction of the sample and the subsequent retuning of the resonator.

The obtained dependence of the breakdown threshold on the position of the sample in the resonator indicates that the cause of the observed effect is heating of the free carriers by the electric microwave field. The breakdown of the metallic electron-hole drops is more readily expected in a microwave magnetic field.

b) Dependence of P_d on the microwave pulse delay time t_d. Figure 4 shows the dependence of the threshold breakdown power on the microwave-pulse delay time relative to the laser pulse at a fixed microwave pulse duration $\tau = 0.2$ μsec and at an initial density $\bar{n} = 10^{15}$ cm^{-3} of the nonequilibrium carriers in the sample. We see that the curve has two branches, the first in the region 3–40 μsec and the second at $t_d = 50$–150 μsec, the second branch having a minimum at $t_d = 100$ μsec.

The character of the breakdown is somewhat different on the indicated branches. This difference pertains both to the sharpness of the breakdown threshold and to the time of its development. In the first section of the region where this breakdown exists, the threshold is indistinct, and the amplitude of the breakdown spike increases relatively slowly with increasing microwave pulse power. The duration of the leading and trailing edges for this branch is $\tau' \approx \tau'' \approx 0.2$ μsec. For the second branch, the breakdown threshold is very strongly pronounced, and the rise and fall-off times of the breakdown spikes are $\tau' \approx 0.1$ μsec and $\tau'' \approx 0.2$ μsec.

We note that the first branch of the $P_d(t_d)$ curve exists only at sufficiently high laser-pump levels. At an excitation intensity corresponding to the initial carrier density $\bar{n} \leq 5 \times 10^{14}$ cm^{-3} in the sample, only the second branch with a minimum appears on the $P_d(t_d)$ curve.

It was found that the breakdown effect under discussion is observed in all the investigated samples with different contents of residual impurity, including ultrapure germanium ($N_A + N_D = 10^{10}$ cm^{-3}),[3] and has the same characteristics (values of P_d, τ', and τ'') for all the samples.

c) Dependence of P_d on the microwave pulse duration τ. Figure 5 shows the dependence of the breakdown threshold power on the microwave pulse duration at an initial electron-hole pair density in the sample $\bar{n} = 10^{15}$ cm^{-3} and a delay time $t_d = 100$ μsec. The $P_d(\tau)$ curve is drooping and flattens into a horizontal straight line at $\tau = 0.5$ μsec.

Experiments were also performed on breakdown by two successive microwave pulses of equal duration and amplitude. It turns out that if the time interval between the microwave pulses is shorter than 10–15 μsec, then the breakdown is produced only by the first pulse, but if the second pulse is delayed relative to the first by more than 10–15 μsec, then the second pulse also produces breakdown.

In addition to the exciton breakdown, we observed

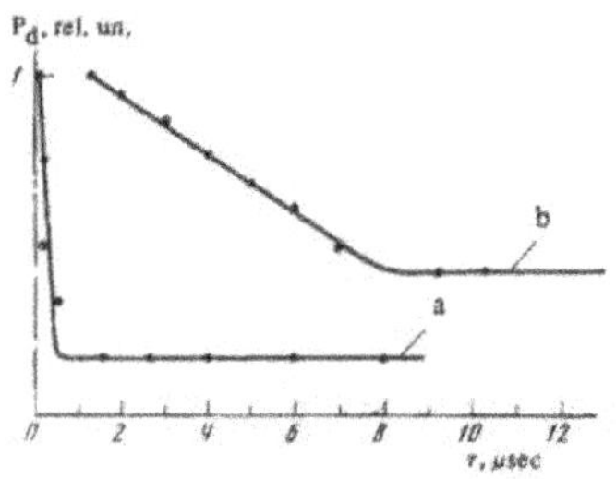

FIG. 5

FIG. 6

FIG. 5. Dependence of the breakdown threshold P_d on the microwave pulse duration τ: a—exciton breakdown, b—impurity breakdown. The exciton-breakdown data were obtained at a microwave pulse delay time (relative to the exciting laser pulse) $t_d = 100$ μsec, and at initial electron-hole pair density $\bar{n} = 10^{15}$ cm^{-3}. The impurity density in the sample is 5×10^{12} cm^{-3}.

FIG. 6. Dependence of the exciton breakdown threshold P_d on the microwave-pulse delay time t_d for different optical-excitation levels (initial concentrations of the generated electron-hole pairs). ●—$\bar{n} = 10^{15}$ cm^{-3}, X—$\bar{n} = 5 \times 10^{14}$ cm^{-3}, O—$n = 2.5 \times 10^{14}$ cm^{-3}. Microwave pulse duration $\tau = 0.2$ μsec.

also an optically-induced impurity breakdown. The characteristics of the impurity breakdown differed significantly from those of the exciton breakdown and varied from sample to sample. Thus, in samples with residual-impurity concentration 10^{11}–5×10^{12} cm^{-3} the impurity-breakdown threshold was independent of the delay time t_d and existed up to $t_d = 10$ msec. In ultrapure germanium, however, the impurity of breakdown was observed only at delay times up to 250–300 μsec following the laser pulse. In addition, the duration of the breakdown spike for the impurity was much larger than for the exciton breakdown, and amounted to ~2 μsec with a rise time ~1 μsec.

We investigated the dependence of the impurity-breakdown threshold on the microwave pulse duration (curve b on Fig. 5). The strong difference between the plots of $P_d(\tau)$ for the exciton and impurity breakdowns makes it possible to separate the two effects by varying τ. For example, a microwave pulse of duration less than 1 μsec (as seen from Fig. 5) produces only exciton breakdown. We note also that exciton breakdown is observed only at $T \leq 2.5$°K, whereas impurity breakdown is observed also at higher temperatures (up to 4.2°K).

d) Dependence of P_d on the optical-excitation intensity. Greatest interest attaches to the second branch of exciton breakdown. As will be shown below, the existence of a minimum of the breakdown threshold can be interpreted on the basis of the model with breakdown of an exciton gas in equilibrium with an electron-hole drop. We have therefore investigated in detail the dependence of the exciton breakdown on the initial concentration of the generated electron-hole pairs, which is determined by the intensity of the optical excitation. Figure 6 shows a family of $P_d(t_d)$ curves for different optical-excitation levels I. We see that $P_{d\,min}$ shifts towards shorter delays with decreasing light intensity. Figure 7 shows a plot of instant $t^*_d(I)$ when $P_{d\,min}$ occurs against the laser-pulse intensity.

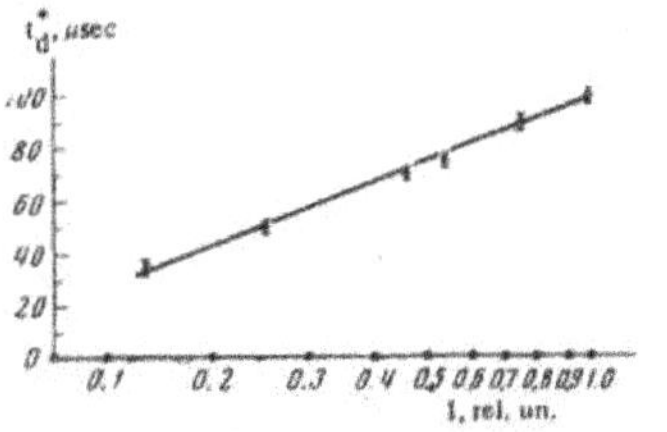

FIG. 7. Dependence of the time t_d^* of the minimum exciton breakdown threshold on the laser-excitation intensity I. The abscissa scale is logarithmic. The point I = 1 corresponds to an initial electron-hole pair density $\bar{n} = 10^{15}$ cm^{-3}.

4. THEORY OF MICROWAVE BREAKDOWN OF EXCITONS. DISCUSSION OF EXPERIMENTAL RESULTS

The most characteristic of the discussed qualitative factors is the considerable time delay of the breakdown relative to the perturbing pulse. The optimal conditions for exciton breakdown are produced in the sample within ~100 μsec after the appearance of the excitons in the sample (Fig. 6), a time that exceeds by more than one order of magnitude the lifetime of the free excitons [20]. On the other hand, it is known from numerous recent experimental investigations [9-15] that at low temperatures the excitons in Ge stick together to form certain complexes, electron-hole drops (EHD), or biexcitons. The lifetimes of these complexes are much larger than the lifetime of the free excitons. There is therefore practically no doubt that the singularities of the observed microwave breakdown of excitons are connected in some manner with the presence of these complexes. In other words, the observed breakdown developes in a system comprising electrons + excitons + EHD or in a system comprising electrons + excitons + biexcitons. We shall discuss the experimental data described above on the basis of both the EHD and biexciton models.

We start with the EHD model, since, as we shall show, it makes it possible to describe in a natural fashion most of the results. In this model we start from the fact that the bulk of the nonequilibrium carriers in the sample are bound into drops of average radius R and of particle density n_0. The drop concentration N is connected with the average density of the carriers introduced in the sample by the obvious relation

$$\bar{n} = {}^4/_3 \pi R^3 N n_0. \tag{2}$$

In addition to the EHD, the sample contains also excitons with concentration n and free carriers with concentration n_e. We, however, did not take them into account in (2), inasmuch as $\bar{n} \gg n \gg n_e$ at the sufficiently low temperatures of interest to us. During the first several microseconds after the exciting pulse, a dynamic equilibrium is established between these three groups of bound and free carriers, namely, the free carriers are bound into excitons and stick to the drops, the excitons dissociate into free carriers or are absorbed by the drops, and the drops capture and evaporate the excitons and free carriers. All the concentrations, (n, n_e, and $\bar{n}$) decrease with time as the result of recombination, but the equilibrium between them is preserved for a long time.

The main cause of the breakdown is undoubtedly the heating of the free carriers by the microwave field. The magnitude of this heating can be easily estimated in the usual manner by equating the energy absorbed from the field to the phonon radiation losses:

$$\frac{e^2\mathscr{E}^2}{m\omega^2}\nu = \frac{\hbar\bar{\omega}\nu}{1+2N_{\bar{\omega}}}. \tag{3}$$

Here $\mathscr{E}$ is the field intensity in the sample, ω is the field frequency, e and m are the charge and effective mass of the electron, ν is the frequency of the collisions with the phonons, $\bar{\omega}$ is the average frequency of the emitted and absorbed phonons, and is equal, by virtue of the energy and momentum conservation, to $\bar{\omega} = \sqrt{2\bar{\epsilon} m s^2}/\hbar$, where $\bar{\epsilon}$ is the average electron energy, s is the speed of sound, and $N_{\bar{\omega}}\,[e^{\hbar\bar{\omega}/kT} - 1]^{-1}$ is the equilibrium phonon distribution function. In the case of weak overheating we have $(\bar{\epsilon} \sim kT)\hbar\bar{\omega} \ll kT$ by virtue of $ms^2 \ll kT$, and it is then easy to obtain from (3)

$$\bar{\epsilon} = \frac{kT}{ms^2}\frac{e^2\mathscr{E}^2}{m\omega^2}, \tag{4}$$

and in stronger fields, when $\hbar\bar{\omega} > kT$, we have

$$\bar{\epsilon} = \frac{1}{2ms^2}\left(\frac{e^2\mathscr{E}^2}{m\omega^2}\right)^2. \tag{5}$$

We are interested mainly in the region (5), since $\bar{\epsilon} \sim \epsilon_0 \gg kT$ at breakdown (ϵ_0 is the exciton binding energy). In this region $\bar{\epsilon}$ increases very rapidly with increasing field ($\sim\mathscr{E}^4$). It is easy to estimate also the fields corresponding to the start of the overheating, putting for this purpose $\bar{\epsilon} \sim kT$ in (4). Then $e^2\mathscr{E}^2 \approx (ms\omega)^2$, which yields $\mathscr{E} \approx 2$ V/cm at $m \approx 10^{-28}$ g, $s = 5 \times 10^5$ cm/sec, and $\omega = 7 \times 10^{10}$ sec^{-1}. In fields 10 V/cm, the average electron energy becomes ~10°K, which should correspond to breakdown of the free excitons present in the sample.

The key factor in the explanation of the entire aggregate of the described experimental data is that the EHD, without being noticeably overheated in the considered field and without being destroyed by bombardment with free carriers, are at the same time effective trapping centers for electrons and holes. To verify this, let us estimate the time Γ_e^{-1} of trapping of an electron by one of the drops

$$\Gamma_e^{-1} = [4\pi N R^2 v_\epsilon]^{-1} = \frac{n_0}{3\bar{n}}\frac{\bar{R}}{v_\epsilon}; \tag{6}$$

Here $v_\epsilon = \sqrt{\bar{\epsilon}/m}$ is the average random electron velocity in a microwave field of intensity $\mathscr{E}$. Substituting in (6) $n_0 = 2 \times 10^{17}$ cm^{-3}, $\bar{n} \approx 10^{15}$ cm^{-3}, $\bar{R} \approx 10^{-3}$ cm and $v_\epsilon \approx 10^7$ cm/sec, we obtain $\Gamma_e \sim 10^{-8}$ sec.

Thus, in a sufficiently pure sample or under conditions when the other capture centers (due to impurities and defects) are filled after the exciting pulse, the mechanism considered here turns out to be decisive for the free-carrier density. The electric breakdown of the excitons is determined in this case by the ratio of the rate of multiplication of the free carriers as the result of impact ionization $\alpha(\mathscr{E}) n n_e$ ($\alpha = \langle \sigma_i v \rangle$ is the impact-ionization coefficient and σ_i is the cross section for the disintegration of the exciton by an electron) to the rate of capture by the drops, and also of the reciprocal process of the binding into excitons βn_e^2. The equations describing the change in the concentrations n and n_e thus take the form

$$\frac{dn_e}{dt} = \alpha n n_e - \beta n_e^2 - \left(\Gamma_e + \frac{1}{\tau_e}\right) n_e = \alpha n n_e - \beta n_e^2 - \tilde{\Gamma}_e n_e, \tag{7a}$$

$$\frac{dn}{dt} = -\alpha n n_e + \beta n_e^2 - \Gamma(n - n_T) - \frac{n}{\tau_{ex}}. \tag{7b}$$

In the right-hand side of (7) we have introduced, in addition to the already mentioned terms, also terms that take into account the recombination of the exciton

n/τ_{ex}, their capture by the drops Γ_n, and evaporation of the excitons from the drops Γn_T, and also other electron-capture mechanisms not connected with the drops, n_e/τ_e, so that $\bar{\Gamma}_e = \Gamma_e + 1/\tau_e$. Obviously, $\Gamma = v_T\Gamma_e/v_\epsilon$, where v_T is the thermal velocity of the excitons. (We assume that both the excitons and the free carriers are trapped by the drop in each collision.) The quantity n_T, which characterizes the rate of evaporation from the EHD, is determined by the temperature of the carriers in the drop, and if the drops are not overheated relative to the crystal lattice, it is equal to the concentration of the excitons that are in thermodynamic equilibrium with the EHD at the specified temperature T. In (7) we did not take into account the thermal dissociation of the excitons, assuming it to have low probability and not to play a fundamental role in the breakdown. In addition, we did not introduce explicitly any source of free carriers, since the latter are apparently of non-equilibrium origin.

The coefficients α and β depend on the carrier energy distribution and consequently on the field. The coefficient α increases rapidly (exponentially) with increasing $\mathscr{E}$, while β decreases. On the other hand the coefficients $\bar{\Gamma}_e$ and Γ depend explicitly on the time, for owing to the recombination of the carriers in the drop, with a time constant τ_0, the total surface area of all the drops decreases like $\exp(-2t/3\tau_0)$. In the absence of an electric field, the number of free carriers is quite small, and the solution of (7b), within a very short time after the exciting pulse, $\sim(\Gamma + 1/\tau_{ex})^{-1}$, assumes a quasistationary form

$$n(t) = \frac{\Gamma(t)\tau_{ex}}{1+\Gamma(t)\tau_{ex}} n_T. \qquad (8)$$

When a field pulse is applied at a certain instant of time, the criterion for the breakdown is obviously the condition $dn_e/dt > 0$, i.e., by virtue of (7a) we have

$$\alpha n(t_0) - \beta n_e(t_0) - \bar{\Gamma}_e(t_0) > 0. \qquad (9)$$

Since the breakdown develops within a very short time, the additional evaporation of the excitons from the EHD within this time can be neglected, i.e., to describe the breakdown spike one can omit from (7b) the term $\Gamma(n_T - n)$ and put $\bar{\Gamma}_e(t) \approx \bar{\Gamma}_e(t_0) = \text{const}$. Then the system (7) can be integrated and reduced to the form

$$\frac{d\ln n_e}{dt} = \frac{\alpha\bar{\Gamma}_e}{\alpha+\beta}\chi(n_e), \qquad (10)$$

while the function $\chi(n_e)$ is determined by the transcendental equation

$$(n_e - n_{e\,max})\frac{(\alpha+\beta)^2}{\alpha\bar{\Gamma}_e} = -\chi + \ln(1+\chi). \qquad (11)$$

The integration constant $n_{e\,max}$ corresponds to the solution of (11) with $\chi = 0$, i.e., by virtue of (10), to the maximum value of n_e in the breakdown spike, and can be expressed in terms of the initial values $n(t_0) = \bar{n}$ and $n_e(t_0) = \bar{n}_e$ by comparing (7a) with (10)–(11) at $t = t_0$:

$$n_{e\,max} = \bar{n}_e + \frac{\alpha\bar{\Gamma}_e}{(\alpha+\beta)^2}\{\chi(\bar{n}_e) - \ln(1+\chi(\bar{n}_e))\} = \bar{n}_e + \left\{\frac{1}{(\alpha+\beta)\bar{n}_e}\frac{dn_e}{dt}\right.$$
$$\left.\left. - \frac{\alpha\bar{\Gamma}_e}{(\alpha+\beta)^2}\ln\left(1 + \frac{\alpha+\beta}{\alpha\bar{\Gamma}_e}\frac{1}{n_e}\frac{dn_e}{dt}\right)\right\}\right|_{t=t_0} = \bar{n}_e + \frac{\alpha\bar{n} - \beta\bar{n}_e - \bar{\Gamma}_e}{\alpha+\beta} \qquad (12)$$
$$- \frac{\alpha\bar{\Gamma}_e}{(\alpha+\beta)^2}\ln\left[-\frac{\beta}{\alpha} + \left(1+\frac{\beta}{\alpha}\right)\frac{\alpha\bar{n} - \beta\bar{n}_e}{\bar{\Gamma}_e}\right].$$

The time of cascade development can be obtained by direct integration of (10):

$$t_{max} - t_0 = \frac{\alpha+\beta}{\alpha\bar{\Gamma}_e}\int_{\bar{n}_e}^{n_{e\,max}} \frac{dn_e}{n_e\chi(n_e)}. \qquad (13)$$

The subsequent decrease occurs with a time constant $\bar{\Gamma}_e/(1 + \beta/\alpha)$ since it is easy to verify from (10) and (11) that $\chi(n_e) \to -1$ at $dn_e/dt < 0$ and $(n_{e\,max} - n_e) \times(\alpha + \beta)^2/\alpha\bar{\Gamma}_e \gg 1$.

To simplify the subsequent formulas and estimates, we shall assume that $\beta n_{e\,max} \ll \bar{\Gamma}_e$, which is apparently correct in the case of interest to us, for if the process of binding into excitons, which is quadratic in n_e, were to play an essential role, the breakdown could not be abrupt. Assuming therefore $\beta/\alpha \ll 1$, we can rewrite (12) in the compact form

$$n_{e\,max} = \bar{n}_e + \bar{n}\left[1 - \frac{\bar{\Gamma}_e}{\alpha\bar{n}}\left(1 + \ln\frac{\alpha\bar{n}}{\bar{\Gamma}_e}\right)\right]. \qquad (14)$$

Near the threshold, when $\alpha n_{e\,max} \ll \bar{\Gamma}_e$, relation (11) reduces to

$$\chi^2 \approx \frac{2\alpha}{\bar{\Gamma}_e}(n_{e\,max} - n_e), \qquad (11a)$$

and Eq. (10)

$$\frac{d\ln n_e}{dt} = \sqrt{2\alpha\bar{\Gamma}_e(n_{e\,max} - n_e)} \qquad (10a)$$

has the solution

$$n_e = n_{e\,max}/\mathrm{ch}^2\left[\sqrt{2\alpha\bar{\Gamma}_e n_{e\,max}}(t - t_{max})\right]. \qquad (15)$$

The breakdown spike is in this case almost symmetrical, and its duration $(\alpha\bar{\Gamma}_e n_{e\,max})^{-1/2}$ decreases with increasing $n_{e\,max}$, i.e., of the field. When the breakdown is fully developed and $\alpha\bar{n} \gg \bar{\Gamma}_l$, $n_{e\,max} \sim \bar{n}$, and the expansion (11a) is no longer valid, the spike becomes strongly asymmetrical, namely, its trailing edge falls off, as already noted, with a time constant $\bar{\Gamma}_e^{-1}$, and the leading front increases much more rapidly within a time $\sim(\alpha n)^{-1}$.

The foregoing solution is valid in a time interval $\ll \Gamma^{-1} \sim v_\epsilon\Gamma_e^{-1}/v_T$. At larger times it is no longer possible to neglect in (7b) the evaporation of the excitons from the drops. Accordingly, after the number of free carriers has decreased very strongly following the breakdown, the number of excitons again begins to increase slowly, and after a time $\sim\Gamma^{-1}$, which exceeds by one or two orders of magnitude the duration of the breakdown spike, the breakdown can repeat, in agreement with the experimental results. The considered picture makes it possible to describe in a natural manner also other observed regularities in the breakdown. The breakdown criterion (9) as $\beta \to 0$, with allowance made for (8) and for the explicit form of the dependence of Γ on Γ_e, on the time and the connection between Γ and Γ_e, can be easily reduced to the form

$$\alpha n_T\tau_{ex} = \frac{v_\epsilon}{v_T} + \frac{\tau_{ex}}{\tau_e} + 2\sqrt{\frac{v_\epsilon}{v_T}\frac{\tau_{ex}}{\tau_e}}\,\mathrm{ch}\left(\frac{2}{3}\frac{t-t_0}{\tau_0}\right), \qquad (16)$$

where

$$t_0 = \frac{3}{2}\tau_0\ln\left(\sqrt{\frac{v_\epsilon}{v_T}\tau_e\tau_{ex}}\Gamma_0\right), \qquad (17)$$

$\Gamma_0 = \Gamma(t = 0)$ is the value at the initial instant, directly after the formation of the drops, which is obviously proportional $I^{2/3}$, where I is the intensity of the exciting pulse. Formulas (16) and (17) agree fully with the experimental relations (figs. 4, 6, 7). The time constant τ_0 determined by comparing (17) with Fig. 7 turns out to be 35 μsec, in reasonable agreement with the lifetimes of EHD known from other studies.

The presence of the first breakdown region at relatively small delays, $t_d < 40$ μsec, is more readily due to the weak overheating of the sample after the exciting pulse, as a result of which the evaporation of the exciton from the drops increases, and consequently an increase takes place in the value of $\bar{n}$ in the criterion (9). We note that this overheating can be very small ($\lesssim 1^\circ$K), since the dependence of n_T on the temperature is exponential.

Let us discuss now the dependence of the breakdown threshold on the duration of the microwave pulse.

In all the preceding arguments we have assumed that the field does not change during the time of development of the breakdown spike. However, if $\Gamma_e\tau < 1$, i.e., the carriers do not manage to be captured by the drops within the time of action of the pulse, then it is clear that the breakdown criterion is equivalent to the condition $\alpha\bar{n}\tau \sim 1$, i.e., to the requirement that each electron experience during the time of action of the pulse at least one ionization collision. Consequently, with further decrease of τ the impact ionization coefficient α corresponding to the type of the breakdown should increase, and with it also the threshold breakdown field. In experiment (Fig. 5) this growth begins at $\tau < 0.5 \times 10^{-6}$ sec, from which it follows that $\Gamma_e \gtrsim 2 \times 10^6$ sec^{-1}, in qualitative agreement with the estimates based on the trailing edge of the breakdown spike $\Gamma_e \approx 5 \times 10^6$ sec^{-1}. Assuming as an average estimate $\Gamma_e = 3\bar{n}v_\epsilon/n_0R \approx 3 \cdot 10^6$ sec^{-1} and taking $v_e \approx 10^7$ cm/sec, $n_0 \approx 2 \times 10^{17}$ cm^{-3} and $\bar{n} \approx 2 \times 10^{13}$ cm^{-3} (the last figure was calculated under the assumption that each incident photon produces on electron-hole pair with allowance for the subsequent decay of the carriers within 100 μsec with a time constant $\tau_0 \approx 35$ μsec), we can estimate the average drop radius R $\approx 10^{-3}$ cm and the number of drops per unit volume $N \approx 5 \times 10^4$ cm^{-3}. Both estimates agree in order of magnitude with the results of the direct measurement of N and R by the light-scattering method[22,23]. Thus, the model in which the electron-hole drops play the role of trapping centers for the free carriers explains practically all the experimentally observed regularities of exciton microwave breakdown in germanium at $T \lesssim 2^\circ$K.

We are unable at present to suggest a satisfactory alternate description for all these facts, based on the assumption that the excitons are bound not into electron-hole drops but into biexcitons. The model proposed in[15] attributed the breakdown delay to the fact that after the exciting pulse the electron collision frequency ν, which enters in (1), is determined by electron-biexciton collisions and $\nu \gg \omega$. As the biexciton concentration decreases this frequency decreases and the energy absorbed by each electron increases, so that the large heating, at a given microwave power, occurs at the instant determined by the condition $\nu \approx \alpha$. However, under the conditions $\nu \gg \omega$ we have $\Delta\epsilon'' \gg \Delta\epsilon'$, which contradicts the experimental results, as does also the fact that $\Delta\epsilon' \sim n_e/\nu^2 \sim n_e/n_B^2$ should increase as n_B decreases with time. In addition, direct measurements by the cyclotron-resonance method[24] show that the inverse condition, $\nu \ll \omega$, is satisfied.

An attempt to explain the delay of the breakdown in analogy with the procedure used above for the EHD model, assuming that the biexcitons capture free carriers, also leads to difficulties, even if complexes of the negative-ion type were to exist. The point is that the biexciton ionization energy (the energy for its disintegration into an electron, hole, and an exciton) exceeds the exciton ionization energy by certainly less than a factor of 2, and the average electron energy $\bar{\epsilon}$ increases with increasing field, as we have seen, in proportion to $\mathcal{E}^4$. Therefore the breakdown field for the bielectrons therefore exceeds the breakdown field for excitons by not more than 20%. In fact, the breakdown of the biexcitons is more likely to set in earlier, inasmuch as under the experimental conditions their concentration should be higher by several orders of magnitude than the exciton concentration. The condition for biexciton breakdown is $\alpha_B n_B n_e - \gamma n_B n_e > 0$, where α_B is the coefficient of biexciton impact ionization, n_B is the biexciton concentration, and γ is a coefficient that determines the rate of capture of the electrons by the biexcitons. This criterion reduces simply to the obvious condition $\alpha_B > \gamma$ and does not depend on the concentration n_B, and consequently on the time, i.e., it cannot explain the delay of the breakdown. No less important a difficulty in the considered "gas" model of the breakdown is that within the framework of this model there is no visible explanation of the fact that the breakdown spike develops and terminates within a very short time ($\lesssim 1$ μsec), and then does not repeat after a rather long time (~ 10 μsec), in spite of the continuing action of the microwave pulse.

Thus, the theory of microwave breakdown of excitons in the presence of EHD explains the entire aggregate of the presented experimental data, including also the presence and position of the minimum on the plot of the breakdown threshold against the microwave-pulse delay time, the waveform of the breakdown spike, the dependence of the threshold power on the microwave pulse duration and on the optical-excitation power, and the possibility of a repeated breakdown only after a sufficiently long time.

Using the investigated dependence of the breakdown threshold on the microwave pulse duration, we were able to calculate the radius and the number of the EHD, which hitherto were known only from experiments on light scattering[22,23]. A similar analysis of the breakdown effect assuming biexciton existence does not make it possible to explain the observed regularities of the breakdown.

The authors are grateful to A. M. Prokhorov, V. S. Vavilov, V. S. Bagaev, and T. N. Galkina for useful discussions, and to V. A. Sanina, A. S. Seferov, and S. P. Smolin for help with the experiments.

1)We note that approximately the same characteristic times were observed in the kinetics of recombination radiation of germanium [20].

2)We note that by virtue of (1) the contribution made to ϵ' by the free carriers is much lower at higher frequencies. This may be due to the fact that in [8] they observed $\Delta\epsilon' > 0$, i.e., the effect was not determined by the free carriers.

3)This sample was manufactured by the General Electric Company [21].

1L. V. Keldysh, Proc. 9-th Internat. Conf. on Semiconductor Physics, Mir (1968), p. 1387.

2L. V. Keldysh, in: Éksitony v poluprovodnikakh (Excitons in Semiconductors), Nauka (1971), p. 5.

3M. Combescot and P. Noziers, J. Phys. C. Sol. Stat. Phys., 5, 2369 (1972).

[4]W. Brinkman, T. Rice, P. Anderson and S. Chui, Phys. Rev. Lett., 28, 961 (1972).
[5]Ya. E. Pokrovskiĭ and K. I. Svistunova, ZhETF Pis. Red. 9, 435 (1969) [JETP Lett. 9, 261 (1969)].
[6]V. S. Bagaev, T. I. Galkina, and O. V. Gogolin, in: Éksitony v poluprovodnikakh (Excitons in Semiconductors), Nauka (1971), p. 19.
[7]V. S. Vavilov, V. A. Zayats, and V. N. Murzin, ZhETF Pis. Red. 10, 304 (1969) [JETP Lett. 10, 192 (1969)].
[8]I. C. Hensel and T. G. Phillips
[9]C. Benoit a lâ Guillaume and M. Voos, Phys. Rev. B7, 4, 1723 (1973).
[10]V. M. Asnin, B. V. Zubov, T. M. Murina, A. M. Prokhorov, A. A. Rogachev, and N. I. Savlina, Zh. Eksp. Teor. Fiz. 62, 737 (1972) [Sov. Phys.-JETP 35, 390 (1972)].
[11]V. M. Asnin, A. A. Rogachev, and N. I. Savlina, Tekh. Poluprovdn. 5, 802 (1971) [Sov. Phys.-Semicond. 5, 712 (1971)].
[12]P. S. Gladkov, B. G. Zhurkin, and N. A. Penin, Fiz. Tekh. Poluprovodn. 6, 1919 (1972) [Sov. Phys.-Semicond. 6, 1649 (1973)].
[13]B. M. Ashkinadze and F. K. Sultanov, ZhETF Pis. Red. 16, No. 5, 271 (1972 [JETP Lett. 16, 190 (1972)].
[14]T. Sanada, T. Ohyama and E. Otsuka, Sol. St. Comm., 12, 1201 (1972).
[15]A. A. Manenkov, V. A. Milyaev, G. N. Mikhaĭlova, and S. P. Smolin, ZhETF Pis. Red. 16, 454 (1972) [JETP Lett. 16, 322 (1972)].
[16]P. Suleebka and R. Snrau, J. Phys. D. Appl. Phys., 5, 97 (1972).
[17]Yu. K. Danileĭko, A. A Manenkov, A. M. Prokhorov, and V. A. Khaimov-Mal'kov, Zh. Eksp. Teor. Fiz. 58, 31 (1970) [Sov. Phys.-JETP 31, 18 (1970)].
[18]A. D. MacDonald, Microwave Breakdown in Gases, Wiley, 1966.
[19]B. V. Novikov, E. F. Gross, and M. A. Drygin, ZhETF Pis. Red. 8, 15 (1968) [JETP Lett. 8, 8 (1968)].
[20]B. V. Zubov, V. P. Kalinushkin, T. M. Murina, A. M. Prokhorov, and A. A. Rogachev, Fiz. Tekh. Poluprovdn. 7, 1614 (1973) [Sov. Phys.-Semicond. 7, 1077 (1974)].
[21]R. N. Hall and T. G. Solys, IEEE Trans., NS-18, 160 (1971).
[22]Ya. E. Pokrovskiĭ and K. I. Svistunova, ZhETF Pis. Red. 13, 297 (1971) [JETP Lett. 13, 212 (1971)].
[23]V. S. Bagaev, N. A. Penin, N. N. Sibel'din, and V. A. Tsvetkov, Fiz. Tverd. Tela 15, 177 (1973) [Sov. Phys.-Solid State, 15, 121 (1973)].
[24]P. S. Gladkov, Candidate's dissertation, Moscow, FIAN (1972).

Translated by J. G. Adashko.
222

Absorption of ultrasound by electron-hole drops in a semiconductor

L. V. Keldysh and S. G. Tikhodeev

P. N. Lebedev Physics Institute, USSR Academy of Sciences
(Submitted April 11, 1975)
ZhETF Pis. Red. **21**, No. 10, 582–585 (May 20, 1975)

The interaction of ultrasound with electron-hole drops (EHD) in a semiconductor is considered. The sound absorption coefficient and the drift velocity of the EHD dragged by the sound are obtained. In the case of small drops, the temperature dependence of the absorption coefficient has a sharp maximum, and the EHD drift velocity reaches the speed of sound even at very small amplitudes of the sound wave.

PACS numbers: 62.80., 71.85.C

At a sufficiently high density of the electrons and holes in a semiconductor at low temperatures, a first-order phase transition is possible, in which the gas of the nonequilibrium carriers stratifies into two phases—gaseous and liquid—in the form of an electron-hole drop (EHD) having an equilibrium carrier density n_0 and an average radius R, and distributed in the volume of the semiconductor with a density N_0.[1] This liquid is a degenerate electron-hole plasma and interacts effectively with ultrasound, while the mechanism whereby the ultrasound is absorbed in the semiconductor filled with the EHD depends on the ratio of the wavelength λ and dimensions of the EHD.

At $\lambda \ll R$, the absorption of sound is the absorption of a spatially-homogeneous electron-hole Fermi liquid and is described by known formulas,[2] except that account must be taken of the fact that not the entire volume of the semiconductor is filled with drops.

On the other hand, if $\lambda \gg R$, the absorption mechanism is essentially different. In this approximation, each electron-hole pair in the EHD is acted upon by equal forces in the field of the inhomogeneous deformation,[3] so that the EHD are set to oscillate under the influence of the ultrasound and dissipate energy by interacting with the thermal phonons of the lattice. The friction of the drop against the lattice at low drop velocities $V \ll S$ (S is the speed of sound of the semiconductor) will be viscous, with a kinematic friction coefficient[4]

$$\gamma = \frac{2}{3(2\pi)^3}\,\frac{m_e^2 D_e^2 + m_h^2 D_h^2}{\hbar^2 (m_e + m_h)\rho S}\left(\frac{k_0 T}{\hbar S}\right)^5 \int_0^{\xi_0} \xi^5 \frac{\exp \xi}{(\exp\xi - 1)^2}\, d\xi\,, \tag{1}$$

where $\xi_0 = 2\hbar s(3\pi^2 n_0)^{1/3}/k_0 T$, T is the crystal temperature, ρ is its density, m_e, m_h and D_e, D_h are the mass and the deformation potential of the electron and of the hole, respectively.

In a cubic single-valley semiconductor with nondegenerate bands, a small drop ($R \ll \lambda$) in the field of a longitudinal acoustic wave of frequency ω and wave vector **k** directed along the OX axis, we have $\epsilon = \epsilon_0 \sin(\omega t - kx)$ ($\epsilon = T\eta\epsilon_{ij}$ is the deformation of the crystal), is acted upon by the force

$$\mathrm{F} = \frac{4}{3}\pi R^3 n_0 D\mathbf{k}\epsilon_0 \cos(\omega t - k\xi), \tag{2}$$

where ξ is the x coordinate of the drop, and $D = D_e + D_h$.

The equation of motion of the drop in the field of the force (2) will take the form

$$\frac{d^2\xi}{dt^2} + \gamma\frac{d\xi}{dt} = \frac{Dk\epsilon_0}{m}\cos(\omega t - k\xi), \text{ where } m = m_e + m_h. \tag{3}$$

We seek the solution of (3) in the form $\xi = Vt + x(t)$, where V is the constant drift velocity of the EHD and $x(t)$ are small oscillations (with amplitude $x_0 \ll \lambda$). Then $\cos(\omega t - k\xi) = \cos\Omega t + kx\sin\Omega t$, ($\Omega = \omega - kV$), and (3) takes the form

$$\frac{d^2 x}{dt^2} + \gamma\frac{dx}{dt} + \gamma V = \frac{Dk\epsilon_0}{m}\cos\Omega t + \frac{Dk^2\epsilon_0}{m}x\sin\Omega t. \tag{4}$$

Solving (5), we obtain

$$V = \frac{D^2\epsilon_0^2}{2m^2}k^3\frac{1}{\Omega(\Omega^2+\gamma^2)}\,, \tag{5}$$

$$x = \frac{Dk\epsilon_0}{(\Omega^2+\gamma^2)m}\left(\frac{\gamma}{\Omega}\sin\Omega t - \cos\Omega t\right). \tag{6}$$

At $\omega \sim \gamma$, the condition $x_0 \ll \lambda$ yields $V \ll S$ at ultrasound intensities ($I = \rho S^3\epsilon_0^2/2$):

$$I \ll \frac{m^2\rho S^7}{D^2} \sim 10\ \mathrm{W/cm^2} \tag{7}$$

At $D \sim 10$ eV, $m \sim 10^{-27}$ g, $\rho \approx 5$ g-cm^{-3}, and $S \approx 5\times10^5$ cm-sec^{-1}, Eq. (6) enables us to determine the energy absorbed by the drop per unit time at $V \ll S$

$$\overline{\mathrm{F}\frac{dx}{dt}} = -\frac{\frac{4}{3}\pi R^3 n_0 D^2 k^2 \epsilon_0 \gamma}{2m(\gamma^2+\omega^2)}$$

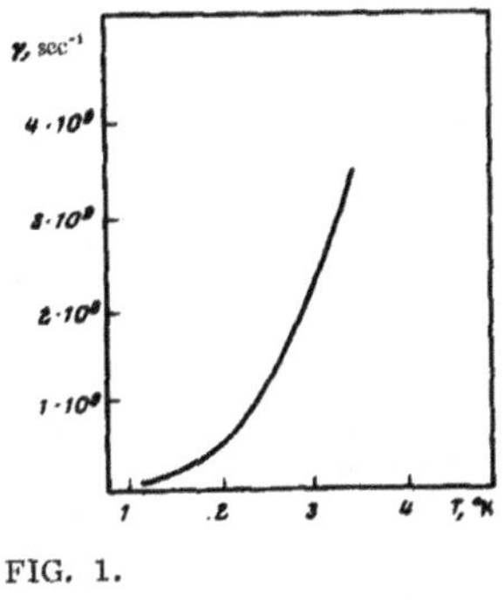

FIG. 1.

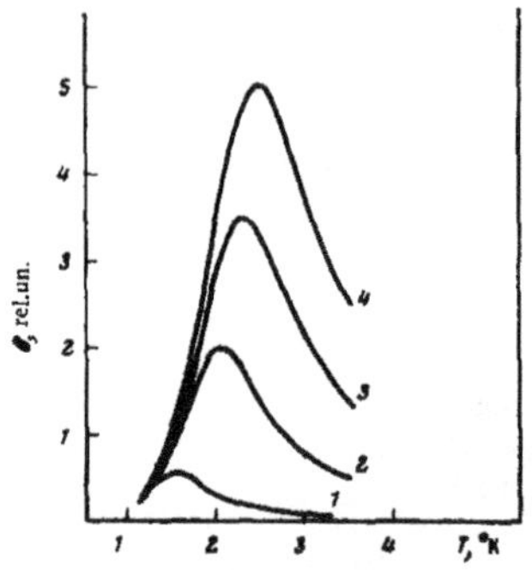

FIG. 2. Temperature dependence of the ultrasound absorption coefficient at frequencies ω equal to: 1) 10^8, 2) 4×10^8, 3) 7×10^8, 4) 10^9 sec^{-1}.

and the ultrasound absorption coefficient due to the entire aggregate of the drops is

$$\delta = \left|\frac{d\ln I}{dx}\right| = \bar{n}\frac{D^2}{\rho m S^5}\frac{\gamma\omega^2}{\gamma^2+\omega^2}, \tag{8}$$

where $\bar{n}=(4/3)\pi R^3 n_0 N_0$ is the average density of the condensate over the entire semiconductor. At $\lambda \gg R$, the absorption coefficient does not depend on the dimensions of the drops. In order of magnitude, we have $\delta \sim 1-10$ cm^{-1} at $\omega \sim 10^9$ sec^{-1} and $\bar{n}\sim 10^{14}$ cm^{-3}. Figures 1 and 2 show the temperature dependences of γ and δ (at different frequencies), calculated from formulas (1) and (8) respectively (at $D=5$ eV, $m=0.5\times10^{-27}$ g, $n_0=2\times10^{17}$ cm^{-3}, $S=5\times10^5$ cm-sec^{-1}, and $\rho=5$ g/cm^3). The quantity δ is given in relative units, since its absolute value is proportional to the excitation level of the semiconductor, which depends on the experiment. As seen from these figures, the most characteristic feature of the absorption of ultrasound by small EHD at low ultrasound intensities is its sharply pronounced resonant dependence on the temperature.

At intensities $I > (\gamma^2/\omega^2)(m^2\rho S^7/2D^2)$, the drifts will be completely dragged by the ultrasonic wave: $x = St + x_0$. Equation (3) yields in this case

$$x_0 = \frac{1}{k}\arccos\frac{2\gamma S m}{D k \epsilon_0} = \frac{1}{k}\arccos\left(\frac{\gamma}{\omega}\frac{mS^2}{D}\sqrt{\frac{\rho S^3}{2I}}\right). \tag{9}$$

Although formally we cannot use formula (1) at $V \sim S$, there are grounds for assuming that γ remains of the same order in this case.[5]

The considered mechanism of ultrasound absorption is of course not confined to cubic single-valley crystals. It should exist in semiconductors with arbitrary band structure and anisotropy, and can occur for all types of acoustic waves, not only pure longitudinal ones.

[1] Ya. Pokrovskii, Phys. Stat. Sol. (a)11, 385 (1972).
[2] A. I. Akhiezer, M. I. Kaganov, and G. Ya. Lyubarskiĭ, Zh. Eksp. Teor. Fiz. **32**, 837 (1957) [Sov. Phys.-JETP **5**, 685 (1957)].
[3] V. S. Bagaev, T. I. Galkina, O. V. Gogolin, and L. V. Keldysh, ZhETF Pis. Red. **10**, 309 (1969) [JETP Lett. **10**, 195 (1969)].
[4] L. V. Keldysh, in: Éksitony v poluprovodnikakh (Excitons in Semiconductors), Nauka (1974).
[5] S. G. Tikhodeev, Kratkie soobshcheniya po fizike, FIAN SSSR (in press).

Phonon wind and dimensions of electron-hole drops in semiconductors

L. V. Keldysh

P. N. Lebedev Physics Institute, USSR Academy of Sciences
(Submitted December 1, 1975)
Pis'ma Zh. Eksp. Teor. Fiz. **23**, No. 2, 100–103 (20 January 1976)

It is shown that a flux of nonequilibrium phonons produced upon recombination of electrons and holes leads to instability of rather large volumes of an electron-hole liquid and to their breakup into smaller drops.

PACS numbers: 71.85.Ce

Electron-hole drops (EHD) are sources of intense fluxes of nonequilibrium phonons produced in nonradiative recombination of electrons and holes. After being absorbed or scattered by the electron-hole liquid, these phonons transfer to it part of their quasimomentum, and this is equivalent from the macroscopic point of view to the action of a certain volume force $\mathbf{f}(\mathbf{r})$ proportional to the local phonon energy flux $\mathbf{w}(\mathbf{r})$ [1]

$$\mathbf{f}(\mathbf{r}) = A\mathbf{w}(\mathbf{r}) \ . \tag{1}$$

The EHD volume element δV located in the vicinity of the point $\mathbf{r}$ produces at a point $\mathbf{r}'$ a flux

$$\delta \mathbf{w}(\mathbf{r}') = \frac{B}{4\pi}\,\frac{\mathbf{r}'-\mathbf{r}}{|\mathbf{r}'-\mathbf{r}|^3}\,\delta V \ , \tag{2}$$

where B is the energy radiated in phonons by a unit EHD volume in a unit time. It follows from (1) and (2) that the two volume elements δV and $\delta V'$ located in the vicinities of the points $\mathbf{r}$ and $\mathbf{r}'$, respectively, repel each other with a force

$$\mathbf{F} = \frac{AB}{4\pi}\,\frac{\mathbf{r}'-\mathbf{r}}{|\mathbf{r}'-\mathbf{r}|^3}\,\delta V \delta V' \ . \tag{3}$$

In other words, the volume forces produced in the electron-hole liquid by the "phonon wind" are exactly the same as would be produced if this liquid were uniformly charged, with charge density equal, by virtue of formula (1) for $\mathbf{f}$ and $\mathbf{w}$, to

  86

$$\rho^2 = \frac{AB}{4\pi} = \begin{cases} \dfrac{a^{(abs)}}{4\pi} \dfrac{n_0 E_g}{\tau_0} \dfrac{d^2 m^2}{\hbar^3 \rho_c s^2} |\mathbf{k}|, & \hbar|\mathbf{k}| < 2p_F \quad (4a) \\ \dfrac{a^{(sc)}}{4\pi} \dfrac{n_0 E_g}{\tau_0} \left(\dfrac{d^2 m^2}{\hbar^3 \rho_c}\right)^2 \left(\dfrac{|\mathbf{k}|^4}{\omega_{\mathbf{k}}^3}\right), & \hbar|\mathbf{k}| \gg 2p_F \quad (4b) \end{cases},$$

where n_0 and τ_0 are the equilibrium density and the lifetime of the electrons and holes in the EHD, E_g is the width of the forbidden band, d and m are certain mean values of the deformation potentials and the effective masses, ρ_c is the density of the crystal, s is the speed of sound, $a^{(abs)}$ and $a^{(sc)}$ are numerical coefficients (each $\sim 10^{-2}$) that depend on the details of the band structure and the anisotropy of the electron-phonon interaction, $\mathbf{k}$ and $\omega_{\mathbf{k}}$ are the wave vector and frequency of the phonons, and p_F is the Fermi momentum of the electrons and holes. The difference between (4a) and (4b) is due to the fact that at $\hbar|\mathbf{k}| < 2p_F$ the principal role is played by phonon absorption processes, and at $\hbar|\mathbf{k}| \gg 2p_F$ the scattering predominates.

An obvious consequence of these arguments is the conclusion that sufficiently large volumes of the electron-hole liquid cannot be stable and must inevitably break up into smaller EHD. Consider two typical situations: a spherical EHD of radius R, which grows gradually from an exciton cloud, and a plane layer of electron-hole liquid of thickness L, which can be produced and apparently is really produced in experiments in the case of very brief but quite intense lasing of a semiconductor. Elementary calculations of the oscillations of the shape of the surface of a uniformly charged incompressible liquid, analogous to[2], show that when a certain critical radius R_c is reached the EHD becomes unstable to deformations of the quadrupole type, which cause it to be divided into two parts, and

$$R_c = \left(\frac{15}{2\pi}\frac{\alpha}{\rho^2}\right)^{1/3}, \tag{5}$$

where α is the surface-tension coefficient of the EHD.

The problem of the oscillations of a flat layer under the influence of phonon wind is similar to the well-known problem of capillary-gravitational waves,[2] subject only to the fundamental difference that the force $f = 4\pi\rho^2 L$ is directed along the outward normal to the liquid surface, and it is this which leads at $L > L_c = 0.385 R_c$ to instability of all the surface waves with wave vectors q satisfying the condition $\alpha \mathbf{q}^2 + 2\pi\rho^2 q^{-1} - 4\pi\rho^2 L < 0$. At $L \gg L_c$ the growth increment is maximal for $q = R_L^{-1}$, where

$$R_L = \left(\frac{3}{4\pi}\frac{\alpha}{\rho^2 L}\right)^{1/2} = \left(\frac{1}{10}\frac{R_c}{L}\right) R_c . \tag{6}$$

The development of this instability should lead to a breakup of the initial layer into EHD with radii $\sim R_L$, which move apart under the influence of the phonon wind with velocity $v = 4\pi\rho^2 L (M n_0 \gamma)^{-1}$, where M is the sum of the effective masses of the electron and the hole, and γ^{-1} is the time of velocity relaxation (deceleration) of the EHD.

Formulas (4) show that the intensity of the phonon wind depends strongly on the phonon frequency distribution. Thus, for the parameters of germanium ($d=4$ eV, $m=4\times10^{-28}$ g, $\rho_c\approx5$ g/cm^3, $s=s_t\approx3\times10^5$ cm/sec, $n_0=2\times10^{17}$ cm^{-3}, $\tau_0=4\times10^{-5}$ sec, $E_g=0.74$ eV), depending on $|\bar{\mathbf{k}}|$, the values of ρ vary in the interval $(0.1-1.0)\times10^3$ g$^{1/2}$cm$^{-3/2}$ sec^{-1}. Assuming $\alpha=2\times10^{-4}$ g/sec^2 we obtain $R_c=(0.7-3.5)\times10^{-3}$ cm and $R_L=(0.15-1.5)L^{-1/2}\times10^{-4}$ cm. For silicon we obtain values of R_L and R_c smaller approximately by one order of magnitude.

The presented values of R_c are in reasonable agreement with the observed[4] limit $R\leq1\times10^{-3}$ cm, and do not contradict the large EHD observed in[5], since the uniaxial deformations used in[5] strongly attenuate the phonon wind.[1] There are also experimental indications in[6] that an electron-hole liquid is scattered in the form of minute drops following a strong pulsed excitation. The foregoing considerations allow us to describe this scattering. It is easy to show that an EHD separated from the initial layer at the instant of time t_0 and at the point z_0 ($z_0\leq L$, where z is the distance from the surface of the sample), moves subsequently in accordance with the law

$$z(t)=z_0\left[1+\frac{4\pi\rho^2\tau_0}{Mn_0\gamma}\left(1-e^{-\frac{t-t_0}{\tau_0}}\right)\right]. \tag{7}$$

On the other hand, if the pulse of the exciting radiation is focused, so that the initially produced liquid is in the shape of a sphere of radius $R\gg R_c$, then this volume breaks up into drops with radii that differ from (6) only in the replacement $L\to R/3$, and are scattered in accordance with the law

$$r(t)=r_0\left[1+\frac{4\pi\rho^2\tau_0}{Mn_0\gamma}\left(1-e^{-\frac{t-t_0}{\tau_0}}\right)\right]^{1/3},\quad r_0\leqslant R. \tag{8}$$

Formulas such as (7) and (8) describe also the case when the initial pulse produces a cloud of minute EHD, if the second terms in the right-hand sides of these formulas are multiplied by the fraction $\bar{n}/n_0$ of the volume occupied by the liquid phase in the cloud. At high temperatures, when most nonequilibrium carriers recombine in the gas phase, τ_0 in Eqs. (7) and (8) must be replaced by the effective lifetime τ_{eff} of the system of nonequilibrium carriers.

[1]V.S. Bagaev, L.V. Keldysh, N.N. Sibel'din, and V.A. Tsvetkov, Preprint FIAN, 1975; Zh. Eksp. Teor. Fiz. **70**, 702 (1976) [Sov. Phys.-JETP **43**, No. 2 (1976)].
[2]L.D. Landau and E.M. Lifshitz, Mekhanika sploshnykh sred (Fluid Mechanics) 61, GITTL, Moscow, 1953, [Pergamon, 1958].
[3]L.M. Sander, H.B. Shore, and L.J. Sham, Phys. Rev. Lett. **31**, 533 (1973); H. Büttner and E. Gerlach, J. Phys. **C6**, L433 (1973); T.M. Rice, Phys. Rev. **B9**, 1540 (1974); T.L. Reinecke and S.C. Ying, Solid State Commun. **14**, 381 (1974); V.S. Bagaev, N.N. Sibel'din, and V.A. Tsvetkov, Pis'ma Zh. Eksp. Teor. Fiz. **21**, 180 (1975) [JETP Lett. **21**, 80 (1975)]; T.L. Reinecke and S.C. Ying, Phys. Rev. Lett. **35**, 311 (1975).
[4]A.S. Alexeev, T.A. Astemrrov, V.S. Bagaev, T.I. Galkina, N.A. Penin, N.N. Sybeldin, and V.A. Tsvetkov, Proc. Twelfth Intern. Conf. on Physics of Semiconductors, Stuttgart, 1974, p. 91; V.S. Bagaev, N.V. Zamkovets,

L.V. Keldysh, N.N. Sibel'din, and V.A. Tsvetkov, Preprint FIAN, 1975; Zh. Eksp. Teor. Fiz. **70**, 1501 (1976) [Sov. Phys.-JETP **43**, No. 4 (1976)].
[5]R.S. Markiewicz, J.P. Wolfe, and C.D. Jeffries, Phys. Rev. Lett. **32**, 1357 (1974); J.P. Wolfe, R.S. Markiewicz, C. Kittel, and C.D. Jeffries, Phys. Rev. Lett. **34**, 275 (1975); J.P. Wolfe, W.L. Hansen, E.E. Haller, R.S. Markiewicz, C. Kittel, and C.D. Jeffries, Phys. Rev. Lett. **34**, 1292 (1975).
[6]J.M. Hvam and I. Balslev, Phys. Rev. **B11**, 5052 (1975); J.M. Worlock, Soviet-American Symposium on the Theory of Light Scattering in Solids, Moscow, 1975.

Dragging of excitons and electron-hole drops by phonon wind

V. S. Bagaev, L. V. Keldysh, N. N. Sibel'din, and V. A. Tsvetkov

P. N. Lebedev Physics Institute, USSR Academy of Sciences
(Submitted August 19, 1975)
Zh. Eksp. Teor. Fiz. **70**, 702–716 (February 1976)

Dragging of excitons by nonequilibrium short-wave phonons is studied theoretically. It is shown that in multivalley semiconductors with degenerate valence bands (such as Ge or Si) the cross section for scattering of short-wave phonons by indirect excitons may exceed by several orders of magnitude the scattering cross section calculated for the case of isotropic and nondegenerate bands. The kinetics of electron-hole drop (EHD) growth is also considered with allowance for the diffusion of excitons to the drop surface and dragging of the excitons by the phonon wind. By invoking the dragging effect one can explain a number of experiments in which the EHD "diffusion" coefficient is measured: moreover, the dragging effect may restrict the growth of the EHD. Results are presented of experimental observation of EHD motion induced by the phonon wind.

PACS numbers: 71.80.+j

Despite the appreciable progress in the understanding of the properties of electron-hole drops (EHD), [1–3] many facts connected with the kinetics of drop formation and growth and with the spatial distribution and motion remain unclear. These include experiments aimed at determining the coefficient of EHD "diffusion"[2,4–7] and the limitation of the EHD radius at temperatures close to threshold.[8,9] The EHD diffusion coefficient was measured by different workers,[2,4–7,9–12] but there are colossal discrepancies (six orders of magnitude) between the results of different measurements. In most studies the diffusion coefficient was determined by measuring the dimensions of the region occupied by the EHD. In[2,4,7], where a relatively low excitation intensity was used, a value $D < 1\ \mathrm{cm}^2/\mathrm{sec}$ was obtained for the diffusion coefficient. A value $D \approx 150$ is cited in[5] and values and 25–500 cm²/sec are cited in[6], depending on the experimental conditions, with an increase of D observed with increasing excitation level, while in[11] it is shown that when the excitation level is raised from 14 to 600 mW the "diffusion" coefficient increases from 0.8 to 80 cm²/sec. The EHD diffusion coefficient was estimated from experiments on the motion of the EHD in a uniform deformation field,[9,10]; in these experiments they measured directly the EHD momentum relaxation time. The obtained EHD diffusion coefficient was $D \approx 10^{-4}$–10^{-3} cm²/sec, depending on the EHD radius assumed in the estimate. Since the dimensions of the EHD are sufficiently well known from light-scattering experiments,[2,3,8,9,12] it can be assumed that this value is not far from the true one. An estimate of the diffusion coefficients from mobility measurements in experiments aimed at determining the EHD charge yields approximately the same value.[11,12]

Thus, experiments on the motion of EHD[9,10] are apparently the most direct way of measuring the diffusion coefficients. The results of experiments in which the size of the region occupied by the EHD was measured[2,4–7,11,12] suggest that some EHD drift mechanism exists in which the drift velocity increases with increasing excitation intensity. This assumption is confirmed by results of studies[13,14] that have demonstrated that the region occupied by the EHD is spherical and has a sharp edge—a fact difficult to explain from the point of view of EHD diffusion. It was suggested in[11,12,15,16] that the EHD are dragged by excitons in the presence of an exciton density gradient. Estimates show, however, that this effect should be very small, especially at low temperatures, when the exciton density is negligible, whereas experiment yields a contrary result.[6,17] Dragging of EHD in an electric field by a current of free electrons and holes was considered in[18]. However, the measurements cited above were performed in the absence of an electric field, and furthermore the estimates made in[18] are not sufficiently well founded, since the concentration of the free carriers is usually much less than is assumed in that paper.

We consider in the present paper the dragging of excitons and of EHD by the phonon wind. The generation and recombination of the non-equilibrium carriers are accompanied by dissipation of energy, which goes mainly to the phonons, so that intense currents of nonequilibrium exist in the system and can drag the excitons and the EHD, acting by the same token on their spatial distributions. These currents flow both from the region where the carriers are generated, on account of their thermalization, and directly from each EHD, on account of the energy released in the nonradiative recombination channels. We note that in this case we are dealing mainly with short-wave acoustic phonons, which are generated mainly by thermalization and by nonradiative recombination, and have sufficiently large lifetimes, while the optical phonons decay very rapidly into acoustic ones. We shall consider below the process of exciton dragging by short-wave phonons and shall show that in semiconductors such as Ge and Si, owing to the peculiarities of their band structure, the cross section for the scattering of such phonons by indirect excitons is larger by several orders of magnitude than the scattering cross section calculated for nondegenerate isotropic bands. No rigorous analysis was made for EHD, but owing to the large relaxation time of the EHD momentum[9,10] we can expect the drop-dragging effect to be more appreciable. Results of an experimental observation of EHD motion under the influence of phonon wind will be presented.

The phonon-wind assumption helps explain the results of experiments on the spatial distribution of EHD[2,4–7,11–14] and makes it possible to understand the mechanism that moves the drops to the p-n junction in measurements of current pulses.[5,15,19–21] In addition, the phonon wind, propagating from each EHD and dragging excitons with it, may be the cause of the limitation, observed in[8,9,17], of the growth of the EHD radius.

THEORY

We consider the dragging of excitons by a current of nonequilibrium phonons flowing from a region where nonequilibrium carriers are generated by an external excitation source. As these phonons collide with the excitons and are absorbed or scattered by them, they drag the excitons with them. The force exerted by this "phonon wind" on the exciton can be determined in terms of the average momentum transferred to each exciton by the phonons per unit time:

$$\mathbf{f}=d\mathbf{p}/dt. \quad (1)$$

By virtue of the energy and momentum conservation laws and with allowance for the fact that the velocity of sound s is much less than the thermal velocity of the excitons v_T, the momentum transferred to the exciton in each collision should be of the order of the thermal momentum of the exciton $p_T=(2MkT)^{1/2}$, even though the average quasimomentum of the phonon is $|\hbar\mathbf{k}| \gg p_T$ (here M is the effective mass of the exciton). Introducing a certain average effective cross section σ for the scattering of an exciton by a nonequilibrium phonon (a detailed calculation and an estimate of σ are given below, formula (20)), and denoting the phonon flux density by $\mathbf{w}$, we obtain

$$\mathbf{f}=p_T\sigma\mathbf{w}/\hbar\bar{\omega}, \quad (2)$$

where $\hbar\bar{\omega}$ is the average energy of these phonons.

Let, for example, the nonequilibrium carriers be generated on a flat sample surface (with linear dimensions $\gg L_D=(D\tau)^{1/2}$, where L_D is the exciton free diffusion length, D is their diffusion coefficient, and τ is their lifetime) in a narrow surface layer of thickness $\ll L_D$. Then, in order to find the distribution of the excitons $n(x)$ over the sample thickness it is necessary to solve the continuity equation with boundary conditions $n(x)=0$ as $x\to\infty$ and $n(x)=n(0)$ at $x=0$. In the stationary case this equation takes the form

$$\operatorname{div}\mathbf{S}+n(x)/\tau=0, \quad (3)$$

where the expression for the exciton flux density, with account taken of both the diffusion and the directional drift of the excitons under the influence of the force **f**, can be written in the form

$$S(x)=-D\frac{\partial n(x)}{\partial x}\mathbf{i}+\frac{D}{kT}\mathbf{f}n(x), \quad (4)$$

here D/kT is the exciton mobility and **i** is a unit vector along the x axis. The solution of Eq. (3) is

$$n(x)=n(0)\exp\{-x/L_{\rm eff}\}, \quad (5)$$

where

$$L_{\rm eff}=L_D/\{[1+(fL_D/2kT)^2]^{1/2}-fL_D/2kT\}. \quad (6)$$

Expression (5) differs from the usual solution of the diffusion equation in that L_D is replaced by $L_{\rm eff}$, the latter being dependent on the phonon flux density. At $fL_D/kT \gg 1$ (a condition apparently satisfied in Ge at $w\gg 10^{-2}$ W/cm^2, in view of the estimates given above) we have $L_{\rm eff}\approx fL_D^2/kT\sim w$.

The relation $L_{\rm eff}\sim w$, however, is not universal but is determined by the geometry of the excited region. Thus, if the radiation is focused into a narrow strip (of width $\ll L_D$ and length $L\gg L_D$), then the phonon flux density, and with it also the force $f=p_T\sigma W/\pi r L\hbar\bar{\omega}$, decreases with increasing distance from the excited region, and the spatial distribution of the excitons takes the form

$$n(r)={\rm const}\left(\frac{r}{L_D}\right)^{\nu}K_\nu\left(\frac{r}{L_D}\right), \quad (7)$$

where r is the distance from the excited strip, W is the total energy carried away by the nonequilibrium phonons per unit time, and $y=fr/kT=p_T\sigma W/\pi L\hbar\bar{\omega}kT$ and $K_\nu(r/L_D)$ is the known Macdonald function. At $\nu\lesssim 1$, the distribution (7) differs little from the usual distribution that does not take into account the phonon dragging by the excitons, and has a width $\sim L_D$. At $\nu\gg 1$ we have in the region $r<\nu L_E$

$$n(r)\approx{\rm const}\cdot\exp[-r^2/2\nu L_D^2], \quad (8)$$

i.e.,

$$L_{\rm eff}=L_D(2p_T\sigma W/\pi L\hbar\bar{\omega}kT)^{1/2}\sim W^{1/2}.$$

Similarly, when the exciting radiation is focused into a spot of diameter $\ll L_D$, the decrease of the concentration with increasing distance, a sufficiently large phonon-flux power

$$W\gg 4\pi kT\hbar\bar{\omega}L_D/p_T\sigma \quad (9)$$

follows the law

$$n(r)\approx{\rm const}\cdot\exp[-(r/L_{\rm eff})^3], \quad (10)$$

where

$$L_{\rm eff}=(3p_T\sigma L_D^2W/2\pi\hbar\bar{\omega}kT)^{1/3}\sim W^{1/3}.$$

It is seen from these expressions that at sufficiently high excitation levels the volume of the excited region always increases in proportion to W, and therefore, regardless of the geometry of the excited region, there is a certain maximum exciton density n_m that can be attained under conditions of stationary but not volume excitation.

We proceed now to calculated σ. There can be two mechanisms of momentum transfer from the nonequilibrium phonons to the excitons: phonon absorption and phonon scattering by excitons. The cross section, averaged over the Maxwellian distribution, for the absorption of an acoustic phonon of frequency ω is

$$\sigma_{\rm abs}(\omega)=\frac{d^2}{2\hbar\rho s^2v_T}\exp\left\{-\frac{(\hbar\omega-2Ms^2)^2}{8Ms^2kT}\right\}, \quad (11)$$

where d is the deformation potential and ρ is the density of the crystal. Thus, only phonons with energy $\hbar\omega\lesssim 2^{3/2}(Ms^2kT)^{1/2}$ are effectively absorbed. If we assume that the nonequilibrium phonons are sufficiently rapidly thermalized (within a time $\ll L_D/s\sim 10^{-7}$ sec), then at low temperatures $kT\lesssim 8Ms^2$ the absorption cross section (11) reaches values $\sim 10^{-14}$ cm^2 and the dragging effect becomes noticeable starting with fluxes $w\sim 10^{-2}$ W/cm^2. This assumption, however, is hardly justified at low temperatures. During the course of thermalization of nonequilibrium electrons and holes, the phonons produced in the main are optical and short-wave acoustic with energies $\hbar\omega\gtrsim 10$ meV. The optical longitudinal acoustic phonons seem to decay rapidly into softer transverse acoustic phonons, but further energy relaxation is already greatly hindered by the energy and quasimomentum conservation laws. Indeed, there are many experiments[22,23] that show that at low temperatures there exists in germanium a ballistic (collisionless) regime of propagation of strongly nonequilibrium ($\hbar\omega\gg kT$) phonons over distances[1)] ~ 1 cm. It is therefor more legitimate to assume that $\bar{\omega}$ is of the order of the limiting frequency of the transverse acoustic phonons. By virtue of (11), however, such phonons are not absorbed at all by the excitons, and can only be scattered by them.

Phonon scattering is a two-phonon process and its cross section, at first glance, should be rather small in comparison with the cross section (11) for single-phonon scattering. Indeed, if we use for the description of the electron-phonon interaction the usual linear-in-strain Hamiltonian

$$H^{(1)}_{e\text{-ph}}=d\varepsilon \quad (12)$$

(ε is the relative strain), then in second-order perturbation theory we obtain phonon-scattering cross section values $\sigma_{sc}\sim d^4M^5p_T/(\rho s\hbar|\mathbf{k}|)^2s\sim(10^{-18}-10^{-19})$ cm^2; these values are too small to make the dragging effect noticeable.

In crystals of the germanium and silicon type, however, which have a multivalley conduction band and a degenerate valence band, there is a specific feature that increases the cross section for the scattering of phonons by indirect excitons by several orders of magnitude. The point is that the ground level of the exciton in such a crystal is split into two closely lying levels.[24,25] The physical cause of this splitting can be explained very roughly in the following manner: the electric field produced by the electron entering the exciton and belonging to some definite valley of the conduction band is anisotropic, by virtue of the anisotropy of the effective masses of the electron. It therefore lowers locally (in the vicinity of the given electron) the crystal symmetry and splits the valence band. Thus, the splitting δ of the exciton level is similar in character to the splitting produced in degenerate band by relative strains, which also lower the symmetry of the crystal, and it is convenient to characterize this splitting by a certain equivalent strain[25]

$$\varepsilon_c = \delta/d_v. \qquad (13)$$

We present for simplicity a purely schematic consideration, disregarding the tensor structure of the strain ε and of the deformation potentials d_v. When a true strain ε, which generally speaking does not coincide at all in direction with ε_c, is produced in a crystal, the splitting of the exciton levels is given by[25]

$$\delta E = d_v(\varepsilon_c^2 + \varepsilon^2 + c\varepsilon\varepsilon_c)^{1/2}, \qquad (14)$$

where c is a constant that depends on the orientation of ε. The connection between the exciton energy and the strains, and hence with the phonons, is thus strongly nonlinear even at quite small ε. Confining ourselves to the second-order terms of interest to us, we can express the Hamiltonian of the exciton-phonon interaction in place of (12) in the form

$$H_{e\text{-ph}} = H^{(1)} + H^{(2)} = \left(d + \frac{1}{2}cd_v\right)\varepsilon + \frac{1}{2}d_v\frac{\varepsilon^2}{\varepsilon_c}\left(1 - \frac{c^2}{8}\right). \qquad (15)$$

It is the large value $\varepsilon_c^{-2} = (d_v/\delta)^2 \sim 10^6$ which determines the anomalously large probabilities of the two-phonon processes in our case, compared with the case of simple nondegenerate bands. Expanding in the usual manner[23] the strains in terms of the normal vibrations (phonons), we obtain in first order of perturbation theory in $H^{(2)}$ the probability of scattering of a phonon with wave vector $\mathbf{k}$ into a state $\mathbf{k}'$ by an exciton with momentum $\mathbf{p}$:

$$w_{sc}(\mathbf{p}; \mathbf{k}\to\mathbf{k}') = \frac{2\pi}{\hbar}C\left(\frac{d_v}{2\varepsilon_c}\right)^2 \frac{\hbar^2|\mathbf{k}|^2|\mathbf{k}'|^2}{4\rho^2V^2\omega_\mathbf{k}\omega_{\mathbf{k}'}}\delta\left(\frac{p^2 - (\mathbf{p} + \hbar\mathbf{k} - \hbar\mathbf{k}')^2}{2M} + \hbar\omega_\mathbf{k} - \hbar\omega_{\mathbf{k}'}\right). \qquad (16)$$

The constant $C \sim 1$ was introduced here to take phenomologically into account (in the mean) the differences between difference modes and the direction (c in (14), the possible scattering with intervalley transitions of the exciton, the anisotropy of the effective masses, etc.; V is the normalization valume, and $\omega_\mathbf{k}$ is the frequency of the phonon with wave vector $\mathbf{k}$. Using (16), we calculate now not the mean value of the scattering cross section, but directly the force $\mathbf{f}$ exerted on the exciton by the phonon flux. In accord with the definition (1),

$$\mathbf{f}(\mathbf{r}) = \int \frac{d^3k\,d^3k'\,d^3p\,V^2}{(2\pi)^9\hbar^3}\hbar(\mathbf{k}-\mathbf{k}')w_{sc}(\mathbf{p}, \mathbf{k}\to\mathbf{k}')\left(\frac{2\pi\hbar^2}{MkT}\right)^{3/2} \times \exp\left\{-\frac{p^2}{2MkT}\right\}F(\mathbf{k}, \mathbf{r}). \qquad (17)$$

We average here over the Maxwellian distribution for the excitons; $F(\mathbf{k}, \mathbf{r})$ is the phonon distribution function in the wave vectors at the point $\mathbf{r}$. Recognizing that the momentum $\hbar\varkappa = \hbar(\mathbf{k} - \mathbf{k}')$ transferred in the scattering is on the order of $p_T \ll |\hbar\mathbf{k}|$ and that $M|\nabla_\mathbf{k}\omega_\mathbf{k}|^2 \sim Ms^2 \ll kT$, we can integrate in (17) with respect to $\mathbf{p}$ and $\varkappa$ in explicit form:

$$\mathbf{f}(\mathbf{r}) = \frac{2CkT}{3\pi^2\hbar v_T}\left(\frac{Md_v}{2\varepsilon_c\hbar\rho}\right)^2 \int \frac{d^3k}{(2\pi)^3}\nabla_\mathbf{k}\omega_\mathbf{k}\cdot\hbar\omega_\mathbf{k}\frac{|\mathbf{k}|^4}{\omega_\mathbf{k}^3}F(\mathbf{k}, \mathbf{r}). \qquad (18)$$

We note now that the phonon energy flux density at the point $\mathbf{r}$ is of the form

$$\mathbf{w}(\mathbf{r}) = \int \frac{d^3k}{(2\pi)^3}\nabla_\mathbf{k}\omega_\mathbf{k}\cdot\hbar\omega_\mathbf{k}F(\mathbf{k}, \mathbf{r}). \qquad (19)$$

Therefore, taking outside the integral sign in (18) a certain mean value of the factor that distinguishes between the integrands in (18) and (19), we can rewrite $\mathbf{f}(\mathbf{r})$ in a more compact form:

$$\mathbf{f}(\mathbf{r}) = \frac{2CkT}{3\pi^2\hbar v_T}\left(\frac{Md_v}{2\varepsilon_c\hbar\rho}\right)^2 \overline{\left(\frac{|\mathbf{k}|^4}{\omega_\mathbf{k}^3}\right)}\mathbf{w}(\mathbf{r}). \qquad (20)$$

Strictly speaking, in the spirit of the simplifications already made on going from (18) to (20), we have neglected also the difference that the crystal anisotropy can produce between the directions of the vectors $\mathbf{f}$ and $\mathbf{w}$. The principal uncertainty in (20) is connected with the lack of any information whatever concerning the distribution function of the nonequilibrium phonons radiated in the generation region, and hence the impossibility of determining reliably the value of $\overline{(|\mathbf{k}|^4/\omega_\mathbf{k}^3)}$. The most natural assumption, however, is that these phonons are more or less uniformly distributed over the entire Brillouin zone in the most long-lived transverse acoustic modes.

Using (20), we can write down exact formulas for the effective drift length of the excitons. At sufficiently high power we obtain in the planar case from (6)

$$L_{\text{eff}} \approx \frac{2CL_D^2}{3\pi^2\hbar v_T}\left(\frac{Md_v}{2\varepsilon_c\hbar\rho}\right)^2 \overline{\left(\frac{|\mathbf{k}|^4}{\omega_\mathbf{k}^3}\right)}w, \qquad (21)$$

in the case of excitation by a narrow strip, the index of the Macdonald function is

$$\nu = \frac{C}{3\pi^3\hbar v_T}\left(\frac{Md_v}{2\varepsilon_c\hbar\rho}\right)^2 \overline{\left(\frac{|\mathbf{k}|^4}{\omega_\mathbf{k}^3}\right)}\frac{W}{L} \qquad (22)$$

and

$$L_{\text{eff}} = \frac{L_D Md_v}{2\pi\varepsilon_c\hbar\rho}\left[\frac{2C}{3\pi\hbar v_T}\overline{\left(\frac{|\mathbf{k}|^4}{\omega_\mathbf{k}^3}\right)}\frac{W}{L}\right]^{1/2} \qquad (23)$$

and finally, in the case of excitation in a point

$$L_{\text{eff}}=L_D\left[\frac{C}{\pi^3\hbar v_T}\left(\frac{Md_v}{2\varepsilon_c\hbar\rho}\right)^2\overline{\left(\frac{|\mathbf{k}|^4}{\omega_{\mathbf{k}}^3}\right)}\frac{W}{L_D}\right]^{1/3}. \tag{24}$$

The limiting exciton concentration attainable under conditions of stationary excitation is given by

$$n_m\sim\frac{\hbar v_T}{D\Delta E}\left(\frac{2\varepsilon_c\hbar\rho}{Md_v}\right)^2\overline{\left(\frac{|\mathbf{k}|^4}{\omega_{\mathbf{k}}^3}\right)}^{-1}, \tag{25}$$

where ΔE is the energy going into heat for each pair of nonequilibrium carriers produced by the excitation.

The nonequilibrium phonon fluxes produced at each of the EHD can, by dragging the excitons with them, hinder the exciton diffusion towards the surface of the drop and by the same token decrease the growth rate of the EHD. Assuming that in each recombination act an energy $\approx E_g$ (the width of the forbidden band) is released in the EHD and that the nonradiative recombination channels (say, the Auger channel) predominate, we find that the energy flux from a drop of radius R is

$$W=\frac{4\pi}{3}R^3\frac{n_0}{\tau_0}E_g, \tag{26}$$

where n_0 is the density of the electron-hole liquid and τ_0 is the lifetime of the carriers in the EHD. The energy flux density at a distance r from the center of the drop is

$$\mathbf{w}(\mathbf{r})=W\mathbf{r}/4\pi r^3\hbar\bar{\omega}. \tag{27}$$

It is convenient to express the force exerted by the phonon wind on the excitons as the gradient of a certain effective potential $U(\mathbf{r})$, i.e., $\mathbf{f}=-\text{grad}\,U(\mathbf{r})$. Then, using (20), (26), and (27), we obtain

$$U(\mathbf{r})=\frac{2C}{9\pi^2}\frac{kT}{\hbar}\left(\frac{Md_v}{2\varepsilon_c\hbar\rho}\right)^2\overline{\left(\frac{|\mathbf{k}|^4}{\omega_{\mathbf{k}}^3}\right)}\frac{n_0E_g}{v_T\tau_0}\frac{R^3}{r}=\frac{R^3kT}{R_0^2r}, \tag{28}$$

where

$$R_0=\frac{3\pi}{(2C)^{1/2}}\frac{2\varepsilon_c\hbar\rho}{Md_v}\left[\overline{\left(\frac{|\mathbf{k}|^4}{\omega_{\mathbf{k}}^3}\right)}^{-1}\frac{\hbar v_T\tau_0}{n_0E_g}\right]^{1/2}. \tag{29}$$

We consider now the spatial distribution of the excitons $n(\mathbf{r})$ in the vicinity of a drop with allowance for the potential $U(\mathbf{r})$, and the diffusion of the excitons in the drop. It is necessary for this purpose to solve the continuity equation with the corresponding boundary conditions. Since we are dealing with the solution of the problem in a small region near an individual drop (it is assumed that the drop dimensions are small in comparison with the distance between them), within which the quasi-stationary distribution of the excitons is established within a very short time, it follows that this equation goes over into the stationary diffusion equation div $\mathbf{S}=0$, where $\mathbf{S}(\mathbf{r})$ is defined by E2. (4). The boundary conditions are $n(r\to\infty)=n$, where n is the average exciton density in the volume of the crystal, and

$$S(R)=[n(R)-n_T(R)]v_T, \tag{30}$$

where $S(R)$ is the exciton flux density on the surface of the drop, $N(R)$ is the exciton gas density near the surface of the drop,

$$n_T(R)=g\left(\frac{M_dkT}{2\pi\hbar^2}\right)^{3/2}\exp\left\{-\frac{\Delta}{kT}+\frac{2\alpha}{n_0RkT}\right\} \tag{31}$$

is the thermodynamic-equilibrium exciton gas density over a drop of radius R, and $V_T=(kT/2\pi M)^{1/2}$ is the thermal velocity of the excitons; In (31), Δ is the work function of the excitons from the EHD as $R\to\infty$, M_d is the effective mass of the density of states of the excitons, α is the coefficient of surface tension of the electron-hole liquid, and g is the multiplicity of the degeneracy of the ground state of the exciton. Solving the diffusion equation with allowance for the boundary conditions, we find the exciton flux on the surface of a drop of radius R, expressed in terms of the average exciton density

$$S(n,R)=4\pi R^2\left[1+\frac{v_TR}{D}\frac{kT}{U(R)}(1-e^{-U(R)/kT})\right]^{-1}[ne^{-U(R)/kT}-n_T(R)]v_T. \tag{32}$$

Expression (32) enables us to write down an equation that describes the kinetics of the EHD growth and recombination, with allowance for the diffusion of the excitons to the surface of the drop and to their dragging by the phonons:

$$\frac{d}{dt}\left(\frac{4}{3}\pi R^3n_0\right)=S(n,R)-\frac{4}{3}\pi R^3\frac{n_0}{\tau_0}. \tag{33}$$

In the stationary case, the left-hand side of (33) is equal to zero. Then solving (32) and (33) simultaneously, we obtain

$$n=e^{U(R)/kT}\left\{n_T(R)+\frac{n_0R^2}{3D\tau_0}\frac{kT}{U(R)}\left[1+\frac{D}{v_TR}\frac{U(R)}{kT}-e^{-U(R)/kT}\right]\right\}. \tag{34}$$

This expression gives the radius of the EHD that are in equilibrium with an exciton gas of density n.

It is easy to verify that at $R\ll R_0$ Eqs. (32) and (34) go over into the corresponding expressions obtained without allowance for the effect of exciton dragging by the phonons.[17] When R approaches R_0, however, the density of the excitons that are in equilibrium with drops of radius R begins to increase very rapidly ($\sim\exp(R^2/R_0^2)$), in other words, the growth of the EHD radius slows down sharply with increasing excitation level, and the dependence of the radius on the temperature becomes very weak (since $R_0\sim T^{1/4}$), in qualitative agreement with the results of[8,9,17]. Substitution of the numerical values shows that in the temperature region $T\gtrsim 3\,^\circ$K the right-hand side of (34) is well approximated by the first term, so that

$$n\approx n_T(R)\exp\frac{U(R)}{kT}. \tag{35}$$

Thus, $U(R)$ represents the effective decrease of the work function of the EHD, which becomes noticeable at $R\approx R_0$ and increases in proportion to R^2. The lack of information on the distribution function of the phonons emitted by the EHD does not make it possible to estimate reliably the value of R_0. Substituting then in (29), by way of estimate, $|\bar{\mathbf{k}}| \sim 10^8$ cm^{-1} and $\bar{\omega}_{\mathbf{k}} \sim 10^{13}$ sec^{-1}, and using for all the remaining quantities more or less well known values ($\rho \approx 5$ g/cm^3, $E_g \approx 0.74$ eV, $d_v \approx 4$ eV, $2\delta = 2\varepsilon_c d_v \approx 1$ meV, $M \approx 10^{-27}$ g, $v_T \approx 5\times10^5$ cm/sec, $n_0 \approx 2\times10^{17}$ cm^{-3}, and $\tau_0 \approx 4\times10^{-5}$ sec), we obtain $R_0 \approx 1.5\times10^{-3}$ cm, which is close to the limiting value $\sim 10^{-3}$ which we observed for the EHD radius,[8,9,17] although this agreement should not be given too large a significance. The point is that the quantities n_m and R_0 are determined, in essence, by the same combination of the parameters. Using (25) and (29), we obtain

$$R_0 \sim \left(\frac{n_m}{n_0}\frac{\tau_0}{\tau}\frac{\Delta E}{E_g}\right)^{1/2} L_D, \tag{36}$$

therefore a measurement of n_m would make it possible to determine also R_0. We know of no direct experiments of this kind, but the numerous communications[27–29] reporting that an exciton density $n \gtrsim 10^{15}$ has been attained under conditions of stationary optical excitation suggest, after comparison with (36) (assuming $\Delta E \sim 1$ eV) that the action of the "phonon wind" on the excitons is strongly exaggerated[2)] in the foregoing estimates. If we assume by way of estimate $|\bar{\mathbf{k}}| \sim 3\times10^7$ cm^{-1}, then at these values of $|\mathbf{k}|$ we have $\omega_{\mathbf{k}}/|\mathbf{k}| \approx s_t \approx 3\times10^5$ cm/sec (s_t is the velocity of the transverse sound) and $\overline{(|\mathbf{k}|^4/\omega_{\mathbf{k}}^3)} \approx |\mathbf{k}|/s_t^3 \approx 10^{-9}$ sec^3/cm^4; then (29) yields $R_0 \sim 1.5\times10^{-2}$ cm. This is in reasonable agreement with the value of R_0 determined from (36) at $n_m \sim 10^{15}$ cm^{-3}.

We make a few more remarks concerning the results. Although the "phonon wind" greatly hinders the growth of the EHD by condensation of the exciton from the gas phase, this does not exclude in any way the possibility of observing drops with $R \gg R_0$; for example, if the carrier generation takes place directly in the drop, or if a carrier density $n \gtrsim n_0$ is produced immediately in some part of the sample. Nor do the conclusions contradict the results of[30–32], in which EHD with a radius of hundreds of microns were observed. The point is that these drops were produced in uniaxially deformed samples and had significantly lower carrier densities n_0, and as a consequence[31] larger lifetimes ($\sim 5\times10^{-4}$ sec). However, as is easily seen from (29), a decrease of n_0 by one order of magnitude and an equal increase of τ_0 increase R_0 by one order. In addition, if the strain produced in the samples greatly had exceeded ε_c, then the cross section for the scattering of phonons by excitons should decrease strongly.

It is presently impossible to investigate sufficiently fully the dragging of EHD by a flux of strongly-nonequilibrium phonons, owing to the lack of detailed calculations of the energy and of the spectrum of the excitations of the EHD, with account taken of the real band structure in weakly deformed germanium and silicon. The experiments[10] show, however, that in the case of small strains the dependence of the energy per particle pair in the EHD on the strain is strongly nonlinear. We can point to at least one mechanism that produces the strong nonlinearity in this case. It is known[33] that the dependence of the energy of the holes on the strain in these crystals has schematically a structure of the type of (14), provided we replace $\delta = d_v\varepsilon_c$ in this formula by the kinetic energy. Inasmuch as only holes with energies close to the Fermi energy ε_{Fh} take part in the phonon absorption and scattering, by virtue of the Pauli principle, we can introduce in analogy with (15) also a nonlinear hole–phonon interaction Hamiltonian

$$H_{h-ph} = d\varepsilon + \tilde{c}\frac{d_v^2}{\varepsilon_{Fh}}\varepsilon^2.$$

By a procedure similar to the derivation of (20), but with the Fermi degeneracy taken into account, we can obtain for the volume density of the force $n_0\mathbf{f}(\mathbf{r})$ (n_0 is the EHD density, $\mathbf{f}(\mathbf{r})$ is the force acting on one electron-hole pair), due to the scattering of the phonons by holes, the expression

$$n_0\mathbf{f}(\mathbf{r}) = \frac{2\tilde{C}}{3\pi}\left(\frac{d_v^2 m_h^2}{2\pi\hbar^3\rho}\right)^2 \overline{\left(\frac{|\mathbf{k}|^4}{\omega_{\mathbf{k}}^3}\right)}\mathbf{w}(\mathbf{r}), \tag{20a}$$

where m_h is the effective mass of the hole, the constant $\tilde{C} \sim 1$, and the averaging procedure has the same meaning as in (20).

Just as for the excitons, the dragging of EHD by a current of phonons with relatively large wavelengths, $|\mathbf{k}| \lesssim n_0^{1/3}$, is determined, naturally, by the absorption process and not by the scattering. The expression for the force is then

$$n_0\mathbf{f}_{abs}(\mathbf{r}) = \frac{d^2m^2}{8\pi\hbar^2\rho}\int_{|\mathbf{k}|\leq k_0}\frac{d^3k}{(2\pi)^3}\mathbf{k}|\mathbf{k}|F(\mathbf{k},\mathbf{r}) = \frac{d^2m^2}{8\pi\hbar^3\rho s^2}\overline{(|\mathbf{k}|)}_{|\mathbf{k}|<k_0}\mathbf{w}(\mathbf{r}),$$

$$k_0 = 2\pi\left(\frac{3}{\pi}n_0\right)^{1/3}.$$

This formula takes into account the absorption of the phonons by the electrons bound in the EHD (m is the electron effective mass). To find the total force it is necessary to add a similar term for the holes.

THE EXPERIMENT

The dragging of excitons and EHD by the phonon wind can be detected by the displacement of the region occupied by the excitons or EHD under the influece of a current of nonequilibrium phonons produced with the aid of an additional excitation source. The experiments were performed with the setup used by us previously to measure the scattering of light by EHD.[34] A block diagram of the setup is shown in Fig. 1. The excitation source was a laser of ~10 mW power, operating at a wavelength 1.52 μ. The radiation of this laser was focused into a spot of diameter ~200 μ on the front surface of the sample. The exciting radiation was modu-

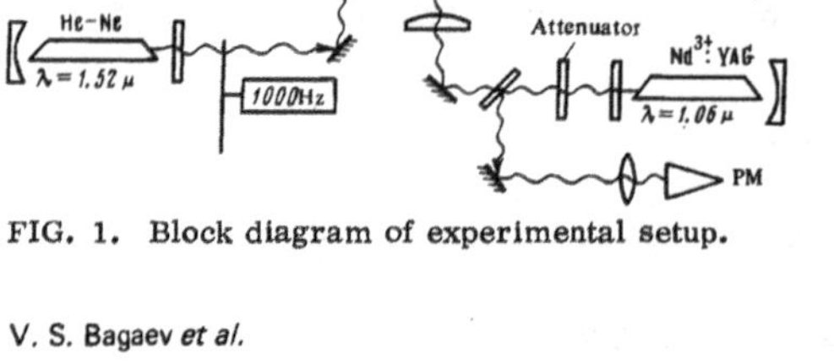

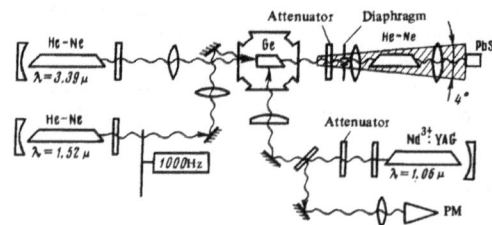

FIG. 1. Block diagram of experimental setup.

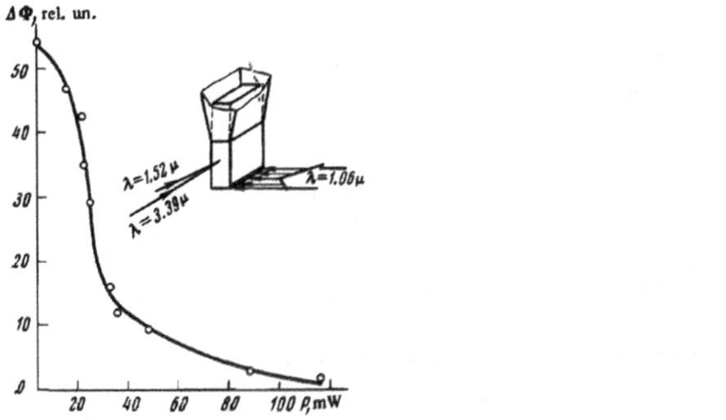

FIG. 2. Dependence of the absorption signal $\Delta\Phi$ on the power P at $x=0$ and $T=1.95$ °K.

lated at a frequency 1 kHz. We measured the absorption of 3.39-μ laser radiation by the nonequilibrium carriers bound into excitons or EHD. This radiation was also focused on the front surface of the sample, into a spot of ~300 μ diameter. The beams of both lasers could be collocated (insert in Fig. 2) or separated by a distance x (insert in Fig. 3), with the position of the 3.39-μ laser beam remaining unchanged in all the measurements. The absorption signal was recorded with a goniometer consisting of a quantum amplifier and a PbS photoreceiver cooled to ~100°K. This registration method made it possible to detect the EHD motion not only from the change of the absorption signal, but also from the change of the pattern of the diffraction by the edge of the region occupied by the EHD.[35]

The nonequilibrium phonons were generated upon thermalization of the nonequilibrium carriers, as well as in the nonradiative recombination of the excitons and in the EHD produced with the aid of stationary illumination of the sample by an Nd^{3+} YAG laser. This radiation was focused with a cylindrical lens on the lateral face of the sample into a narrow strip ~5 mm long, parallel to the light beams of the two other lasers (Figs. 2 and 3). The distance from this strip to the 3.39-μ laser beam was ~7 mm. The radiation of the Nd^{3+} YAG laser was attenuated with calibrated light filters and monitors with a power meter ("attenuator" and "PM" in Fig. 1, respectively).

The measurements were made on Ge samples with residual-impurity density less than 10^{12} cm^{-3}. The samples were mechanically polished and measured 15 ×5×2 mm. The sample plane through which the 3.39-μ radiation passed was inclined 2° to the opposite plane, to eliminate parasitic interference. Just as in the scattering measurement,[8] the samples were soldered into the bottom of the helium container of the cryostat in such a way that the working part of the sample was in vacuum while the other part was immersed in liquid helium.

The dragging of the EHD by the phonon wind was measured at $T\approx 1.95$ °K. At this temperature almost all the nonequilibrium carriers were bound into EHD. The radii of the EHD and their concentration in the crystal, obtained from measurements of the light scattering, were under the conditions of our experiments $R=5$ μ and $N\approx 2\times 10^7$ cm^{-3}, respectively. Figure 2 shows the dependence of the absorption signal (which is proportional to the volume of each phase in that part of the crystal through which the probing 3.39-μ laser beam passed) on the power P of the laser radiation that produced the flux of nonequilibrium phonons, for the case when the probing and exciting laser beams were collocated ($x=0$). The absorption signal decreased with increasing power P. The connection between the observed effect and the motion of the EHD is confirmed by the next experiment.

To observe directly the EHD motion due to the phonon wind, the exciting laser spot was displaced a distance x from the probing laser spot. At $x>0.3$ mm, no absorption signal was observed with the 1.06-μ laser radiation blocked (i.e., at $P=0$). When the exciting laser beam was displaced upward relative to the probing laser beam, no absorption signal produced even when the illumination that produced the nonequilibrium phonons was turned on. When the exciting laser beam was shifted downward on the sample in such a way that it passed between the probing beam and the region where the phonons were generated (insert in Fig. 3), an absorption signal appeared at sufficiently large P and its magnitude depended on the power P and on the distance x.

Figure 3 shows the dependence of the absorption signal on the distance x at a phonon-producing laser power $P\approx 40$ mW. It is seen from the figure that the absorption signal, and hence also the number of particles bound into EHD, decreases with increasing x, in accord with expression (5). We emphasize that the spatial distribution of the excitons and the carriers bound into drops of the liquid phase is described by the same expressions (5), (7), (8), and (10), except that in the EHD case it is necessary to replace L_D in the formulas (21), (23), and (24) for L_{eff} by the quantity $(\tau_0\tau_r kT/M)^{1/2}$ (τ_r is the EHD momentum relaxation time) and L_{eff} must be expressed in terms of the force with the aid of (20) (now f is the force exerted on a pair of particles in the EHD by the phonon wind). From the slope of the straight line (Fig. 3) we obtain $L_{eff}\approx 0.5$ mm. According to (6), at sufficiently high intensity of the phonon wind acting on the EHD, we have

$$L_{eff}=\tau_0\frac{\tau_r}{M}f. \qquad (37)$$

It is seen from this formula that L_{eff} has the meaning of the distance traversed by the EHD during the lifetime

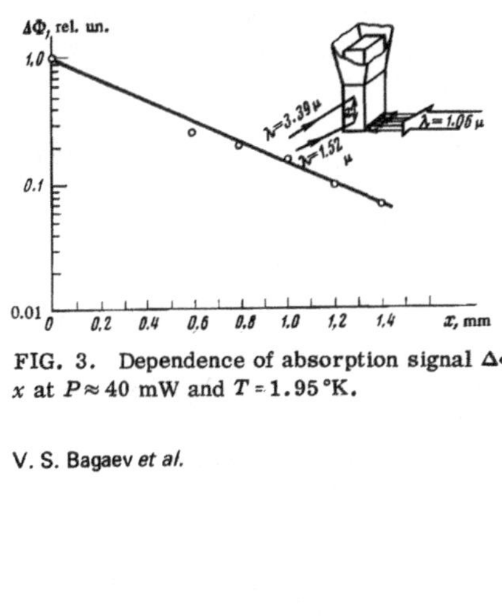

FIG. 3. Dependence of absorption signal $\Delta\Phi$ on the distance x at $P\approx 40$ mW and $T=1.95$ °K.

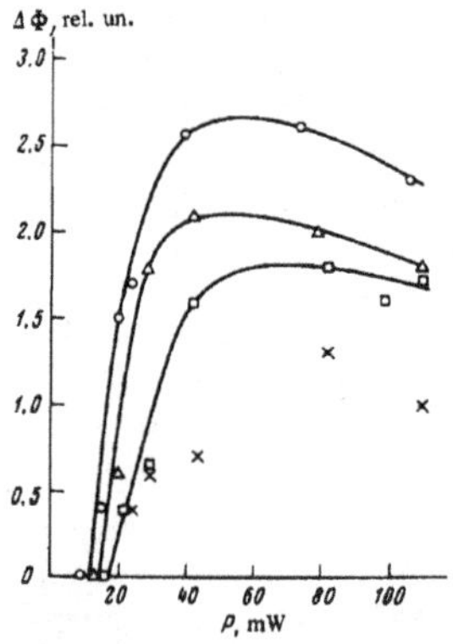

FIG. 4. Plots of the absorption signal ΔΦ vs. the power P for different distances x: ○—$x=0.6$ mm, △—0.8 mm, □—1.0 mm, ×—1.4 mm; $T=1.95$ °K.

τ_0, since $v=\tau_r f/M$ is the EHD drift velocity. Assuming as an estimate that $\tau_0=40$ μsec and $\tau_r\approx 2\times10^{-8}$ sec, [9] we find that $v\approx 1.3\times10^3$ cm/sec and $f\approx 2.5\times10^{-17}$ dyne.

Using (20a), we can now estimate the value of $(\overline{|\mathbf{k}|^4/\omega_{\mathbf{k}}^3})$ for the phonons that interact effectively with the EHD. Assuming that half of the power $P\approx 40$ mW incident on the sample goes to generate the directed flux of nonequilibrium phonons, we obtain the energy flux density $w\approx 0.2$ W/cm². Substituting in (20a) $m_h \approx 0.4m_0$ (m_0 is the mass of the free electron) and the numerical values cited above for the remaining quantities, we obtain $(\overline{|\mathbf{k}^4|\omega_{\mathbf{k}}^3})\sim 6\times10^{-10}$ sec³/cm⁴. This value does not differ greatly from the value 10^{-9} sec³/cm⁴ used above to estimate R_0 and n_m from formulas (29) and (36).

The dependence of the absorption power on the power P at different distances x are shown in Fig. 4. It is seen that the absorption signal appears only when a certain minimal power is reached, and this minimum power increases with increasing distance between the exciting and probing light beams. With further decrease of P, the absorption signal first increases, and then goes through a maximum and starts to decrease. To analyze this behavior we must use expression (5). Integrating (5) with respect to x from zero to ∞, we obtain the total number of carriers bound into drops of the condensed phase; in the stationary case this number is equal to $G\tau_0$ (G is the rate of generation). After determining in this manner the average density of the carriers bound into EHD at the point $x=0$ and substituting it in (5) we obtain for the distribution of the liquid-phase particles (and hence also of the absorption signal) over the sample volume:

$$N_{\Sigma}(x)=\frac{GM}{\tau_r f}\exp\left(-\frac{M}{\tau_0\tau_r f}x\right). \qquad (38)$$

Given x, the function $N_{\Sigma}(x)$ (as a function of f) has a maximum at

$$f_m=Mx/\tau_0\tau_r, \qquad (39)$$

and since in our case $f\sim P$, the last formula yields the position of the maxima of the curves shown in Fig. 4, and shows that when the distance x is increased the maximum should shift towards higher powers, in agreement with the experimental results. Relation (38) has a simple physical meaning: at $f<f_m$ the EHD drift velocity is so low that during the time of motion from $x=0$ to the observation point the EHD volume decreases strongly as a result of recombination; at $f>f_m$, owing to the appreciable velocity, the EHD spread over the large volume of the sample and this decreases the average density of the liquid-phase particles at the observation point x. The maximum value of the density at the point x is reached when the distance x is negotiated by the drops during the lifetime.

Expression (38) does not describe the cutoff of the absorption signal at small P. We recall, however, that (38) is valid only at a sufficiently high power of the phonon flux, when L_{eff} is determined by (37). At small P it is necessary to use for L_{eff} expression (6), which is valid in the general case, and from which it is seen that so long as the intensity of the phonon wind is insignificant, L_{eff} does not exceed appreciably the dimensions of the generation region. It follows from (38) that at the point $x=0$ the absorption should decrease in inverse proportion to the power P. However, the power dependence observed by us for the absorption (Fig. 1) is stronger ($\Delta\Phi\sim p^{-1.6}$). It is possible that the decrease of the absorption at $x=0$ is due not only to the increase of the EHD but also to the influence of the phonon wind on the kinetics of the exciton condensation and the generation of the EHD.

The diffraction measurement results agree qualitatively with the data obtained from the absorption measurements, although their quantitative reduction is difficult.

Reduction of the measurements[6,11,13,14] of the spatial distribution of the EHD with the aid of expressions (10), (23), and (24) has shown that these expressions are in qualitative agreement with experiment. The force acting on a pair of particles in the EHD, estimated from these measurements, agrees within an order of magnitude with the estimate given above.

Measurements made at 4.2 °K, when all the nonequilibrium carriers produced by the exciting 1.52-μ radiation were bound into excitons, have shown that the effect of dragging of the excitons by nonequilibrium phonons is much less than in the case of EHD. A noticeable displacement of the exciton cloud is observed only at powers $P>100$ mW. However, the sensitivity of the apparatus turned out not to be high enough for quantitative measurements. Assuming that $L_{eff}\approx L_D\approx 1$ mm at $P\approx 100$ mW, we obtain with the aid of (21) the value $(\overline{|\mathbf{k}^4|/\omega_{\mathbf{k}}^3})\sim 6\times10^{-10}$ sec³/cm⁴, in agreement with the value estimated from the measurements of the EHD dragging.

Thus, a theoretical analysis and the results of experiments that revealed motion of EHD induced by phonon wind over distances exceeding 2 mm shows that the dragging of EHD and excitons by nonequilibrium phonons is an effect essential for the explanation of the

measurements of the spatial EHD distribution and can influence the kinetics of their growth.

We thank G. E. Pikus for a useful discussion and N. V. Znakomets for help with the experiments.

[1)] We note also that elastic scattering, say by impurities or defects, which can appreciably limit the thermal conductivity, do not play a noticeable role in our analysis, since the dragging effect, as seen from (2)–(11), is determined by the phonon frequency distribution and by the total energy flux, the latter being governed by the excitation source.

[2)] It is possible, however, that this exciton density was reached in[27–29] only in a small vicinity of the exciting light spot. In addition, it is possible that shorter-wavelength phonons interact with the excitons near the drops (the phonons move away from the drop that emits them within a time $\sim 10^{-8}$ sec).

[1] L. V. Keldysh, Tr. IX Mezhdunarodnoĭ konferentsii po fizike poluprovodnikov (Proc. 9-th Internat. Conf. on Semiconductor), Nauka, 1968, p. 1387; in: Éksitony v poluprovodnikakh (Excitons in Semiconductors) ed. by B. M. Vul, Nauka, 1971, p. 5.

[2] Ya. Pokrovskii, Phys. Status Solidi [a] 11, 385 (1972).

[3] V. S. Bagaev, Springer Tracts in Modern Phys. 73, 72 (1975).

[4] Ya. E. Pokrovskiĭ and K. I. Svistunova, Fiz. Tverd. Tela 13, 1485 (1971) [Sov. Phys. Solid State 13, 1241 (1971)].

[5] C. Benoit à la Guillaume, M. Voos, and F. Salvan, Phys. Rev. Lett. 27, 1214 (1971).

[6] R. W. Martin, Phys. Status Solidi [b] 61, 223 (1974).

[7] J. C. Hensel and T. G. Phillips, Proc. Twelfth Intern. Conf. on Physics of Semiconductors, Stuttgart, 1974, p. 51.

[8] V. S. Bagaev, N. A. Penin, N. N. Sibel'din, and V. A. Tsvetkov, Fiz. Tverd. Tela, 15, 3269 (1973) [Sov. Phys. Solid State 15, 2179 (1974)].

[9] A. S. Alekseev, T. A. Astemirov, V. S. Bagaev, T. I. Galkina, N. A. Penin, N. N. Sybeldin, and V. A. Tsvetkov, Proc. Twelfth Intern. Conf. on Physics of Semiconductors, Stuttgart, 1974, p. 91.

[10] A. S. Alekseev, V. S. Bagaev, and T. I. Galkina, Zh. Eksp. Teor. Fiz. 63, 1020 (1972) [Sov. Phys.-JETP. 36, 536 (1973)].

[11] Ya. E. Pokrovsky and K. I. Svistuna, Proc. Twelfth Intern. Conf. on Physics of Semiconductors, Stuttgart, 1974, p. 71.

[12] Ya. E. Polrovskiĭ and K. I. Svistunova, Fiz. Tverd. Tela 16, 3399 (1974) [Sov. Phys. Solid State 16, 2202 (1975)].

[13] M. Voos, K. L. Shaklee, and J. M. Worlock, Phys. Rev. Lett. 33, 1161 (1974).

[14] B. J. Feldman, Phys. Rev. Lett. 33, 359 (1974).

[15] O. Christensen and J. M. Hvam, Proc. Twelfth Intern. Conf. on Physics of Semiconductors, Stuttgart, 1974, p. 56.

[16] I. Balslev and J. M. Hvam, Phys. Status Solidi [b] 65, 531 (1974).

[17] V. S. Bagaev, N. V. Zamkovets, L. V. Keldysh, N. N. Sibel'din, and V. A. Tsetkov, Preprint FIAN No. 139 (1975).

[18] V. B. Fuks, Pis'ma Zh. Eksp. Teor. Fiz. 20, 33 (1974) [JETP Lett. 20, 14 (1974)].

[19] V. M. Asnin, A. A. Porachev, and N. I. Sablina, Pis'ma Zh. Eksp. Teor. Fiz. 11, 162 (1970) [JETP Lett. 11, 99 (1970)].

[20] C. Benoit à la Guillaume, M. Voos, F. Salvan, J. Laurant, and A. Bonnot, Compt. Rend. 272, 236B (1971).

[21] J. M. Hvam and O. Christensen, Solid State Commun. 15, 929 (1974).

[22] M. Pomeranz, R. J. von Gutfeld, Tr. IX Mezhdunarodnoĭ konferentsii po fizike poluprovodnikov (Proc. 9-th Internat. Conf. on Semiconductor Physics), Nauka, 1968, p. 732.

[23] R. C. Dynes, V. Narayanamuri, and M. Chin, Phys. Rev. Lett. 26, 181 (1971).

[24] T. P. McLean and R. Loudon, J. Phys. Chem. Solids 13, 1 (1960).

[25] G. L. Bir and G. E. Pikus, Fiz. Tverd. Tela 17, 696 (1975) [Sov. Phys. Solid State 17, 448 (1975)].

[26] R. Peierls, Quantum Theory of Solids, Oxford, 1955 (Russ. Transl., IIL, 1956).

[27] A. S. Alekseev, V. S. Bagaev, T. I. Galkina, O. V. Gogolin, and N. A. Penin, Fiz. Tverd. Tela 12, 3516 (1970) [Sov. Phys. Solid State 12, 2855 (1971)].

[28] T. K. Lo, B. J. Feldman, and C. D. Jeffries, Phys. Rev. Lett. 31, 224 (1973).

[29] G. A. Thomas, T. M. Rice, and J. C. Hensel, Phys. Rev. Lett. 33, 219 (1974).

[30] R. S. Markiewicz, J. P. Wolfe, and C. D. Jeffries, Phys. Rev. Lett. 32, 1357 (1974).

[31] J. Wolfe, R. Markiewicz, C. Kittel, and C. Jeffries, Phys. Rev. Lett. 34, 275 (1975).

[32] J. P. Wolfe, W. L. Hansen, E. E. Haller, R. S. Markiewicz, C. Kettel, and C. D. Jeffries, Phys. Rev. Lett. 34, 1292 (1975).

[33] G. L. Bir and G. E. Pikus, Simmetriya i deformatsionnye effekty v popuprovodnikakh (Symmetry and Deformation Effects in Semiconductors), Nauka, 1972, chapter V.

[34] V. S. Bagaev, N. V. Zamkovets, N. A. Penin, N. N. Sibel'-din, and V. A. Tsvetkov, Prib. Tekh. Éksp. No. 2, 242 (1974).

[35] N. N. Sibel'din, V. S. Bagaev, V. A. Tsvetkov, and N. A. Penin, Preprint FIAN, No. 117, 1972.

Translated by J. G. Adashko

Kinetics of exciton condensation in germanium

V. S. Bagaev, N. V. Zamkovets, L. V. Keldysh, N. N. Sibel'din, and V. A. Tsvetkov

P. N. Lebedev Physics Institute, USSR Academy of Sciences
(Submitted November 6, 1975)
Zh. Eksp. Teor. Fiz. 70, 1501–1521 (April 1976)

A light-scattering method is used to investigate the dimension and the concentration of electron-hole drops (EHD) in germanium as functions of the temperature, excitation intensity, and waveform of the exciting light pulse. It is shown that the dependences of the condensed-phase volume on the temperature and intensity of the excitation are determined mainly by the number of produced drops (especially near the condensation threshold). It is observed that nuclei of the EHD are produced mainly when the excitation is turned on, i.e., on the front of the exciting pulse. Theoretical expressions are obtained for the description of the EHD density as a function of the temperature and of the excitation level at various rise times of the exciting-pulse front. The experimental results are discussed on the basis of a condensation theory that takes into account the surface tension of the electron-hole drop and the diffusion of the excitons to the surface of the EHD.

PACS numbers: 71.80.+j

INTRODUCTION

Since the publication of the first papers on electron-hole drops (EHD), [1–7] many theoretical and experimental investigations were made of the properties of the electron-hole liquid. There is, however, a considerable gap in the understanding of the physical picture of the exciton condensation, a gap concerning the kinetics of the formation and growth of the EHD. In addition, some presently known experimental facts which are apparently also connected with the kinetics of condensation have not been satisfactorily explained in the literature. Foremost among these is the difference, observed by a number of workers, in the dependence of the intensity of the EHD on the excitation power and the difference in the values of the binding energies of the carriers in EHD, as determined from optical and thermal measurements.

There are a number of cited results on the kinetics of exciton condensation, [8–14] obtained from measurements of the dependence of the volume of the liquid phase on the generation rate. These experimental data were analyzed on the basis of the equations of the condensation kinetics given in, [7,8] and it was assumed in most of their interpretations that the EHD concentration in the crystal does not depend on the experimental conditions, but is determined by the number of condensation centers. It is possible that this is justified under certain conditions, although it was shown in[15,16] that the EHD concentration depends strongly on the temperature and on the excitation intensity. In[17,18], optical hysteresis of the intensity of the EHD emission was observed as well as a very sharp growth of the volume of the liquid phase near the condensation threshold with increasing generation level. A sharp growth of the liquid-phase volume was observed also in impurity-containing samples. [9,10] These results are difficult to explain by assuming a fixed concentration of the EHD, as was noted in[9].

In the analysis of the causes of the discrepancies between the values of the binding energies of the carriers in EHD, determined from spectral and thermal measurements, account must be taken of the strong temperature dependence of the EHD concentration. Spectral measurements of the binding energy per pair of particles in the liquid phase were carried out in[13,18–22]. It appears that the presently most reliable binding energy is 2.06 ± 0.15 meV. [18,22] Temperature measurements[13,17,23,24] yield values approximately 30% lower. It is possible that it is precisely the dependence of the rate of liquid-phase nucleus phase formation on the temperature which leads to this difference.

The binding energy was also determined from measurements of the EHD recombination and evaporation kinetics. [23,25] These measurements yield approximately the same value as the temperature measurements. It should be noted in this connection that the results of studies of the recombination kinetics[23,25,26] were reduced with the aid of kinetics equations[7,8] that do not take into account the diffusion of the excitons towards the EHD surface, nor the surface tension of the electron-hole liquid. The contribution of the diffusion may turn out to be appreciable at large EHD radii. It is possible that failure to take this circumstance into account leads to a stronger growth of the EHD radius near the condensation temperature threshold, as observed in[25]. Theoretical calculations[27–30] and measurement results[16,31] show that the surface-tension coefficient of the electron-hole liquid is quite high ($\sim 10^{-4}$ dyn/cm), and therefore the role of the surface tension can become noticeable at small drop radii. In[32], the influence of surface tension on the kinetics of exciton condensation was considered, but the exciton diffusion was not taken into account there, too.

We used a light-scattering method to investigate the dependence of the dimension and concentration of the EHD in the case of volume quasistationary optical excitation on the following experimental conditions: the temperature, the excitation intensity, and the wave form of the light pulse. Our results show that the concentration of the EHD in the crystal increases with increasing excitation level and with decreasing temperature, and near the threshold temperature the dependence of the drop concentration on the generation level

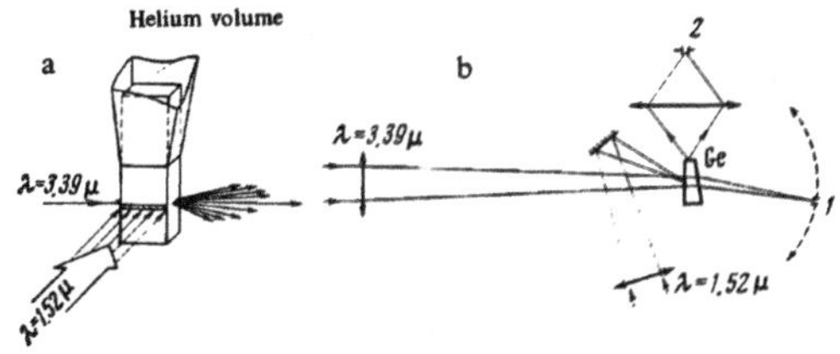

FIG. 1. System for the superposition of the laser beams in the measurement of the concentration (a) and of the total number of EHD (b): 1) entrance diaphragm of the quantum amplifier, 2) monochromator slit.

is stronger. Hysteresis-type phenomena were observed, i.e., under quasi-stationary excitation, the drop concentration increased with increasing slope of the light-pulse front.

The experimental results show that at low temperatures the stationary dimensions of the EHD are determined by the generation level and by the drop concentration. At high temperatures, when the recombination in the system comprising the exciton gas and the EHD proceeds mainly via the gas phase, the radius of the drops depends little on the excitation level and is independent of their concentration.

We have considered theoretically the model of liquid-phase nuclei formation and of their growth to form drops of stationary dimension; this model takes into account the surface tension of the electron-hole liquid and the diffusion of the excitons to the EHD drop. The presence of surface energy brings about a situation wherein EHD with radii not smaller than a certain temperature-dependent value can be in stable equilibrium with the exciton gas. This model agrees satisfactorily with experiment at low temperatures, but the behavior of the EHD radius in the high-temperature region still lacks an unambiguous explanation. Examination of the kinetics of formation of liquid-phase nuclei at various excitation conditions has made it possible to obtain for the dependence of the EHD concentration in the crystal on the temperature and on the generation levels expressions that describe correctly the experimental results.

EXPERIMENTAL PROCEDURE

The measurements were performed with a setup described in detail in[33]. The excitation source was a helium-neon laser of ~10 mW power, operating at a wavelength 1.52 μ. In the measurements we used two schemes for focusing the exciting radiation: in the measurements of the drop concentration, the exciting radiation was focused by a cylindrical lens on the lateral surface of the sample into a narrow strip (Fig. 1a), and in the measurement of the total number of the scattering particles it was focused on the front face of the crystal into a spot of ~200 μ diameter (Fig. 1b). The exciting radiation was modulated with a mechanical chopper of frequency 1 kHz. In the case when a large slope of the fronts of the exciting light pulse was required, the laser radiation was sharply focused on the modulator disk.

We investigated the scattering of 3.39-μ radiation from a helium-neon laser. The beam from this laser was focused with a long-focus lens on the front surface of the crystal in the measurements in accordance with the scheme of Fig. 1a, or on the entrance diaphragm of a quantum amplifier (Fig. 1b). The beams of both lasers were collocated in accordance with the maximum of the absorption signal, and during the measurement of the total number of the drops their collocation was also monitored against the symmetry of the diffraction pattern produced when the 3.39-μ radiation was diffracted by the region occupied by the EHD.[34]

The radiation scattered by the EHD was amplified with an optical quantum amplifier operating with a helium-neon mixture and was registered with a PbS receiver cooled to ~100 °K. The angular distribution of the intensity of the scattered light was recorded on the chart of an automatic plotter.

Simultaneously with the scattering, we record the spectrum of the recombination radiation of germanium. To observe the photoluminescence we used a standard setup containing a high-transmission MDR-2 monochromator. The radiation was registered with a cooled PbS photoresistor.

The measurements were performed on germanium samples with residual-impurity content not larger than 10^{12} cm^{-3}. The samples measured 15×5×2 mm and were sealed into the helium volume of the cryostat in such a way that the working half of the sample was in vacuum. The samples were mechanically polished, and the plane of the surface from which the scattered radiation emerged was inclined 2° to the opposite plane, in order to eliminate parasitic interference.

RESULTS AND DISCUSSION

The main results of our experiments are shown in Figs. 2–10. Although all the measurements were performed under conditions close to stationary (long-duration exciting pulses), phenomena of the hysteresis type were observed, i.e., a dependence of the result on the slope of the excitation-pulse front. We shall report first the results obtained under conditions closest to quasi-stationary, when the buildup of the excitation was quite slow (rise time $t_0 \approx 100$ μsec, much longer than the lifetime of either the excitons or the EHD).

Figure 2 shows the temperature dependences of the intensities I_{exc} (at the maximum of the spectral line) of the recombination radiation of the free excitons and I_d, of the EHD and also I_{abs} and I_{sc}, of the absorption and scattering signals respectively (I_{sc} is the light flux scattered at an angle 8° (in vacuum) into the aperture of the receiving system). It is seen that when the threshold temperature is reached ($T_c \approx 3.55$ °K at the employed excitation level), scattering and drop-luminescence signals appear simultaneously and the absorption signal begins to grow rapidly.

Figure 3 (upper curve) and Fig. 4 (lower curve) show plots of the EHD radius and of the EHD concentration against the temperature at a fixed excitation level.

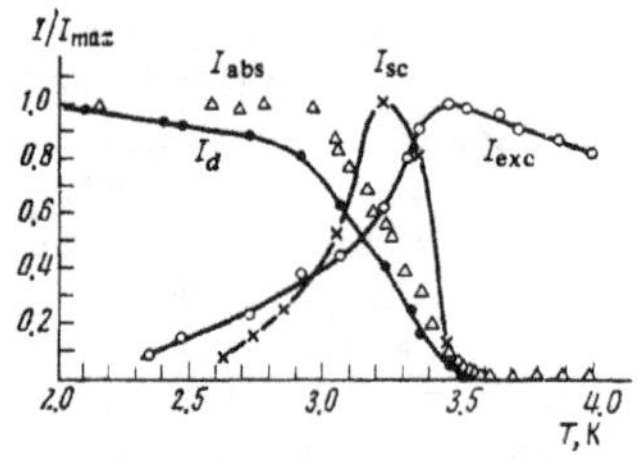

FIG. 2. Temperature dependences of absorption and scattering signals and of the intensities of the recombination radiation of the EHD and of the free excitons (the curves are normalized to their maximum values).

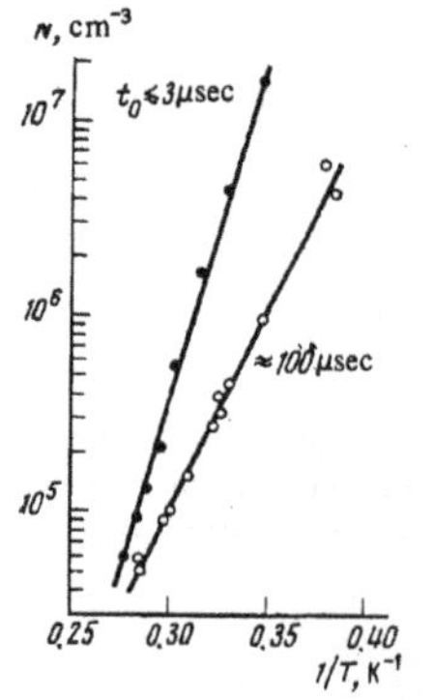

FIG. 4. Temperature dependences of the EHD concentration for long and short exciting-pulse fronts.

These measurements are performed with the beams superimposed in accordance with the scheme of Fig. 1a, and the EHD concentration was calculated from the measured values of their radii and from the magnitude of the scattering and absorption signals, as described in[35]. The most characteristic feature of these results is the strong increase of the EHD concentration and the decrease of their radius with decreasing temperature. We note also that the angular dependence of the scattering signal agrees well with the assumption that the radii of all the observed drops are practically the same, although the limitations of the measurement procedure do not exclude a possibility of the presence in the sample of a certain amount of EHD with dimensions much larger or much smaller than those observed.

The EHD dimensions were practically independent of the distance to the illuminated surface of the sample, although the volume of the liquid phase (which is proportional to the absorption signal) decreased with increasing distance (Fig. 5). When this distance was increased from 0.5 to 1.5 mm, the EHD dimensions decreased by not more than 10–15%, and this decrease was larger at lower temperatures. Consequently, the decrease of the volume of the liquid phase with increasing distance from the illuminated surface at high temperatures (curve 3 of Fig. 5) is more readily connected with a decrease of the drop concentration. In[36], a stronger change in the drop dimension with depth was observed. It is possible that this discrepancy is due to the fact that when an incandescent lamp is used as an excitation source the distribution of the nonequilibrium carriers bound in the EHD over the thickness of the sample is relatively inhomogeneous even at low temperatures.

The measurement setup illustrated in Fig. 1a is the most convenient for the interpretation of the data on light scattering by EHD. However, the power of the employed excitation source in this geometry of the excited region made it possible to produce only relatively low concentrations of impurity carriers, so that the threshold condensation temperature in the experiments described above did not exceed 3.55 °K, and furthermore, it was impossible to trace the dependences of the concentration and of the EHD dimensions on the excitation level.

To investigate these dependences in the most interesting temperature region, $2.5\,°K \lesssim T \lesssim 4.2\,°K$, the beams were superimposed in accordance with the scheme of Fig. 1b with relatively sharp focusing of the exciting beam. Then, however, the scattering experiments determine not the EHD concentration, but the total number of the drops along the path of the probing beam, which varies with changing temperature and with changing excitation level not in proportion to the concentration, inasmuch as the spatial distribution of the EHD over the depth of the sample changes (Fig. 5). Therefore the EHD concentration (N, in relative units) was calculated as the ratio of the intensity of the recombination radiation of the EHD to the cube of the EHD radius (the monochromator slit cut out a strip of width ≈ 0.3 mm near the illuminated surface of the sample). The absolute values of the total number of electron-hole pairs and of the total number of EHD in

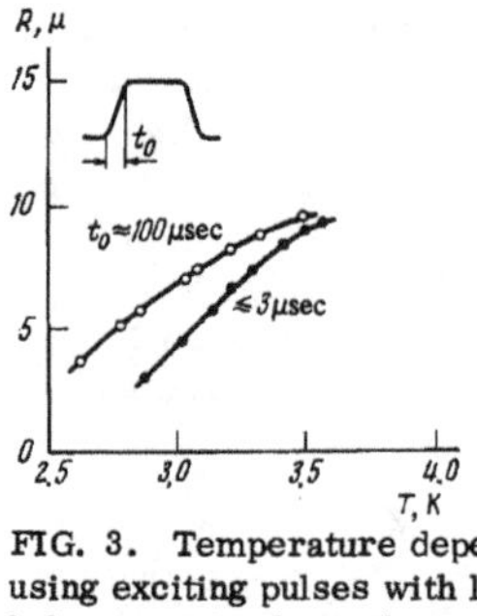

FIG. 3. Temperature dependences of EHD radius, measured using exciting pulses with long and short fronts. In the upper left corner is shown the wave form of the exciting pulse.

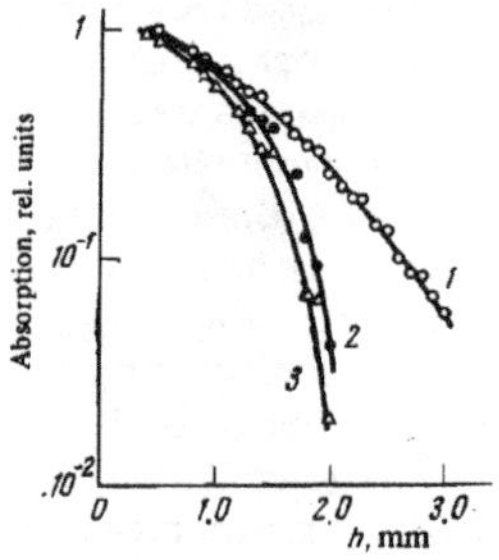

FIG. 5. Distribution of the absorption over the sample thickness: 1) $T=2.8\,°K$, $P=8$ mW; 2) $T=2.8\,°K$, $P=2.4$ mW; 3) $T=3.8\,°K$, $P=8$ mW (P is the power of a laser with wavelength 1.52 μ, h is the distance from the illuminated surface.)

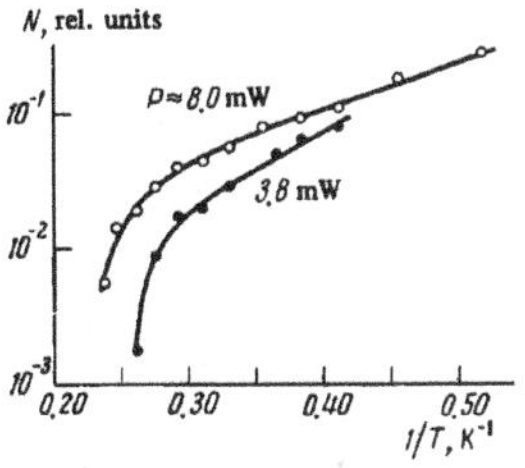

FIG. 6. Temperature dependences of the EHD concentration at two different excitation intensities. Front duration $t_0 \approx 100$ μsec.

the sample at the maximum excitation level, estimated from measurements of the absorption scattering, were respectively ~ 10^{12} and ~ 10^4 at 1.9 °K, and their concentrations were ~ 2×10^{15} and ~ 2×10^7 cm^{-3}.

It is seen from Fig. 6 that the very sharp growth of the EHD concentration with decreasing temperature, observed at $T \gtrsim 3$ °K, slows down greatly at lower temperatures. The growth of the EHD concentration with increasing excitation level g also slows down with decreasing temperature (Fig. 7a). The $N(g)$ dependence is superlinear at temperatures relatively close to the threshold, and sublinear at $T \lesssim 3$ °K.

We note finally one more striking result (Fig. 8): at temperatures $T \gtrsim 3$ °K the EHD radii are equal to ≈ 10 μ and are practically independent of either the temperature or the excitation level.

So far we have described the results obtained with a slow (quasi-stationary) growth of the light intensity in the exciting pulses. The results of experiments in which the sample was excited with pulses of practically the same duration (~ 500 μsec) but with much shorter rise times ($t_0 \lesssim 3$ μsec) are shown in Fig. 3 (lower curve), Fig. 4 (upper curve), and in Figs. 7b, 9, and 10. Although practically all the general tendencies remain the same as before, it is seen from these figures that given the temperature and the excitation level, the EHD concentration is higher, and their radius is smaller, than in the case of excitation with smoothly growing pulses. In addition, the EHD concentration increases more rapidly with decreasing temperature and with increasing excitation level. The function $N(g)$ turns out to be significantly superlinear (Fig. 7b) in the entire temperature interval ($T \gtrsim 2.5$ °K). This ex-

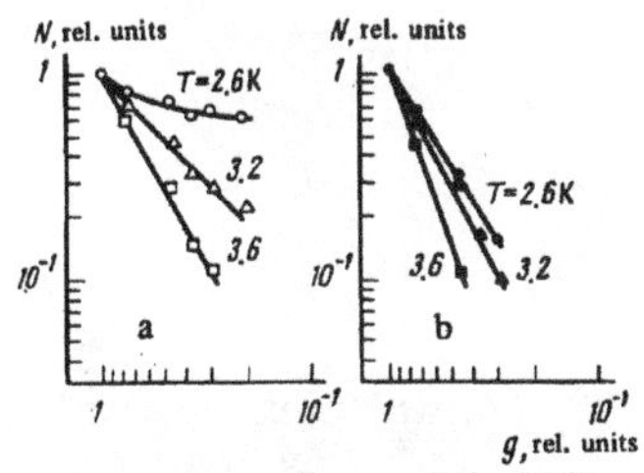

FIG. 7. Dependence of the EHD concentration on the excitation level at three different temperatures: a) $t_0 \approx 100$ μsec, b) $t_0 \lesssim 3$ μsec.

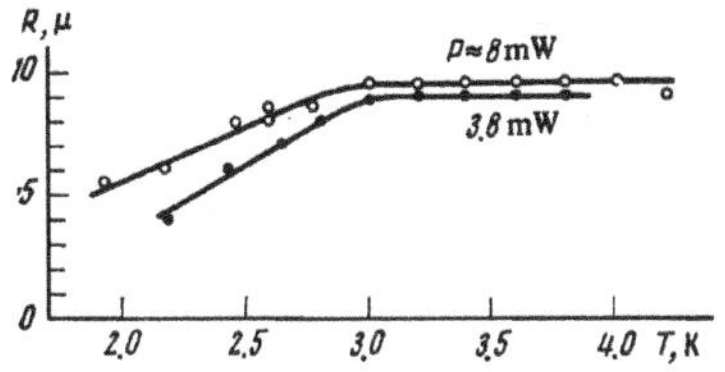

FIG. 8. Temperature dependence of the EHD radius at two different excitation intensities ($t_0 \approx 100$ μsec).

plains, in particular, the only qualitative difference between the results obtained using exciting pulses with long and short pulses, i.e., the directly contradictory dependences of the EHD radii on the excitation level, namely, when the exciting pulse has a long rise time the radius increases with increasing excitation level (Fig. 8), and in the case of a short rise time it decreases (Fig. 10). We emphasize that at temperatures $T \gtrsim 3.5$ °K the EHD radius tends to the same value (~ 10 μ) as in the case of a long rise time of the exciting light pulse.

As will be shown below, all these differences can be explained in the following manner: When the exciting pulses have short rise times, a strongly supersaturated exciton "vapor" is produced in the sample within a short time, as a result of which a large number of EHD nuclei begin to grow practically simultaneously. On the other hand in the case of a slow rise of the excitation, the relatively small number of EHD that begin to grow ahead of the others manage to absorb all the excess of the subsequently introduced nonequilibrium carriers, preventing by the same token the formation of new drops.

An analysis of the kinetics of the formation of the condensates calls for a consideration of two essentially different questions: the onset of the EHD nuclei and the growth of each individual drop until equilibrium is established between it and the surrounding exciton gas. We start with the second of these problems, since it is simpler.

The change in the number of particles in a spherical drop of radius R is determined by the equation

$$\frac{d}{dt}\left(\frac{4}{3}\pi R^3 n_0\right) = S - \frac{4}{3}\pi R^3 \frac{n_0}{\tau_0}, \qquad (1)$$

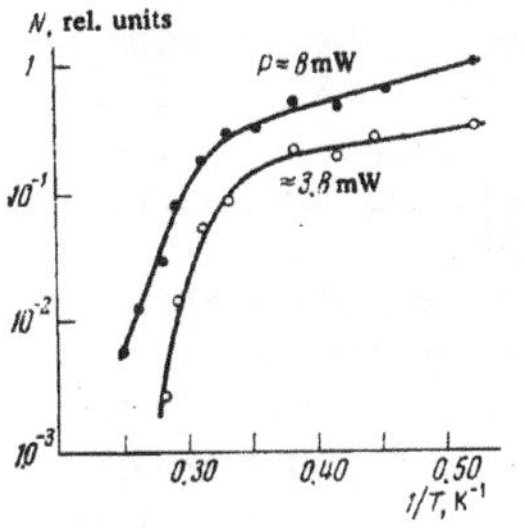

FIG. 9. Temperature dependence of EHD concentration at two different excitation intensities ($t_0 \lesssim 3$ μsec).

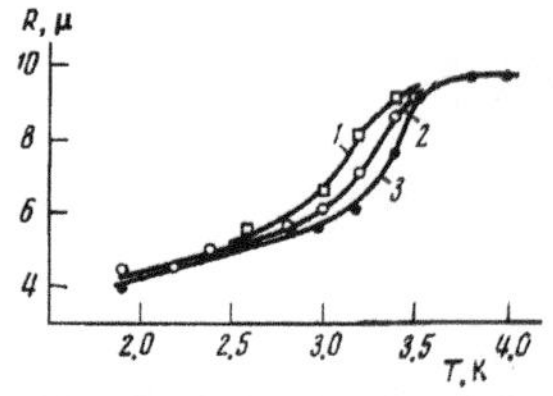

FIG. 10. Dependences of the EHD radius on the temperature at three different excitation intensities ($t_0 \lesssim 3$ μsec): 1) $P \approx 2.4$ mW, 2) $P = 3.8$ mW, 3) $P = 8$ mW.

where n_0 and τ_0 are the equilibrium density and lifetime of the carriers in the condensed phase, and S is the flux of the excitons entering or leaving the drop:

$$S = 4\pi R^2 \gamma [n(R) - n_T(R)] v_T. \tag{2}$$

Here $n(R)$ is the concentration of the excitons near the surface of the drop, $v_T = (kT/2\pi M)^{1/2}$ is the thermal velocity of the excitons, M is their effective mass, γ is the "sticking coefficient" of the excitons in the drop, i.e., the probability that an exciton landing on the surface of the drop will be captured rather than reflected and

$$n_T(R) = \nu \left(\frac{M_d kT}{2\pi\hbar^2}\right)^{3/2} \exp\left(-\frac{\Delta}{kT} + \frac{2\alpha}{n_0 R kT}\right) \tag{3}$$

is the thermodynamic-equilibrium concentration of the excitons above a drop of radius R. In formula (3), Δ is the work function of the excitons from the EHD as $R \to \infty$, M_d is the effective mass of the density of state of the excitons, α is the coefficient of the surface tension of the electron-hole liquid, and ν is the multiplicity of the degeneracy of the ground state of the exciton (with respect to spin, the number of equivalent extrema in the Brillouin zone, etc.). The term $-4\pi R^2 \gamma n_T v_T$ in (2) describes the evaporation of the excitons from the drop, and the term $2\alpha/n_0 RkT$ in the argument of the exponential in (3) describes the increase of the evaporation rate due to the surface tension.

Equations of the type (1) and (3) were proposed by Pokrovskiĭ and Svistunova in[8] and used in a number of studies,[7,11,13,23,25,26] in all of which it was assumed, however, that the exciton concentration $n(R)$ near the surface of the drop coincides with their average concentration in the volume, and no account was taken of the role of the surface tension. Silver[32] used a somewhat more general approach in the analysis of the structure of the two-phase system of the non-equilibrium carriers, based on equations of the Fokker-Planck type for the EHD radius distribution function, and demonstrated within the framework of this approach the need for taking the surface tension into account. However, the difference between $n(R)$ and n (averaged over the volume of the exciton concentration) was not taken into account in his paper. At the same time, the presence of an exciton-concentration drop is necessary for the existence of the flux S.

The dependence of the exciton concentration $n(\mathbf{r}, t)$ on the coordinate $\mathbf{r}$ and on the time t is determined by the diffusion equation

$$\frac{\partial n(\mathbf{r},t)}{\partial t} + \operatorname{div}[-D\nabla n(\mathbf{r},t)] + \frac{n(\mathbf{r},t)}{\tau} = g(\mathbf{r},t), \tag{4}$$

where D is the exciton diffusion coefficient, τ is their lifetime, and $g(\mathbf{r}, t)$ is the rate of generation of nonequilibrium carriers by the excitation source (it is assumed that the time of binding the carriers into excitons is short enough). Conditions (2) on the surface of each EHD serve as the boundary conditions for (4). Recognizing that the distance between the drops is much smaller than the exciton diffusion length $L_D = \sqrt{D\tau}$, we can average Eq. (4) over the volume elements containing many EHD but small relative to L_D. Then

$$\frac{\partial \bar{n}}{\partial t} + \operatorname{div}(-D\nabla\bar{n}) + NS(\bar{n}, R) + \frac{\bar{n}}{\tau} = g(\mathbf{r},t), \tag{5}$$

where $\bar{n}(\mathbf{r}, t)$ is the exciton concentration averaged in the indicated manner and N is the drop concentration. Equation (5) describes the time variation of the average exciton concentration $\bar{n}(\mathbf{r}, t)$ as the result of their generation, recombination, and diffusion, as well as capture and evaporation by the drops (the term $NS(\bar{n}, R)$). In the case of a spatially homogeneous excitation of the sample we have $\operatorname{div}(-D\nabla\bar{n}) = 0$ and the diffusion fluxes of the excitons propagate only towards each of the drops.

To find $S(\bar{n}, R)$ it is necessary to solve (4) in the vicinity of each given drop with boundary conditions (2) and $n(r \to \infty) = \bar{n} \equiv n$ (it is assumed that the dimension of the drops is small in comparison with the distance between them). Since we are now dealing with the solution of the problem in a small region $r \ll N^{-1/3}$ (r is the distance from the center of the drop), within which a nonstationary exciton distribution is established in a rather short time, the terms with $\partial n/\partial t$, n/τ, and g can be omitted from (4), accurate to the small terms

$$NR^3 \ll 1, \quad (NL_D^3)^{-1} \ll 1. \tag{6}$$

Thus

$$D \operatorname{div} \operatorname{grad} n(r) = \frac{D}{r^2}\frac{d}{dr}\left(r^2 \frac{dn(r)}{dr}\right) = 0.$$

Inasmuch as $D\nabla n|_{r=R} = S/4\pi R^2$, the solution of this equation with allowance for the boundary conditions takes the form $n(r) = n - S/4\pi Dr$. Combining it with (2) and (1) we obtain

$$S(n,R) = 4\pi\gamma R^2 \frac{n - n_T(R)}{1 + \gamma v_T R/D} v_T \tag{7}$$

and

$$\frac{dR}{dt} = \frac{n - n_T(R)}{n_0} \frac{\gamma v_T}{1 + \gamma v_T R/D} - \frac{R}{3\tau_0}. \tag{8}$$

For small drops, Eqs. (8), (5), and (3) differ from those usually employed[7,8,11,13,23,25,26] only in that the surface tension is taken into account in the evaporation rate (3). However, when the dimensions of the drops

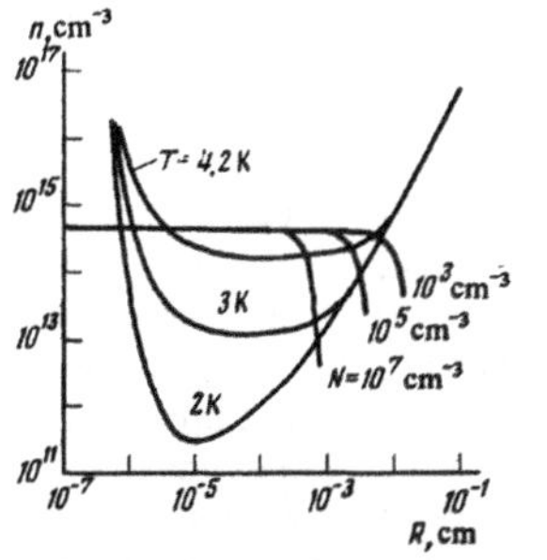

FIG. 11. Plots of n against R, calculated from formula (9) at three different temperatures and from formula (12) for three EHD concentrations at $\gamma=1$, $\tau_0=20$ μsec, $D=1500$ cm^2/sec at $T=3$ °K[13] and with account taken of the fact that $D\sim T^{-1/2}$, $\tau=5$ μsec, and $g=10^{20}$ cm^{-3}sec^{-1} (the numerical values of the remaining parameters are given in the text).

become comparable with the exciton mean free path ($v_T R/D \gtrsim 1$), their further growth is greatly slowed down by the strong decrease of the exciton concentration near the surfaces of the drops in comparison with their average concentration n in the volume.

In the stationary case, Eq. (8) determines directly the connection between the average exciton concentration and the EHD radius:

$$n=n_T(R)+\frac{n_0 R}{3\gamma v_T \tau_0}\left(1+\gamma\frac{v_T R}{D}\right), \tag{9}$$

which is plotted in Fig. 11.

In the plane of the variables n and R (Fig. 11), the points that do not land on the $n=f(R)$ curve described by expression (9) cannot correspond to stationary states of the system. For points lying above this curve, the radius of the drop increases with time ($dR/dt>0$), and for points lying below this curve it decreases ($dR/dt<0$). Therefore the stable states are only those on the right-hand ascending branch of the $n=f(R)$ curve at $R>R_{min}$, whereas on the descending (left) part any small deviations of n or R lead to the fact that R begins to vary with time, going farther and farther away from the equilibrium curve.[32]

Thus, for each given temperature there exists a certain minimal radius defined by the relation

$$R_{min}^2\left(R_{min}+\frac{D}{2\gamma v_T}\right)-\frac{3\alpha D\tau_0}{n_0 kT}\frac{n_T(R_{min})}{n_0}=0, \tag{10}$$

so that in the stationary state the drops cannot have a dimension smaller than $R_{min}(T)$. In the case $R \ll l \sim D/v_T$ is the exciton mean free path) we have

$$R_{min}\approx\left(6\gamma\frac{\alpha v_T \tau_0}{n_0 kT}\frac{n_T}{n_0}\right)^{1/3}, \tag{11}$$

where n_T is defined by (3). If furthermore $2\alpha/n_0 RkT \ll 1$, then this quantity can be neglected in the argument of the exponential in (3) and $n_T\approx n_{0T}$, i.e., it is nearly equal to the thermodynamic-equilibrium concentration of the excitons over the "liquid" in the case of a flat interface.

The true value of R is determined by the EHD concentration by virtue of relations (5) and (9). If the excitation can be regarded as stationary and spatially-homogeneous with sufficient degree of accuracy, then with (7) and (8) taken into account, (5) can be transformed into (with allowance for the fact that $\partial n/\partial t = \partial R/\partial t = \mathrm{div}(-D\nabla n)=0$)

$$\frac{n}{\tau}+\frac{4}{3}\pi R^3 N\frac{n_0}{\tau_0}=g, \tag{12}$$

which reflects simply the balance of the number of generated and recombining carriers. Plots of the average concentration of the excitons n against the EHD radius, calculated with the aid of (12) for three different drop concentrations N, are also shown in Fig. 11. The points of intersection of these curves with the stable branch of the $n=f(R)$ curves described by expression (9) yield the radius of the EHD at a given drop concentration and at a given generation level, and the intersection with the unstable branch gives the "critical" radius of the nuclei.

It seems natural to assume that under conditions of stationary excitation, owing to the production of more and more new EHD, the value of N will increase with time, while n and R will decrease and tend gradually to their limiting values that are compatible with (9) and (12) and with the condition $R \geqslant R_{min}$, namely

$$N_{max}(g,T)=\frac{3}{4\pi}\frac{\tau_0}{n_0 R_{min}^3}\left(g-\frac{n_{min}(T)}{\tau}\right), \tag{13}$$

$$n_{min}(T)=n(R_{min}), \tag{14}$$

so that in this stationary state the EHD radius and concentration, as well as the density of the exciton gas, are determined by expressions (10), (13), and (14).

This is precisely the conclusion drawn by Silver.[32] However, the data by others, which are cited above, do not agree with this conclusion. Indeed, as seen from Figs. 3, 8, and 10, at the same temperature $T \lesssim 3$ °K, the EHD radii vary noticeably as functions of the excitation level and the rise time of the exciting pulse, in contrast to R_{min}, which depends only on T. At $T \gtrsim 3.5$ °K, the radii in general are practically independent of T ($R\approx 10$ μ), whereas R_{min} should increase monotonically with temperature. Finally, numerical values of the observed EHD radii also exceed noticeably the values of R_{min} estimated from formula (11), especially at low temperatures ($T\approx 2$ °K). Assuming $\gamma\approx 1$, $\alpha\sim 2\times10^{-4}$ erg/cm^2,[16,31]1) $M=M_d\approx 4\times10^{-28}$ g, $\tau=40$ μsec, $n_0\approx 2\times10^{17}$ cm^{-3}, $\nu=16$, and $\Delta\approx 2.1$ meV we obtain the values $R_{min}\approx 5$ μ and ≈ 0.2 μ at $T=4.2$ °K and 2 °K, respectively. Thus, if the observed values of R can be close to R_{min} then this can occur only at the highest temperatures near 4 °K. There is no doubt, however, that the condition $R>R_{min}$ is satisfied by all the experimental values of R.

All these facts indicate that under the conditions of our experiments, in spite of the long duration of the exciting pulses, greatly exceeding the lifetimes of both

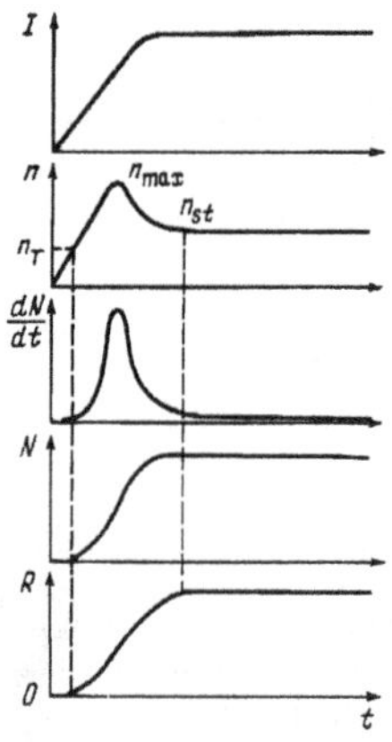

FIG. 12. Plots of the exciton concentration n, of the EHD N, of the rate of nucleus formation dN/dt, and of the drop radius R on the time t with increasing excitation intensity I in the pulse.

the excitons and the EHD, the system is still far from the stationary state described by formulas (13) and (14). This is indicated primarily by the strong dependence of the EHD number on the duration of the front of the exciting pulse. Indeed, the front duration during which the excitation varies with time, is not more than 20% of the pulse duration during which the excitation is perfectly stationary, and the EHD concentration in the case of a short rise time of the exciting pulse nevertheless greatly exceeds the value observed in the case of a slow growth of the excitation. This seems to mean that the EHD or their nuclei are produced only at the very start of the pulse and that their number then remains practically constant during the entire remainder of the pulse. This result can be explained qualitatively in the following manner:

When the exciting pulse is applied, the exciton concentration n in the sample begins to increase (Fig. 12). When this concentration greatly exceeds n_{0T}, EHD nuclei begin to form, but the number of excitons going to these nuclei is relatively small at first, since both the number of the nuclei N and the radius of each of the nuclei are small, so that the term $NS \sim NR^2$ plays no role in (5). Therefore the exciton concentration continues to increase (within times shorter than τ or, in the case of a long rise time—shorter than the front duration). But with increasing degree of supersaturation of the electron gas, the number of the newly produced nuclei increases rapidly (exponentially), and the radii of the previously produced EHD increase with them, in accord with (8). At a certain instant, the number of excitons going into the drops becomes comparable with the rate of carrier generation, and later it even exceeds it. From this instant on, the exciton concentration begins to decrease, in spite of the constant or even increasing excitation intensity I, since all the produced excitons are required for the growth and for the maintenance of the existence of the already produced EHD. The formation of new nuclei, on the other hand, soon practically ceases, because the exciton concentration decreases to a level to n_{st}, at which the probability of production of nuclei, meaning also the rate of their formation dN/dt, is very small. The previously produced EHD continue to increase for some time, until they reach the dimension determined by formulas (9) and (12), but now already at a given number of EHD.[2)]

At first glance, the following considerations, based on the stabilization of the concentration of the excitons at a level close to n_{0T} contradict certain known results,[13,17] which show that even under stationary excitation, especially at temperatures $T > 3$ °K, the number of excitons continues to increase with increasing excitation level after passing through the condensation threshold. This, however, is more readily due to the fact that in all the experiments, as was already emphasized in[32], the excitation is not spatially homogeneous and the condensation threshold is reached first in a very narrow region, where the excitation is maximal. With further rise of the carrier generation level, the number of excitons continues to grow on account of those regions where the condensation threshold has not yet been reached. The increase in the dimensions of the region in which the EHD are concentrated with increasing excitation was noted by many workers.[38–42] It is seen also in our experiments on Fig. 5. At the same time, the above-described time evolution of the exciton condensation process agrees not only with our data, but also with the well-known results[17,18] on the observation of hysteresis phenomena in the EHD and exciton radiation, results which can hardly be explained in any other way.

A more detailed examination of this evolution (Appendix) on the basis of the well-known theory of formation of critical nuclei of a liquid phase in supersaturated vapor[43–45] explains practically all the qualitative features of our experimental results.

Indeed, formulas (A.35) and (A.37) show that the drop concentration N increases with decreasing rise time t_0 of the exciting pulse, in agreement with the results shown in Figs. 4, 6, and 9. With decreasing temperature, N first increases very rapidly, and this growth is stronger in the case of a short rise time (A.37) than a long one (A.35). The experimentally observed slopes of log N against $1/T$ (Fig. 4) are well represented by these formulas at $\Delta \approx 20$–25 °K, which agrees with the *spectroscopically* determined work function of the excitons from the EHD, $\Delta \approx 2$ meV. With further decrease of temperature, the growth of N decreases very rapidly ($\beta\xi$ in (A.35) and $\beta\eta$ in (A.37) become of the order of unity), as is clearly seen from Figs. 6 and 9. The dependence of N on the excitation level g is, in accordance with (A.37), stronger (superlinear) for a short front than for a long one—see (A.35). It becomes steeper with increasing temperature; for example, in the case of a short front, as T approaches the condensation threshold at a given g, the delay time t_m of the drop formation increases, and when it becomes larger than t_0 we go over from the short-front situation described by formula (A.37) to the case of "extremely short front," described by formula (A.21); we change from the relation $N \propto g^{3/2}$ to $N \propto g^3$. All these results agree qualitatively with those shown in Fig. 7.

We have not carried out a detailed quantitative re-

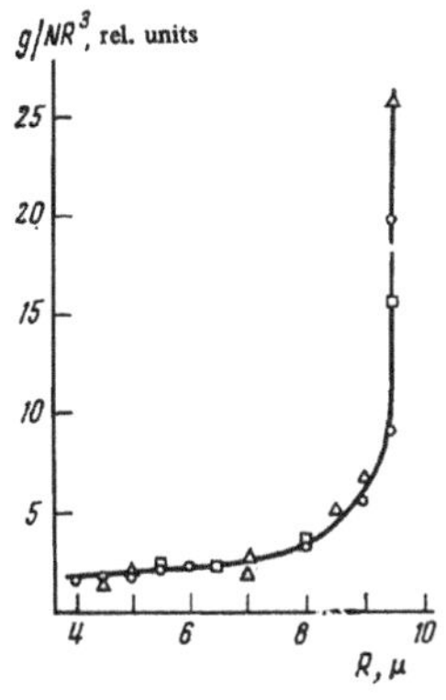

FIG. 13. Plot of g/NR^3 against R at three different excitation levels: o – $P \approx 8$ mW, Δ – $P \approx 3.8$ mW, □ – $P \approx 2.4$ mW.

duction of the experimental results by means of the theoretical formulas, inasmuch as these results are known to be distorted by the aforementioned expansion of the spatial region of concentration with decreasing temperature or with rising excitation level. We note, however, that formulas (A.35) and (A.37) yield the correct order of magnitude of not only the qualitative relations but also of the absolute values of the EHD concentrations. Thus, for example, for the case $t_0 = 3$ μsec, assuming $\Delta \approx 2.1$ mV, $\alpha \approx 2\times10^{-4}$ erg/cm^2, $g\tau \approx 5\times10^{13}$ cm^{-3} (which corresponds to an excitation power ≈ 8 mW under the conditions of our experiment), $\tau = 5$ μsec, and $N_i \approx 10^{12}$ cm^{-3} and taking the values of these parameters given above for Ge, we obtain from (A.37) and (A.38) the values $N \sim 4\times10^5$ and $\sim 10^7$ cm^{-3} at $T = 3.2$ and 2.8 °K, respectively, in reasonable agreement with the result shown in Fig. 4. We note also that these calculated values depend very little on the choice of the number N_i of the condensation centers.

In conclusion, let us discuss the behavior of the EHD radius. Figure 13 shows the dependence of the ratio of the generation rate to the volume of the liquid phase, g/NR^3, on the EHD radius for three different generation levels. At small R (low temperatures), $g/NR^3 \approx$ constant, as follows from (12) in the case when the recombination in the gas phase can be neglected. At large EHD radii (high temperatures), however, the radius exhibits a clearly pronounced tendency to the value ≈ 10 μ (Figs. 3, 8, 10), which depends neither on the temperature nor on the level and the method of excitation. This tendency is particularly clearly manifest in Fig. 13 by the abrupt growth of g/NR^3 at a practically constant value of the radius. This result cannot be described by expressions (9) and (12), which seems to indicate that the theoretical picture described above is incomplete.

The experimental data indicate that there exists a certain mechanism that limits strongly the growth of the EHD radius at $R \sim 10$ μ, at least under the conditions of our experiments. At the present time we are unable to explain this fact finely. A number of possible mechanisms that limit the EHD radius have been considered in[46,47].

We are indebted to S. G. Tikhodeev for a discussion of the results.

APPENDIX

The well-known formula[45] for the rate of production of critical nuclei in supersaturated vapor with particle density n can be represented in the form

$$\frac{dN}{dt} = A\left(\frac{\beta\Delta}{kT}\right)^{1/2} v_T n_0^{1/3} N_i \frac{n}{n_0} \exp\left\{-\lambda(T) \Big/ \ln^2\frac{n}{n_{0T}}\right\}, \qquad \text{(A.1)}$$

where

$$\lambda(T) = \frac{2\pi}{3}\left(\frac{2\beta\Delta}{kT}\right)^3. \qquad \text{(A.2)}$$

The dimensionless coefficient β is connected with the surface-tension coefficient α by the relation

$$\alpha = \beta\Delta n_0^{2/3}, \qquad \text{(A.3)}$$

N_i is the concentration of the impurities that serve as condensation centers, $n_{0T} = n_T(R = \infty)$ is the equilibrium density of the saturated vapor over the planar phase-separation boundary, and A is a numerical factor of the order of unity. Strictly speaking, the numerical factor in the pre-exponential term of (A.1) can be omitted, since inclusion of this factor is an exaggeration of the accuracy with which this formula is derived. We shall therefore omit from all the final formulas the numerical pre-exponential factors.

Equations (5), (7), (8), and (A.1) form a complete system of equations describing exciton condensation. We consider the case of a long (with duration much larger than τ) pulse of spatially homogeneous excitation $g(t)$, beginning with the instant $t = 0$ ($g(t<0) = 0$). At the start of the pulse, in the absence of EHD, the exciton concentration increases monotonically in accordance with the regular law

$$n(t) = \int_0^t \exp\left\{-\frac{t-t'}{\tau}\right\} g(t')\,dt'. \qquad \text{(A.4)}$$

Formation of the EHD begins at the instant of time t_1 at which $n(t_1)$ reaches the value n_{0T}, and their contribution to the change in the number of free excitons can be represented in the form

$$-N(t)S(n,R) = -4\pi v_T \int_{t_1}^{t} \frac{dN(t')}{dt'} \frac{R^2(t,t')}{1+v_T R(t,t')/D}[n - n_T(R(t,t'))]\,dt', \qquad \text{(A.5)}$$

where $R(t, t')$ is the value, at the instant t, of the radius of the drop produced at the instant t', i.e., the solution of equation (8) with initial condition $R(t') = 0$. With (A.5) taken into account, Eq. (5) takes the form

$$\frac{\partial n}{\partial t} = g(t) - \frac{n}{\tau} - 4\pi v_T \int_{t_1}^{t} \frac{dN(t')}{dt'} \frac{R^2(t,t')}{1+v_T R(t,t')/D}[n(t) - n_T(R(t,t'))]\,dt'. \qquad \text{(A.6)}$$

Recognizing that $\lambda \gg 1$, we can easily see that so long as the supersaturation of the exciton gas is small, $(n - n_{0T})/n_{0T} \ll 1$, the contribution of the last term in the right-hand side of (A.6) is negligibly small, but with increase of supersaturation, at $(n - n_{0T})/n_{0T} \gtrsim 1$, it begins to grow very rapidly, and at a certain instant

t_m the right-hand side of (A.6) reverses sign, i.e., the exciton concentration reaches the maximum value n_m, which is connected with t_m by the relation

$$g(t_m)-\frac{n_m}{\tau}-4\pi v_T\int_{t_1}^{t_m}\frac{dN(t')}{dt'}\frac{R^2(t_m,t')}{1+v_TR(t_m,t')/D}[n_m-n_T(R(t_m,t'))]dt'=0. \tag{A.7}$$

By virtue of the very strong dependence of dN/dt on n, the main contribution to the integral in (A.7), as well as in all the subsequently encountered integrals containing dN/dt, is made by a narrow vicinity of the instant of time t_m. Therefore, expanding the argument of the exponential in (A.1) in terms of the small difference $t_m - t' = s \ll t_m$, we obtain

$$\frac{dN}{dt'}\approx A\left(\frac{\beta\Delta}{kT}\right)^{1/2}v_Tn_0^{1/3}N_i\frac{n_m}{n_0}\exp\left\{-\frac{\lambda}{\mathscr{L}^2}+\frac{\lambda}{n_m\mathscr{L}^3}n_m''s^2\right\}, \tag{A.8}$$

where $\ln(n_m/n_{0T})\equiv \mathscr{L}$, $n_m''=(\partial^2 n/\partial t^2)_{t=t_m}$ appeared. To find n_m'' we differentiate (A.6) once more with respect to time at the instant $t=t_m$, recognizing that $(\partial n/\partial t)_{t_m}=0$ and $R(t_m, t_m)=0$:

$$n_m''=\left(\frac{dg}{dt}\right)_{t_m}-4\pi v_T\int_{t_1}^{t_m}\frac{dN(t')}{dt'}\frac{d}{dt_m}\times\left\{\frac{R^2(t_m,t')}{1+v_TR(t_m,t')/D}[n_m-n_T(R(t_m,t'))]\right\}dt'. \tag{A.9}$$

Using (A.8), we usually obtain the total number of the produced EHD:

$$N\approx\int_{-\infty}^{\infty}\frac{dN(t_m-s)}{ds}ds\approx A\pi^{1/2}\left(\frac{\beta\Delta}{kT}\right)^{1/2}\left(\frac{n_m\mathscr{L}^3}{\lambda|n_m''|}\right)^{1/2}v_Tn_0^{1/3}N_i\frac{n_m}{n_0}\exp\left\{-\frac{\lambda}{\mathscr{L}^2}\right\}$$

$$=\frac{\sqrt{3}}{4}A\frac{kT}{\beta\Delta}\frac{v_TN_i}{n_0^{2/3}}\left(\frac{n_m^3\mathscr{L}^3}{|n_m''|}\right)^{1/2}\exp\left\{-\frac{\lambda}{\mathscr{L}^2}\right\}. \tag{A.10}$$

On the other hand, the calculation of the analogous integrals in (A.7) and (A.9) calls for an additional analysis of the properties of the integrands. The point is that as $t'\to t_m$ the radius R tends to zero, or more accurately to the radius of the critical nucleus

$$R_c(n)=\frac{2\beta\Delta}{kT\ln(n/n_{0T})}n_0^{-1/3}. \tag{A.11}$$

The difference $n-n_T(R)$ tends to zero at the same time. We cannot therefore replace in (A.7) and (A.9) all the factors by their values at $s=0$, with the exception of dN/dt. However, with increasing s, the radius $R(t_m, t_m-s)$ increases in accordance with (8) like

$$\frac{n}{n_0}v_Ts=\int_{(1+\delta)R_c}^{R(t_m,t_m-s)}dR\left(1-\exp\left\{-\frac{R-R_c}{R}\ln\frac{n}{n_{0T}}\right\}\right)^{-1}, \tag{A.12}$$

where $\delta\cdot R_c$ is the initial excess of the radius above the critical value at $t'=t_m-s$ $(0<\delta\lesssim 1)$.

In the derivation of (A.12) we took into account the fact that we are dealing here with very small time intervals and radii, so that the terms in (8), which describe the recombination and diffusion, were left out.

In addition, it was assumed that $\gamma=1$ and that the exciton concentration n at the instant $t'=t_m-s$ does not change significantly in the interval s. It is easy to verify that if the condition

$$\frac{n}{n_0}\frac{v_Ts}{R_c}\gg\frac{1}{\ln(n/n_{0T})} \tag{A.13}$$

is satisfied, the EHD radius begins to depend linearly on s, and the difference become $n-n_T(R)\approx n-n_{0T}$. We shall see later on that for values of s that make the effective contribution to the integrals of (A.7) and (A.9), the condition (A.13) is satisfied practically in all cases when the condensation is observable. Then, taking all the foregoing into account we have

$$R^2(t_m,t_m-s)[n_m-n_T(R(t_m,t_m-s))]\approx\frac{(n_m-n_{0T})^3}{n_0^2}v_T^2s^2;$$

$$g(t_m)-\frac{n_m}{\tau}-\pi^{1/2}A\left(\frac{\beta\Delta}{kT}\right)^{1/2}v_T^4N_in_0^{1/3}n_m\left(\frac{n_m-n_{0T}}{n_0}\right)^3\times\left(\frac{n_m\mathscr{L}^3}{\lambda|n_m''|}\right)^{3/2}\exp\left\{-\frac{\lambda}{\mathscr{L}^2}\right\}=0 \tag{A.14}$$

and

$$n_m''=\left(\frac{dg}{dt}\right)_{t_m}-4\pi A\left(\frac{\beta\Delta}{kT}\right)^{1/2}v_T^4N_in_0^{1/3}\left(\frac{n_m-n_{0T}}{n_0}\right)^3\frac{n_m^2\mathscr{L}^3}{\lambda|n_m''|}\exp\left\{-\frac{\lambda}{\mathscr{L}^2}\right\}. \tag{A.15}$$

Recognizing that the effective values of $s\ll t_m$ and $g(t)$ vary with time much more slowly than the last term in (A.6), we can determine the instant of time t_m by using directly Eq. (A.4) in the form

$$n_m\approx\int_0^{t_m}g(t)\exp\left\{-\frac{t_m-t}{\tau}\right\}dt. \tag{A.16}$$

Formula (A.10) together with equations (A.14)–(A.16) solves completely in principle the problem of the number of produced EHD for sufficiently long pulses (large t_m) of volume excitation. However, the character of the resultant dependences of N on g and T can be quite different under different experimental conditions. We demonstrate this by considering certain limiting cases, assuming that the excitation intensity first increases linearly with time, and then levels off at the stationary value

$$g(t)=\begin{cases}gt/t_0, & 0\le t\le t_0\\ g, & t>t_0\end{cases}. \tag{A.17}$$

1. Extremely short rise time of the exciting pulse: $t_0\to 0$. The EHD production occurs at $t>t_0$, i.e., $(dg/dt)_{t_m}=0$. In this case

$$N=\text{const}\left[\left(\frac{kT}{\beta\Delta}\right)^{1/2}\frac{N_i}{n_m-n_{0T}}\mathscr{L}\right]^{3/4}n_m\exp\left\{-\frac{3}{4}\frac{\lambda}{\mathscr{L}^2}\right\}, \tag{A.18}$$

and n_m is determined by the transcendental equation

$$g-\frac{n_m}{\tau}=\left(\frac{\pi^3A}{8}\right)^{1/4}\left[\left(\frac{\beta\Delta}{kT}\right)^{1/2}N_in_0^{1/3}\right]^{1/4}n_mv_T\left(\frac{n_m-n_{0T}}{\lambda n_0}\right)^{3/4}\mathscr{L}^{9/4}\exp\left\{-\frac{1}{4}\frac{\lambda}{\mathscr{L}^2}\right\}. \tag{A.19}$$

In (A.18) and the sequel, the constant stands for a numerical factor on the order of unity. At low excesses of the excitation level above threshold, $g_T = n_{0T}/\tau$, so long as

$$\left(\frac{\beta\Delta}{kT}\right)^{1/2} N_t n_0^{1/3} (v_T\tau)^4 \left[\frac{kT}{\beta\Delta}\ln\frac{g\tau}{n_{0T}}\right]^6 \left(\frac{g\tau-n_{0T}}{n_0}\right)^3 \exp\left\{-\frac{\lambda}{\ln^2(g\tau/n_{0T})}\right\} \ll 1, \tag{A.20}$$

we have $t_m \gg \tau$ and $n_m \approx g\tau$, so that the number of the produced EHD is given by formula (A.18), in which n_m is replaced everywhere by $g\tau$.

At a higher excitation level or at lower temperatures, when an inequality inverse to (A.20) is satisfied, t_m becomes smaller than τ, and $n_m \ll g\tau$, while formulas (A.18) and (A.19) can be reduced to

$$N = \text{const}\cdot\left(\frac{g}{N_T v_T}\right)^3 \left(\frac{n_0}{N_T}\right)^2 \frac{1}{\zeta^6} \exp\left\{5\frac{\Delta}{kT}[1-(0.762\pi)^{1/3}\beta\zeta]\right\}, \tag{A.21}$$

where

$$N_T = \nu(M_d kT/2\pi\hbar^2)^{3/2}, \tag{A.22}$$

and the dimensionless parameter ζ is defined by the relation

$$\zeta = \left(\frac{7}{\lambda}\right)^{1/3} \mathscr{L} = \left(\frac{21}{16\pi}\right)^{1/3} \frac{kT}{\beta\Delta}\mathscr{L} \tag{A.23}$$

and is determined from the equation

$$\frac{1}{\zeta^2} - \zeta - \frac{9}{7}\left(\frac{21}{16\pi}\right)^{1/3} \frac{kT}{\beta\Delta}\ln\zeta = \left(\frac{21}{16\pi}\right)^{1/3} \frac{kT}{7\beta\Delta}\ln\left[\left(\frac{\beta\Delta}{kT}\right)^{1/2} N_t n_0^{1/3} \left(\frac{n_{0T}}{n_0}\right)^3 \left(\frac{n_{0T}v_T}{g}\right)^4\right], \tag{A.24}$$

which follows directly from (A.19) and (A.23) at $t_m \ll \tau$.

2. Production of EHD on the front of the exciting pulse ($t_m < t_0$). In this case it follows from (A.16) and (A.17) that $(dg/dt)_{t_m} = g/t_0$ and

$$n_m = \frac{g\tau^2}{t_0}\left(\frac{t_m}{\tau} - 1 + e^{-t_m/\tau}\right). \tag{A.25}$$

After determining n''_m from (A.15) and introducing, to simplify the subsequent formulas, the notation

$$\chi = 1 - e^{-t_m/\tau}, \tag{A.26}$$

$$G = g\tau^2/n_{0T}t_0, \tag{A.27}$$

$$\Lambda = 64A\left(\frac{\beta\Delta}{kT}\right)^{1/2} N_t n_0^{1/3} (v_T\tau)^4, \tag{A.28}$$

we reduce equations (A.14) and (A.25) to the form

$$\chi = \frac{1}{8}\left[\frac{2\pi}{\lambda G}\frac{n_m}{n_{0T}}\mathscr{L}^3\right]^{1/2} \frac{F(n_m, G)}{\{[1+F(n_m,G)]^{1/2}-1\}^{3/2}}, \tag{A.29}$$

$$G^{-1}\frac{n_m}{n_{0T}} = -[\chi + \ln(1-\chi)], \tag{A.30}$$

where

$$F(n_m, G) = \frac{\pi}{4\lambda}\frac{\Lambda}{G^2}\left(\frac{n_m}{n_{0T}}\right)^2 \left(\frac{n_m - n_{0T}}{n_0}\right)^3 \mathscr{L}^3 \exp\left(-\frac{\lambda}{\mathscr{L}^2}\right). \tag{A.31}$$

Obviously, by virtue of the definition of χ, the condition $t_m < t_0$ corresponds to $\chi < 1 - e^{-t_0/\tau}$. This condition can be satisfied, naturally, only after a certain minimum excitation level G_{min} is reached, defined by the relation

$$1 - e^{-t_0/\tau} = \frac{1}{8}\left[\frac{2\pi}{\lambda G_{min}}\frac{n(G_{min}, t_0)}{n_{0T}}\ln^3\frac{n(G_{min}, t_0)}{n_{0T}}\right]^{1/2} \times \frac{F(n(G_{min}, t_0), G_{min})}{\{[1+F(n(G_{min}, t_0), G_{min})]^{1/2}-1\}^{3/2}}$$

in which $n(G_{min}, t_0)$ is given by formula (A.25) with $G = G_{min}$ and $t_m = t_0$. At $G < G_{min}$ we have $t_m > t_0$ and the process of EHD production, in spite of the finite rise time of the exciting pulse, is described by the formulas considered above for the case $t_0 \to 0$. At excitation levels noticeably exceeding G_{min}, we have $t_m < t_0$ and it is easy to show that the inequality $F(n_m, G) \ll 1$, is satisfied, and when this inequality is taken into account (A.10) and equations (A.29) and (A.30) can be reduced to the form:

$$N = \text{const}\cdot\left[\left(\frac{\beta\Delta}{kT}\right)^{1/2}\frac{N_t n_m}{n_{0T}^2}G\right]^{1/2} \frac{n_0^{2/3}}{v_T\tau}\left(\frac{n_m}{n_{0T}} - 1\right)^{-3/2} \exp\left\{-\frac{\lambda}{2\mathscr{L}^2}\right\} \tag{A.32}$$

$$G^{-1}\frac{n_m}{n_{0T}} + \left[\frac{Gn_{0T}}{\Lambda n_m}\left(\frac{n_0}{n_m - n_{0T}}\right)^3\right]^{1/2} \exp\left\{\frac{\lambda}{2\mathscr{L}^2}\right\} = -\ln\left\{1 - \left[\frac{Gn_{0T}}{\Lambda n_m}\left(\frac{n_0}{n_m - n_{0T}}\right)^3\right]^{1/2} \exp\left\{\frac{\lambda}{2\mathscr{L}^2}\right\}\right\}. \tag{A.33}$$

As written, formulas (A.32) and (A.33) are still not clear enough. They can however, be, noticeably simplified in two limiting cases, which we shall somewhat arbitrarily call the case of the short ($t_m \ll \tau$) and long ($t_m \gg \tau$) rise times of the exciting pulse. These cases are realized at $G \gg G_c$ and $G \ll G_c$, respectively, where G_c is determined by the equality

$$\frac{\lambda}{\ln^2 G_c} - 3\ln(G_c - 1) = \ln\left[\Lambda\left(\frac{n_{0T}}{n_0}\right)^3\right]. \tag{A.34}$$

It is obvious that the case of a long front is not realized at all at t_0 and the case of a short front can be realized also at $t_0 \lesssim \tau$, provided that the excitation level is high enough.

a) Case of long front ($G_{min} < G < G_c$). In this case

$$N = \text{const}\cdot\frac{gn_0^2}{N_T^2 v_T^3 \tau t_0}\exp\left\{\frac{3\Delta}{kT}\left[1 - 2\left(\frac{\pi}{6}\right)^{1/3}\beta\xi\right]\right\}, \tag{A.35}$$

where $\xi \equiv (4/\lambda)^{1/3}\mathscr{L}$ is defined by the equation

$$\frac{1}{\xi^2} - \xi = \frac{1}{8}\left(\frac{6}{\pi}\right)^{1/3}\frac{kT}{\beta\Delta}\ln\left[\frac{\Lambda}{G}\left(\frac{n_{0T}}{n_0}\right)^3\right]. \tag{A.36}$$

Formula (A.35) is accurate to within a factor $(1 - n_{0T}/n_m)^3$ close to unity. At relatively high temperatures and at not too high excitation level, when $(\Lambda/G)(n_{0T}/n_0)^3 \gg 1$, we have $\beta\xi \ll 1$ and N increases quite rapidly with decreasing temperature. At low temperatures, however, this growth slows down.

b) Case of short front ($G > G_c$). In this case

$$N = \text{const}\cdot\left(\frac{g}{N_T v_T^2 t_0}\right)^{3/2}\left(\frac{n_0}{N_T}\right)^2 \exp\left\{\frac{7}{2}\frac{\Delta}{kT}\left[1 - 2\left(\frac{2\pi}{15}\right)^{1/3}\beta\eta\right]\right\} \tag{A.37}$$

The parameter $\eta \equiv (5/\lambda)^{1/3} \mathscr{L}$ is defined by the equation

$$\frac{1}{\eta^2} - \eta = \frac{1}{10}\left(\frac{15}{2\pi}\right)^{1/3} \frac{kT}{\beta\Delta} \ln\left[\frac{2\Lambda}{G^2}\left(\frac{n_{0T}}{n_0}\right)^3\right]. \tag{A.38}$$

We discuss now the reason of applicability of the approximation used above. The entire analysis was based on the assumption that the time interval

$$s_{\text{eff}} = (n_m \mathscr{L}^3/\lambda |n_m''|)^{1/3} \tag{A.39}$$

is short enough

$$s_{\text{eff}} \ll t_m, \tag{A.40}$$

but is at the same time long enough for the drops to grow to dimensions $R \gg R_c$ (the condition (A.13)).

We consider first the case of an extremely short front. Using (A.15), (A.16), (A.19), and (A.39) we easily obtain the ratio

$$\frac{s_{\text{eff}}}{t_m} = \frac{\tau_*}{t_m}(e^{t_m/\tau} - 1)\left(\frac{1}{\lambda}\mathscr{L}^3\right)^{1/3}, \tag{A.41}$$

which shows that (A.40) is practically always satisfied by virtue of $\mathscr{L}^3 \ll \lambda$.

In the case of a front of finite duration we obtain from (A.15), (A.26), (A.39), and (A.29), assuming again $F \ll 1$, the estimate

$$s_{\text{eff}} \sim \tau(1 - e^{-t_m/\tau}),$$

from which it follows that the condition $s_{\text{eff}} \ll t_m$ is satisfied for a long front, and for a short front it is at the limit, i.e., $s_{\text{eff}} \sim t_m$. Thus, in this last case formulas (A.37) and (A.38) can claim only to describe the qualitative relations and the orders of magnitude.

The second criterion (A.13) reduces in the cases of long, short, and extremely short fronts, to the respective forms

$$\frac{n_0}{n_m}\frac{\lambda^{1/3}}{\mathscr{L}^2} \ll n_0^{1/3} v_T \tau, \tag{A.42}$$

$$\left(\frac{g\tau^2}{n_m t_0}\right)^{1/2} \frac{n_0}{n_m}\frac{\lambda^{1/3}}{\mathscr{L}^2} \ll n_0^{1/3} v_T \tau, \tag{A.43}$$

$$\frac{g\tau n_0}{n_m^2}\frac{\lambda^{5/6}}{\mathscr{L}^5} \ll n_0^{1/3} v_T \tau. \tag{A.44}$$

Satisfaction of these conditions does not follow automatically from any of the assumptions made above. However, in view of the fact that their right-hand sides contain the very large quantity $n_0^{1/3} v_T \tau \sim 10^7$ (for germanium), it is practically always satisfied when condensation is observable, i.e., at $n_m \gtrsim 10^{11}$–10^{12} cm^{-3}.

[1)]A value $\alpha \approx 1.6 \times 10^{-4}$ erg/cm^2 is cited in[16,31]. However, formula (2) of[31] is incorrect. To obtain the correct formula it is necessary to replace $\ln(T_0/T)$ in (2) by the quantity $[(T_0 - T)/T_0 + 3/2(kT/\varepsilon_0)\ln(T_0/T)]$. The reduction of the experimental result obtained in[31] by means of the corrected formulas does not change the qualitative conclusions of the paper, and the corrected value of the surface-tension coefficient is $\alpha = 1.8 \times 10^{-4}$ erg/cm^2. We are grateful to D. Hensel for calling our attention to this circumstance.

[2)]Similar arguments were advanced in[37].

[1]L. V. Keldysh, Proc. Trudy, Ninth Intern. Conf. on Physics of Semiconductors, Moscow, 1968, vol., Nauka, Leningrad (1969), p. 1387.

[2]V. M. Asinin and A. A. Pogachev, Pis'ma Zh. Eksp. Teor. Fiz. **9**, 415 (1969) [JETP Lett. **9**, 248 (1969)].

[3]Ya. E. Pokrovskiĭ and K. I. Svistunova, *ibid.*, 435 [261].

[4]V. S. Vavilov, V. A. Zayats, and V. N. Murzin, *ibid.* **10**, 304 (1969) [**10**, 192 (1969)].

[5]B. S. Bagaev, T. I. Galkina, O. V. Gogolin, and L. V. Keldysh, *ibid.*, 309 [195].

[6]L. V. Keldysh, Usp. Fiz. Nauk 100, 514 (1970) [Sov. Phys. Usp. **13**, 292 (1970)].

[7]L. V. Keldysh, in: Éksitony v poluprovodnikakh (Excitons in Semiconductors), Nauka, 1971, p. 5.

[8]Ya. E. Pokrovskiĭ and K. I. Svistunova, Fiz. Tekh. Poluprovodn. **4**, 491 (1970) [Sov. Phys. Semicond. **4**, 409 (1970)].

[9]A. S. Alekseev, V. S. Bagaev, T. I. Galkina, O. V. Gogolin, and N. A. Penin, Fiz. Tverd. Tela **12**, 3516 (1970) [Sov. Phys. Solid State **12**, 2855 (1971)].

[10]V. S. Bagaev, T. I. Galkina, and O. V. Gogolin, Proc. Tenth Intern. Conf. on Physics of Semiconductors, Cambridge, Mass., 1970, publ. by US Atomic Energy Commission, Washington, DC (1970), p. 500.

[11]Ya. Pokrovsky, A. Kaminsky, K. Svistunova, *ibid.*, p. 504.

[12]V. S. Vavilov, V. A. Zayats, and V. N. Murzin, *ibid.*, p. 509.

[13]Ya. E. Pokrovskii, Phys. Status Solidi [a] **11**, 385 (1972).

[14]B. V. Zubov, V. P. Kalinushkin, T. M. Murina, A. M. Prokhorov, and A. A. Pogachev, Fiz. Tekh. Poluprovodn. **7**, 1614 (1973) [Sov. Phys. Semicond. **7**, 1077 (1974)].

[15]V. S. Bagaev, N. A. Penin, N. N. Sibel'din, and V. A. Tsvetkov, Fiz. Tverd. Tela **15**, 3269 (1973) [Sov. Phys. Solid State **15**, 2179 (1974)].

[16]A. S. Alekseev, T. A. Astemirov, V. S. Bagaev, T. I. Galkina, N. A. Penin, N. N. Sybeldin, and V. A. Tsvetkov, Proc. Twelfth Intern. Conf. on Physics of Semiconductors, Stuttgart, 1974, p. 91.

[17]T. K. Lo, B. J. Feldman, and C. D. Jeffries, Phys. Rev. Lett. **31**, 224 (1973).

[18]T. K. Lo, B. J. Feldman, R. M. Westervelt, J. L. Staehli, C. D. Jeffries, and E. E. Haller, Preprint, 1975.

[19]V. S. Bagaev, T. I. Galkina, and O. V. Gogolin, in: Eksitony v poluprovodnikakh (Excitons in Semiconductors), Nauka, 1971, p. 19.

[20]C. Benoit a la Guillaume and M. Voos, Phys. Rev. [B] **7**, 1723 (1973).

[21]G. A. Thomas, T. G. Phillips, T. M. Rice, and J. C. Hensel, Phys. Rev. Lett. **31**, 386 (1973).

[22]T. K. Lo, Solid State Commun. **15**, 1231 (1974).

[23]J. C. Hensel, T. G. Phillips, and T. M. Rice, Phys. Rev. Lett. **30**, 227 (1973).

[24]J. C. McGroddy, M. Voos, and O. Christensen, Solid State Commun. **13**, 1801 (1973).

[25]C. Benoit a la Guillaume, M. Capizzi, B. Etienne, and M. Voos, Solid State Commun. **15**, 1031 (1974).

[26]R. M. Westervelt, T. K. Lo, J. L. Staehli, and C. D. Jeffries, Phys. Rev. Lett. **32**, 1051 (1974).

[27]T. M. Rice, Phys. Rev. [B] **9**, 1540 (1974).

[28]L. M. Sander, H. B. Shore, and L. J. Sham, Phys. Rev. Lett. **31**, 533 (1973).

[29]H. Büttner and E. Gerlach, J. Phys. C**6**, L 433 (1973).

[30]T. L. Reinecke and S. C. Ying, Solid State Commun. **14**, 381 (1974).

[31]V. S. Bagaev, N. N. Sibel'din and V. A. Tsvetkov, Pis'ma Zh. Eksp. Teor. Fiz. 21, 180 (1975) [JETP Lett. **21**, 80 (1975)].

[32]R. N. Silver, Phys. Rev. [B] **11**, 1569 (1975).

[33]V. S. Bagaev, N. V. Zamkovets, N. A. Penin, N. N. Sibel'din, and V. A. Tsvetkov, Zweite Intern. Tagung Laser und ihre Anwendungen, Dresden, DDR, 1973, *k.* 107; Prib. Tekh. Eksp. No. **2**, 258 (1974).

[34]N. N. Sibel'din, V. S. Bagaev, V. A. Tsvetkov, and N. A. Penin, Preprint FIAN No. 117, Moscow, 1972.
[35]N. N. Sibel'din, V. S. Bagaev, V. A. Tsvetkov, and N. A. Penin, Fiz. Tverd. Tela **15**, 177 (1973) [Sov. Phys. Solid State **15**, 121 (1973)].
[36]Ya. E. Pokrovskiĭ and K. I. Svistunova, Pis'ma Zh. Eksp. Teor. Fiz. **13**, 297 (1971) [JETP Lett. **13**, 201 (1971)].
[37]R. N. Silver, Preprint, 1975.
[38]R. W. Martin, Phys. Status Solidi [b] **61**, 223 (1974).
[39]Ya. E. Pokrovsky and K. I. Svistunova, Proc. Twelfth Intern. Conf. on Physics of Semiconductors, Stuttgart, 1974, p. 71.
[40]Ya. E. Pokrovskiĭ and K. I. Svistunova, Fiz. Tverd. Tela **16**, 3399 (1974) [Sov. Phys. Solid State **16**, 2202 (1975)].
[41]B. J. Feldman, Phys. Rev. Lett. **33**, 359 (1974).
[42]M. Voos, K. L. Shaklee, and J. M. Worlock, Phys. Rev. Lett. **33**, 1161 (1974).
[43]R. Becker and W. Döring, Ann. Phys. (Leipz.) **24**, 719 (1935).
[44]Ya. B. Zel'dovich, Zh. Eksp. Teor. Fiz. **12**, 525 (1942).
[45]Ya. I. Frenkel', Kineticheskaya teoriya zhidkosteĭ (Kinetic Theory of Liquids), AN SSSR, 1945, Chap. VII.
[46]V. S. Bagaev, L. V. Keldysh, N. N. Sibel'din, and V. A. Tsvetkov, Zh. Eksp. Teor. Fiz. **70**, 702 (1976) [Sov. Phys. JETP **43**, 362].
[47]L. V. Keldysh, Pis'ma Zh. Eksp. Teor. Fiz. **23**, 100 (1976) [JETP Lett. **23**, 86 (1976)].

Translated by J. G. Adashko

Electron liquid in a superstrong magnetic field

L. V. Keldysh and T. A. Onishchenko

P. N. Lebedev Physics Institute, USSR Academy of Sciences
(Submitted June 8, 1976)
Pis'ma Zh. Eksp. Teor. Fiz. **24**, No. 2, 70–73 (20 July 1976)

It is shown that the energy of an electron-hole plasma in ultrastrong magnetic field has, as a function of the density, a minimum in the strong-compression region. The equilibrium density is $n_0 \sim (eHa_{0\,2}/\hbar c)^{8/7} a_0^{-3}$, and the energy per particle is given by $E/n_0 \sim -n_0^{1/4} e^2 a_0^{-1/4}$, where H is the field intensity and a_0 is the effective Bohr radius.

PACS numbers: 52.25.–b

We calculate in this paper the energy of high-density electron-plasma in the limit of strong magnetic fields.

We use Coulomb units $e = \hbar = m_0 = 1$ and the unit of the magnetic field intensity is $e^3/\hbar^3 m_0^2 c$.

We assume that the magnetic length is $\lambda = H^{-1/2} \ll 1$ and that the particle concentration satisfies the condition $1/\lambda^2 \ll n \ll 1/\lambda^3$. Then, by virtue of the right-hand side of the inequality, all the electrons and holes are at the lower Landau level, and the transitions between the different Landau levels can be neglected, while the left half of the written inequality is the compressibility condition of the system, inasmuch as the effective volume of the atom (exciton) is $\propto\lambda^2$.[1] In the considered density range, the system is therefore a degenerate Fermi liquid with Fermi momentum $p_0 = 2\pi^2 n\lambda^2 \gg 1$. An analysis of the perturbation-theory series shows that, just as in any compressed system with Coulomb interaction, the principal contribution (in terms of the parameter p_0^{-1}) to the correlation energy is made in this situation by the so-called random-phase-approximation (RPA) diagrams with momentum transfers $|\mathbf{k}| \ll n^{1/3}$. However, as will be shown below, a peculiarity of our problem is the presence of a concentration region in which the main contribution to the energy is made by $|\mathbf{k}| \sim n^{1/4} \gg p_0$, so that the dependence of the correlation energy on the density is qualitatively altered and a minimum of the energy appears.

For simplicity we calculate the energy of the ground state of an electron gas in a magnetic field $\mathbf{H} \parallel OZ$, and write out the final results for an electron-hole plasma with isotropic masses of the electrons and holes. The correlation of the system

$$E_{corr} = \frac{i}{2}\int_0^1 \frac{da}{a}\int \frac{d^3k\,d\omega}{(2\pi)^4}\left[\frac{\frac{4\pi a}{k^2}\Pi(a,\omega,\mathbf{k})}{1-\frac{4\pi a}{k^2}\Pi(a,\omega,\mathbf{k})} - \frac{4\pi a}{k^2}\Pi_0(\omega,\mathbf{k})\right]$$

is expressed in terms of the exact polarization operator $\Pi(a,\omega,\mathbf{k})$ and its first-order approximation $\Pi_0(\omega,\mathbf{k})$ in the coupling constant α. The contribution of diagrams other than the RPA diagrams is small in the parameter $1/n\lambda^2$, and at momentum transfers on the order of $n^{1/4}\Pi(a,\omega,\mathbf{k})$ it reduces to $\Pi_0(\omega,\mathbf{k}) \cdot E_{corr}$ is expressed by the well-known formula

$$E_{corr} = \frac{1}{2}\int \frac{d^3k\,d\omega}{(2\pi)^4}\left\{\ln\left[1 - \frac{4\pi}{k^2}\Pi_0(i\omega,\mathbf{k})\right] + \frac{4\pi}{k^2}\Pi_0(i\omega,\mathbf{k})\right\},$$

$$\Pi_0(i\omega,\mathbf{k}) = -\frac{1}{4\pi^2\lambda^2}\exp\left(-\frac{k_\perp^2\lambda^2}{2}\right)\frac{1}{k_z}\ln\left[\frac{\omega^2+\left(\frac{1}{2}k_z^2+k_zp_0\right)^2}{\omega^2+\left(\frac{1}{2}k_z^2-k_zp_0\right)^2}\right].$$

Integrating over the transverse momentum with accuracy to $1/n\lambda^2$ and expressing ω in terms of a new independent variable v

$$v = \frac{1}{\pi\lambda^2 k_z^3}\ln\frac{\omega^2+\left(\frac{1}{2}k_z^2+k_zp_0\right)^2}{\omega^2+\left(\frac{1}{2}k_z^2-k_zp_0\right)^2},$$

we obtain

$$E_{corr} = -\frac{4p_0^5}{\pi^3 y}f(y), \text{ where } y = 1/8\,\pi\lambda^2p_0^3$$

$$f(y) = \int_0^\infty dv\left[(1+v)\ln(1+v) - v\right]\int_0^{u(v/y)} dx\,x^7\exp\left(-\frac{1}{y}vx^3\right)$$

$$\times\left[1-\exp\left(-\frac{1}{y}vx^3\right)\right]^{-3/2}\left[(1+x)^2\exp\left(-\frac{1}{y}vx^3\right) - (1-x)^2\right]^{-1/2}.$$

$u(v/\gamma)$ is the solution of the equation

$$v/\gamma = (2/u^3)\ln\left|\frac{1+u}{1-u}\right| .$$

At $\gamma \gg 1$ the calculations yield

$$\frac{E_{corr}}{n} = -\frac{2^5\pi^{3/4}}{5[\Gamma(1/4)]^2}n^{1/4} = -1.1\,n^{1/4} .$$

The main contribution accumulates in this case in a momentum-transfer region of the order of $n^{1/4}$, and the contribution of momenta smaller than p_0 is of the order of $\max\{n^4\lambda^{10};n\lambda^2\}\ln\gamma$. At $\gamma \ll 1$ the main contribution to E_{corr} is due to momenta smaller than p_0, and

$$\frac{E_{corr}}{n} = -\frac{1}{64\pi^6 n^2\lambda^6}\ln\frac{1}{\gamma} ,$$

which agrees with the results of [2] apart from the factor $(\ln 2\pi^3 n\lambda^2)(\ln 8\pi^7 n^3\lambda^8)^{-1}$.

The exchange part of the energy is of the order of $-n\lambda^2\ln(1/n\lambda^3)$, and is negligible in the considered density interval. The energy of the ground state

$$\frac{E}{n} = \frac{2}{3}\pi^4 n^2\lambda^4 - \frac{2^5\pi^{3/4}}{5\left[\Gamma\left(\frac{1}{4}\right)\right]^2}n^{1/4}$$

has a minimum $(\dot{E}/n)_{min} \approx -0.42H^{2/7}$ at $n_0 \approx 0.030H^{8/7}$. For an electron-hole plasma with $m_e = m_h = m$ at $\gamma \gg 1$ we have

$$\frac{E_{corr}}{n} = -\frac{2^5\pi^{3/4}}{5\left[\Gamma\left(\frac{1}{4}\right)\right]^2}2^{5/4}\frac{m^{1/4}}{\epsilon^{5/4}}n^{1/4} \approx -2.7\frac{m^{1/4}}{\epsilon^{5/4}}n^{1/4} ;$$

$$\left(\frac{E}{n}\right)_{min} \approx -1.0H^{2/7}m^{3/7}\epsilon^{-10/7}$$

at $n = n_e = n_h \approx 0.034m^{5/7}H^{8/7}\epsilon^{-5/7}$, where ϵ is the dielectric constant of the medium. If m_e and m_h are essentially different, then E_{corr} is determined mainly by the large mass, in contrast to the binding energy of the exciton.

Just as in the absence of a magnetic field [3], the spectrum has a gap as a result of the logarithmically diverging diagrams near the Fermi surface. [4,5] The magnets of the gap, however, is negligible because of the high density of the plasma. If m_e and m_h are so different that the assumptions of the present paper are satisfied only for light nuclei, and the mass of the heavy particles is much larger than $n^{1/3}$, then the heavy particles form a Wigner crystal.

We note in conclusion that the result has a general character in view of the universality of the asymptotic form of the polarization operator at large momentum transfers.

The RPA diagrams with momentum transfer on the order of $n^{1/4}$ make a contribution of the order $n^{1/4}$ per particle to E_{corr} in any high-density system in which the characteristic momenta of the particles are much smaller than $n^{1/4}$, regardless of the dimensionality of the system, and in the presence of bound states the characteristic momentum is the reciprocal dimension of the exciton.

[1]R. J. Elliott and R. Loudon, J. Phys. Chem. Solids **15**, 196 (1960).
[2]N. J. Horing, R. W. Danz, and M. L. Glasser, Phys. Rev. **A6**, 2391 (1972).
[3]L. V. Keldysh and Yu. V. Kopaev, Fiz. Tverd. Tela **6**, 2791 (1964) [Sov. Phys. Solid State 6, 2219 (1965)].
[4]A. A. Abrikosov, J. Low Temp. Phys. **2**, 37 (1970).
[5]S. A. Brazovskiĭ, Zh. Eksp. Teor. Fiz. **62**, 820 (1972) [Sov. Phys. JETP **35**, 433 (1972)].

Electron-hole liquid in strongly anisotropic semiconductors and semimetals

E. A. Andryushin, V. S. Babichenko, L. V. Keldysh, T. A. Onishchenko, and A. P. Silin

P. N. Lebedev Physics Institute, USSR Academy of Sciences

(Submitted July 2, 1976)
Pis'ma Zh. Eksp. Teor. Fiz. **24**, No. 4, 210–214 (20 August 1976)

It is shown that in semiconductors and semimetals with strongly anisotropic electron spectra the electron-hole liquid has a minimal energy at densities corresponding to strong compression. The binding energy of the liquid greatly exceeds in this case the binding energy of the excitons.

PACS numbers: 71.80.+j

It was shown in a preceding paper[1] that in sufficiently strong magnetic fields, $H \gg 1$, the energy of the electron-hole liquid, as a function the density n of the electron-hole pairs, has a minimum at $n \sim H^{8/7} \gg 1$. We use here the "Coulomb" system of units $e^2/\epsilon = \hbar = m = 1$ and measure H in units of $e^3 m^2 c/\epsilon^2 \hbar^2$, where ϵ is the dielectric constant of the crystal (without allowance for the contribution of the free electrons and holes), and m is one of the effective masses, the choice of which will be specified concretely in each individual case.

Thus, the electron-hole plasma turns out to be strongly compressed by the joint action of the magnetic field and the Coulomb interaction. It will be shown below that this tendency to "self-compression" of the electron-hole liquid, i.e., the fact that its energy has a minimum at $n \gg 1$, is a characteristic feature of a large number of model systems even without a magnetic field. A common feature of these models is a strong anisotropy of the electron spectrum. From a more formal point of view, a criterion for the selection of such systems is the existence for them of a region of concentrations n in which the inequalities

$$1 \ll p_F \ll n^{1/4}, \tag{1}$$

are satisfied (p_F is the Fermi momentum). In the well-known formula for the correlation energy per particle pair

$$E_{corr} = \frac{1}{2n} \int_0^1 \frac{d\lambda}{\lambda} \int \frac{d^3k\,d\omega}{(2\pi)^4}\left[\frac{4\pi\chi(\mathbf{k}, i\omega; \lambda)}{1 + 4\pi\chi(\mathbf{k}, i\omega; \lambda)} - 4\pi\chi^{(0)}(\mathbf{k}, i\omega; \lambda)\right], \tag{2}$$

where $\chi(\mathbf{k}, i\omega; \lambda)$ is the polarizability of the system of electrons and holes with concentration n and charge $\pm\sqrt{\lambda}$ for each particle (+for the holes), if (1) is satisfied, then the principal contribution to χ in terms of the parameter p_F^{-1} is made by the diagram $\chi^0(k, i\omega; \lambda)$ of lowest order in λ, while the main contribution to the integral with respect to the momentum transfer k is made by the region $k \propto n^{1/4} \gg p_F$. It is important that the effective radius of this interaction is $k^{-1} \propto n^{-1/4} \gg n^{-1/3}$, where $n^{-1/3}$ is the average distance between particles. In this momentum range we have

$$\chi^{(0)}(\mathbf{k}, i\omega; \lambda) \approx \lambda \sum_i \frac{2n_i\,\epsilon_i(\mathbf{k})}{k^2[\epsilon_i^2(\mathbf{k}) + \omega^2]}, \tag{3}$$

where the subscript i labels different types of charged particles, and $\epsilon_i(\mathbf{k})$ are their energies as functions of the momentum. Assuming all the dispersion laws of ϵ_i to be quadratic, we easily obtain from (2) and (3)

$$E_{corr} \approx -An^{1/4}, \tag{4}$$

where A is a coefficient that depends on the electron and hole mass ratio, on their anisotropy, on the number of equivalent minima in the electron and hole bands, and on other details of the electron spectrum. An explicit form of this coefficient is given below for serveral typical systems satisfying the criterion (1). For all these systems, the exchange energy in the considered concentration range is small in comparison with E_{corr}, and the total energy $E(n)$ per pair of particles, which is equal to the sum of the Fermi energy and E_{corr}, reaches a minimum value E_{min} at a concentration n_{min} satisfying the criterion (1).

I. *Quasi-one-dimensional systems*, i.e., systems of parallel conducting filaments. The transitions of electrons and holes from one filament to another are assumed to have negligibly low probability, and the density of the number of filaments per unit of the employed scale for the surface perpendicular to them is $N \gg 1$. This case is closest to the case of a strong magnetic field, considered in [1], and the results are practically the same, apart from the substitutions $N \rightleftharpoons H$. We present them here, however, in a somewhat more general form, taking into account the difference between the electron and hole masses m_e and m_h, and assuming $m = m_e m_h/(m_e + m_h) = 1$ and $N^{-1/2} \lesssim \sigma \equiv m_e/m_h \leq 1$. The proportionality coefficient in (4) for this case is

$$A_{\rm I} = \left(\frac{4}{\pi}\right)^{1/4} \frac{64\pi}{5[\Gamma(1/4)]^2} f_{\rm I}(\sigma),$$

where

$$f_{\rm I}(\sigma) = \frac{[\Gamma(1/4)]^2}{8\cdot 2^{1/4}\pi^{3/2}} \int_0^\infty dx \left[\frac{1}{(1+\sigma)x^2 + \dfrac{1}{1+\sigma}} + \frac{1}{\dfrac{1+\sigma}{\sigma}x^2 + \dfrac{\sigma}{1+\sigma}}\right]^{5/4}, \tag{5}$$

$$f_{\rm I}(1) = 1 \qquad f_{\rm I}(\sigma << 1) = (1/2)^{3/2} \frac{1}{\sigma^{1/4}}, \tag{6}$$

$$p_F = \frac{\pi}{2} Nn; \quad n_{min} = \left[\frac{3}{2\pi^2} A_{\rm I}\right]^{4/7} N^{8/7}; \quad E_{min} = -\frac{7}{8}\left(\frac{3}{2}\right)^{1/7} \frac{1}{\pi^{2/7}} A_{\rm I}^{8/7} N^{2/7}. \tag{7}$$

II. *Quasi-two-dimensional (layered) systems*, i.e., systems of parallel conducting planes with distances $c \ll 1$ between them and with negligibly low probability of carrier transfer from one plane to another.

Assuming $2m_e m_h/(m_e+m_h) = 1$ and $c \lesssim \sigma \lesssim 1$, we obtain for the case of spectra that are isotropic in the plane of the layer

$$A_{\rm II} = \left(\frac{\pi}{2}\right)^{1/4} \frac{256\pi^2}{5[\Gamma(1/4)]^2} f_{\rm II}(\sigma) \approx 3.27 f_{\rm II}(\sigma), \tag{8}$$

where

$$f_{\rm II}(\sigma) = \frac{5[\Gamma(1/4)]^2 \left(\dfrac{1}{\sqrt{\sigma}} + \sqrt{\sigma}\right)^{3/2}}{32\sqrt{2}\pi} \int_0^1 \frac{dx}{x^{1/4}}$$

$$\times \sqrt{1 - \left(\sigma + \frac{1}{\sigma}\right)x + \sqrt{1 - 2x\left(\frac{1}{\sqrt{\sigma}} - \sqrt{\sigma}\right)^2 + x^2 \frac{1}{\sigma} - \sigma^2}}$$

$$= \begin{cases} 1, & \sigma = 1 \\ \dfrac{1}{4(4\sigma)^{1/4}}, & \sigma << 1 \end{cases}, \tag{9}$$

$$p_F = \sqrt{2\pi nc}\,; \quad n_{min} = \left(\frac{A_{\rm II}}{4\pi}\right)^{1/3} c^{-4/3}, \quad E_{min} = -\frac{3}{4}\left(\frac{A_{\rm II}}{4\pi}\right)^{1/3} c^{-1/3} \approx -\frac{1.57}{c^{1/3}} f_{\rm II}^{4/3}(\sigma). \tag{10}$$

III. *Multivalley semiconductors and semimetals*, i.e., having several equivalent (by virtue of the crystal symmetry) minima in the electron and (or) hole spectra. The fact that the presence of many valleys increases the role of E_{corr} and the binding energy of the electron-hole liquid has already been noted in a number of papers.[2–5] From the formal point of view, the use of a number of valleys in the electron and hole bands, ν_e and ν_h, as large parameters is analogous to the expansion, known from the theory of phase transition, in terms of the reciprocal of the number of components of the order parameter (the $1/n$ expansion). It justifies the replacement of χ by $\chi^{(0)}$ in (2), and on the other hand it ensures at $n \ll \nu^4$ satisfaction of the condition $p_F \sim (n/\nu)^{1/3} \ll n^{1/4}$ and hence the validity of the asymptotic form (3). It is convenient to define the system of units in this case by the relation

$$m^{1/4} = \frac{[\Gamma(1/4)]^2}{8\pi 2^{1/4}(2\pi)^{3/2}} \int_0^\infty dx \overline{\left[\frac{1}{\nu_e} \sum_{k=1}^{\nu_e} \frac{(m_e^{-1})_{ij}^{(k)} n_i n_j}{x^2 + [(m_e^{-1})_{ij}^{(k)} n_i n_j]^2} + \frac{1}{\nu_h} \sum_{k=1}^{\nu_h} \frac{(m_h^{-1})_{ij}^{(k)} n_i n_j}{x^2 + [(m_h^{-1})_{ij}^{(k)} n_i n_j]^2} \right]^{5/4}} = 1, \tag{11}$$

where $\overline{[\cdots]}$ denotes averaging over the direction of the unit vector $\mathbf{n}$, and $(m_{e,h}^{-1})_{ij}^{(k)}$ is the reciprocal-mass tensor of thek-th valley in terms of the axes of the reciprocal lattice of the crystal. Then

$$A_{III} = \frac{32\sqrt{2}(2\pi)^{3/4}}{5[\Gamma(1/4)]^2}; \qquad n_{min} = \left[\frac{5A_{III}M}{4\pi^{2/3}(3\pi)^{2/3}}\right]^{12/5};$$

$$E_{min} = -\frac{5}{8}\left(\frac{5}{4\cdot 3^{2/3}\pi^{4/3}}\right)^{3/5} A_{III}^{8/5} M^{3/5}, \tag{12}$$

where

$$M^{-1} = M_e^{-1} + M_h^{-1}; \qquad M_{e,h} = [\det(m_{e,h}^{-1})]^{-1/3} \nu_{e,h}^{2/3}.$$

IV. We present also the result for a semimetal with ν electron valleys and one hole valley in a strong magnetic field directed in equivalent fashion relative to all the electron ellipsoids. Let $m_{e,h\parallel}$ be the effective masses of motion in the field direction, $\sigma = m_{e\parallel}/m_{h\parallel}$, $m^{-1} = m_{e\parallel}^{-1} + m_{h\parallel}^{-1} = 1$. Then

$$A_{IV}(\sigma) = \left(\frac{4}{\pi}\right)^{1/4} \frac{64\pi}{5[\Gamma(1/4)]^2} f_{IV}(\sigma), \qquad f_{IV}(\sigma) = f_I(\sigma),$$

$$n_{min} = \left[\frac{A_{IV}(\sigma)(1+\sigma)\nu^2}{1+\nu^2\sigma}\right]^{4/7} \left(\frac{eH}{c}\right)^{8/7} \left(\frac{3}{16\pi^4}\right)^{4/7},$$

$$E_{min} = -\frac{7}{8}\left[\frac{3}{16\pi^4} A_{IV}(\sigma) \frac{(1+\sigma)\nu^2}{1+\sigma\nu^2}\right]^{1/7} \left(\frac{eH}{c}\right)^{2/7}.$$

We note in conclusion, as one of the consequences of our results, that semiconductors with a narrow forbidden band E_g pertaining to one of the classes considered above become unstable relative to a first-order phase transition into the semimetallic state as E_{min} approaches E_g, i.e., long before the exciton instability sets in.

[1]L. V. Keldysh and T. A. Onishchenko, Pis'ma Zh. Eksp. Teor. Fiz. **24**, 70 (1976) [JETP Lett. **24**, (1976)].

[2]V. S. Bagaev, T. I. Galkina, O. V. Gogolin, and L. V. Keldysh, ibid. **9**, 435 (1969) [**9**, 261 (1969)].

[3]M. Combescot and P. Nozieres, J. Phys. **C5**, 2369 (1972). M. Combescot, Phys. Rev. **B10**, 5045 (1974).

[4]W. F. Brinkan and T. M. Rice, Phys. Rev. **B7**, 1503 (1973).

[5]P. Vashishta, P. B. Bhattacharyya, and K. S. Singwi, Nuovo Cimento **23B**, 172 (1974); P. Vashishta, S. G. Das, K. S. Singwi, Phys. Rev. Lett. **33**, 911 (1974).

Electron-hole liquid and the metal-dielectric phase transition in layered systems

E. A. Andryushin, L. V. Keldysh, and A. P. Silin

P. N. Lebedev Physical Institute, USSR Academy of Sciences
(Submitted April 13, 1977)
Zh. Eksp. Teor. Fiz. **73**, 1163–1173 (September 1977)

A phase transition of the gas-liquid type in a system of electrons and holes in a quasi-two-dimensional (layered) semiconductor is considered. The phase diagram, critical temperature and density for the transition are obtained. It is shown that near the critical point the transition is of a purely plasma nature. In other words, bound states of the exciton type are absent in both (gas and liquid) phases.

PACS numbers: 71.35.+z, 71.30.+h

It was shown in a previous work[1] that the role of correlation effects in electron-hole plasma (EHP) turns out to be anomalously large in semiconductors and semimetals possessing extremely strong anisotropies of the electron spectrum, in particular, layered systems. A significant decrease in the energy associated with inter-electron correlations creates a tendency to "self-compression" of such a plasma, i.e., the formation of an electron-hole liquid (EHL) with particular density $n \gg a_x^{-3}$ and binding energy per electron-hole pair $|E_{\min}| \gg E_x$, where E_x and a_x are the binding energy and the effective radius of the hydrogen-like exciton. Formation of an EHL at low temperatures takes place through a first-order phase transition, in which the concentration of the carriers, free and bound in excitons, reaches some critical value $n_{gas}(T)$ that depends on the temperature—the density of the saturated vapor. This situation can be realized both at thermodynamic equilibrium for a semiconductor with a sufficiently narrow forbidden band E_g upon increase in the temperature or decrease in E_g, and under essentially nonequilibrium conditions, in which the critical concentration is achieved by introduction of selected carriers in intense excitation of the semiconductor.

In the first case, the formation of the EHL means a discontinuous change in the width of the forbidden band E_g to some negative value (overlapping of the bands), i.e., a transition of the initial semiconductor to a semimetal state. The second case corresponds to the so-called condensation of nonequilibrium carriers (or excitons) into EHL drops. With the same accuracy with which the nonequilibrium carriers can be assumed to be thermalized, the thermodynamics of both these transitions is identical and we shall consider them here using as an example one of the types of systems previously described[1]—layered systems. Just as in Ref. 1, we shall mean by a layered semiconductor or semimetal an idealized model in which the motion of the electrons and holes is two-dimensional, i.e., it takes place only in the plane of the layers without transitions between them. For simplicity, the dispersion laws of the electrons $\varepsilon_e(\mathbf{p})$ and holes $\varepsilon_h(\mathbf{p})$ will be assumed to be quadratic and isotropic in the plane of the layer: $\varepsilon_{e,h}(\mathbf{p}) = p^2/2m_{e,h}$. We shall also assume the permittivity tensor of the crystal to be isotropic in the plane of the layers. This tensor is characterized by two principal values $\varepsilon_\perp$ and $\varepsilon_\parallel$ for the directions along and perpendicular to the layers, respectively.

We emphasize that we are speaking of a permittivity without any contribution of free carriers. We shall use a system of units defined by the relations

$$e^2/(\varepsilon_\parallel \varepsilon_\perp)^{1/2} = \hbar = 2m = 1,$$

where e and $\hbar$ are the charge on the electron and Planck's constant, and m is the reduced mass, $m^{-1} = m_e^{-1} + m_h^{-1}$. In this work, along with the parameter of the ratio of the effective masses of the electron and the hole, $\sigma = m_e/m_h$ that is usually employed, we will find it convenient to use another parameter $s = (1-\sigma)/(1+\sigma)$.

The system of electrons and holes in the considered model is described by the Hamiltonian

$$H - \mu_e n_e - \mu_h n_h = \sum_{\mathbf{p}s_1l}\left[\left(\frac{1+s}{2}p^2 - \mu_e\right)a^+_{\mathbf{p}s_1l}a_{\mathbf{p}s_1l} + \left(\frac{1-s}{2}p^2 - \mu_h\right)b^+_{\mathbf{p}s_1l}b_{\mathbf{p}s_1l}\right]$$
$$+\frac{1}{2}\sum_{\substack{\mathbf{p}_1\mathbf{p}_2\mathbf{q}\\ s_1s_2l_1l_2}} V_{q,l_1-l_2}[a^+_{\mathbf{p}_1s_1l_1}a^+_{\mathbf{p}_2s_2l_2}a_{\mathbf{p}_2+\mathbf{q}s_2l_2}a_{\mathbf{p}_1-\mathbf{q}s_1l_1}$$
$$+ b^+_{\mathbf{p}_1s_1l_1}b^+_{\mathbf{p}_2s_2l_2}b_{\mathbf{p}_2+\mathbf{q}s_2l_2}b_{\mathbf{p}_1-\mathbf{q}s_1l_1} - 2a^+_{\mathbf{p}_1s_1l_1}b^+_{\mathbf{p}_2s_2l_2}b_{\mathbf{p}_2+\mathbf{q}s_2l_2}a_{\mathbf{p}_1-\mathbf{q}s_1l_1}], \quad (1)$$

where $a^+_{\mathbf{p}s_{1,2}l}$ and $b^+_{\mathbf{p}s_{1,2}l}$ are the electron and hole creation operators in the layer l ($l = 0, \pm1, \pm2, \ldots$), and spin projections $s_{1,2}$ and momentum $\mathbf{p}$ in the plane of the layer; $\mathbf{p}_{1,2}$ and $\mathbf{q}$ are two-dimensional vectors; the normalized area of the layer is set equal to unity:

$$V_{q,l} = 2\pi q^{-1}\exp(-q|l|c^*); \quad (2)$$

$c^* = c(\varepsilon_\perp/\varepsilon_\parallel)^{1/2}$; c is the distance between the neighboring planes; μ_e and μ_h are the chemical potentials of the electrons and holes, calculated from the bottom of the respective bands and connected by the condition of electric neutrality

$$\sum_{\mathbf{p}s_1}\langle a^+_{\mathbf{p}s_1l}a_{\mathbf{p}s_1l}\rangle = \sum_{\mathbf{p}s_2}\langle b^+_{\mathbf{p}s_2l}b_{\mathbf{p}s_2l}\rangle = nc; \quad (3)$$

n is the volume density of the number of electron-hole pairs; nc is the surface density in a single layer.

The introduction of the chemical potentials μ_e and μ_h is necessary for the description of nonequilibrium systems, when the electrons and holes assumed to be dif-

 0038-5646/77/4603-0616$02.40

ferent types of particles and their concentration n is specified arbitrarily. In a thermodynamically equilibrium situation, the chemical potential of the electrons in both bands should be the same, which adds an additional condition for μ_e and μ_h:

$$\mu_e + \mu_h = -E_g^{(0)}, \tag{4}$$

where $E_g^{(0)}$ is the width of the forbidden band, i.e., the distance between the initial values of the energies of the electrons and holes in (1). It should be noted that $E_g^{(0)}$ is the unrenormalized width of the forbidden band, a width that enters into the bare dispersion law. The account given below of the interaction of electrons and holes leads to a significant difference between $E_g^{(0)}$ and the real width E_g of the forbidden band right up to the transition from the semiconductor unrenormalized continuous spectrum ($E_g^{(0)} > 0$) to the semimetal ($E_g < 0$) after renormalization.

The investigation of the thermodynamics of the considered model is most conveniently carried out by starting from the dependence of the particle number density on the temperature and the chemical potential

$$n_{e,h}(\mu_{e,h}, T) = \frac{2T}{c} \sum_k \int \frac{d^2p}{(2\pi)^2} G_{e,h}(\mathbf{p}, \varepsilon_k) e^{i\varepsilon_k \tau}$$
$$= \frac{2T}{c} \sum_k \int \frac{d^2p}{(2\pi)^2} \frac{e^{i\varepsilon_k \tau}}{i\varepsilon_k + \mu_{e,h} - \varepsilon_{e,h}(\mathbf{p}) - \Sigma_{e,h}(\mathbf{p}, \varepsilon_k)}, \tag{5}$$

where $G_{e,h}$ is the Green's function of the electrons and holes in the Matsubara technique[2] (see also Ref. 3), $\Sigma_{e,h}$ are the self-energy parts, $\varepsilon_k = \pi T(2k+1)$, and k is an integer and $\tau \to +0$.

Further consideration is based on the assumption that the distances between the layers c are small in comparison with the radius of the two-dimensional exciton, i.e., with account of the scales employed,

$$c \ll 1. \tag{6}$$

We shall show below (formulas (43)–(49)) that upon satisfaction of the condition (6), in the region of concentrations and temperatures defined by the inequalities

$$c^{-1} \ll n \ll c^{-2}, \tag{7}$$
$$T \ll n^{1/2}, \tag{8}$$

the self-energy parts $\Sigma_{e,h}$ do not depend in first approximation on either the temperature or the arguments $\mathbf{p}$ and ε_k, and have the form

$$\Sigma_{e,h} = -a(\pm s)(c/c^*)^{1/4} n^{3/4}, \tag{9}$$

where the plus sign corresponds to electrons, and the minus to holes,

$$a(s) = \frac{2^{11/8}\pi^{1/4}}{[\Gamma(1/4)]^2} \int_0^\infty dx \frac{1+s}{x^2+(1+s)^2} \left[\frac{1+s}{x^2+(1+s)^2} + \frac{1-s}{x^2+(1-s)^2}\right]^{1/4}. \tag{10}$$

Thus, in this region of concentrations and temperatures, the interaction in the electron-hole plasma leads to a narrowing of the forbidden band without any appreciable change in any other of the parameters of the electron spectrum. We shall also show that the entire region of existence of the electron-hole liquid, including the critical point, falls in the interval of concentrations and temperatures satisfying the conditions (7) and (8). In order to establish this, we consider first the thermodynamics of the electron-hole plasma, starting out from the formulas (9) and (10), and then give the basis of the formulas themselves. Carrying out the summation and integration in (5) with account of the independence of Σ of $\mathbf{p}$ and ε_k, we obtain

$$n_{e,h} = n = \frac{T}{\pi c(1 \pm s)} \ln\left(1 + \exp\frac{\mu_{e,h} - \Sigma_{e,h}}{T}\right). \tag{11}$$

Solving the relation (11) for $\mu_{e,h}$ and combining the equations thus obtained, we find the dependence of the chemical potential of a single electron-hole pair $\mu = \mu_e + \mu_h$ on n and T:

$$\mu(n, T) = \Sigma_e + \Sigma_h + T \ln\left\{\left(\exp\frac{\pi nc(1+s)}{T} - 1\right)\left(\exp\frac{\pi nc(1-s)}{T} - 1\right)\right\}$$
$$= -A\left(\frac{c}{c^*}\right)^{1/4} n^{3/4} + T \ln\left[2e^{\pi nc/T}\left(\operatorname{ch}\frac{\pi nc}{T} - \operatorname{ch}\frac{\pi ncs}{T}\right)\right], \tag{12}$$

where the coefficients $A = a(s) + a(-s)$ is connected in the following fashion with the function $f(\sigma)$ introduced previously[1]:

$$A = 2^{23/4}\pi^{9/4} f(\sigma)/[\Gamma(1/4)]^4. \tag{13}$$

The relation (12) is essentially the equation of state of the electron-hole plasma, written in the variables μ and n instead of the more customary pressure and volume. Several typical curves, described by these relations for the simplest case $\sigma = 1$ ($s = 0$), are given in Fig. 1. It is seen that they have the typical van der Waals character, i.e., at temperatures $T < T_c$ there are two stable branches $n_{\text{gas}}(\mu, T)$ and $n_{\text{liq}}(\mu, T)$ and one thermodynamically unstable branch $(\partial\mu/\partial n)_T < 0$ of the solutions. The lesser of the two stable solutions n_{gas} corresponds to the gaseous phase, the greater—n_{liq}—corresponds to the electron-hole liquid. The critical point (T_c, n_c) is determined by the two equations

$$(\partial\mu/\partial n)_{T_c, n_c} = (\partial^2\mu/\partial n^2)_{T_c, n_c} = 0, \tag{14}$$

which reduce to a single transcendental equation for the parameter $z_c = \pi c n_c / T_c$:

$$z_c = \frac{3(\operatorname{ch} z_c - \operatorname{ch}(sz_c))(\exp(z_c) - \operatorname{ch}(sz_c) - s\operatorname{sh}(sz_c))}{4[(1+s^2)(\operatorname{ch} z_c \operatorname{ch}(sz_c) - 1) - 2s\operatorname{sh} z_c \operatorname{sh}(sz_c)]}. \tag{15}$$

The critical parameters of the liquid are expressed in terms of z_c by the formulas

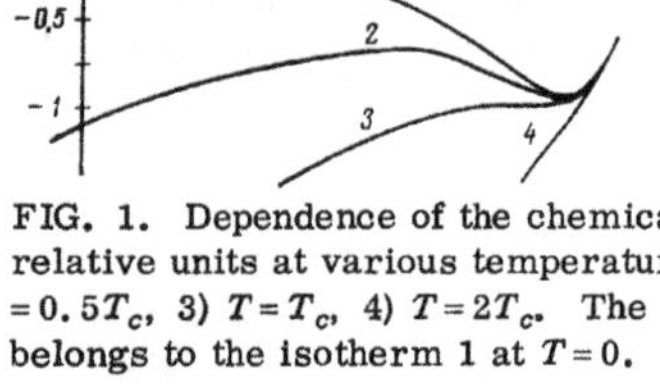

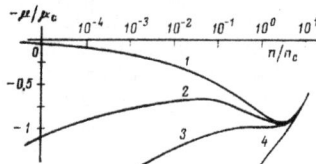

FIG. 1. Dependence of the chemical potential on the density in relative units at various temperatures: curves 1) $T = 0$, 2) $T = 0.5T_c$, 3) $T = T_c$, 4) $T = 2T_c$. The semi-axis $n = 0$, $\mu \leqslant 0$ also belongs to the isotherm 1 at $T = 0$.

$$n_c c=\left[\frac{A}{4\pi}\frac{\operatorname{ch} z_c-\operatorname{ch}(z_c s)}{\exp(z_c)-\operatorname{ch}(z_c s)-s\operatorname{sh}(z_c s)}\right]^{4/3}\frac{1}{c^{*1/3}}, \quad (16)$$

$$T_c=\frac{\pi}{z_c}\left[\frac{A}{4\pi}\frac{\operatorname{ch} z_c-\operatorname{ch}(z_c s)}{\exp(z_c)-\operatorname{ch}(z_c s)-s\operatorname{sh}(z_c s)}\right]^{4/3}\frac{1}{c^{*1/3}}, \quad (17)$$

$$\mu_c=-T_c\left[\left(3+4\frac{\operatorname{sh} z_c-s\operatorname{sh}(sz_c)}{\operatorname{ch} z_c-\operatorname{ch}(sz_c)}\right)z_c-\ln 2-\ln(\operatorname{ch} z_c-\operatorname{ch}(sz_c))\right]. \quad (18)$$

By virtue of (15), z_c depends only on a single parameter—the ratio of the masses of the electron and hole σ (or s). The dependence of z_c on σ is very weak:

$$z_c(\sigma)=\begin{cases}0.5501, & \sigma=1\\ 0.4877, & \sigma=0.\end{cases} \quad (19)$$

With accuracy to one hundredth, the function $z_c(\sigma)$ can be obtained in explicit form by expansion of the right side of (15) in a series in z_c up to terms of third order:

$$z_c(\sigma)\approx\frac{3}{2(1+s^2)}\left[\left(1+\frac{8}{9}(1+s^2)\right)^{1/2}-1\right]. \quad (20)$$

With the same accuracy, the formulas (16)–(18) reduce to the form

$$n_c c\approx\left[\frac{3A}{32\pi}z_c\left(1+\frac{1+s^2}{12}z_c^2\right)\right]^{4/3}\frac{1}{c^{*1/3}}, \quad (21)$$

$$T_c=\frac{\pi}{z_c}\left[\frac{3A}{32\pi}z_c\left(1+\frac{1+s^2}{12}z_c^2\right)\right]^{4/3}\frac{1}{c^{*1/3}}, \quad (22)$$

$$\mu_c=-T_c\left[\frac{32}{3}\left(1-\frac{1+s^2}{12}z_c^2\right)-z_c-\ln z_c^2(1-s^2)\left(1+\frac{1+s^2}{12}z_c^2\right)\right]. \quad (23)$$

At temperatures less than critical, one of the two stable solutions $n_{gas}(\mu, T)$ and $n_{liq}(\mu, T)$ is generally metastable. Only for a single value of the chemical potential $\mu(T)$, determined for each temperature T, is the existence of both phases possible with balanced densities $n_{gas}(T)=n_{gas}(\mu(T), T)$ and $n_{liq}(T)=n_{liq}(\mu(T), T)$. The quantities $\mu(T)$, $n_{gas}(T)$ and $n_{liq}(T)$ are determined by the conditions that the chemical potentials and pressures be equal in both phases:

$$\mu(T)=\mu(n_{gas}(T), T)=\mu(n_{liq}(T), T), \quad (24)$$

$$\int_{n_{gas}}^{n_{liq}} dn\,\mu(n, T)=\mu(T)(n_{liq}(T)-n_{gas}(T)), \quad (25)$$

with the accuracy shown above, near the critical point, $T_c-T\ll T_c$:

$$n_{liq}(T)-n_c=n_c-n_{gas}(T)\approx 2\cdot 6^{1/2}\left(1-\frac{1+s^2}{3}z_c^2\right)n_c\left(\frac{T_c-T}{T_c}\right)^{1/2}, \quad (26)$$

$$\mu(T)-\mu_c=\left\{2-\ln z_c^2(1-s^2)+\frac{1}{12}(1+s^2)z_c^2\right\}(T_c-T). \quad (27)$$

At low temperatures $T\ll T_c$:

$$n_{gas}c=n_0 c\left\{1-\frac{25\pi^2}{2^{10}(1-s^2)}\left(\frac{15z_c}{4}\right)^{2/3}\left[1+\frac{2}{9}(1+s^2)z_c^2\right]\left(\frac{T}{T_c}\right)^2\right\}$$

$$\approx 0.166f^{4/3}(\sigma)\left\{1+0.582\frac{z_c^{2/3}}{1-s^2}\left[1+\frac{2}{9}(1+s^2)z_c^2\right]\left(\frac{T}{T_c}\right)^2\right\}\frac{1}{c^{*1/3}}, \quad (28)$$

$$\mu(T)=\mu_0\left\{1+\frac{25\pi^2}{2^{12}(1-s^2)}\left(\frac{15z_c}{4}\right)^{2/3}\left[1+\frac{2}{9}(1+s^2)z_c^2\right]\left(\frac{T}{T_c}\right)^2\right\}$$

$$\approx -1.57f^{4/3}(\sigma)\left\{1+0.146\frac{z_c^{2/3}}{1-s^2}\left[1+\frac{2}{9}(1+s^2)z_c^2\right]\left(\frac{T}{T_c}\right)^2\right\}\frac{1}{c^{*1/3}}, \quad (29)$$

where

$$n_0 c=\left(\frac{A}{5\pi}\right)^{4/3}c^{*-1/3}=\frac{0.166f^{4/3}(\sigma)}{c^{*1/3}}, \quad (30)$$

$$\mu_0=E_{min}=-3\pi\left(\frac{A}{5\pi}\right)^{4/3}c^{*-1/3}=-\frac{1.57f^{4/3}(\sigma)}{c^{*1/3}} \quad (31)$$

are respectively the equilibrium density and minimum energy of the ground state of the EHL at $T=0$. In this limiting case, the expressions (28), (29) go over into the formula (10) of Ref. 1.[1)] It is seen from Eqs. (21)–(23) and (30), (31) that the region of existence of the EHL satisfies the inequalities (7) and (8).

It should be noted that the expansion used in (20) and subsequently assures an accuracy to within a few percent in all the formulas. This accuracy will obviously be quite sufficient when a comparison is made with the experimental data, the more so that, in the region of real values of the distance between the layers c, the expressions (21)–(23), (31) and (32) will be satisfied only qualitatively. The problems of the applicability of the present consideration are discussed at the end of the article.

Figure 2 shows a phase diagram in reduced units kn/n_c and T/T_c, constructed numerically according to (24), (25) for the case $\sigma=1$. The form of the phase diagram does not depend on the parameter c and depends weakly on the parameter σ. It follows from the calculation that for all values of the parameters, the following relation between the critical temperature T_c and the binding energy of the EHL $|E_{min}|$ at $T=0$ is well satisfied:

$$T_c\approx 0.1|E_{min}|. \quad (32)$$

This relation turns out to be quite general. Thus, it holds approximately for the experimentally observed phase diagrams of the EHL in germanium and silicon, which are never layered materials. The experimental data are for germanium[4]: $T_c=6.5\pm 0.1$ K, $|E_{min}|=5.66\pm 0.15$ meV; for silicon[5]: $T_c=28\pm 2$ K, $|E_{min}|=22.6\pm 0.2$ meV. Calculations of the critical point in germanium and silicon were also carried out in other researches.[6] A similar relation evidently exists for liquid-gas transitions generally; see the note on page 288 of the book of Landau and Lifshitz.[7]

However, we now return to layered system. In the considered model, the gas phase is a nondegenerate plasma, and the excitons in it do not play any significant role. Even at very low temperatures $T\ll T_c$, their concentration $\propto\exp[(\mu_0+1)/T]$ is much less than the concentration of the free carriers, which is of the order of

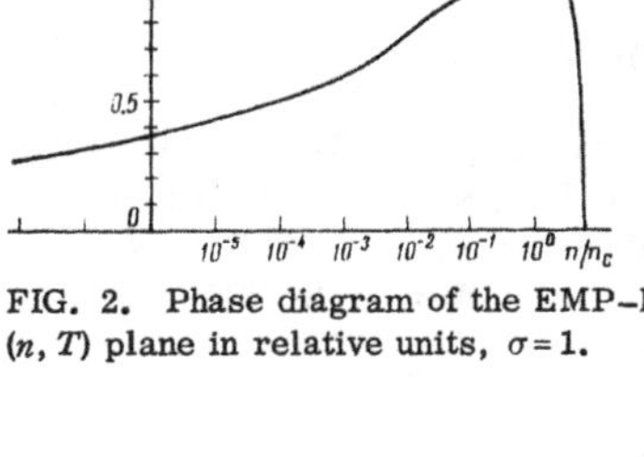

FIG. 2. Phase diagram of the EMP–EHL transition on the (n, T) plane in relative units, $\sigma=1$.

$\exp(\mu_0/2T)$, by virtue of the condition $|\mu_0| \sim c^{-1/2} \gg 2$. At temperatures that are close to critical, the density of the carriers in the gas phase is so large that the existence of excitons in it is generally impossible.

We also give the formulas for the symmetric case $\sigma = 1$. The equation for z_c is materially simplified:

$$\tfrac{4}{3} z_c = \exp(z_c) - 1, \tag{33}$$

its solution is $z_c = 0.5501$ and Eqs. (21)–(23) reduce to

$$n_c c = \left(\frac{A}{6\pi} z_c \exp(-z_c)\right)^{4/3} \frac{1}{c^{*1/3}} \approx \frac{0.028}{c^{*1/3}} \approx 0.17 n_0 c, \tag{34}$$

$$T_c = \frac{\pi}{z_c}\left(\frac{A}{6\pi} z_c \exp(-z_c)\right)^{4/3} \frac{1}{c^{*1/3}} \approx \frac{0.161}{c^{*1/3}} \approx 0.103 |E_{min}|, \tag{35}$$

$$\mu_c = -2[3 + 4z_c - \ln(4z_c/3)] T_c \approx -11.0 T_c. \tag{36}$$

Up to now we have had in mind a nonequilibrium situation, in which the total number of electrons and holes in the sample is given by the external source of the excitation, although in all the other parameters, these carriers are in equilibrium with the crystal lattice and with one another. In the case of complete thermodynamic equilibrium, i.e., in the absence of an excitation, the formulas (12)–(36) obtained above remain in force, but their meaning is modified somewhat. The condition that the chemical potentials of the electrons be the same in the valence band and the conduction band (4) leads to the replacement in Eq. (12) of μ by $E_g^{(0)}$—the unrenormalized width of the forbidden band:

$$E_g^{(0)} = A\left(\frac{c}{c^*}\right)^{1/4} n^{1/4} - T \ln 2 \exp\left(\frac{\pi nc}{T}\right)\left(\operatorname{ch}\frac{\pi nc}{T} - \operatorname{ch}\frac{\pi ncs}{T}\right). \tag{37}$$

Equation (37) determines the dependence of the electron and hole concentration n on the temperature T and on $E_g^{(0)}$.

The actual (renormalized) width of the forbidden band E_g, by virtue of (9) and (10), is equal to

$$E_g(T) = E_g \; - A(c/c^*)^{1/4} n^{1/4}(T). \tag{38}$$

The gas-liquid phase transition means a discontinuous increase in $n(T)$ and therefore a decrease in $E_g(T)$, i.e., semiconductor—semimetal transition at sufficiently small values of the bare forbidden band. In the plane of the variables $(T, E_g^{(0)})$ (Fig. 3) the curve $E_g^{(0)} = -\mu(T)$, described by the formulas (22)–(25), (27), (29), (31), divides the regions of existence of the semimetal (SM) and semiconductor (SC) phases. In the region $T < T_c$, $|\mu_0| \leq E_g^{(0)} < |\mu_c|$ the carrier concentrations n and the width of the forbidden band E_g depend in discontinuous fashion on T and $E_g^{(0)}$:

$$n \leq n_{gas}(T), \; E_g(T) \geq |\mu(T)| - A(c/c^*)^{1/4} n_{gas}^{1/4}(T)$$

for the SC phase and

$$n \geq n_{liq}(T), \; E_g(T) \leq |\mu(T)| - A(c/c^*)^{1/4} n_{liq}^{1/4}(T)$$

for the SM phase. In the actual phase diagram, in place of $E_g^{(0)}$ (Fig. 3), we should have some thermodynamic parameter whose change causes $E_g^{(0)}$ to change, for example, the external pressure or the concentration of one of the components in the case of solid solutions.

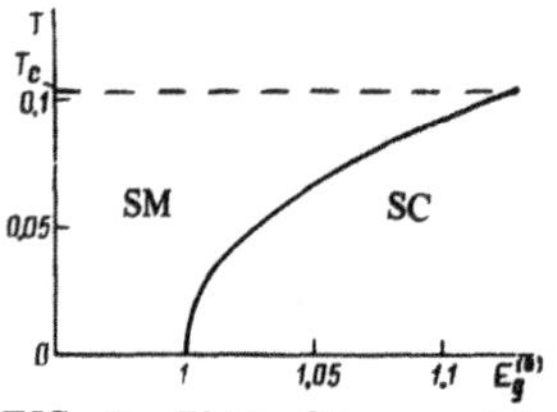

FIG. 3. Phase diagram of the semiconductor–semimetal transition on the plane of the variables $(T, E_g^{(0)})$, with both variables in units of $|E_{min}| = |\mu_0|$. The continuous curve separates the semiconducting (SC) from the semimetal (SM) phase.

We now proceed to the derivation of Eqs. (9) and (10), on which all the previous consideration was based. The irreducible self-energy part Σ_e (Σ_h can be obtained from ε_e by the obvious substitution $s \to -s$ in all formulas), is expressed by the well known formula

$$\Sigma_e(\mathbf{p}, \varepsilon_{k_1}) = -T \sum_{k_2} \sum_{l_1} \int \frac{d^2q}{(2\pi)^2} \tilde{V}_{l-l_1}(\mathbf{q}, \omega_{k_2}) \times G_l(\mathbf{p}+\mathbf{q}, \varepsilon_{k_1}+\omega_{k_2}) \Gamma^{(e)}_{l-l_1}(\mathbf{p}, \mathbf{q}; \varepsilon_{k_1}, \omega_{k_2}), \tag{39}$$

where the screened interaction $\tilde{V}$ is determined by the equation

$$\tilde{V}_{l-l_1}(\mathbf{q}, \omega_{k_2}) = V_{q,l-l_1} + \sum_{l_3 l_4} V_{q,l-l_3} \Pi_{l_3-l_4}(\mathbf{q}, \omega_{k_2}) \tilde{V}_{l_4-l_1}(\mathbf{q}, \omega_{k_2}). \tag{40}$$

The polarization operator

$$\Pi_{l_3-l_4}(\mathbf{q}, \omega_{k_2}) = \sum_{\alpha=e,h} 2T \sum_{k_1} \int \frac{d^2p}{(2\pi)^2} G_\alpha(\mathbf{p}, \varepsilon_{k_1}) G_\alpha(\mathbf{p}+\mathbf{q}, \varepsilon_{k_1}+\omega_{k_2}) \times \Gamma^{(\alpha)}_{l_3-l_4}(\mathbf{p}, \mathbf{q}; \varepsilon_{k_1}, \omega_{k_2}), \tag{41}$$

$\varepsilon_{k_1} = \pi T(2k_1+1)$, $\omega_{k_2} = 2\pi k_2 T$; $k_{1,2}$ are integers; $\Gamma^{(\alpha)}_{l-l_1}$ is the vertex part, defined as the sum of all the irreducible diagrams with a single incoming (into the layer l_1) interaction line and two out external electron or hole lines on the same layer l.

In the region of high concentrations $nc \gg 1$ of interest to us, classification of the terms of the series of perturbation theory in powers of the density n is possible. The set of principal diagrams forms, as is well known, the so-called diagrams of the random phase approximation (RPA), corresponding to the use for $\Gamma^{(\alpha)}_{l-l_1}$ and $\Pi_{l_3-l_4}$ of their first nonvanishing approximations:

$$\Gamma^{(\alpha)}_{l-l_1}(\mathbf{p}; \mathbf{q}; \varepsilon, \omega) = \delta_{l_1 l},$$

$$\Pi_{l_3-l_4}(\mathbf{q}, \omega_{k_2}) = -4\delta_{l_3 l_4} \sum_{\alpha=e,h} \int \frac{d^2p}{(2\pi)^2} f_\alpha(\mathbf{p}) \frac{\varepsilon_\alpha(\mathbf{p}+\mathbf{q}) - \varepsilon_\alpha(\mathbf{p})}{\omega_{k_2}^2 + [\varepsilon_\alpha(\mathbf{p}+\mathbf{q}) - \varepsilon_\alpha(\mathbf{p})]^2}, \tag{42}$$

where $f_{e,h}(\mathbf{p})$ is the Fermi distribution function. The form of the expression (42) is the usual one (see, for example, Ref. 3, p. 250); in the summation, however, it is now assumed that Σ_α is independent of its arguments, an assumption discussed below.

The expression (39) for Σ reduces in this approximation to

$$\Sigma_e(\mathbf{p}, \varepsilon_{k_1}) = -\int \frac{d^2q}{(2\pi)^2} \frac{2\pi}{q} f_e(\mathbf{p}+\mathbf{q}) + T \sum_{k_2} \int \frac{d^2q}{(2\pi)^2} \left[\tilde{V}_0(\mathbf{q}, \omega_{k_2}) - \frac{2\pi}{q}\right] G_e(\mathbf{p}+\mathbf{q}, \varepsilon_{k_1}+\omega_{k_2}), \tag{43}$$

where V_0 is obtained by solution of the Eq. (40):

$$V_0(\mathbf{q},\omega_{k_2})=\frac{2\pi}{q}\left\{1-\frac{4\pi}{q}\Pi(\mathbf{q},\omega_{k_2})\operatorname{cth} qc+\frac{4\pi^2}{q^2}\Pi^2(\mathbf{q},\omega_{k_2})\right\}^{-1/2}. \quad (44)$$

The first term in the right side of (43) is the exchange correction to the dispersion of the electrons; the second is the correlation correction. The exchange part is obviously $\sim nc/\bar{p}$ and, as we shall see below, is small in comparison with the correlation part at concentrations satisfying the inequalities (7) and (8). For estimates of the correlation contribution Σ_{corr}, we consider (44) in more detail.

For the behavior of the polarization operation, a knowledge of the quantity

$$(q/\bar{p})^2+(\omega_k/q\bar{p})^2$$

is essential, where $\bar{p}$ is the average momentum of the particles, $\bar{p}\sim\max[(2\pi nc)^{1/2}, T^{1/2}]$. In the two limiting cases, we have

$$\Pi(\mathbf{q},\omega_k)\approx-\frac{1}{\pi}\left[\frac{f_e(0)}{1+s}+\frac{f_h(0)}{1-s}\right]\sim-\frac{nc}{\bar{p}^2},\quad \left(\frac{q}{\bar{p}}\right)^2+\left(\frac{\omega_k}{q\bar{p}}\right)^2\ll 1, \quad (45)$$

$$\Pi(\mathbf{q},\omega_k)\approx-4\left[\frac{1}{(1+s)(q^2+(2\omega_k/(1+s)q)^2)}+\frac{1}{(1-s)(q^2+(2\omega_k/(1-s)q)^2)}\right]nc,\quad \left(\frac{q}{\bar{p}}\right)^2+\left(\frac{\omega_k}{q\bar{p}}\right)^2\gg 1. \quad (46)$$

In the region of (7), (8) we have $\bar{p}\ll c^{-1/2}$. Therefore, the transferred momenta $q\gtrsim c^{-1}$ correspond to the asymptotic form (46) of the polarization operator,

$$q^{-1}\Pi\lesssim nc^4\ll 1$$

and the contribution of the region $qc\gtrsim 1$ to the integral for the correlation part Σ_e is very small. The basic contribution is made by the region $qc\ll 1$. Using this inequality and the condition $|\Pi|\lesssim 1$, which is obvious by virtue of (45), (46) and the definition of p, we get

$$\Sigma_{e\,corr}(\mathbf{p},\varepsilon_{k_1})\approx-T\sum_{k_2}\int\frac{d^2q}{(2\pi)^2}\frac{2\pi}{q}\left[\left(1-\frac{4\pi}{cq^2}\Pi(\mathbf{q},\omega_{k_2})\right)^{-1/2}-1\right]\times G_e(\mathbf{p}+\mathbf{q},\varepsilon_{k_1}+\omega_{k_2}). \quad (47)$$

Substituting (45) and (46) successively in (47), we can easily establish the fact that at $\bar{p}\gg n^{1/4}$ the basic contribution to (47) is made by $q\lesssim(p/\bar{p})^{1/2}\ll\bar{p}$, to which corresponds the asymptotic form (45), and at $\bar{p}\ll n^{1/4}$ by the momenta $q\sim n^{1/4}\gg\bar{p}$, which corresponds to the asymptotic form (46). But the inequality $\bar{p}\ll n^{1/4}$, by virtue of the definition of $\bar{p}$, is equivalent to the inequalities (7) and (8), so that in the region of concentrations and temperatures of interest to us we should use the polarization operator in the form (46). Moreover, in (47) we can omit $\mathbf{p}$ in the argument of G_e because of the inequality $q\gg\bar{p}$ and, consequently, for real momenta of the particles Σ_{corr} does not depend on $\mathbf{p}$. In the summation over the frequencies, the basic contribution under these conditions is made by $|\omega_{k_2}|\lesssim n^{1/2}$, as is easily understood from (45) and (46). Therefore, by virtue of (8) and of the definition of $\omega_{k_2}=2\pi k_2 T$, the summation over k_2 can be replaced by the integration

$$T\sum_{k_2}\to\int\frac{d\omega}{2\pi},$$

which is equivalent to neglect of the temperature dependence, i.e., the replacement of Σ_{corr} by its value at $T=0$.

Finally, the dependence on ε_{k_1} in the argument G_e in (47) also turns out to be unimportant for $|\varepsilon_{k_1}|\lesssim n^{1/2}$, i.e., right up to energies significantly greater than the mean energy of the particles $\bar{p}^2$. With account of all these simplifications, (47) takes the form

$$\Sigma_{e\,corr}=-\int\frac{d^2q}{(2\pi)^2}\int_{-\infty}^{+\infty}\frac{d\omega}{2\pi}\frac{2\pi}{q}\left[\left\{1+4\pi n\left(\frac{1+s}{\omega^2+[(1+s)^2q^2/2]^2}+\frac{1-s}{\omega^2+[(1-s)^2q^2/2]^2}\right)\right\}^{-1/2}-1\right]\frac{1}{i\omega-q^2(1+s)/2+\mu_e-\Sigma_e}$$

$$=\int_0\frac{dq\,d\omega}{\pi}\left[\left\{1+4\pi n\left(\frac{1+s}{\omega^2+[(1+s)^2q^2/2]^2}+\frac{1-s}{\omega^2+[(1-s)^2q^2/2]^2}\right)\right\}^{-1/2}-1\right]\frac{-q^2(1+s)/2+\mu_e-\Sigma_e}{\omega^2+[q^2(1+s)/2+\Sigma_e-\mu_e]^2} \quad (48)$$

or, after the introduction of the new integration variables

$$x=\frac{2\omega}{q^2},\quad y=(16\pi n)^{-1/4}q,$$

$$\Sigma_{e\,corr}=\frac{(16\pi n)^{1/4}}{\pi}\int_0^\infty\int dx\,dy\left[\left\{1+\frac{1}{y^4}\left(\frac{1+s}{x^2+(1+s)^2}+\frac{1-s}{x^2+(1-s)^2}\right)\right\}^{-1/2}-1\right]\frac{1+s-(\mu_e-\Sigma_e)/2y^2(\pi n)^{1/2}}{x^2+[1+s-(\mu_e-\Sigma_e)/2y^2(\pi n)^{1/2}]^2}. \quad (49)$$

The system of equations (11) and (49) together determines the dependence of $\mu_{e,h}$ and $\Sigma_{e,h}$ on n. With relative accuracy $\sim n^{1/4}$ the terms $(\mu-\Sigma)/n^{1/2}$ from the right side of (49) and, after integration over y, (49) reduces to (9) and (10). In connection with this solution, it is necessary to make the following remark. The approximation that we have used is, strictly speaking, not the random phase approximation, although it is similar to it. In the approach that is usual for the RPA, one must use in (5), (39) and (41) the zeroth approximation for the Green's function

$$G_{e,h}^{(0)}(\mathbf{p},\varepsilon_k)=(i\varepsilon_k-\varepsilon_{e,h}(\mathbf{p})+\mu_{e,h})^{-1}. \quad (50)$$

Neglect of $\Sigma_{e,h}$ in the integrand of (49) actually corresponds to this. However, the total Green's functions (in the principal approximation in n)

$$G_{e,h}(\mathbf{p},\varepsilon_k)=(i\varepsilon_k-\varepsilon_{eh}(\mathbf{p})+\mu_{e,h}-\Sigma_{e,h})^{-1} \quad (51)$$

must be used in (5), (11), (41) and (42). This difference is very important, since the Matsubara diagram technique is developed at fixed μ and not at fixed n. Therefore, substitution of $G^{(0)}$ in (5) and (41) would have yielded, at large negative values of μ of interest to us, exponentially small current carrier densities $n_0\sim e^{\mu/T}$ and, correspondingly, exponentially small screening of the interaction $\Pi\sim n_0$. In other words, we would have found for each μ only a single solution corresponding at low temperatures to the gas phase and generally not suitable at $T\sim T_c$. The analysis given above shows the existence for the nonlinear system of Eqs. (39)–(41), (51) of a second solution of the form (41), (9), corresponding to high concentrations n at negative μ, i.e., to a liquid phase.

In conclusion, we discuss the limits of applicability of the results. The model used in the present work and described by the Hamiltonian (1) is itself extremely idealized. In real layered semiconductors and semimetals, the probability of an electron transition from one layer to another differs from zero; this leads to the dependence of the energies of the electrons and holes on the quasimomentum components p_z perpendicular to the planes of the layers. If the transition probability is sufficiently small, this dependence is of the form $W\cos(p_z c)$, where W is proportional to the overlap integral of the wave functions of the electrons on neighboring layers. The consideration given above, which does not take into account the dependence of $\varepsilon_{e,h}$ on p_z, is therefore valid only so long as all the energies entering into the calculations significantly exceed the width W of transverse-motion band, i.e., $W \ll c^{-1/2}$. In the case of opposite sign of this inequality, the motion of the electrons is essentially three-dimensional and for the description of the EHL we must use the approach based on the strong anisotropy of the effective masses for motion in the layer (m_t) and perpendicular to it (m_l).[1] The role of the small parameter of the theory is played in this case by $m_l/m_t \ll 1$ instead of c.

1) The coefficient A_{II} of Ref. 1, which determines the correlation contribution to the total energy, differs by the factor $\frac{4}{5}$ from the A coefficient used in the present work in the self-energy parts of the electrons and holes. Moreover, in formulas (10) from Ref. 1, there are errors. The correct form of these formulas agrees with formulas (30) and (31) of this paper.

[1] E. A. Andryushin, V. S. Babichenko, L. V. Keldysh, T. A. Opishchenko, and A. P. Silin, Pis'ma Zh. Eksp. Teor. Fiz. **24**, 210 (1976) [JETP Lett. **24**, 185 (1976)].

[2] T. Matsubara, Progr. Theor. Phys. **14**, 351 (1955).

[3] A. A. Abrikosov, L. P. Gor'kov, and I. E. Dzyaloshinskiĭ, Metody kvantovoĭ teorii polya v statisticheskoĭ fizike (Quantum Field Theory Methods in Statistical Physics) Fizmatgiz, 1962.

[4] T. K. Lo, Solid State Commun. **15**, 1231 (1974); C. A. Thomas, T. M. Rice, and J. C. Hensel, Phys. Rev. Lett. **33**, 219 (1974).

[5] A. F. Dite, V. D. Kulakovskiĭ, and V. B. Timofeev, Zh. Eksp. Teor. Fiz. **72**, 1156 (1977) [Sov. Phys. JETP **45**, 604 (1977)].

[6] R. N. Silver, Phys. Rev. **B8**, 2403 (1973); M. Combescot, Phys. Rev. Lett. **32**, 15 (1974); P. Vashishta, S. G. Das, and K. S. Singwi, Phys. Rev. Lett. **33**, 911 (1964); G. Mahler, Phys. Rev. **B11**, 4050 (1975); T. L. Reinecke and S. C. Ying, Phys. Rev. Lett. **35**, 311 (1975).

[7] L. D. Landau and E. M. Lifshitz, Statisticheskaya fizika (Statistical Physics) Ch. 1, Nauka, 1976.

Translated by R. T. Beyer

 0038-5646/77/4603-0621$02.40

Polaritons in thin semiconducting films

L. V. Keldysh
P.N. Lebedev Physics Institute, USSR Academy of Sciences

(Submitted 12 July 1979)
Pis'ma Zh. Eksp. Teor. Fiz. **30**, No. 4, 244–249 (20 August 1979)

It is shown that the oscillator strengths of the excitons in semiconducting films, whose thickness is less than the effective exciton radius and whose dielectric constant is much greater than that of the medium surrounding the film, increase sharply with decreasing thickness of the film. The reflectivity of the film increases in the neighborhood of the excitonic lines and conditions are established for propagation of the electromagnetic waves-polaritons that are localized near the film.

PACS numbers: 71.35. + z, 71.36. + c, 73.60.Fw, 78.20.Dj

It was shown in a recent letter[1] that in sufficiently thin semiconductor films whose dielectric constant ϵ greatly exceeds the dielectric constant of the substrate the Coulomb interaction between the charges will increase (compared with bulk samples of the same semiconductor) if the distance between them exceeds the thickness d of the film d. For $d \ll a_0 \equiv \omega\hbar^2(me^2)^{-1}$ (e is the electron charge, $\hbar$ is the Planck constant, and m is the reduced mass of the electron and of the hole), this circumstance will increase the binding energy $\mathscr{E}_0(d)$ and decrease the effective radius a of the Vanier–Mott excitons

$$\mathcal{E}_0(d) = \frac{e^2}{\epsilon d}\left\{ \ln\left[\left(\frac{2\epsilon}{\epsilon_1 + \epsilon_2}\right)^2 \frac{d}{a_0}\right] - 0.8 \right\}, \qquad (1)$$

$$a = \frac{1}{2}\sqrt{a_0 d}\,. \qquad (2)$$

The subscripts 1 and 2 here and elsewhere denote quantities associated with the half-space on either side of the film. The excitons in this situation are almost two-dimensional.

The intensity of the optical transition for the dipole-active excitons examined below is determined by $|\phi(0)|^2$, where $\phi(r)$ is the wave function of the relative motion of the electron and hole.[2] It is clear that $|\phi(0)|^2$ is inversely proportional to the "exciton volume," i.e., because of Eq. (2) $|\phi(0)|^2 \sim a^{-2} \sim (a_0 d)^{-1}$. Thus, the binding energy of excitons and the oscillator strength corresponding to them increase with decreasing d. Taking this into account, the contribution of the excitonic transition to the electromagnetic response of the film to the field with the frequency ω and projection $\mathbf{k}$ of the wave vector on the plane of the film can be described by the induced current density $\mathbf{j}$:

$$j_\alpha(\mathbf{k}, \omega; z) = -i\omega\Lambda_{\alpha\beta}\,\chi_{\mathbf{k}\omega}\,\bar{E}_\beta 2\sin^2\frac{\pi z}{d}, \tag{3}$$

$$\chi_{\mathbf{k}\omega} = \frac{1}{\epsilon}\left(\frac{e^2}{dE_\mathbf{k}}\right)^2 m\,|V_{cv}|^2 \frac{2\mathcal{E}_\mathbf{k}}{\mathcal{E}_\mathbf{k}^2 - (\hbar\omega)^2)}, \tag{4}$$

$$\bar{\mathbf{E}} = 2d^{-1}\int_0^d \mathbf{E}_{\mathbf{k}\omega}(z)\sin^2\frac{\pi z}{d}\,dz. \tag{5}$$

Here $\mathbf{E}$ is the electric field, z is the coordinate normal to the plane of the film that occupies the band $0 \leqslant z \leqslant d$, V_{cv} is the matrix element of the velocity operator for a transition from the valence band to the conduction band, the functions $(2/d)^{1/2} \times \sin(\pi z/d)$describe the size-quantized transverse motion of the electrons and holes relative to the film, and $\mathcal{E}_k$ is the resonance energy of the excitonic transition. The dimensionless coefficients $\Lambda_{\alpha\beta} \sim 1$, which characterize the polarization properties of this transition, depend on the symmetry of the c and v bands. For simplicity, we assume that the materials of the film and of the substrate are optically isotropic. Thus, $\Lambda_{\alpha\beta}$ will contain only two different quantities: $\Lambda_\perp \equiv \Lambda_{zz}$ and $\Lambda_\parallel$. The matrix elements V_{cv} for typical semiconductors are associated with the masses and the forbidden bands. In the Kane model,[3] $m|V_{cv}|^2 = \mathcal{E}_k$ with an accuracy to a factor close to 1, which can be included in the definition of Λ. Thus,

$$\chi_{\mathbf{k}\omega} = \frac{2}{\epsilon}\left(\frac{e^2}{d}\right)^2 [\mathcal{E}_\mathbf{k}^2 - (\hbar\omega)^2]^{-1}. \tag{6}$$

The most characteristic properties of Eqs. (3)–(6) are the $\chi_{k\omega} \sim d^{-2}$ dependence and the coupling, nonlocal in z, of the induced current with the field.

Following the usual procedure for the optics of thin films,[4,5] we can solve with an accuracy to small terms $\sim [(\omega/c)d]^2$ and $(kd)^2$ the Maxwell equations together with Eqs. (3) and (5) and the equation of continuity in the $0 \leqslant z \leqslant d$ band, and use the resulting solution to match the solutions in the half-spaces 1 ($z<0$) and 2 ($z>d$). Henceforth the presence of a film will be taken into account only by these boundary conditions, which, if we allow for the nonlocalizability, have the form:

$$\begin{aligned}\mathbf{E}_2 - \mathbf{E}_1 &= i\tilde{\epsilon}_\parallel\,\mathbf{n}d\left(\mathbf{k}\frac{\mathbf{E}_1+\mathbf{E}_2}{2}\right) - i\frac{\omega}{c}d\left[\mathbf{n}\frac{\mathbf{H}_1+\mathbf{H}_2}{2}\right]\\ &+ i\frac{ck}{\epsilon\omega}d\,\frac{\epsilon+2\pi\Lambda_\perp\chi}{\epsilon+6\pi\Lambda_\perp\chi}\left(\mathbf{k}\left[\mathbf{n}\frac{\mathbf{H}_1+\mathbf{H}_2}{2}\right]\right),\end{aligned} \tag{7}$$

$$\mathbf{H}_2 - \mathbf{H}_1 = - i\mathbf{n}\, d\left(\mathbf{k}\frac{\mathbf{H}_1 + \mathbf{H}_2}{2}\right) + i\tilde{\epsilon}_{\parallel}\frac{\omega}{2}\, d\left[\mathbf{n}\frac{\mathbf{E}_1 + \mathbf{E}_2}{2}\right]. \tag{8}$$

Here $\mathbf{E}_{1,2}$ and $\mathbf{H}_{1,2}$ are the boundary values of the electronic and magnetic fields in the half-spaces 1 and 2, c is the velocity of light, $\mathbf{n}$ is the unit vector of the normal to the film, and $\tilde{\epsilon}_{\parallel} = \epsilon + 4\pi\Lambda_{\parallel}\chi$. Henceforth, we drop the small terms $\sim(\omega/c)d$ and kd [the second term in Eq. (7) and the first term in Eq. (8)] but retain the terms $\sim\epsilon(\omega/c)d$ and ϵkd. The third term in Eq. (7) should be taken into account only in the narrow frequency range in which $|1 + 6\pi\Lambda_{\perp}\chi\epsilon^{-1}| \ll \epsilon^{-1}$.

The coefficient of reflection of light from the structure under consideration at normal incidence

$$R = \frac{(\sqrt{\epsilon_1} + \sqrt{\epsilon_2})^2 + \left(\dfrac{\omega}{c} d\tilde{\epsilon}_{\parallel}\right)^2}{(\sqrt{\epsilon_1} + \sqrt{\epsilon_2})^2 + \left(\dfrac{\omega}{c} d\tilde{\epsilon}_{\parallel}\right)^2} \tag{9}$$

has a maximum near $\omega = \omega_0 = \hbar^{-1}\mathscr{E}_{k=0}$, which increases as d^{-2} due to Eqs. (4) and (6). For oblique incidence when the electric vector lies in the incident plane a second maximum R appears near the frequency

$$\omega_l = \left[\omega_0^2 + 12\pi\Lambda_{\perp}\left(\frac{e^2}{\epsilon\hbar d}\right)^2 \frac{m|V_{cv}|^2}{\hbar\omega_0}\right]^{1/2}. \tag{10}$$

If the film is sufficiently perfect and homogeneous in thickness, then a fast increase of $\chi_{\mathbf{k}\omega}$ with decreasing d will lead to the appearance of pronounced poles (maxima) and zeros in the quantities $\epsilon_{\parallel}$ and $\epsilon + 6\pi\Lambda_{\perp}\chi$, which are necessary for manifestation of the polariton effects. Of course, we can discuss only the modes such as $E_{1,2}(\mathbf{r}) \sim \exp[i\mathbf{kr} - \kappa_{1,2}|z|]$, which are localized near the film,[5-8] because the film determines only their spectrum. The dispersion of these waves is characterized by the quantities Δ_E and Δ_M

$$\Delta_E = 4\pi\frac{e^2}{\hbar c}\frac{e^2}{\epsilon d}\frac{m|V_{cv}|^2}{\hbar\omega_0}, \qquad \Delta_M = \frac{4\pi}{\epsilon}\left(\frac{e^2}{\epsilon d\hbar\omega_0}\right)^2 m|V_{cv}|^2. \tag{11}$$

I. Transverse electric polaritons. $\mathbf{E}\|[\mathbf{kn}]$. These exists in the frequency range $2\Lambda_{\parallel}\hbar^{-1}\Delta_E/\sqrt{|\epsilon_1 - \epsilon_2|} \gg \omega_0 - \omega > 0$. The dispersion law is

$$\omega = \omega_0 - \frac{2\Lambda_{\parallel}\Delta_E}{\hbar\sqrt{|\epsilon_1 - \epsilon_2|}}\left\{\frac{c^2k^2}{\omega_0^2} - \left[\left(\frac{c^2k^2}{\omega_0^2} - \epsilon_1\right)\left(\frac{c^2k^2}{\omega_0^2} - \epsilon_2\right)\right]^{1/2} - \frac{\epsilon_1 + \epsilon_2}{2}\right\}^{1/2}, \tag{12}$$

$$\kappa_{1,2} = \frac{\omega_0}{2c}\left(\frac{\Lambda_{\parallel}\Delta_E}{\hbar\omega_0 - \hbar\omega} \pm \frac{\epsilon_2 - \epsilon_1}{2}\right). \tag{13}$$

II. Transverse magnetic polaritons $\mathbf{H}\|[\mathbf{kn}]$. There are two branches of such waves near the frequencies ω_0 and ω_l in frequency intervals of width $\sim\hbar^{-1}\Delta_M$. In the simplest case, $\epsilon_2 = \epsilon_1$, their dispersion laws are

$$\hbar\omega = \hbar\omega_0 + \Lambda_{\parallel}\Delta_M \frac{d\left(k^2 - \epsilon_1\frac{\omega_0^2}{c^2}\right)^{1/2}}{2\frac{\epsilon_1}{\epsilon} + d\left(k^2 - \epsilon_1\frac{\omega_0^2}{c^2}\right)^{1/2}}, \tag{14}$$

$$\hbar\omega = \hbar\omega_l + \frac{\epsilon_1}{2}\Lambda_{\perp}\Delta_M \frac{k^2 d}{\left[k^2 - \epsilon_1\frac{\omega_l^2}{c^2}\right]^{1/2}}. \tag{15}$$

If $\Lambda_{\perp} = 0$, then the branch (15), as well as the reflection peak near ω_l, will be missing.

[1]L.V. Keldysh, Pis'ma Zh. Eksp. Teor. Fiz. **29**, 716 (1979) [JETP Lett. **29**, 658 (1979)].
[2]R.J. Elliot, Phys. Rev. **108**, 1384 (1957); R.S. Knox, Teoriya eksitonov (Theory of Excitons) Mir, M., 1966 (Academic Press, N.Y., 1963).
[3]E.O. Kane, J. Phys. Chem. Solids **1**, 249 (1957).
[4]M. Born and E. Wolf, Osnovy optiki (Principles of Optics), Nauka, M., 1973 (Pergamon Press, 1959).
[5]V.M. Agranovich and V.L. Ginzburg, Kristallooptika s uchetom prostranstvennoĭ disperesii i teoriya eksitonov (Spatial Dispersion in Crystal Optics and the Theory of Excitons), Nauka, M., 1979.
[6]R. Fuchs and K.H. Kliever, Phys. Rev. **A140**, 2076 (1965); **144**, 495 (1966).
[7]R. Ruppin and R. Englman, Rept. Progr. Phys. **33**, 149 (1970).
[8]V.V. Bryksin, D.N. Mirlin, and Yu.A. Firsov, Usp. Fiz. Nauk **113**, 29 (1974) [Sov. Phys. Usp. **17**, 305 (1974)].

Coulomb interaction in thin semiconductor and semimetal films

L. V. Keldysh
P. N. Lebedev Physics Institute, USSR Academy of Sciences

(Submitted 28 April 1979)
Pis'ma Zh. Eksp. Teor. Fiz. **29**, No. 11, 716–719 (5 June 1979)

It is shown that the Coulomb interaction in thin films increases strongly with decreasing film thickness, if the film dielectric constant is much larger than that of the substrate. The variation of exciton binding energy and of shallow impurity levels with film thickness and the ratio of the dielectric constants is determined.

PACS numbers: 73.60.Fw, 71.35. + z, 71.55.Dp, 77.20. + y

Large values of the dielectric constant $\epsilon \sim 10$–100 are characteristic of semiconductors and semimetals. Therefore, the Coulomb interaction of free electrons and holes in these substrates is strongly attenuated, and hydrogen-like binding states such as shallow impurity levels and Wannier-Mott excitons have small binding energies E_0 and macroscopically-large effective radii a_0

$$E_0 = \frac{e^4 m}{2\epsilon^2 \hbar^2} \lesssim 10^{-2}\ \text{eV}, \qquad a_0 = \frac{\epsilon \hbar^2}{m e^2} \gtrsim 10^{-6}\ \text{cm}. \tag{1}$$

Here e and $\hbar$ are the electron charge and Planck's constant, and m is the effective mass. The very fact of the existence of these levels is manifested only at sufficiently low temperatures. However, in thin films the interaction between charges increases with decreasing thickness d, since for distances between charges $\gtrsim d$ the field produced by these charges in the medium surrounding the film begins to play a perceptible role, and if the dielectric constant for this medium is much less than ϵ, the interaction turns out to be significantly larger than in a homogeneous medium with film dielectric constant ϵ.

Assume the film occupies a region of space $-d/2 \leqslant z \leqslant d/2$. The half-space $z < -d/2$ (substrate layer) is filled with a medium having a dielectric constant ϵ_1, and the half-space $z > d/2$ with a medium having a dielectric constant ϵ_2. The energy of the interaction between the charges e and e' located at the points (ρ, z) and $(0, z')$ $(z \geqslant z'$; $\rho = (x,y)$ (are the coordinates in the film plane) is equal to

$$V(\vec{\rho}, z, z') = \frac{4\pi e e'}{\epsilon} \int \frac{d^2k}{(2\pi)^2} e^{i\mathbf{k}\vec{\rho}} \frac{\mathrm{ch}\left[|\mathbf{k}|\left(\frac{d}{2} - z\right) + \eta_2\right] \mathrm{ch}\left[|\mathbf{k}|\left(\frac{d}{2} + z'\right) + \eta_1\right]}{|\mathbf{k}|\, \mathrm{sh}[|\mathbf{k}|\, d + \eta_1 + \eta_2]}$$

$$\eta_{1,2} = \frac{1}{2} \ln \frac{\epsilon + \epsilon_{1,2}}{\epsilon - \epsilon_{1,2}}.$$

We will consider only the most interesting case $\epsilon_{12} \ll \epsilon$ and $d \ll a_0$. For $\rho \gg d$ there exists in the interval **k** such that $|\mathbf{k}|d \ll 1$. Moreover, V is independent of z and z', and can be reduced as follows:

$$V(\vec{\rho}) = \frac{2ee'}{\epsilon d} \int_0^\infty \frac{J_0(t)\,dt}{t + \frac{\epsilon_1 + \epsilon_2}{\epsilon} \frac{\rho}{d}} = \frac{\pi e e'}{\epsilon d} \left[\mathcal{H}_0\left(\frac{\epsilon_1 + \epsilon_2}{\epsilon} \frac{\rho}{d}\right) - N_0\left(\frac{\epsilon_1 + \epsilon_2}{\epsilon} \frac{\rho}{d}\right) \right], \tag{2}$$

where $N_0(x)$ and $\mathcal{H}_0(x)$ are the Neumann and Struve functions. In the interval $d \ll \rho \ll \epsilon d/(\epsilon_1 + \epsilon_2)$

$$V(\vec{\rho}) \approx \frac{2ee'}{\epsilon d} \left[\ln\left(\frac{2\epsilon}{\epsilon_1 + \epsilon_2} \frac{d}{\rho}\right) - C \right], \tag{3}$$

$C \simeq 0.577$ is the Euler constant, and for $\rho \gg [\epsilon/(\epsilon_1 + \epsilon_2)]\, d$

$$V(\vec{\rho}) \approx \frac{2ee'}{(\epsilon_1 + \epsilon_2)\rho}. \tag{4}$$

When $d \ll a_0$ the distance between the dimensionally quantized energy levels h^2/md^2 is much larger than the interaction energy of Eqs. (3)–(4). Therefore, the transverse motion of charges relative to the film does not change when interaction is taken into account, and the problem concerning their relative motion becomes two-dimensional

$$-\frac{\hbar^2}{2m} \vec{\Delta}_{\vec{\rho}} \psi_n(\vec{\rho}) + V(\vec{\rho})\psi_n(\vec{\rho}) = E_n \psi_n(\vec{\rho}). \tag{5}$$

For films satisfying the condition $a_0 \gg d \gg (\epsilon_1 + \epsilon_2/2\epsilon)^2 a_0$, the effective radii of the fundamental and first excited binding states fall within a range of distances where $V(\boldsymbol{\rho})$ has the form of Eq. (3), and by substituting $\boldsymbol{\rho} = a\boldsymbol{\xi}$ we transform Eq. (5) as follows:

$$\Delta_{\vec{\xi}} \psi_n(\vec{\xi}) - \ln|\vec{\xi}|\, \psi_n(\vec{\xi}) = \gamma_n \psi_n(\vec{\xi}), \tag{6}$$

where

$$a = \frac{1}{2} \sqrt{\frac{\epsilon \hbar^2}{me^2} d} = \frac{1}{2} \sqrt{a_0 d}, \tag{7}$$

$$E_n = -\frac{e^2}{\epsilon d} \left\{ \ln\left[\left(\frac{2\epsilon}{\epsilon_1 + \epsilon_2}\right)^2 \frac{d}{a_0}\right] - 2C - 2\gamma_n \right\}. \tag{8}$$

Equations (7) and (8) determine the effective radii and binding energies of the hydrogenic states in a thin film. Since Eq. (6) contains no parameters, the terms $\gamma_n \sim 1$ are small in comparison with the logarithm which, thus, determines the binding energy of both the ground and the lowest excited states. The spacing between these levels is substantially less than their binding energy.

If the parameters of the film and substrate satisfy the condition

$$\frac{(\epsilon_1 + \epsilon_2)^2}{\epsilon} \times \frac{m_o}{m} >> 10,$$

where $m_0 \simeq 9 \times 10^{-28}$ g, then the condition

$$d << \left(\frac{\epsilon_1 + \epsilon_2}{\epsilon}\right)^2$$

a_0 is consistent with the requirement that the film be macroscopic, i.e., that it contains a large number of atomic layers. For films like this the radius of the bound states lies in the region where $V(\rho)$ has the form of Eq. (4), and the binding energy ceases to increase for a further decrease in d as it acquires a form which is characteristic of the two-dimensional Coulomb problem[1,2] with a dielectric constant $(\epsilon_1 + \epsilon_2)/2$[3]

$$E_n = \left(\frac{2\epsilon}{\epsilon_1 + \epsilon_2}\right)^2 \frac{4E_o}{(2n+1)^2} . \tag{9}$$

Equation (9) has, however, hardly any region of applicability for $n = 0$, since the small effective masses required for this are usually associated with the narrow forbidden bands E_g in the electron spectrum and with its strong nonparabolicity[4]

$$-E_p = E_{go}\left[1 + \frac{p^2}{mE_{go}}\right]^{1/2} .$$

Because of the effect of dimensional quantization, E_g increases in the films along with the effective mass value, so that the condition for the applicability of Eq. (3) for $V(\rho)$ is automatically satisfied. Taking nonparabolicity into account

$$a = \frac{1}{2}\sqrt{a_o d}\left[1 + \frac{1}{mE_{go}}\left(\frac{\pi\hbar}{d}\right)^2\right]^{-1/4} = \frac{1}{2}\sqrt{\epsilon d}\left[\left(\frac{me^2 d}{\hbar^2}\right)^2 + \pi^2 \frac{e^4 m}{\hbar^2 E_{go}}\right], \tag{10}$$

$$E_o(d) = -\frac{e^2}{\epsilon d}\ln\left[\left(\frac{2\epsilon}{\epsilon_1 + \epsilon_2}\right)^2 \sqrt{\left(\frac{d}{a_o}\right)^2 + 2\pi^2 \frac{E_o}{E_{go}}}\right]. \tag{11}$$

The ratio of m and E_{g0} is such[4] that usually $e^4 m/\hbar^2 E_{g0} \sim 1$, so that Eqs. (10) and (11) are valid for all $d \ll a_0$, providing $4\epsilon/(\epsilon_1 + \epsilon_2)^2 \gg 1$.

[1]R.J. Elliot, Polarons and Excitons, Oliver and Boyd, London (1963); H.I. Ralph, Solid State Com. 3, 303 (1965).

[2]V.L. Ginzburg and V.V. Kelle, Pis'ma Zh. Eksp. Teor. Fiz. **17**, 428 (1973) [JETP Lett. **17**, 306 (1973)].
[3]Yu.E. Lozovik and V.I. Yudson, Phys. Lett. **56A**, 393 (1976).
[4]E.O. Kane, J. Phys. Chem. Solids **1**, 249 (1957).

Restructuring of polariton and phonon spectra of a semiconductor in the presence of a strong electromagnetic wave

A. L. Ivanov and L. V. Keldysh

P. N. Lebedev Physics Institute, USSR Academy of Sciences
(Submitted 30 July 1982)
Zh. Eksp. Teor. Fiz. **84**, 404–421 (January 1982)

We investigate the propagation of a strong electromagnetic wave in a direct-band semiconductor following resonant excitation of excitons. It is assumed that the given polariton wave attenuates because of scattering by longitudinal acoustic phonons, and the case is analyzed when the scattering due to the absorption of the phonon wave by the excitons greatly exceeds the scattering at which the wave excitons emit phonons. Using a diagram technique for nonequilibrium processes, a system of equations is obtained for the description of the behavior of the initial polariton wave, of the scattered electrons, and of the scattering phonons. The polariton- and phonon-spectrum restructuring accompanied by an abrupt decrease of the electromagnetic-wave absorption coefficient is considered within the framework of the model indicated.

PACS numbers: 71.35. + z, 71.36. + c

The interaction of electromagnetic radiation with excitons in direct-band semiconductors has been attracting considerable interest both theoretically and experimentally. As a rule, the problems investigated are connected with the luminescence and kinetics of exciton systems[1–6] and with the description of the propagation of electromagnetic waves in resonant excitation of excitons.[7–12] For a correct treatment of these problems, the polariton concept is introduced[13] because in direct-band semi-conductors at low temperatures there is realized a strong exciton-photon coupling that leads to a mixing of the exciton and photon states and to a restructuring of the photon and exciton spectra.

Among the problems of luminescence and kinetics of exciton systems is included the investigation of thermalization of strongly excited exciton systems, and in particular the consideration of the possibility of Bose condensation of excitons both on the upper polariton branch[5,6] and on the lower one.[1–4] It is appropriate to note that one of the principal difficulties of experimentally observing Bose condensation of excitons in the course of their thermalization is as a rule the long lifetime of the excitons in the semiconductor.

The problem of propagation of electromagnetic radiation in resonant excitation of excitons, a problem connected with the calculation of the absorption coefficient of the electromagnetic wave,[7,8] with the experimental confirmation of the polariton character of the electromagnetic waves in the semiconductors,[9,10,12] and with others, has also a bearing on the manifestation of Bose condensation of excitons. In fact, an electromagnetic wave $\mathbf{k}_0$ produced by an external source and propagating in a crystal has a finite amplitude, and thus the mode $\mathbf{k}_0$ is macroscopically filled in the usual sense:

$$\langle \hat{a}_{\mathbf{k}_0}{}^{+}\hat{a}_{\mathbf{k}_0}\rangle \propto \langle \hat{b}_{\mathbf{k}_0}{}^{+}\hat{b}_{\mathbf{k}_0}\rangle \propto V,$$

where $\hat{a}^{+}_{\mathbf{k}_0}$ and $\hat{b}^{+}_{\mathbf{k}_0}$ are the photon and exciton creation operators in the mode $\mathbf{k}_0$ and V is the volume of the crystal. In this case there can take place a relatively short-time thermalization with possible formation of a Bose condensate of non-equilibrium-excited excitons, as is proposed in the first group of problems, and a much faster scattering of the excitons from the initial wave and establishment of quasiequilibrium in such an exciton system with a condensate in the mode $\mathbf{k}_0$.

The purpose of the present article is a consistent description of the propagation of a macroscopically filled polariton wave $\mathbf{k}_0$ with allowance for the specific damping mechanism. Just as in Refs. 7 and 8, we shall consider the case when the cause of the damping of the wave is scattering of the exciton component by a longitudinal acoustic phonons. In addition, we propose that the scattering as a result of absorption of an acoustic-phonon wave by the exciton is much larger than the scattering due to the emission of a phonon wave by the exciton. This assumption makes it possible, on the one hand, to investigate a number of singularities in the propagation of a high-power polariton wave with the indicated damping mechanism, and on the other hand apply the results to the phenomenon of ordinary equilibrium Bose condensation of an ideal Bose gas in a phonon thermostat into the mode $\mathbf{k}_0 = 0$. We shall take into account hereafter only scattering connected with absorption of acoustic phonons, and defer the analysis of the indicated assumptions and the extent to which the model is realistic to the end of the article.

In the general case the problem consists of a joint examination of a macroscopically filled wave $\mathbf{k}_0$, scattered excitons $\mathbf{k}$, and scattering acoustic phonons $\mathbf{k}-\mathbf{k}_0$. With the aid of the diagram technique for nonequilibrium processes[14] used in Ref. 15 for the case when an explicit account is taken of the finite lifetime of interacting quasiparticles, we obtain kinetic equations for the distribution functions of the scattered excitons $N_k(\mathbf{r}t)$, for the scattering phonons $n_{\mathbf{k}-\mathbf{k}_0}(\mathbf{r}t)$, as well as the restructuring of the spectra of these quasiparticles, and describe the propagation of the initial wave. The latter will depend substantially on the state of the wave $\mathbf{k}_0$. Namely, if the wave $\mathbf{k}_0$ is in a Glauber state, i.e., coherent, it is described by the Maxwell and Schrödinger field equations with respect to the mean values of the corresponding field operators, while if the wave $\mathbf{k}_0$ is completely incoherent, i.e., a noise, it is described by a kinetic equation with respect to

 0038-5646/83/010234-11$04.00

$N_0(\mathbf{r}t)$—the density of the excitons $\mathbf{k}_0$ produced by the wave. The density $N_0(\mathbf{r}t)$ is connected with the intensity of the polariton wave $I_0(\mathbf{r}t)$ by relation (78) below. In the general case it is necessary to consider simultaneously in the indicated manner the coherent and noise components of the wave.

An important parameter of the problem is the quantity $\Gamma(\mathbf{k},\mathbf{k}_0)$—the characteristic energy uncertainty in the elementary act of scattering of the exciton by the initial wave. We shall obtain the following expression for this quantity:

$$\Gamma(\mathbf{k},\mathbf{k}_0)\approx\{[\gamma(\mathbf{k})+\gamma_{\rm ph}^{A}(\mathbf{k}-\mathbf{k}_0)]^2+4N_0m_{\mathbf{k}-\mathbf{k}_0}\}^{1/2}, \quad (1)$$

where $\gamma(\mathbf{k})=1/\tau$ is the reciprocal lifetime of the scattered exciton $\mathbf{k}$ and is connected with the possibility of further scattering by equilibrium acoustic phonons, $\gamma_{\rm ph}^{A}(\mathbf{k}-\mathbf{k}_0)=1/\tau_{\rm ph}^{A}$ is the reciprocal lifetime of the scattering phonon $\mathbf{k}-\mathbf{k}_0$ because of the lattice anharmonicity of the crystal, and $m_{\mathbf{k}-\mathbf{k}_0}$ is defined by (17) below. The quantity $\Gamma(\mathbf{k},\mathbf{k}_0)$ determines the number of polariton modes $\mathbf{k}$ in which scattering of excitons by the initial wave takes place, and also the number of the phonon-subsystem $\mathbf{k}-\mathbf{k}_0$ modes whose phonons participate most effectively in the scattering process. Depending on the intensity of the polariton wave, we can separate different cases.

Low intensities:

$$N_0<\gamma(k_0)\gamma_{\rm ph}^{A}(k_0)/m_{k_0}. \quad (2)$$

These low intensities correspond to the linear theory[8]: the absorption coefficient of the wave $\sigma_{\rm sp}$ is proportional to the semiconductor temperature T and is independent of the wave intensity.

Medium intensities:

$$\gamma(k_0)\gamma_{\rm ph}^{A}(k_0)/m_{k_0}<N_0<[\gamma(k_0)+\gamma_{\rm ph}^{A}(k_0)]^2/4m_{k_0}. \quad (3)$$

This case corresponds to nonliner-wave propagation, and the absorption coefficient $\sigma_{\rm sp}\propto 1/N_0$ and can reach values that are smaller by several orders than for low intensities. At low temperatures, the inequality $\gamma_{\rm ph}^{A}(k_0)\ll\gamma(k_0)$ is satisfied, and in this case we have phonon nonlinearity, wherein the phonon subsystem is greatly depleted in the course of scattering of the initial wave: the act of exciton scattering into the mode $\mathbf{k}$ is annihilation of the phonon $\mathbf{k}-\mathbf{k}_0$. Such a strong decrease of the occupation numbers of the modes of the phonon subsystem is naturally accompanied by a decrease of the wave absorption coefficient. At high temperatures $\gamma(k_0)\ll\gamma_{\rm ph}^{A}(k_0)$, and the nonlinear decrease of the absorption can be attributed to nonequilibrium Bose condensation. This phenomenon consists in the possibility of filling the polariton modes $\mathbf{k}$, in which the excitons are scattered up to values of the equilibrium phonon occupation numbers of the modes $\mathbf{k}-\mathbf{k}_0$. The scattering of the initial wave is then effectively suppressed and the absorption is determined exclusively by the further scattering of the excitons $\mathbf{k}$.

High intensities:

$$[\gamma(k_0)+\gamma_{\rm ph}^{A}(k_0)]^2/4m_{k_0}<N_0. \quad (4)$$

At such high intensities, a restructuring of the polariton and phonon spectra takes place (see Fig. 1). It consists of unification and splitting of the indicated terms, in analogy with the formation of the polariton dispersion curve from the exciton and photon spectra. The restructured spectrum corresponds to a new excitation consisting of polariton and phonon components, which we shall call phonoriton for short. The wave absorption coefficient is nonliner, as before, and is proportional to $N_0^{-1/2}$. The inequality (4) is the condition that the splitting of the spectrum $2(m_{k0}N_0)^{1/2}$ exceed the reciprocal lifetime of the obtained excitation $[\gamma(k_0)+\gamma_{\rm ph}^{A}(k_0)]$, i.e., the indicated restructuring of the spectrum can be observed. On the other hand, this condition corresponds to the case when the quantity $\Gamma(\mathbf{k},\mathbf{k}_0)$ is determined principally by the contribution of the last term under the square root in expression (1), i.e., the action of a macroscopically filled mode $\mathbf{k}_0$. This means that the scattered excitons returns with high probability to the initial mode $\mathbf{k}_0$, this being a consequence of the macroscopic filling of the mode $\mathbf{k}_0$. Such a possibility of exciton scattering followed by return to the wave, accompanied by the absorption of a phonon $\mathbf{k}-\mathbf{k}_0$ with its subsequent emission, corresponds precisely to the restructuring of the polariton and phonon spectra. We shall show below that for the restructured spectrum the characteristic uncertainty of the energy in the elementary scattering act will in fact be independent of the intensity of the initial wave. Moreover, a consistent analysis of the problem presupposes the use of the restructured spectrum at any wave intensity, in particular at intensities defined by inequalities (2) and (3). We note that condition (4) is satisfied at $T=4.2$ K for $N_0\sim10^{15}-10^{17}$ cm^{-3}, depending on the type of semiconductor and on its quality.

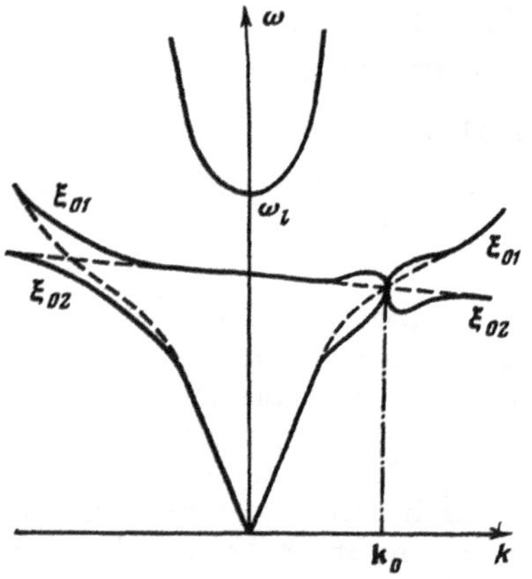

FIG. 1.

We begin the analysis of the formulated problem with the case when the initial wave $\mathbf{k}_0$ is fully incoherent. For a consistent solution, we need introduce the retarded $(G^{(R)},D^R)$, advanced (G^A,D^A), and statistical (G^+,D^+) Green's functions[14] of the excitons and phonons, respectively. If there is no interaction between the quasiparticles, these functions take the form

$$G_0^{R,A}(k)=\frac{1}{\omega-\omega_{\mathbf{k}}\pm i\delta},\qquad G_0^+(k)=-2\pi iN_{\mathbf{k}}\delta(\omega-\omega_{\mathbf{k}}); \quad (5)$$

$$D_0^{R,A}(k)=\left[\frac{1}{\omega-\omega_k^{ph}\pm i\delta}-\frac{1}{\omega+\omega_k^{ph}\pm i\delta}\right],$$

$$D_0^+(k)=-2\pi i[(1+n_k)\delta(\omega+\omega_k^{ph})+n_k\delta(\omega-\omega_k^{ph})];\quad k=(\omega,\mathbf{k}). \tag{6}$$

Here $\hbar\omega_k^{ph}=\hbar uk$ (u is the speed of sound in the crystal) is the energy of the acoustic phonon **k** and ω_k is the energy of the exciton **k** and is determined by the polariton dispersion curve. Since the mode k_0 is macroscopically filled, the statistical Green's function (5) takes at $\mathbf{k}=\mathbf{k}_0$ form

$$G_0^+(k_0)=-2\pi i N_0 V\delta(\varepsilon-\omega_{k_0}),$$
$$N_{k_0}=(2\pi)^3N_0\delta(\mathbf{k}-\mathbf{k}_0),\qquad k_0=(\varepsilon,\mathbf{k}_0). \tag{7}$$

We note that the exciton Green's functions represent in fact the exciton part of the Green's function of the polariton, and should therefore be multiplied by a factor [see (79) below] that depends on **k**. However, as will be noted below, this factor, in the interesting region of values of **k** near the polariton splitting of the spectrum, is equal to unity with high accuracy, therefore the polariton character of the excitons can be taken into account only by assuming $\omega_\mathbf{k}$ to be the polariton energy. In addition, to simplify the description we shall consider the case when the propagating wave $\mathbf{k}_0$ is linearly polarized, and the dipole moment of the exciton transition **d** has a preferred direction. These assumptions allow us to disregard the tensor character of the Green's functions of the excitons and photons.

In this case, when quasiparticle interaction takes place, the Green's functions are determined by the corresponding Dyson equations and we can make the following assumption concerning their form[15]:

$$G^R(k)=(\omega-\tilde\omega_k+i\tilde\gamma)^{-1},\quad G^+(k)=-2\pi i N_{k\omega}\tilde\delta[\tilde\gamma|\omega-\tilde\omega_k]. \tag{8}$$

Here

$$\tilde\delta[\tilde\gamma|\omega-\tilde\omega_k]=\tilde\gamma(k)/\pi[(\omega-\tilde\omega_k)^2+\tilde\gamma^2(k)] \tag{9}$$

is the energy difference smeared by a δ function, and the generalized occupation numbers $N_{\mathbf{k}\omega}$ are connected with the usual numbers $N_\mathbf{k}$ by the relation

$$N_k=\int N_{k\omega}\tilde\delta[\tilde\gamma|\omega-\tilde\omega_k]\,d\omega. \tag{10}$$

The smearing $\tilde\gamma(k)$ of the δ function and the changed frequencies $\tilde\omega_\mathbf{k}$ are connected with the self-energy part $\Sigma_f^R(\omega,\mathbf{k})$ in the following manner:

$$\tilde\omega_k=\omega_k+\mathrm{Re}\,\Sigma_f^R(\omega_k,\mathbf{k}), \tag{11}$$

$$\tilde\gamma(k)=-\mathrm{Im}\,\Sigma_f^R(\omega,\mathbf{k}). \tag{12}$$

Similar relations hold for the phonon Green's functions.

To describe the scattered excitons, the scattering phonons, and the excitons of the initial wave we shall separate explicitly only the first stage of the scattering, which consists of absorption of one acoustic phonon, by the exciton of the wave or of emission of an acoustic phonon by the scattered exciton with transition of this exciton into the mode k_0. In addition, as already noted, the scattered excitons and scattering phonons have characteristic lifetimes τ and τ_{ph}^A respectively. The self-energy parts in the Dyson equations investigated below will be defined by an integral convolution of the corresponding complete Green's functions, i.e., we assume the vertex part to be equal to unity. In this approximation there are no anomalous Green's functions, for in this case they enter into the Dyson equations together with a factor $m_{\mathbf{k}=0}=0$ for the acoustic (and also optical) phonons [for a definition of $m_\mathbf{k}$ see (17)].

We consider first the dynamic part of the problem, namely questions connected with the shift or restructuring of the quasiparticle spectra and with the change of the characteristic lifetimes, i.e., of the widths of the energy levels, when the interaction is turned on. As indicated, this information is provided by the Dyson equations for the advanced or retarded Green's functions of the corresponding quasiparticles. For scattered excitons we have

$$G^R(k)=G_0^R(k)+G_0^R(k)\Sigma_f^R(k)G^R(k), \tag{13}$$

$$G^R(k)=[\omega-\omega_k+i\gamma(\mathbf{k})-\Sigma^R(k)]^{-1}. \tag{14}$$

The self-energy part $\Sigma_f^R(k)$ consists of two terms, the first of which $\Sigma^R(k)$ is connected with the first stage of the scattering, and the second term $\Sigma_\tau^R(k)$ is the self-energy part connected with the probable further scattering of the exciton **k** not into the mode $\mathbf{k}_0$. These self-energy parts are determined by the relations

$$\Sigma_\tau^R(k)=-i/\tau=-i\gamma(\mathbf{k}), \tag{15}$$

$$\Sigma^R(k)=i\int[D^R(k-q)G^R(q)+D^+(k-q)G^R(q)+D^R(k-q)G^+(q)]m_{\mathbf{k}-\mathbf{k}_0}\frac{d^4q}{(2\pi)^4}=\frac{N_0m_{\mathbf{k}-\mathbf{k}_0}}{\omega-\tilde\omega^{ph}_{\mathbf{k}-\mathbf{k}_0}-\tilde\omega_{\mathbf{k}_0}+i(\tilde\gamma_{N_0}+\tilde\gamma_{ph})}. \tag{16}$$

The factor $m_\mathbf{k}$ is connected here with the value of the matrix element $M_\mathbf{k}$ of exciton scattering by a phonon in the following manner:

$$m_k=\frac{2\pi}{\hbar}V|M_k|^2. \tag{17}$$

From the relations (12), (15), and (16) we can obtain an expression for the width of the energy level of the scattered exciton:

$$\tilde\gamma(k)\equiv\tilde\gamma(\omega,\mathbf{k})=\gamma(\mathbf{k})+\pi N_0m_{\mathbf{k}-\mathbf{k}_0}\tilde\delta[\tilde\gamma_{ph}+\tilde\gamma_{N_0}|\omega-\tilde\omega^{ph}_{\mathbf{k}-\mathbf{k}_0}-\tilde\omega_{\mathbf{k}_0}]. \tag{18}$$

Similarly, for the scattering phonons $\mathbf{k}-\mathbf{k}_0$ we have

$$D^R(k-k_0)=[(\omega-\varepsilon)-\omega^{ph}_{\mathbf{k}-\mathbf{k}_0}+i\gamma_{ph}^A(\mathbf{k}-\mathbf{k}_0)-\Sigma^R(k-k_0)]^{-1}, \tag{19}$$

$$\Sigma^R_{\tau_{ph}^A}(k-k_0)=-i\frac{1}{\tau_{ph}^A}=-i\gamma_{ph}^A(\mathbf{k}-\mathbf{k}_0),$$
$$\Sigma^R(k-k_0)=\frac{N_0m_{\mathbf{k}-\mathbf{k}_0}}{(\omega-\varepsilon)-(\tilde\omega_\mathbf{k}-\tilde\omega_{\mathbf{k}_0})+i(\tilde\gamma_{N_0}+\tilde\gamma)} \tag{20}$$

$$\tilde\gamma_{ph}(k-k_0)=\gamma_{ph}^A(\mathbf{k}-\mathbf{k}_0)+\pi N_0m_{\mathbf{k}-\mathbf{k}_0}\tilde\delta[\tilde\gamma_{N_0}+\tilde\gamma|(\omega-\varepsilon)-(\tilde\omega_\mathbf{k}-\tilde\omega_{\mathbf{k}_0})]. \tag{21}$$

The last relation connects the width of the energy level of the scattering phonon $\mathbf{k}-\mathbf{k}_0$ with the corresponding values for the scattered excitons and the excitons of the initial mode $\mathbf{k}_0$. An analysis of the Dyson equation for the retarded Green's function of the excitons $G^{(R)}(k_0)$ of a macroscopically filled mode yields the remaining relations needed to obtain the closed system of equations (18), (21), and (23) with respect to the widths $\tilde{\gamma}$, $\tilde{\gamma}_{N_0}$ and $\tilde{\gamma}_{\mathrm{ph}}$:

$$\Sigma^R(k_0)=\int m_{\mathbf{k}-\mathbf{k}_0}[n_{\mathbf{k}-\mathbf{k}_0,\omega-\varepsilon}-N_{\mathbf{k},\omega}] \times\frac{1}{\tilde{\omega}_{\mathbf{k}}-\varepsilon-\tilde{\omega}^{\mathrm{ph}}_{\mathbf{k}-\mathbf{k}_0}+i(\tilde{\gamma}+\tilde{\gamma}_{\mathrm{ph}})}\frac{d^3k}{(2\pi)^3}, \tag{22}$$

$$\tilde{\gamma}_{N_0}(k_0)=\pi\int m_{\mathbf{k}-\mathbf{k}_0}[n_{\mathbf{k}-\mathbf{k}_0,\omega-\varepsilon}-N_{\mathbf{k},\omega}] \times\tilde{\delta}[\tilde{\gamma}+\tilde{\gamma}_{\mathrm{ph}}|\tilde{\omega}_{\mathbf{k}}-\varepsilon-\tilde{\omega}^{\mathrm{ph}}_{\mathbf{k}-\mathbf{k}_0}]\frac{d^3k}{(2\pi)^3}. \tag{23}$$

A consistent treatment of the problem within the framework of the assumption (8) concerning the form of the Green's function shows that the characteristic energy uncertainty in the act of the scattering of the excitons of the initial wave $\mathbf{k}_0$ is determined by the sum of the widths of the levels of all the quasiparticles that take part in the scattering:

$$\Gamma(\mathbf{k},\mathbf{k}_0)=\tilde{\gamma}_{N_0}(\tilde{\omega}_{\mathbf{k}_0},\mathbf{k}_0)+\tilde{\gamma}(\tilde{\omega}_{\mathbf{k}},\mathbf{k})+\tilde{\gamma}_{\mathrm{ph}}(\tilde{\omega}^{\mathrm{ph}}_{\mathbf{k}-\mathbf{k}_0},\mathbf{k}-\mathbf{k}_0), \tag{24}$$

After solving the indicated system of equations relative to the widths of the energy levels and substituting them in (24), we obtain for $\Gamma(\mathbf{k},\mathbf{k}_0)$ the approximate relation (1). As indicated earlier, the physically unjustified strong dependence of the level widths, of the frequency shifts, and also of the value of $\Gamma(\mathbf{k},\mathbf{k}_0)$ on the intensity of the wave $\mathbf{k}_0$ is eliminated in practice by introducing a renormalization of the phonon and polariton spectra. Namely, taking into consideration the relations (16) and (20) for the self-energy parts, the expressions for the retarded Green's functions of the scattered excitons and of the scattering phonons can be transformed into

$$G^R(k)=\frac{\omega-\omega^{\mathrm{ph}}_{\mathbf{k}-\mathbf{k}_0}-\tilde{\omega}_{\mathbf{k}_0}+i\gamma^A_{\mathrm{ph}}}{(\omega-\omega_{\mathbf{k}}+i\gamma)(\omega-\omega^{\mathrm{ph}}_{\mathbf{k}-\mathbf{k}_0}-\tilde{\omega}_{\mathbf{k}_0}+i\gamma^A_{\mathrm{ph}})-N_0m_{\mathbf{k}-\mathbf{k}_0}}, \tag{25}$$

$$D^R(k-k_0)=\frac{(\omega-\varepsilon)-(\omega_{\mathbf{k}}-\tilde{\omega}_{\mathbf{k}_0})+i\gamma}{[(\omega-\varepsilon)-\omega^{\mathrm{ph}}_{\mathbf{k}-\mathbf{k}_0}+i\gamma^A_{\mathrm{ph}}][(\omega-\varepsilon)-(\omega_{\mathbf{k}}-\tilde{\omega}_{\mathbf{k}_0})+i\gamma]-N_0m_{\mathbf{k}-\mathbf{k}_0}}. \tag{26}$$

It is known that the poles of the Green's functions yield information on the spectrum and damping of the considered excitation, therefore the dispersion curves of the restructured spectrum are determined by the equation

$$(\omega-\omega_{\mathbf{k}}+i\gamma)(\omega-\omega^{\mathrm{ph}}_{\mathbf{k}-\mathbf{k}_0}-\tilde{\omega}_{\mathbf{k}_0}+i\gamma^A_{\mathrm{ph}})-N_0m_{\mathbf{k}-\mathbf{k}_0}=0, \tag{27}$$

which have the following roots:

$$\xi_1=\xi_{01}-i\Gamma_1,$$

$$\xi_{01}=\tfrac{1}{2}\{(\omega_{\mathbf{k}}+\omega^{\mathrm{ph}}_{\mathbf{k}-\mathbf{k}_0}+\tilde{\omega}_{\mathbf{k}_0})+[(\omega_{\mathbf{k}}-\omega^{\mathrm{ph}}_{\mathbf{k}-\mathbf{k}_0}-\tilde{\omega}_{\mathbf{k}_0})^2+4N_0m_{\mathbf{k}-\mathbf{k}_0}]^{1/2}\},$$

$$2\Gamma_1=\gamma+\gamma^A_{\mathrm{ph}}+(\gamma-\gamma^A_{\mathrm{ph}})\frac{\omega_{\mathbf{k}}-\omega^{\mathrm{ph}}_{\mathbf{k}-\mathbf{k}_0}-\tilde{\omega}_{\mathbf{k}_0}}{[(\omega_{\mathbf{k}}-\omega^{\mathrm{ph}}_{\mathbf{k}-\mathbf{k}_0}-\tilde{\omega}_{\mathbf{k}_0})^2+4N_0m_{\mathbf{k}-\mathbf{k}_0}]^{1/2}};$$

$$\xi_2=\xi_{02}-i\Gamma_2, \tag{28}$$

$$\xi_{02}=\tfrac{1}{2}\{(\omega_{\mathbf{k}}+\omega^{\mathrm{ph}}_{\mathbf{k}-\mathbf{k}_0}+\tilde{\omega}_{\mathbf{k}_0})-[(\omega_{\mathbf{k}}-\omega^{\mathrm{ph}}_{\mathbf{k}-\mathbf{k}_0}-\tilde{\omega}_{\mathbf{k}_0})^2+4N_0m_{\mathbf{k}-\mathbf{k}_0}]^{1/2}\},$$

$$2\Gamma_2=\gamma+\gamma^A_{\mathrm{ph}}-(\gamma-\gamma^A_{\mathrm{ph}})\frac{\omega_{\mathbf{k}}-\omega^{\mathrm{ph}}_{\mathbf{k}-\mathbf{k}_0}-\tilde{\omega}_{\mathbf{k}_0}}{[(\omega_{\mathbf{k}}-\omega^{\mathrm{ph}}_{\mathbf{k}-\mathbf{k}_0}-\tilde{\omega}_{\mathbf{k}_0})^2+4N_0m_{\mathbf{k}-\mathbf{k}_0}]^{1/2}}.$$

Here, as well as hereafter, we neglect terms of order $(\gamma^A_{\mathrm{ph}})^2/N_0m_{\mathbf{k}-\mathbf{k}_0}$ and $\gamma^2/N_0m_{\mathbf{k}-\mathbf{k}_0}$ in the calculations. The dispersion curves $\xi_{01}(k)$ and $\xi_{02}(k)$ shown in Fig. 1 are the restructured phonon and polariton spectra. Their deviation from the original ones is most substantial near values of $\mathbf{k}$ defined by the expression

$$\omega_{\mathbf{k}}-\omega^{\mathrm{ph}}_{\mathbf{k}-\mathbf{k}_0}-\tilde{\omega}_{\mathbf{k}_0}=0, \tag{29}$$

i.e., for a given scattering direction near two points of intersection of the unperturbed terms. In the vicinity of one of them, a characteristic term splitting $\Delta\xi=\xi_{01}-\xi_{02}=2(N_0m_{\mathbf{k}-\mathbf{k}_0})^{1/2}$, takes place, and in the vicinity of the point $\mathbf{k}_0$, at values of $\mathbf{k}$ satisfying the condition

$$|\mathbf{k}-\mathbf{k}_0|<4N_0m_{\mathbf{k}-\mathbf{k}_0}/k_0(u+V_{\mathbf{k}_0}{}^p)^2 \tag{30}$$

($V^p_{\mathbf{k}_0}$ is the polariton group velocity of the wave $\mathbf{k}_0$) the spectrum restructuring is such that $\xi_{i0}\propto|\mathbf{k}_0-\mathbf{k}|^{1/2}$. The final expressions for the Green's functions (25) and (26) can be written in the form

$$G^R(k)=\frac{\varphi_1{}^R(k)}{\omega-\xi_1}+\frac{\varphi_2{}^R(k)}{\omega-\xi_2}, \tag{31}$$

$$D^R(k-k_0)=\frac{\psi_1{}^R(k-k_0)}{(\omega-\varepsilon)-(\xi_1-\tilde{\omega}_{\mathbf{k}_0})}+\frac{\psi_2{}^R(k-k_0)}{(\omega-\varepsilon)-(\xi_2-\tilde{\omega}_{\mathbf{k}_0})}, \tag{32}$$

where the functions $\varphi^R_i(k)$ and $\psi^R_i(k-k_0)$ are given by

$$\varphi^R_{1,2}(k)=\pm\frac{\omega-\omega^{\mathrm{ph}}_{\mathbf{k}-\mathbf{k}_0}-\tilde{\omega}_{\mathbf{k}_0}+i\gamma^A_{\mathrm{ph}}(\mathbf{k}-\mathbf{k}_0)}{[(\omega_{\mathbf{k}}-\omega^{\mathrm{ph}}_{\mathbf{k}-\mathbf{k}_0}-\tilde{\omega}_{\mathbf{k}_0})^2+4N_0m_{\mathbf{k}-\mathbf{k}_0}]^{1/2}}, \tag{33}$$

$$\psi^R_{1,2}(k-k_0)=\pm\frac{(\omega-\varepsilon)-(\omega_{\mathbf{k}}-\tilde{\omega}_{\mathbf{k}_0})+i\gamma(\mathbf{k})}{[(\omega_{\mathbf{k}}-\omega^{\mathrm{ph}}_{\mathbf{k}-\mathbf{k}_0}-\tilde{\omega}_{\mathbf{k}_0})^2+4N_0m_{\mathbf{k}-\mathbf{k}_0}]^{1/2}}. \tag{34}$$

The energy $\tilde{\omega}_{k_0}$ of the excitons of the mode $\mathbf{k}_0$ is given by Eq. (11), where $\Sigma^R(k_0)$ is given by expression (55) below.

We proceed now to obtain and analyze the kinetic equations for scattered excitons that scatter phonons and excitons of the mode $\mathbf{k}_0$. The kinetic equations are derived from the Dyson equations for the corresponding statistical Green's functions by a known method,[14] but in this case the derivation contains a number of peculiarities, and we shall therefore carry it out in sufficient detail.

The Dyson equation for the Green's function of the scattered excitons if of the form

$$G^+(\tilde{x},x')=G_0^+(\tilde{x},x')$$
$$+\int G_0^+(\tilde{x},x_2)\Sigma_f^A(x_2,x_1)G^A(x_1,x')\,d^4x_2\,d^4x_1$$
$$+\int G_0^R(\tilde{x},x_2)\,[\Sigma_f^R(x_2,x_1)G^+(x_1,x')$$
$$+\Sigma_f^+(x_2,x_1)G^A(x_1,x')]\,d^4x_2\,d^4x_1, \tag{35}$$

where $x_i=(t_i,\mathbf{r}_i)$.

From this we obtain

$$G_0^{-1}(\tilde{x})G^+(\tilde{x},x')$$
$$=\int[\Sigma_f^R(\tilde{x},x_1)G^+(x_1,x')+\Sigma_f^+(\tilde{x},x_1)G^A(x_1,x')]\,d^4x_1, \tag{36}$$
$$[G_0^{-1}(x')]^*G^+(\tilde{x},x')$$
$$=\int[G^+(\tilde{x},x_1)\Sigma^A(x_1,x')+G^R(\tilde{x},x_1)\Sigma^+(x_1,x')]\,d^4x_1.$$

We now take the difference of these two equations, make the change of variables $x=\frac{1}{2}(x'+\tilde{x}), \eta=\tilde{x}-x'$ and take the total Fourier transform[5] with respect to η of both sides of the obtained equations. After the foregoing operations the left-hand side of the equation takes the form

$$-\sum_{i=1}^{2}\left[\frac{\partial}{\partial t}+\mathbf{V}_{\mathbf{k}}^{(i)}\nabla_{\mathbf{r}}\right]N_{\mathbf{k}}^{(i)}(\mathbf{r}t), \tag{37}$$

where $\mathbf{V}_{\mathbf{k}}^{(i)}=\partial\xi_{0i}/\partial\mathbf{k}$. The distribution function of the scattered excitons $N_{\mathbf{k}}(\mathbf{r}t)$ breaks up into two components $N_{\mathbf{k}}^{(i)}(\mathbf{r}t)$, each of which is a distribution function of excitons with a corresponding restructured spectrum $\omega=\xi_{0i}(\mathbf{k})$, with the index i numbering the branch of the restructured spectrum. Such a breakdown is formally connected with the definition of the distribution fuction in terms of the statistical Green's function:

$$N_{\mathbf{k}}(\mathbf{r}t)=i\int G^+(x,k)\frac{d\omega}{2\pi}, \tag{38}$$

since the function $G^+\propto G^R-G^A$ and according to (31) it contains two terms proportional to $\tilde{\delta}[\Gamma_i|\omega-\xi_{0i}]$.

The right-hand side of the equation assumes after the indicated transformations the form

$$\int\frac{d\omega}{2\pi}\{[\Sigma^+(x,k)+\Sigma_\tau(x,k)][G^R(x,k)-G^A(x,k)]$$
$$-G^+(x,k)[\Sigma^R(x,k)+\Sigma_\tau^R(x,k)-\Sigma^A(x,k)-\Sigma_\tau^A(x,k)]\}. \tag{39}$$

The self-energy parts are defined in the following manner:

$$\Sigma_{(k)}^R=\sum_{i=1}^{2}\Sigma_i^R(k)=\psi_1^R(k-k_0)\frac{N_0m_{\mathbf{k}-\mathbf{k}_0}}{\omega-\xi_1}+\psi_2^R(k-k_0)\frac{N_0m_{\mathbf{k}-\mathbf{k}_0}}{\omega-\xi_2}, \tag{40}$$

$$\Sigma^+(k)=i\int m_{\mathbf{k}-\mathbf{q}}D^+(k-q)G^+(q)\frac{d^4q}{(2\pi)^4}$$
$$=\sum_{i=1}^{2}n_{\mathbf{k}-\mathbf{k}_0}^{(i)}[\Sigma_i^R(k)-\Sigma_i^A(k)], \tag{41}$$

$$\Sigma_\tau^R(x,k)=-i\frac{1}{\tau},\quad \Sigma_\tau^A(x,k)=i\frac{1}{\tau},$$
$$\Sigma_\tau^+(x,k)=-i\frac{2N_{\mathbf{k}}^0}{\tau}=0. \tag{42}$$

Here $N_{\mathbf{k}}^0$ are the equilibrium exciton occupation numbers, equal to zero, since we assume complete absence of excitons when there is no polariton wave in the crystal. Relation (40) is obtained from (16) when (32) is taken into account. After certain transformations of (39) using (40)–(42) we obtain the right-hand side of the equation

$$\sum_{i=1}^{2}\Big\{-\frac{2N_{\mathbf{k}}^{(i)}(\mathbf{r}t)}{\tau}$$
$$+N_0m_{\mathbf{k}-\mathbf{k}_0}[\varphi_i(\mathbf{k})n_{\mathbf{k}-\mathbf{k}_0}^{(i)}(\mathbf{r}t)-\psi_i(\mathbf{k}-\mathbf{k}_0)N_{\mathbf{k}}^{(i)}(\mathbf{r}t)]\frac{1}{\Gamma_i(\mathbf{k})}$$
$$+N_0m_{\mathbf{k}-\mathbf{k}_0}[\varphi_j(\mathbf{k})n_{\mathbf{k}-\mathbf{k}_0}^{(i)}(\mathbf{r}t)-\psi_j(\mathbf{k}-\mathbf{k}_0)N_{\mathbf{k}}^{(i)}(\mathbf{r}t)]_{i\neq j}$$
$$\times 2\pi\tilde{\delta}[\Gamma_1+\Gamma_2|\xi_{01}-\xi_{02}]\Big\}. \tag{43}$$

The distribution functions of the phonons of the corresponding branches of the restructured spectrum $n_{\mathbf{k}-\mathbf{k}_0}^{(i)}(\mathbf{r}t)$ are determined by a formula similar to (38), and the factors $\varphi_i(\mathbf{k})$ and $\varphi_i(\mathbf{k}-\mathbf{k}_0)$, which can be called the weighting factors, satisfy the following equations:

$$\psi_1(\mathbf{k}-\mathbf{k}_0)=\varphi_2(\mathbf{k})=\frac{1}{2}[\varphi_2^R(\omega=\xi_{02},\mathbf{k})+\varphi_2^{R*}(\omega=\xi_{02},\mathbf{k})]$$
$$=\frac{(\omega_{\mathbf{k}-\mathbf{k}_0}^{ph}+\tilde{\omega}_{\mathbf{k}_0}-\omega_{\mathbf{k}})+[(\omega_{\mathbf{k}-\mathbf{k}_0}^{ph}+\tilde{\omega}_{\mathbf{k}_0}-\omega_{\mathbf{k}})^2+4N_0m_{\mathbf{k}-\mathbf{k}_0}]^{1/2}}{2[(\omega_{\mathbf{k}-\mathbf{k}_0}^{ph}+\tilde{\omega}_{\mathbf{k}_0}-\omega_{\mathbf{k}})^2+4N_0m_{\mathbf{k}-\mathbf{k}_0}]^{1/2}}, \tag{44}$$

$$\psi_2(\mathbf{k}-\mathbf{k}_0)=1-\varphi_2(\mathbf{k})=\varphi_1(\mathbf{k})$$
$$=\frac{1}{2}[\varphi_1^R(\omega=\xi_{01},\mathbf{k})+\varphi_1^{R*}(\omega=\xi_{01},\mathbf{k})]$$

$$=\frac{(\omega_{\mathbf{k}}-\omega^{ph}_{\mathbf{k}-\mathbf{k}_0}-\tilde{\omega}_{\mathbf{k}_0})+[(\omega_{\mathbf{k}}-\omega^{ph}_{\mathbf{k}-\mathbf{k}_0}-\tilde{\omega}_{\mathbf{k}_0})^2+4N_0 m_{\mathbf{k}-\mathbf{k}_0}]^{1/2}}{2[(\omega^{ph}_{\mathbf{k}-\mathbf{k}_0}+\tilde{\omega}_{\mathbf{k}_0}-\omega_{\mathbf{k}})^2+4N_0 m_{\mathbf{k}-\mathbf{k}_0}]^{1/2}}. \tag{45}$$

From (37) and (43) we obtain kinetic equations for the functions of the scattered excitons $N^0_{\mathbf{k}}(\mathbf{r}t)$:

$$\left[\frac{\partial}{\partial t}+\mathbf{V}^{(i)}_{\mathbf{k}}\nabla_r\right]N^{(i)}_{\mathbf{k}}(\mathbf{r}t)=-\frac{2N^{(i)}_{\mathbf{k}}(\mathbf{r}t)}{\tau}+N_0 m_{\mathbf{k}-\mathbf{k}_0}$$
$$\times\{[\varphi_i(\mathbf{k})n^{(i)}_{\mathbf{k}-\mathbf{k}_0}(\mathbf{r}t)-\psi_i(\mathbf{k}-\mathbf{k}_0)N^{(i)}_{\mathbf{k}}(\mathbf{r}t)]\frac{1}{\Gamma_i}$$
$$+2\pi\tilde{\delta}[\Gamma_1+\Gamma_2|\xi_{01}-\xi_{02}][\varphi_j(\mathbf{k})n^{(i)}_{\mathbf{k}-\mathbf{k}_0}(\mathbf{r}t)-\psi_j(\mathbf{k}-\mathbf{k}_0)N^{(i)}_{\mathbf{k}}(\mathbf{r}t)]_{i\neq j}\}. \tag{46}$$

Similarly we derive kinetic equations for the scattering phonons, the only difference being that now the self-energy part $\Sigma^{+A}_{\tau ph}(x,k-k_0)$ differs from zero and is given by

$$\Sigma^{+}_{\tau^A_{ph}}(x,k-k_0)=-i\frac{2}{\tau^A_{ph}}n_0(\mathbf{k}-\mathbf{k}_0), \tag{47}$$

where

$$n_0(\mathbf{k}-\mathbf{k}_0)=\left[\exp\left(\frac{\hbar\omega^{ph}_{\mathbf{k}-\mathbf{k}_0}}{T}\right)-1\right]^{-1}$$

are the occupations numbers of the unperturbed phonon thermostat. In the kinetic equations this self-energy part constitutes a temperature source of phonons in the phonon subsystem, owing to the presence of the thermostat. These equations take the form

$$\left[\frac{\partial}{\partial t}+\mathbf{V}^{(i)}_{\mathbf{k}}\nabla_r\right]n^{(i)}_{\mathbf{k}-\mathbf{k}_0}(\mathbf{r}t)=\frac{2}{\tau^A_{ph}}[\psi_i(\mathbf{k}-\mathbf{k}_0)n_0(\mathbf{k}-\mathbf{k}_0)-n^{(i)}_{\mathbf{k}-\mathbf{k}_0}(\mathbf{r}t)]$$
$$-N_0 m_{\mathbf{k}-\mathbf{k}_0}\Big\{[\varphi_i(\mathbf{k})n^{(i)}_{\mathbf{k}-\mathbf{k}_0}(\mathbf{r}t)-\psi_i(\mathbf{k}-\mathbf{k}_0)N^{(i)}_{\mathbf{k}}(\mathbf{r}t)]\frac{1}{\Gamma_i}$$
$$+2\pi\tilde{\delta}[\Gamma_1+\Gamma_2|\xi_{01}-\xi_{02}]$$
$$\times[\varphi_j(\mathbf{k})n^{(i)}_{\mathbf{k}-\mathbf{k}_0}(\mathbf{r}t)-\psi_j(\mathbf{k}-\mathbf{k}_0)N^{(i)}_{\mathbf{k}}(\mathbf{r}t)]_{i\neq j}\Big\}. \tag{48}$$

The kinetic equations (46) and (48) must be supplemented by conditions on the occupation numbers of the phonons and excitons in the absence of an initial wave ($N_0=0$); these conditions are the initial conditions of the problem in the case of slow growth of the initial-wave amplitude:

$$n^{(i=1)}_{\mathbf{k}-\mathbf{k}_0}=\begin{cases}0, & \omega_{\mathbf{k}}-\omega_{\mathbf{k}_0}-\omega^{ph}_{\mathbf{k}-\mathbf{k}_0}>0\\ n_0(\mathbf{k}-\mathbf{k}_0), & \omega_{\mathbf{k}}-\omega_{\mathbf{k}_0}-\omega^{ph}_{\mathbf{k}-\mathbf{k}_0}<0\end{cases}, \tag{49}$$

$$n^{(i=2)}_{\mathbf{k}-\mathbf{k}_0}=\begin{cases}n_0(\mathbf{k}-\mathbf{k}_0), & \omega_{\mathbf{k}}-\omega_{\mathbf{k}_0}-\omega^{ph}_{\mathbf{k}-\mathbf{k}_0}>0\\ 0, & \omega_{\mathbf{k}}-\omega_{\mathbf{k}_0}-\omega^{ph}_{\mathbf{k}-\mathbf{k}_0}<0\end{cases};\quad N^{(i)}_{k}=0.$$

In the derivation of (46) and (48) we used the following approximate relations:

$$\int N_{\mathbf{k},u}\tilde{\delta}[\Gamma_i|(\varepsilon-\omega)-(\tilde{\omega}_{\mathbf{k}_0}-u)]\tilde{\delta}[\Gamma_j|u-\xi_{0j}]du$$
$$\approx N_{\mathbf{k}}\tilde{\delta}[\Gamma_i+\Gamma_j|(\omega-\varepsilon)-(\xi_{0j}-\tilde{\omega}_{\mathbf{k}_0})], \tag{50}$$
$$\int n_{\mathbf{k}-\mathbf{k}_0,u}\tilde{\delta}[\Gamma_i|\omega-u-\tilde{\omega}_{\mathbf{k}_0}]\tilde{\delta}[\Gamma_j|u-\xi_{0j}]du$$
$$\approx n_{\mathbf{k}-\mathbf{k}_0}\tilde{\delta}[\Gamma_i+\Gamma_j|\omega-\tilde{\omega}_{\mathbf{k}_0}-\xi_{0j}].$$

Let us explain the physical meaning of the kinetic equations (46) and (48). The left-hand side of the equations has the usual form of Aboltzmann operator acting on the distribution function. The first term of the right-hand side of the equations describes the relaxation of the distribution functions to the corresponding equilibrium values, while the second term in the curly brackets describes roughly speaking the tendency of the phonon and exciton components of the given branch of the restructured spectrum to become balanced. This second term, in turn, consists of two parts, the first describing the aforementioned tendency for that branch of the spectrum whose distribution functions are determined by the given equation, while the second corresponds to be second branch of the spectrum. The second part makes a substantial contribution to the corresponding kinetic equation only in the case when the reciprocal lifetime of the phonoriton excitation $\Gamma=\Gamma_1+\Gamma_2=\gamma(\mathbf{k})+\gamma^A_{ph}(\mathbf{k}-\mathbf{k}_0)$ exceeds the spectral splitting $2(N_0 m_{\mathbf{k}-\mathbf{k}_0})^{1/2}$. In this case the obtained kinetic equations can be substantially simplified by changing over to phonon and exciton distribution functions with a non-restructured spectrum[16]:

$$\left[\frac{\partial}{\partial t}+\mathbf{V}_{\mathbf{k}}{}^p\nabla_r\right]N_{\mathbf{k}}(\mathbf{r}t)=-\frac{2N_{\mathbf{k}}(\mathbf{r}t)}{\tau}$$
$$+N_0 m_{\mathbf{k}-\mathbf{k}_0}[n_{\mathbf{k}-\mathbf{k}_0}(\mathbf{r}t)-N_{\mathbf{k}}(\mathbf{r}t)]$$
$$\times\tilde{\delta}[\gamma(\mathbf{k})+\gamma^A_{ph}(\mathbf{k}-\mathbf{k}_0)|\omega_{\mathbf{k}}-\tilde{\omega}_{\mathbf{k}_0}-\omega^{ph}_{\mathbf{k}-\mathbf{k}_0}], \tag{51}$$
$$\left[\frac{\partial}{\partial t}+\mathbf{u}_{\mathbf{k}-\mathbf{k}_0}\nabla_r\right]n_{\mathbf{k}-\mathbf{k}_0}(\mathbf{r}t)=\frac{2}{\tau^A_{ph}}[n_0(\mathbf{k}-\mathbf{k}_0)-n_{\mathbf{k}-\mathbf{k}_0}(\mathbf{r}t)]$$
$$-N_0 m_{\mathbf{k}-\mathbf{k}_0}[n_{\mathbf{k}-\mathbf{k}_0}(\mathbf{r}t)-N_{\mathbf{k}}(\mathbf{r}t)]$$
$$\times\tilde{\delta}[\gamma(\mathbf{k})+\gamma^A_{ph}(\mathbf{k}-\mathbf{k}_0)|\omega_{\mathbf{k}}-\tilde{\omega}_{\mathbf{k}_0}-\omega^{ph}_{\mathbf{k}-\mathbf{k}_0}].$$

In the case

$$\gamma(\mathbf{k})+\gamma^A_{ph}(\mathbf{k}-\mathbf{k}_0)<2(N_0 m_{\mathbf{k}-\mathbf{k}_0})^{1/2}$$

it follows from the system of quasilinear kinetic equations (46) and (48), under the condition that the quantity $N_0(\mathbf{r}t)$ varies slowly and with allowance for the initial conditions (49), that the reciprocal of the characteristic lifetimes of the phonons and excitons having the restructured spectrum is determined by Γ_i for the i-phonoriton branch, and from these equations it is possible to obtain equations of the form

$$\left[\frac{\partial}{\partial t}+\mathbf{V}^{(i)}_{\mathbf{k}}\nabla_r\right]F^{(i)}_{\mathbf{k}}(\mathbf{r}t)=2\Gamma_i(\mathbf{k})[F^{(i)}_0(\mathbf{r}t)-F^{(i)}_{\mathbf{k}}(\mathbf{r}t)], \tag{52}$$
$$F^{(i)}_0(\mathbf{r}t)=\varphi_i(\mathbf{k})\psi_i(\mathbf{k}-\mathbf{k}_0)n_0(\mathbf{k}-\mathbf{k}_0)$$

relative to the phonoriton distribution function of the corresponding branch:

$$F_{\mathbf{k}}^{(i)}(\mathbf{r}t)=\psi_i(\mathbf{k}-\mathbf{k}_0)N_{\mathbf{k}}^{(i)}(\mathbf{r}t)+\varphi_i(\mathbf{k})n_{\mathbf{k}-\mathbf{k}_0}^{(i)}(\mathbf{r}t). \tag{53}$$

We note that, for a given branch i of the phonoriton spectrum, the homogeneous solutions of the two equations (46) and (48) at constant N_0 have generally speaking two different characteristic damping times, owing to the initial conditions (49), however, the case actually realized will be the indicated one, i.e., that of Eq. (52) and the damping Γ_i. Moreover, to a certain degree the two equations (52) with allowance for (53) turn out to be equivalent to the four kinetic equations (46), (48) with allowance for (49). The latter, however, are more lucid and also follow directly from the diagram technique.

A kinetic equation for the excitons of a macroscopically filled mode $\mathbf{k}_0$ is obtained from the Dyson equation

$$G^+(k_0)=G_0^+(k_0)\left[1+\Sigma^A(k_0)G^A(k_0)\right]+G_0^R(k_0)\Sigma^R(k_0)G^+(k_0), \tag{54}$$

where $\Sigma^R(k_0)$ is defined by the relation

$$\Sigma^R(k_0)=i\sum_{i=1}^{2}\int\frac{d^3k}{(2\pi)^4}m_{\mathbf{k}-\mathbf{k}_0}$$
$$\times\left\{\left[\psi_i(\mathbf{k}-\mathbf{k}_0)N_{\mathbf{k}}^{(i)}-\varphi_i(\mathbf{k})n_{\mathbf{k}-\mathbf{k}_0}^{(i)}\right]\frac{\pi}{\Gamma_i(\mathbf{k})}\right.$$
$$\left.+\left[\psi_j(\mathbf{k}-\mathbf{k}_0)N_{\mathbf{k}}^{(i)}-\varphi_j(\mathbf{k})n_{\mathbf{k}-\mathbf{k}_0}^{(i)}\right]_{i\neq j}\frac{2\pi i}{(\xi_{01}-\xi_{02})+i(\Gamma_1(\mathbf{k})+\Gamma_2(\mathbf{k}))}\right\}. \tag{55}$$

From (54) and (55) we obtain in the manner considered above a kinetic equation for the excitons of the macroscopically filled $\mathbf{k}_0$:

$$\left[\frac{\partial}{\partial t}+\mathbf{V}_{\mathbf{k}_0}^{p}\nabla_{\mathbf{r}}\right]N_0(\mathbf{r}t)=-N_0\sum_{i=1}^{2}\int\frac{d^3k}{(2\pi)^3}m_{\mathbf{k}-\mathbf{k}_0}$$
$$\times\left\{\left[\varphi_i(\mathbf{k})n_{\mathbf{k}-\mathbf{k}_0}^{(i)}(\mathbf{r}t)-\psi_i(\mathbf{k}-\mathbf{k}_0)N_{\mathbf{k}}^{(i)}(\mathbf{r}t)\right]\right.$$
$$\times\frac{1}{\Gamma_i(\mathbf{k})}+2\pi\tilde{\delta}\left[\Gamma_1+\Gamma_2|\xi_{01}-\xi_{02}\right]$$
$$\left.\times\left[\varphi_j(\mathbf{k})n_{\mathbf{k}-\mathbf{k}_0}^{(i)}(\mathbf{r}t)-\psi_j(\mathbf{k}-\mathbf{k}_0)N_{\mathbf{k}}^{(i)}(\mathbf{r}t)\right]_{i\neq j}\right\}, \tag{56}$$

where $\mathbf{V}_{\mathbf{k}Q}^{p}$ is the polariton group velocity.

Thus, the system of five equations (46), (48), and (56) for the five variables $n_{\mathbf{k}-\mathbf{k}_0}^{(i)}(\mathbf{r}t)$, $N_{\mathbf{k}}^{(i)}(\mathbf{r}t)$ and $N_0(\mathbf{r}t)$ is the sought system of kinetic equations. We note that it was obtained in the approximation of a weak exciton-phonon interaction and for slow variation of the amplitude of the initial polariton wave $\mathbf{k}_0$.

In the case of a quasistationary spatially homogeneous problem, the indicated system of equations can be solved, and for the coefficient of the temporal damping $\sigma(\mathbf{k}_0)$ of the polariton wave we have the approximate formula

$$\frac{\sigma(\mathbf{k}_0)}{2\gamma(\mathbf{k}_0)}\approx\left[1+\frac{2(N_0m_{\mathbf{k}_0})^{1/2}}{\gamma_{ph}^{A}(\mathbf{k}_0)+\gamma(\mathbf{k}_0)}\right]\left[1+\frac{N_0m_{\mathbf{k}_0}}{\gamma(\mathbf{k}_0)\gamma_{ph}^{A}(\mathbf{k}_0)}\right]^{-1} \tag{57}$$

According to Ref. 17, the reciprocal lifetime of the longitudinal acoustic phonons $\mathbf{k}-\mathbf{k}_0$ is given by

$$\gamma_{ph}^{A}(\mathbf{k}-\mathbf{k}_0)=\frac{1}{\tau_{ph}^{A}(\mathbf{k}-\mathbf{k}_0)}=\frac{\chi_0T^4|\mathbf{k}-\mathbf{k}_0|}{\hbar^3\rho u^4}, \tag{58}$$

where ρ is the crystal density and χ_0 is a dimensionless constant of the order of unity. As for $m_{\mathbf{k}}$, it is defined in accordance with (17), and for acoustic phonons we have

$$M_{\mathbf{k}}=[2kC^2/9\rho uV]^{1/2}, \tag{59}$$

where C is the deformation potential of the semiconducting crystal.[8]

Analyzing (57) with account taken of (58) and (59), we can distinguish, as indicated above, between three regions of the behavior of the coefficient of the temporal damping $\sigma(\mathbf{k}_0)$, depending on the polariton-wave intensity.

1. Weak intensities (linear theory):

$$N_0<\frac{1}{m_{\mathbf{k}_0}}\gamma(\mathbf{k}_0)\gamma_{ph}^{A}(\mathbf{k}_0)\approx\frac{k_0^2T^5\chi_0}{2\pi^2\hbar^4\rho u^5u_{\mathbf{k}_0}}, \tag{60}$$

$$\sigma(\mathbf{k}_0)=2\gamma(\mathbf{k}_0)\approx\frac{2C^2(Tk_0^2)}{9\pi^2u^2u_{\mathbf{k}_0}\rho\hbar^2}. \tag{61}$$

Here $u_{\mathbf{k}0}$ is the average value of the phonoriton velocity $\mathbf{V}_{\mathbf{k}}^{(i)}$ in the vicinity of the phonoriton splitting of the dispersion curves:

$$u<u_{\mathbf{k}_0}<V_{\mathbf{k}_0}^{p}. \tag{62}$$

2. Medium intensities:

$$\frac{1}{m_{\mathbf{k}_0}}\gamma(\mathbf{k}_0)\gamma_{ph}^{A}(\mathbf{k}_0)<N_0<\frac{1}{4m_{\mathbf{k}_0}}[\gamma(\mathbf{k}_0)+\gamma_{ph}^{A}(\mathbf{k}_0)]^2, \tag{63}$$

$$\sigma\approx\frac{C^2\chi_0}{9\pi^4u^7u_{\mathbf{k}_0}^2\hbar^6\rho^2}\left(\frac{T^6k_0^4}{N_0}\right)\sim\frac{1}{N_0}. \tag{64}$$

Depending on the temperature and consequently on the ratio of $\gamma(\mathbf{k}_0)$ and $\gamma_{ph}^{A}(\mathbf{k}_0)$, Eq. (64) corresponds to different nonlinearity mechanisms. For low temperatures, when

$$T<[2C^2u^2\hbar k_0/9\pi^2u_{\mathbf{k}_0}\chi_0]^{1/3}\equiv T_c, \tag{65}$$

the phonon nonlinearity mechanism comes into play and can be realized even at sufficiently low intensities. Thus, e.g., at $T=4.2$ K, phonon nonlinearity can set in at $N_0\gtrsim10^{10}$ cm^{-3}. High temperatures correspond to nonequilibrium Bose condensation whose meaning was explained at the beginning of the article.

3. High intensities (phonoriton restructuring of the spectra):

$$N_0>\frac{1}{4m_{\mathbf{k}_0}}[\gamma(\mathbf{k}_0)+\gamma_{ph}^{A}(\mathbf{k}_0)]^2, \tag{66}$$

$$\sigma(\mathbf{k}_0)\approx\frac{2}{3}\frac{C\chi_0}{\pi^{5/2}u^{11/2}u_{\mathbf{k}_0}\rho^{3/2}\hbar^{9/2}}\left(\frac{T^5k_0^{5/2}}{N_0^{1/2}}\right)\sim\frac{1}{N_0^{1/2}},\quad T<T_c,$$

$$\sigma(\mathbf{k}_0)\approx\frac{4}{27}\frac{C^3}{\pi^{9/2}u^{7/2}u_{\mathbf{k}_0}^2\rho^{3/2}\hbar^{7/2}}\left(\frac{T^2k_0^{7/2}}{N_0^{1/2}}\right)\sim\frac{1}{N_0^{1/2}},\quad T>T_c. \tag{67}$$

In this case, which is accompanied by restructuring of the phonon and polariton spectra, the damping coefficient (67) takes on values smaller by several orders of magnitude than in the linear theory. We note that such a decrease of the damping coefficient at high intensities of the wave $\mathbf{k}_0$ is determined in the general case simultaneously also by the phonon nonlinearity, i.e., by the depletion of the phonon subsystem that participates in the scattering, and by the nonequilibrium Bose condensation, which manifests itself in this case in an active return of the scattered excitons to the initial mode.

The conversion to the spatial absorption coefficient $\sigma_{sp}(\mathbf{k}_0)$ of the polariton wave $\mathbf{k}_0$ is by means of the formula

$$\sigma_{sp}(\mathbf{k}_0)\approx\sigma(\mathbf{k}_0)/V_{\mathbf{k}_0}^{p}, \tag{68}$$

in which case Eq. (57) goes over into

$$\frac{\sigma_{sp}(\mathbf{k}_0, N_0)}{\sigma_{sp}(\mathbf{k}_0, N_0\to 0)}\cdot$$

$$\approx\left[1+\frac{2(N_0 m_{\mathbf{k}_0})^{1/2}}{\gamma_{ph}^{A}(\mathbf{k}_0)+\gamma(\mathbf{k}_0)}\right]\left[1+\frac{N_0 m_{\mathbf{k}_0}}{\gamma(\mathbf{k}_0)\gamma_{ph}^{A}(\mathbf{k}_0)}\right]^{-1}$$

$$\times\frac{V_{\mathbf{k}_0}^{p}(N_0)}{V_{\mathbf{k}_0}^{p}(N_0\to 0)}\cdot \tag{69}$$

We have used here the notation $V_{\mathbf{k}_0}^{p}(N_0)$ for the polariton group velocity $V_{\mathbf{k}_0}^{p}$, since this velocity will be shown below to be substantially dependent on the wave intensity.

We proceed now to the case when the initial wave $\mathbf{k}_0$ is coherent. Now, as already noted, it will be described by the Maxwell and Schrödinger equations relative to the mean values of the operator of the positive-frequency part of the electromagnetic field $\langle\hat{E}(\mathbf{r}t)\rangle=E(x)$ and the operator of the exciton field $\langle\hat{\Phi}(\mathbf{r}t)\rangle=\Phi(x)$, respectively. To obtain the indicated equations by the diagram technique for the nonequilibrium processes we introduce the following quantities

$$\Phi_+(x)=\Phi(x)=\langle S^{-1}T[\hat{\Phi}_0(x)S]\rangle,$$

$$E_+(x)=E(x)=\langle S^{-1}T[\hat{E}_0(x)S]\rangle, \tag{70}$$

$$\Phi_-(x)=\langle[\tilde{T}S^{-1}\hat{\Phi}_0(x)]S\rangle,$$

$$E_-(x)=\langle[\tilde{T}S^{-1}\hat{E}_0(x)]S\rangle,$$

where $\hat{E}_0(x)$ and $\hat{\Phi}_0(x)$ are operators in the interaction representation,

$$S=\exp\left[-i\int_{-\infty}^{+\infty}\hat{H}_{int}(t)\,dt\right]$$

is the evolution operator, and $\hat{H}_{int}(t)$ is the operator of the exciton-photon and exciton-phonon interactions in the interaction representation. We recall that we are considering exciton-phonon interaction in the dipole approximation.

The Dyson equation for the quantities $E_\pm(x)$ and $\Phi_\pm(x)$ are of the form

$$E_+(x)=E_+^{0}(x)+\int D_0^{+}(x,x_1)\Phi_-(x_1)\,dd^4x_1$$
$$-\int D_0^{c}(x,x_1)\Phi_+(x_1)\,dd^4x_1,$$
$$E_-(x)=E_-^{0}(x)+\int D_0^{-}(x,x_1)\Phi_+(x_1)\,dd^4x_1$$
$$-\int \tilde{D}_0^{c}(x,x_1)\Phi_-(x_1)\,dd^4x_1,$$
$$\Phi_+(x)=\Phi_+^{0}(x)+\int[G_0^{+}(x,x_1)E_-(x_1)-G_0^{c}(x,x_1)E_+(x_1)]$$
$$\times dd^4x_1+\int G_0^{c}(x,x_1)[\Sigma^{c}(x_1,x_2)\Phi_+(x_2)$$
$$-\Sigma^{+}(x_1,x_2)\Phi_-(x_2)]d^4x_1\,d^4x_2$$
$$+\int G_0^{+}(x,x_1)[\tilde{\Sigma}^{c}(x_1,x_2)\Phi_-(x_2)-\Sigma^{-}(x_1,x_2)\Phi_+(x_2)]d^4x_1d^4x_2,$$
$$\Phi_-(x)=\Phi_-^{0}(x)+\int[G_0^{-}(x,x_1)E_+(x_1)-\tilde{G}_0^{c}(x,x_1)E_-(x_1)]\cdot$$
$$\times dd^4x_1+\int\tilde{G}_0^{c}(x,x_1)[\tilde{\Sigma}^{c}(x_1,x_2)\Phi_-(x_2)$$
$$-\Sigma^{-}(x_1,x_2)\Phi_+(x_2)]d^4x_1\,d^4x_2$$
$$+\int G_0^{-}(x,x_1)[\Sigma^{c}(x_1,x_2)\Phi_+(x_2)-\Sigma^{+}(x_1,x_2)\Phi_-(x_2)]d^4x_1d^4x_2. \tag{71}$$

Here all the Green's functions $G(x,x_1)$ and $D(x,x_1)$, the self-energy parts $\Sigma(x,x_1)$, as well as their connections with the corresponding retarded and advanced functions are defined in accordance with Ref. 14. From these equations we obtain a system for $\Phi_\pm(x)$ and $E_\pm(x)$:

$$\left[i\hbar\frac{\partial}{\partial t}+\frac{\hbar^2}{2m}\nabla_r^2-\hbar\omega_t\right]\Phi_+(x)=-dE_+(x)+\int[\Sigma^{c}(x,x_1)$$
$$\times\Phi_+(x_1)-\Sigma^{+}(x,x_1)\Phi_-(x_1)]d^4x_1,$$
$$\left[i\hbar\frac{\partial}{\partial t}+\frac{\hbar^2}{2m}\nabla_r^2-\hbar\omega_t\right]\Phi_-(x)=-dE_-(x)+\int[\Sigma^{-}(x,x_1)\Phi_+(x_1)$$
$$-\tilde{\Sigma}^{c}(x,x_1)\Phi_-(x_1)]d^4x_1, \tag{72}$$

$$\left[\frac{\varepsilon_0}{c^2}\frac{\partial^2}{\partial t^2}-\nabla_r^2\right]E_\pm(x)=-\frac{4\pi}{c^2}d\frac{\partial^2}{\partial t^2}\Phi_\pm(x),$$

where ω_t is the position of the exciton level relative to the valence band, ε_0 is the background dielectric constant of the semiconductor, and m is the translational mass of the exciton. It can be easily shown that only one homogeneous solution with respect to $\Phi_\pm(x)$ and $E_\pm(x)$ of the equations of the system (72) gives a physically plausible result $\Phi_+(x)=\Phi_-(x)=\Phi(x)$ with a damping determined by the self-energy part $\Sigma^R(x,x_1)$ and $E_+(x)=E_-(x)=E(x)$. In this case the system (72) assumes the simpler form

$$\left[i\hbar\frac{\partial}{\partial t}+\frac{\hbar^2}{2m}\nabla_r{}^2-\hbar\omega_t\right]\Phi(\mathbf{r}t)$$

$$=-\mathbf{dE}(\mathbf{r}t)+\int\Sigma^R(\mathbf{r}t,\mathbf{r}_1t_1)\Phi(\mathbf{r}_1t_1)\,d^3r_1\,dt_1, \tag{73}$$

$$\left[\frac{\varepsilon_0}{c^2}\frac{\partial}{\partial t^2}-\nabla_r{}^2\right]\mathbf{E}(\mathbf{r}t)=-\frac{4\pi}{c^2}\frac{\mathbf{d}}{v_0}\frac{\partial^2\Phi(\mathbf{r}t)}{\partial t^2}.$$

In these equations we transformed to the dimensionless variable $\Phi(\mathbf{r}t)$, and v_0 is the volume of the unit cell of the crystal.

As for the behavior of the scattered excitons and the scattering phonons, they are subject to all the conclusions of the preceding case both with respect to the restructuring of the spectra and with respect to the form of the kinetic equations (46) and (48). In all these relations, which pertain to scattered excitons and scattering phonons, it is necessary to make the substitution

$$N_0(\mathbf{r}t)\to|\Phi(\mathbf{r}t)|^2/v_0=|\Phi_{\mathbf{k}_0}(\mathbf{r}t)|^2/v_0. \tag{74}$$

If now we take into account the explicit form of $\Sigma^R(x,x_1)$ in accord with (55), the system (73) assumes its final form

$$\left[i\hbar\frac{\partial}{\partial t}+\hbar\tilde\omega_{\mathbf{k}_0}-\hbar\tilde\omega_t(\mathbf{k}_0)\right]\Phi_{\mathbf{k}_0}(\mathbf{r}t)+\frac{\hbar^2}{2m}[i(\mathbf{k}_0\nabla_r)-k_0{}^2]\Phi_{\mathbf{k}_0}(\mathbf{r}t)$$

$$=-\mathbf{dE}_{\mathbf{k}_0}(\mathbf{r}t)-i\hbar\Phi_{\mathbf{k}_0}(\mathbf{r}t)\sum_{i=1}^{2}\int\frac{d^3k}{(2\pi)^3}\frac{m_{\mathbf{k}-\mathbf{k}_0}}{2}$$

$$\{[\varphi_i(\mathbf{k})n^{i}_{\mathbf{k}-\mathbf{k}_0}(\mathbf{r}t)-\psi_i(\mathbf{k}-\mathbf{k}_0)$$

$$\times N_{\mathbf{k}}^{(i)}(\mathbf{r}t)]\frac{1}{\Gamma_i(\mathbf{k})}+2\pi\delta[\Gamma_1+\Gamma_2|\xi_{01}-\xi_{02}][\varphi_j(\mathbf{k})n^{(i)}_{\mathbf{k}-\mathbf{k}_0}(\mathbf{r}t)$$

$$-\psi_j(\mathbf{k}-\mathbf{k}_0)N_{\mathbf{k}}{}^{(i)}(\mathbf{r}t)]_{i\neq j}\}, \tag{75}$$

$$\frac{\varepsilon_0}{c^2}\tilde\omega_{\mathbf{k}_0}\left[i\frac{\partial}{\partial t}+\tilde\omega_{\mathbf{k}_0}\right]\mathbf{E}_{\mathbf{k}_0}(\mathbf{r}t)+[i(\mathbf{k}_0\nabla_r)-k_0{}^2]\mathbf{E}_{\mathbf{k}_0}(\mathbf{r}t)$$

$$=-\frac{4\pi}{c^2}\frac{\mathbf{d}}{v_0}\tilde\omega_{\mathbf{k}_0}\left[i\frac{\partial}{\partial t}+\tilde\omega_{\mathbf{k}_0}\right]\Phi_{\mathbf{k}_0}(\mathbf{r}t).$$

A transition was carried out here to slow envelopes of the polariton wave $\mathbf{k}_0$:

$$\Phi(\mathbf{r}t)=\Phi_{\mathbf{k}_0}(\mathbf{r}t)\,e^{-i\tilde\omega_{\mathbf{k}_0}t+i\mathbf{k}_0\mathbf{r}},$$
$$\mathbf{E}(\mathbf{r}t)=\mathbf{E}_{\mathbf{k}_0}(\mathbf{r}t)\,e^{-i\tilde\omega_{\mathbf{k}_0}t+i\mathbf{k}_0\mathbf{r}} \tag{76}$$

and second-order derivatives were neglected. The displaced position of the exciton level $\tilde\omega_t(\mathbf{k}_0)$ is determined from (55) by means of Eq. (11), while $\tilde\omega_{\mathbf{k}_0}$ is determined directly from the dispersion polariton equation (75). We have thus obtained a system of nonlinear equations (46), (48), and (75) relative to $\Phi_{\mathbf{k}_0}(\mathbf{r}t)$, $E_{\mathbf{k}0}(\mathbf{r}t)$, $n^{(i)}_{\mathbf{k}-\mathbf{k}_0}(\mathbf{r}t)$ and $N^{(i)}_{\mathbf{k}}(\mathbf{r}t)$. The dispersion equation obtained from the polariton equations (75) takes the same form as in the preceding noise-wave case, where it was determined by the denominator of a retarded Green's function; in particular, the absorption of the coherent wave $\mathbf{k}_0$ was determined by the imaginary part of the same self-energy part Σ^R as in the case of the noise wave. This fact greatly facilitates the analysis of the obtained equations; e.g., it is possible to treat in the same manner the spatially homogeneous problem and obtain the results (57)–(69) subject only to the difference connected with the substitution (74).

We dwell now briefly on the general case, when the initial wave k_0 is partially coherent. The general system of equations for the distribution functions $n^{(i)}_{\mathbf{k}-\mathbf{k}_0}(\mathbf{r}t)$, and $N^{(i)}_{\mathbf{k}_0}(\mathbf{r}t)$, for the noise part of the wave $N_0(\mathbf{r}t)$, and for its coherent parts $\Phi_{\mathbf{k}_0}(\mathbf{r}t)$ and $E_{\mathbf{k}_0}(\mathbf{r}t)$ can also be obtained with the aid of the diagram technique. This closed system will consist of the equations (46), (48), (56), and (75), and in all the relations, including the kinetic equations (46) and (48) pertaining to scattered excitons and to scattering phonons, it is necessary to make the substitution

$$N_0(\mathbf{r}t)\to N_0(\mathbf{r}t)+|\Phi_{\mathbf{k}_0}(\mathbf{r}t)|^2/v_0, \tag{77}$$

where the last expression is the total density of the excitons of the mode $\mathbf{k}_0$. From an analysis of this system it can be seen that the coherent and noise components of the initial wave $\mathbf{k}_0$ retain their relative shares in the total intensity in the wave propagation process,, and formulas (57)–(69) are again valid if the substitution (77) is made in them. We note that in this paper we define total coherence in the sense

$$\langle\hat\Phi(x)\rangle=\Phi(x),$$

$$G_{cs}{}^+(x,x')=G^+(x,x')-\langle\hat\Phi^+(x)\rangle\langle\hat\Phi(x')\rangle=0$$

and do not consider at all the higher-order Green's functiosn.

The results can be applied to the phenomenon of equilibrium Bose condensation into the mode $\mathbf{k}_0=0$ of an ideal Bose gas in a phonon thermostat; in this case, however, the kinetic and field equations are actually not necessary, since the Bose gas is at equilibrium and the entire useful information is contained in the corresponding retarded Green's functions. The possibility of quasiparticle-spectrum restructuring in such a system, in the case of Bose condensation of dipole inactive excitons into $\mathbf{k}_0=0$, was indicated also in Ref. 5.

The connection between the total intensity $I_0(\mathbf{r}t)$ of the polariton wave with the concentration of the excitons of the wave $N_0(\mathbf{r}t)$, i.e., with the intensity of the exciton part of the polariton wave, is determined by the relation

$$N_0(\mathbf{r}t)=\alpha(\mathbf{k}_0)\frac{I_0(\mathbf{r}t)}{\hbar\tilde\omega_{\mathbf{k}_0}V_{\mathbf{k}_0}{}^pS}\approx\frac{W(\mathbf{r}t)}{\hbar\tilde\omega_{\mathbf{k}_0}V_{\mathbf{k}_0}{}^p}, \tag{78}$$

where S is the cross-section area, $W=I_0/S$ is the power of the flux of the initial wave $\mathbf{k}_0$, and the weighting factor $\alpha(\mathbf{k}_0)$, which is analogous to the factors $\varphi_i(\mathbf{k}_0)$ and $\psi_i(\mathbf{k}-\mathbf{k}_0)$ considered above for the case of polariton-phonon splitting of the spectra, is defined by the formula

$$\alpha(\mathbf{k}_0)=\frac{(c'k_0-\omega_t)+[(c'k_0-\omega_t)^2+4\pi\beta\omega_t{}^2]^{1/2}}{2[(c'k_0-\omega_t)^2+4\pi\beta\omega_t{}^2]^{1/2}}, \tag{79}$$

where

$$4\pi\beta = 2\omega_{lt}\varepsilon_0/\omega_t, \qquad c' = c/\varepsilon_0^{1/2}.$$

This factor is equal, with high degree of accuracy, to unity in the region of the polariton dispersion curve of interest to us and discussed below.

We consider now the dependence of the group velocity $V^p_{\mathbf{k}_0} = V^p_{\mathbf{k}_0}(N_0)$ of the initial wave $\mathbf{k}_0$ on its intensity. Since, on the one hand, the propagation of the $\mathbf{k}_0$ wave is described by nonlinear equations, and on the other hand the absorption $\sigma_{sp}(\mathbf{k}_0)$ for weak intensities can reach large values, the group velocity cannot be defined in the usual manner in the form of a derivative of a dispersion function. A more general definition of the group velocity is

$$\mathbf{V}_{\mathbf{k}_0}{}^p = \langle \mathbf{s} \rangle / Q, \tag{80}$$

where $\langle \mathbf{s} \rangle$ is the Poynting vector averaged over the period of the wave and Q is the polariton-wave energy density averaged over the period. With the aid of the Poynting theorem and the definition (80), the following expression was obtained in Ref. 7 for the group velocity in the case of a given frequency of the initial wave

$$V_{\mathbf{k}_0}{}^p(N_0) \approx \sigma(\mathbf{k}_0)\,[k_0\sigma(\mathbf{k}_0)/\tilde{\omega}_{\mathbf{k}_0} + 2k_0']^{-1}. \tag{81}$$

Here k_0 and k_0' are respectively the real and imaginary parts of the wave vector and are determined from the equation $\omega = \tilde{\omega}_{\mathbf{k}_0}$, while the dependence of the group velocity on the wave intensity is contained in the previously obtained dependence of $\sigma(\mathbf{k}_0)$ on the density $N_0(\mathbf{r}t)$. It follows from an analysis of (81) that the values of the velocity $V^p_{\mathbf{k}_0}$ lie between the velocity of light c' in the medium and the polariton velocity, which is determined as $\sigma(\mathbf{k}_0) \to 0$ in the usual manner and which can reach values smaller by several orders of magnitude than the velocity of light in the medium. At high intensities (3), (4) of the initial wave $\mathbf{k}_0$, an abrupt decrease of the damping coefficient $\sigma(\mathbf{k}_0)$ takes place, and consequently a strong decrease of the polariton group velocity. This can be physically attributed to the fact that at such a decrease of the wave damping the exciton-photon interaction becomes stronger than the exciton-phonon interaction, and this enhances the polariton character of the initial wave, and, in particular, to decreases its group velocity.

We now examine how realistic is the assumed neglect of the polariton-wave absorption due to the emission of acoustic phonon wave by the excitons. This assumption holds if, following absorption of the phonon, the exciton of the initial wave lands on the essentially exciton-like part of the polariton dispersion curve, where the exciton state density is large, and when a phonon is emitted it lands on the essentially photon-like part, where the state density is low. These state densities are inversely proportional to the group velocities at the corresponding points of the dispersion curve and consequently the ratio of the state densities of the exciton-like part of the dispersion curve and of the photon-like part can reach values 10^3. This can be realized in semiconductors in which the longitudinal-transverse splitting satisfies the condition

$$\omega_{lt} < 2(u/c)\,\varepsilon_0^{1/2}\,\omega_t, \tag{82}$$

and the case of the ground energy state of the excitons is considered. This condition is satisfied, e.g., for the semiconductors GaSe, GaAs, Cu_2O, CdTe. In addition, to exclude the possible accumulation of excitons scattered from the wave $\mathbf{k}_0$ with emission of phonons, an accumulation that can lead to stimulated phonon emission by the excitons $\mathbf{k}_0$ and to an abrupt increase of this absorption channel, it is necessary that these scattered electrons leave the region of the wave within a time shorter than $\tau(\mathbf{k})$. This means that the transverse dimensions of the wave should be limited to a diameter $l \lesssim (c/\varepsilon_0^{1/2})\tau \sim 1$–$0.1$ mm, i.e., should actually be a rather narrow beam.

For a possible experimental observation of the considered phenomena, besides the indicated conditions, it is necessary that the frequency of the polariton wave $\mathbf{k}_0$, equal to $\omega = \tilde{\omega}_{\mathbf{k}_0}$, be located on the inflection of the lower polariton dispersion wave for best realization of condition (82), neglecting scattering due to phonon emission. In this case the wave intensity (78) should be high enough in accord with (3) and (4). If pulsed emission $\mathbf{k}_0$ is considered, of intensity (3) or (4), experimental observation of this decrease of the absorption coefficient calls for a pulse duration τ_p that satisfies the condition

$$\tau_p \gtrsim \tau_{eff} = [(\gamma(\mathbf{k}_0) + \gamma^A_{ph}(\mathbf{k}_0))^2 + 4N_0 m_{\mathbf{k}_0}]^{1/2}/N_0 m_{\mathbf{k}_0}. \tag{83}$$

The spectral width of the pulse $\Delta\omega_p$ should in turn satisfy the inequalities

$$\Delta\omega_p < \omega_{lt}, \qquad \Delta\omega_p < \gamma(\mathbf{k}_0) + \gamma^A_{ph}(\mathbf{k}_0), \tag{84}$$

which are the conditions for the applicability of the obtained kinetic and field equations. This imposes an additional limitation on the pulse duration: $\tau_p > 1/\Delta\omega_p$.

Within the framework of the indicated mode, it is necessary in a general analysis to take into account the damping of the polariton wave on account of phonon emission. In this case the proposed description must be supplemented by kinetic and field (if the initial $\mathbf{k}_0$ wave is coherent) equations for the scattered excitons and for the emitted phonons, and also introduce the corresponding terms in expressions (56) and (75), which describe the initial wave. Such transitions of excitons with emission of phonons will also be accompanied by a restructuring of the spectra, and the corresponding system of equations actually reflects processes of stimulated Brillouin scattering.

Excitons of the initial wave can also be scattered by impurities as a result of exciton-exciton interaction, etc., therefore in the general case the results obtained above on the decrease of the absorption coefficient of the $\mathbf{k}_0$ wave should be regarded as a suppression of the contribution of one of the scattering mechanisms, namely scattering as a result of absorption of acoustic phonons by the wave excitons. Even in this case, however, when the condition (4) is satisfied, where $\gamma(\mathbf{k})$ is determined by all the possible exciton-scattering mechanisms, a restructuring of the phonon and polariton spectra is also possible. Moreover, it appears that a similar restructuring of the spectra of the scattered waves will take place for any anti-Stokes scattering, when the initial $\mathbf{k}_0$ wave is intense enough. We note case of simultaneous presence in the semiconductor of several microscopically filled polariton waves, a restructuring of the phonon and polariton spectra, much more complicated in structure than considered in the present paper, is possible.

The authors thank V. S. Dneprovskiĭ and S. S. Fanchenko for a discussion of a number of questions touched upon in the article.

[1]P. Wiesner and U. Heim, Phys. Rev. **B11**, 3971 (1975).
[2]H. Sumi, Sol. St. Commun. **17**, 701 (1975).
[3]V. M. Agranovich, S. A. Darmanyan, and V. N. Rupasov, Zh. Eksp. Teor. Fiz. **78**, 656 (1980) [Sov. Phys. JETP **51**, 332 (1980)].
[4]V. E. Bisti, Fiz. Tverd. Tela (Leningrad) **18**, 1056 (1976) [Sov. Phys. Solid State **18**, 603 (1976)].
[5]S. A. Moskalenko, Bose-Einstein Condensation of Excitons and Biexcitons [in Russian], Kishinev, 1970.
[6]D. Hulin, A. Mysyrowicz, Benoit a la Guillaume, Phys. Rev. Lett. **45**, 1970 (1980).
[7]W. Tait, Phys. Rev. **B5**, 648 (1972).
[8]W. Tait and R. Weiher, Phys. Rev. **166**, 769 (1968).
[9]A. Bosacchi, B. Bosacchi, and S. Franchi, Phys. Rev. Lett. **36**, 1086 (1976).
[10]J. Vøigt, Phys. Stat. Sol. (b) **64**, 549 (1974).
[11]A. Antipov, V. Chumash, V. Dneprovski, and V. Fokin, Appl. Phys. **15**, 423 (1978).
[12]R. G. Ulbrich and G. W. Fehrenbach, Phys. Rev. Lett. **43**, 963 (1979).
[13]J. J. Hopfield, Phys. Rev. **112**, 1555 (1958).
[14]L. V. Keldysh, Zh. Eksp. Teor. Fiz. **47**, 1515 (1964) [Sov. Phys. JETP **20**, 1018 (1965)].
[15]I. B. Levinson, *ibid.* **65**, 331 (1973) [**38**, 162 (1974)].
[16]A. L. Ivanov and L. V. Keldysh, Dokl. Akad. Nauk SSSR **264**, 1636 (1982) [Sov. Phys. Dokl. **27**, 482 (1982).
[17]V. L. Gurevich, Kinetics of Phonon Systems [in Russian], Nauka, 1980, §§10, 33, 34.

Translated by J. G. Adashko

Quantum nature of the reflection of an exciton from the surface of an electron-hole drop

N. A. Gippius, V. A. Zavaritskaya, L. V. Keldysh, V. A. Milyaev, and S. G. Tikhodeev
Institute of General Physics, Academy of Sciences of the USSR

(Submitted 1 October 1984)
Pis'ma Zh. Eksp. Teor. Fiz. **40**, No. 10, 416–418 (25 November 1984)

The small values of the sticking fraction for the impact of an exciton on the surface of an electron-hole drop are shown to be attributable to above-barrier quantum reflection of the exciton from the surface of a drop.

At low temperatures and very high levels of excitation of a semiconductor, the system of nonequilibrium carriers in Ge and Si separates into electron-hole drops (EHD) and a gaseous phase consisting of free excitons (FE) and free carriers.[1] The important parameter, which describes the interaction of the FE gas with the EHD is the sticking fraction for the impact of an exciton on the surface of a drop ξ. Hammond and Silver[2] and Voison *et al.*[3] estimated ξ to be ~ 0.05, for Si, whereas Milyaev and Sanina[4] and Sanina[5] found ξ to be ~ 0.1–0.2 for Ge. We will show that this low value is apparently a consequence of above-barrier quantum reflection of the exciton from the surface of an EHD and that it cannot be attributed to the phonon wind, as done in Refs. 2 and 3.

To explain theoretically the mechanism for the formation of EHD in Si, it was necessary to assume in Refs. 2 and 3 that the sticking fraction ξ in Si is very small: ξ

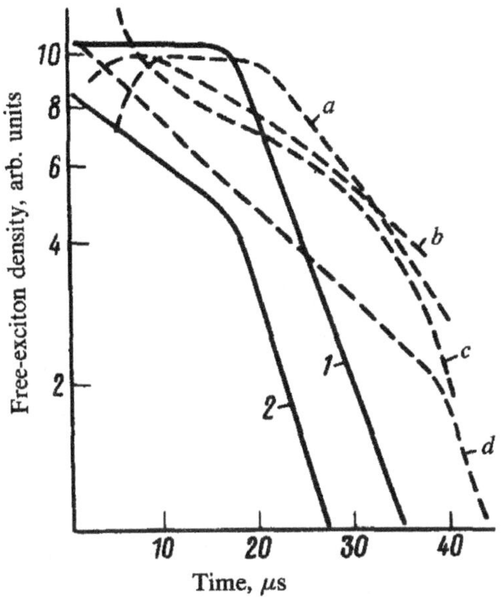

FIG. 1. Kinetics of the density of free excitons in germanium at $T \sim 4$ K. The experimental curves a, b, c, and d are taken from Refs. 8–11, respectively. Curves 1 and 2 of $n(t)$ were computed from Eq. (1) with $\xi = 1(1)$ and $\xi = 0.1(2)$.

~ 0.05. The mechanism for the formation of EHD in Ge was satisfactorily described by assuming that $\xi \sim 1$.[1,6] We note, however, that the kinetics of formation of the liquid phase in Ge is virtually insensitive to variation in ξ over a wide range $(0.01 < \xi < 1)$, because (see Ref. 7) the lifetime of carriers in an EHD in Ge is much longer than in Si. On the other hand, the transient characteristics of the decay of the EHD–FE system in Ge depend strongly on the value of ξ.

Figure 1 shows the decay curves for the density of FE at a temperature of ~ 4 K after pulsed laser excitation. The curves a, b, and c show the decrease in the luminescence signal of the FE with time (the 0.714-eV line).[8–10] The curve d was obtained from an analysis of the cyclotron resonance spectra recorded with a different time delay between the laser pulses.[11]

The time dependences of the density of the FE gas, $n(t)$, and of the average carrier density in the liquid phase, $N(t)$, in the course of the decay are described by

$$\begin{cases} \dfrac{dn}{dt} = -\dfrac{n}{\tau_{ex}} - \Gamma(n - n_T) \\ \\ \dfrac{dN}{dt} = -\dfrac{N}{\tau_0} + \Gamma(n - n_T), \end{cases} \tag{1}$$

where τ_{ex} and τ_0 are the lifetimes of FE and carriers in the EHD, respectively; n_T is the thermodynamic equilibrium density of FE; $\Gamma = \xi 4\pi R^2 N_d V_T$ is the rate of capture of excitons by a drop; R is the radius of the EHD; n_d is their number per cm^3 of the sample; and V_T is the average thermal velocity of FE. It turns out that if $\xi = 1$ and the other parameters in (1) vary within limits that are reasonable for Ge, the numerical solutions of (1) for $n(t)$ are virtually independent of time in the initial stage of decay of the system (curve 1 in Fig. 1). Slowly decaying dependences (curve 2), which are characteristic for all experimental results presented here, can be obtained only when $\xi = 0.1$.

By making use of the analogy between the electrostatic forces and the forces arising when an exciton absorbs nonequilibrium phonons,[12] it is easy to show that the phonon wind has a large effect on the flow of excitons to the EHD only at the lowest temperatures ($T < 0.1$ K for Si and $T = 0.01$ K for Ge), but it cannot explain the small values of ξ in Si and Ge.

The estimate $\xi = 1$, proposed in the preceding studies without allowance for the phonon wind, was based on the classical picture of the descent of an exciton into a potential well—the EHD. However, the classical analysis is applicable only if the de Broglie wavelength satisfies

$$\lambda = \hbar / p \ll a, \tag{2}$$

where p is the momentum of the particle; a is the smallest scale of inhomogeneities of the potential, which in our case is defined as the length of the transitional region from the EHD to the "excitonic atmosphere" and which (see Ref. 1) is approximately $a_{ex} = \epsilon\hbar^2/2\mu e^2$ (the exciton radius); m_e and m_h are the effective masses of the electron and hole; and $\mu = m_e m_h/(m_e + m_h)$ is the reduced mass. In the entire range of temperatures at which EHD exist ($kT \ll E_{ex} = \hbar^2/2\mu a_{ex}^2$), the condition (2) is not satisfied:

$$\frac{\lambda}{a_{ex}} \sim \sqrt{\frac{\mu}{m_e + m_h}} \sqrt{\frac{E_{ex}}{kT}} \gg 1. \tag{3}$$

Estimate (3) and, therefore, the quantum nature of the reflection of excitons from the surface of the EHD are a direct consequence of the relation $m_e \sim m_h$.

Although we do not know the general solution of the quantum problem of scattering of FE, i.e., of a bound state of two particles in an external potential field, we can show that when the kinetic energy of the complex as a whole, E_c, approaches zero, we have simultaneously $\xi \to 0$. This is a standard situation for problems which can be reduced to one-dimensional problems when $\xi \sim \sqrt{E_c}$. By assuming that the energy dependence is similar in nature in the case of reflection of FE, we can roughly estimate

$$\xi \sim \sqrt{kT/E_{ex}},$$

where E_{ex} is the only energy scale in the EHD–free carrier system. For $E_{ex} \sim 100$ K and $T \sim 4$ K, we obtain $\xi \sim 0.2$.

The value $\xi \sim 0.1$ obtained experimentally in Ge and Si can thus be explained by the quantum reflection of excitons from EHD. In addition to the well-known quantum oscillations of luminescence of EHD in strong magnetic fields,[1] the kinetics of recombination of free excitons in the presence of drops is a vivid experimental manifestation of the influence of quantum-mechanical effects on the behavior of a macroscopic system.

[1]Electron-Hole Droplets in Semiconductors, ed. by L. V. Keldysh and C. D. Jeffries, North Holland, New York, 1983.
[2]R. B. Hammond and R. N. Silver, Phys. Rev. Lett. **42**, 523 (1979).

[3]P. Voisin, B. Etienne, and M. Voos, Phys. Rev. Lett. **42**, 526 (1979).
[4]V. A. Milyaev and V. A. Sanina, Preprint FIAN No. 153, Moscow, 1980.
[5]V. A. Sanina, Candidate's Dissertation, Moscow, 1982.
[6]R. M. Westervelt, Phys Status Solidi B **74**, 727 (1976).
[7]S. G. Tikhodeev, Pis'ma Zh. Eksp. Teor. Fiz. **37**, 216 (1983) [JETP Lett. **37**, 255 (1983)].
[8]T. A. Astemirov, V. S. Bagaev, L. I. Paduchikh, and A. G. Poyarkov, Kratkie soobshcheniya po fizike FIAN (Brief Communications in Physics, Physics Institute of the USSR Academy of Sciences) **11**, 3 (1976).
[9]R. M. Westervelt, T. K. Lo, J. L. Staehli, and C. D. Jeffries, Phys. Rev. Lett. **32**, 1051 (1974).
[10]B. M. Ashkinadze and I. M. Fishman, Fiz. Tekh. Poluprovodn. **11**, 408 (1977) [Sov. Phys. Semicond. **11**, 235 (1977)].
[11]A. A. Manenkov, V. A. Milyaev, and V. A. Sanina, Dokl. Akad. Nauk SSSR **250**, 1371 (1980) [Sov. Phys. Dokl. **25**, 116 (1980)].
[12]L. V. Keldysh, Pis'ma Zh. Eksp. Teor. Fiz. **23**, 100 (1976) [JETP Lett. **23**, 86 (1976)].

Translated by M. E. Alferieff
Edited by S. J. Amoretty

High-intensity polariton waves near the threshold for stimulated Brillouin scattering

L. V. Keldysh and S. G. Tikhodeev

P. N. Lebedev Institute of Physics and General Physics Institute, USSR Academy of Sciences

(Submitted 11 December 1985)
Zh. Eksp. Teor. Fiz. **90**, 1852–1870 (May 1986)

We investigate the interaction in a crystal between an intense coherent polariton wave, whose amplitude is close to the threshold for Mandelstam-Brillouin scattering, and a background of scattered-polariton "noise." Correlations between scattered polaritons and phonons emitted during this scattering are taken into account, leading to the creation of mixed phonon-polariton modes. Near threshold, the decay rate of one of these modes reduces to zero, and the number of quanta in the mode grows. Backscattering leads to the appearance of fluctuations in the forward waves, consisting of correlated polariton pairs. We describe the system using diagram techniques devised for nonequilibrium processes, and solve Dyson-type equations in the so-called τ-approximation, in which the usual polarization operators for polaritons and phonons do not depend on the amplitude of the coherent wave while the anomalous phonon-polariton polarization operator is linear in this amplitude. We show that near threshold the τ-approximation ceases to be useful, due to accumulation of quanta in the weakly-damped mode; this leads formally to an increase in the number of diagrams, along with an increase in the order of perturbation theory to which the phonon-polariton interaction must be treated. We show that this problem can be avoided if we include a large number of single-loop diagrams in the expressions for the polarization operators, in which case near threshold all "dressed" diagrams become first order (as in the theory of phase transitions); a full solution of the problem then requires use of the renormalization group. In this paper we set up a self-consistent approximation for treating a simplified one-dimensional system (for example, polaritons in an optical fiber), taking into account only the single-loop diagrams in the polarization operators.

INTRODUCTION

In this paper we will investigate the interaction between a coherent electromagnetic wave (i.e., a polariton) propagating in a crystal with frequency close to the polariton resonance and scattered-polariton "noise," in the case when the amplitude of the coherent wave is close to the threshold for stimulated Mandelstam-Brillouin scattering. In this situation, the intensity of the noise is found to increase strongly, and the "feedback effect" of this noise on the propagation of the coherent wave becomes appreciable. In addition, this effect also gives rise to a significant modification in the spectra both of the scattered polaritons and of the propagating waves. The fact that polariton-acoustic phonon scattering accompanies each scattering event leads to the appearance of a certain coherence between polaritons and phonons. As a result, in place of the original polariton-phonon scattering there appear mixed polariton-phonon modes, just as the photon-exciton interaction gives rise to the polaritons themselves. Near the threshold for stimulated scattering, the decay constant of one of these mixed modes goes to zero; hence, there is an accumulation of quanta in this mode, and the intensity of the mode increases strongly. Under these conditions, the behavior both of the scattered polaritons and the phonons in resonance with them is entirely determined by only one of these weakly-damped modes, so that in practice the scattered polaritons and phonons behave as practically identical particles. The scattering process we investigate here is in effect transformed into one in which a coherent polariton wave decays into two practically identical "mixed" particles. The feedback effect of these scattered particles on the original wave (i.e., back-scattering) leads to the development of intense fluctuations around it, whose spectral width increases as threshold is approached. In polariton language, these fluctuations correspond to polariton pairs whose frequencies and propagation directions are close to the original wave; these pairs arise as a consequence of the intense interaction and secondary scattering of the weakly-damped mixed phonon-polariton modes.

The formal investigation of the physical picture described above is carried out here within the framework of the diagram theory of nonequilibrium processes. For simplicity, the amplitude of the coherent wave is taken to be given, and depends neither on time nor on the coordinates. From a formal point of view, this is equivalent to assuming there is in the medium a distribution of internal sources when sustain this amplitude despite the losses connected with scattering. From the standpoint of physics, this limits the region of applicability of the results we obtain to fields not too close to threshold, for which the intensity of the scattered waves is still small compared to the forward wave, which can therefore be treated approximately as a given "pump." The relevant quantitative criteria will be presented at the end of the article.

In section 1 we formulate rules for the diagram technique, and obtain and solve the equations for the normal

 0038-5646/86/051086-11$04.00

phonon and polariton propagators, the anomalous propagators for correlated phonons and scattered polaritons, and correlated pairs of forward polaritons.

In section 2 we develop the so-called τ-approximation, in which we take into account only low-order terms (not the linear terms mentioned above) in the expansion of the polarization operator in powers of the amplitude of the coherent wave. Within the framework of the τ-approximation we investigate the poles of the Green's function for polaritons which scatter with phonon emission (i.e., the Stokes component of the scattered-polariton spectrum) and for resonance phonons. The analogous problem for the anti-Stokes component was solved in Ref. 1. It is shown that when a certain threshold value of the intensity of the forward wave is attained, the decay constant of one of the mixed phonon-polariton modes changes sign, i.e., stimulated polariton scattering occurs.

In section 3, we investigate the properties of a system near the threshold for stimulated scattering. It is shown that the normal and anomalous propagators for scattered polaritons and phonons near threshold grow (which corresponds to the decrease in the decay constant of one of the mixed modes and the accumulation of quanta in it). Near the threshold, the τ-approximation is found to be inadequate; diagrams which are not taken into account in this approximation for the polarization operator of forward polaritons diverge as threshold is approached. Furthermore, the number of diagrams grows with increasing orders of perturbation theory. This latter difficulty can be overcome if we go beyond the τ-approximation framework and include in the forward-polariton polarization operator the single-loop diagrams, which are very large near threshold. Then the leading diagrams cancel out, and the order of the diagrams does not grow as the order of perturbation theory increases. Consequently, the situation near threshold for stimulated scattering is similar to the situation near a phase transition, and a full solution to the problem requires the use of the renormalization group.

In section 4 we construct a simplified model in which quanta of the coherent mode decay into pairs of quanta with the same free-particle spectrum. In addition, we investigate a one-dimensional system (for example, polariton waves in an optical fiber).

In sections 5 and 6, we set up a self-consistent approximation for the model system near threshold, analogous to the mean-field approximation—including one-loop diagrams for the polarization operator. It is shown that broad-spectrum, low-intensity fluctuations arise near the forward wave. This noisy component near the coherent wave gives rise to an essential change in the lifetime and occupation number of the scattered particles.

In conclusion, we analyze the applicability of the one-dimensional self-consistent approximation to a system of polaritons in an optical fiber.

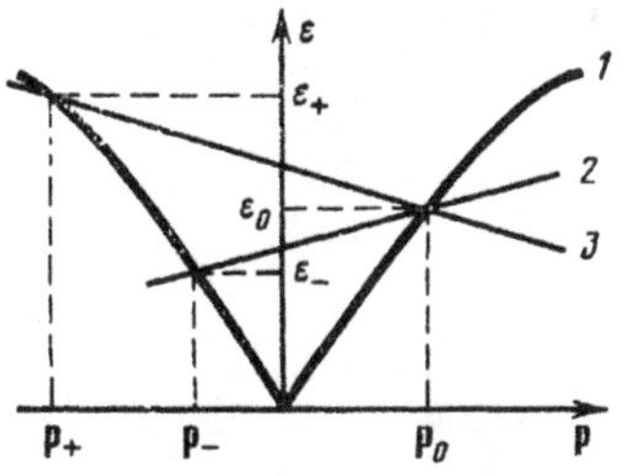

FIG. 1. Polariton branch (1) along with the absorption (2) and emission (3) spectra of phonons. The intersection of the dispersion manifolds with a plane passing through the vector p is shown.

1. SCATTERING OF POLARITONS IN A COHERENT MACROSCOPICALLY-OCCUPIED MODE OFF OF PHONONS: POINT OF VIEW

Let us investigate the interaction between a large-amplitude electromagnetic wave in a direct-gap semiconductor and acoustic phonons. As is well-known,[2] the exciton-photon interaction in direct-gap semiconductors leads to renormalization of the exciton and photon spectra, i.e., to the appearance of polariton branches. We will assume that this renormalization has already been included in the free-particle Hamiltonians for polaritons and phonons. We will treat the influence of the large-amplitude electromagnetic wave by assuming that one of the polariton modes is macroscopically occupied (i.e., its occupation number is large and proportional to the system volume), and furthermore that the mode is in a coherent state. Let $p_0 = (\varepsilon_0, \mathbf{p}_0)$ be the frequency and wave vector of this mode [here $\varepsilon_0 = \varepsilon(\mathbf{p}_0)$, and $\varepsilon(\mathbf{p})$ is the dispersion relation for the polariton]. Then the scattering of polaritons out of this mode with absorption and emission of phonons will also populate other polariton modes. We estimate their characteristic frequency and wave vector $p_\pm = (\varepsilon_\pm, \mathbf{p}_\pm)$ where the index "+" refers to absorption and "−" to emission of a phonon. Let us note that the matrix element for interaction with acoustic phonons grows with the momentum transfer, and is a maximum for backward scattering (see Fig. 1). We therefore take as a characteristic energy of the scattered polaritons the quantities $\varepsilon_\pm = \varepsilon(\mathbf{p}_\pm)$, where the $\mathbf{p}_\pm$, i.e., the characteristic momenta of the scattered polaritons, are solutions to the equation

$$\varepsilon(\mathbf{p}_0) = \varepsilon(\mathbf{p}_\pm) \pm u|\mathbf{p}_\pm - \mathbf{p}_0|, \qquad (1.1)$$

while u is the velocity of sound. The characteristic frequencies and momenta of the phonons are $p_\pm - p_0$. Since $p_\pm$ differ significantly from p_0, the back-scattered polaritons and forward-scattered polaritons are quanta of essentially different kinds. We will underline this state of affairs by using different notations for polariton propagators with $p \sim p_0$ and $p \sim p_\pm$. Because of this notational convention, the results we derive below are applicable to any three-quantum process in which a coherent macroscopically-occupied mode takes part.

In order to describe the system, we will make use of the diagram technique for nonequilibrium processes,[3] in which particle propagators are matrices with temporal indices i, $j = 1, 2$. In applying perturbation theory to the polariton-phonon interaction, we consider only "resonance" diagrams in which the coherent-mode polariton takes part, along with the neighboring modes with $p \sim p_0$, the scattered polaritons with $p \sim p_\pm$ and phonons with $k \sim p_0 - p_\pm$. Measuring frequencies and momenta of these particles from p_0, $p_\pm$, and $p_0-p_\pm$, we can display their free-particle Green's functions in the forms

$$D_0(p)=\begin{pmatrix} 0 & D_0^a \\ D_0^r & S_0 \end{pmatrix}=\begin{pmatrix} 0 & [\varepsilon-\varepsilon_0(\mathbf{p})-i0]^{-1} \\ [\varepsilon-\varepsilon_0(\mathbf{p})+i0]^{-1} & -2\pi i\delta[\varepsilon-\varepsilon_0(\mathbf{p})] \end{pmatrix}, \tag{1.2}$$

$$G_{0,pol}(p)=\begin{pmatrix} 0 & G_{0,pol}^a \\ G_{0,pol}^r & F_{0,pol} \end{pmatrix}=\begin{pmatrix} 0 & [\varepsilon-\varepsilon_\pm(\mathbf{p})-i0]^{-1} \\ [\varepsilon-\varepsilon_\pm(\mathbf{p})+i0]^{-1} & -2\pi i\delta[\varepsilon-\varepsilon_\pm(\mathbf{p})] \end{pmatrix} \tag{1.3}$$

$$G_{0,ph}(p)=\begin{pmatrix} 0 & G_{0,ph} \\ G_{0,ph}^r & F_{0,ph} \end{pmatrix}=\begin{pmatrix} 0 & \mp[\varepsilon-\omega_\pm(\mathbf{p})-i0]^{-1} \\ \mp[\varepsilon-\omega_\pm(\mathbf{p})+i0]^{-1} & -2\pi i(1+2N_\pm)\delta(\varepsilon-\omega_\pm) \end{pmatrix} \tag{1.4}$$

where

$$p=(\varepsilon,\mathbf{p}),\quad \varepsilon_\alpha(\mathbf{p})=\varepsilon(\mathbf{p}+\mathbf{p}_\alpha)-\varepsilon_\alpha,$$

$$\alpha=0,+,-,\quad \omega_\pm(\mathbf{p})=\mp u|\mathbf{p}+\mathbf{p}_0-\mathbf{p}_\pm|+\varepsilon_\pm-\varepsilon_0,$$

$$N_\pm=[\exp(\hbar u|\mathbf{p}_0-\mathbf{p}_\pm|/k_BT)-1]^{-1},$$

and $N_\pm$ is the phonon number density in thermal equilibrium. We have assumed that the temperature is small compared to ε_α/k_B, and that we can neglect the polaritons as a heat source. In order to derive (1.4) from the expressions usually used (see e.g., Ref. 3), we have taken the positive-frequency part for phonon emission and the negative-frequency part for phonon absorption. The indices "$\pm$" on the Green's functions $G_{0,\alpha}$ are not written explicitly, since in the resonance approximation the contributions of absorbed and emitted phonons cannot be confused, and can be computed separately.

Let us denote the matrix D_0 graphically by a wavy line (Fig. 2a), $G_{0,\text{pol}}$ by a straight line (Fig. 2b), and $G_{0,\text{ph}}$ by a dashed line (Fig. 2c). The interaction of polaritons and phonons in the deformation-potential approximation corresponds to the vertices in Fig. 2d. The point in the figures represents the matrix vertex function with temporal indices $i, j, l = 1, 2$:

$$i^{1/2}\mu_k M_{ij}^l(2\pi)^4\delta(p-k-p'), \tag{1.5}$$

where

$$\mu_k={}^1/_2 D(|\mathbf{k}-\mathbf{p}_\pm+\mathbf{p}_0|/\hbar\rho u)^{1/2},\qquad M_{ij}^l=\delta_{1,l}\delta_{i,j}+\delta_{2,l}(\sigma_x)_{ij}, \tag{1.6}$$

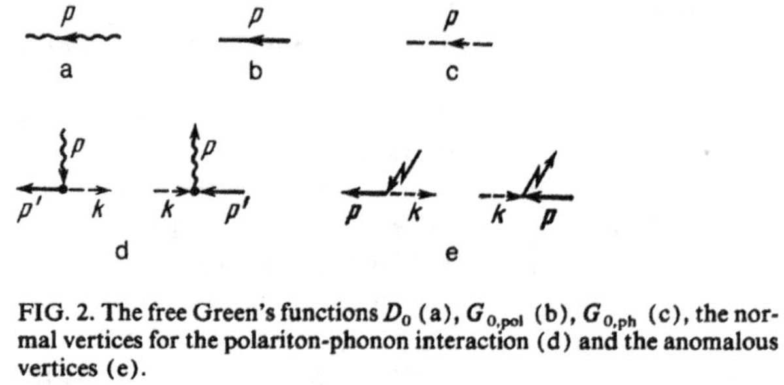

FIG. 2. The free Green's functions D_0 (a), $G_{0,\text{pol}}$ (b), $G_{0,\text{ph}}$ (c), the normal vertices for the polariton-phonon interaction (d) and the anomalous vertices (e).

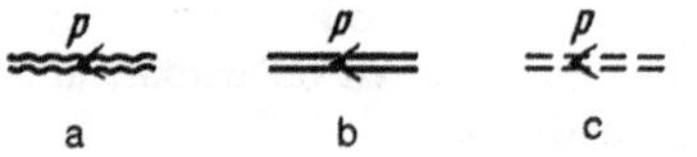

FIG. 3. Green's functions D (a), G_{pol} (b) and G_{ph} (c).

here, D is the polariton deformation potential constant, ρ is the semiconductor density, and σ_x is the first Pauli matrix.

Our treatment of the effect of the coherent mode with $p=p_0$ parallels the treatment of such modes in the diagrammatic analysis of Bose systems,[4] i.e., we add new vertices (of external-field type, see Fig. 2e) which consist of matrices with temporal indices $i, j = 1, 2$:

$$\Phi_{ij}(p)=\Phi_{\mathbf{p}}(2\pi)^4\delta(k+p)(\sigma_x)_{ij}, \tag{1.7}$$

where

$$\Phi_{\mathbf{p}}=\mu_{-\mathbf{p}}(2n_0)^{1/2}=D(n_0|\mathbf{p}+\mathbf{p}_\pm-\mathbf{p}_0|/2\hbar\rho u)^{1/2},$$

n_0 is the spatial density of polaritons in the coherent mode, which we take as an externally-imposed parameter. Likewise, in determining the anomalous vertices corresponding to creation and annihilation of polaritons, we have fixed the phase of the coherent wave at $\varphi=0$. This is legitimate, since later on we will be interested in stationary solutions which do not depend on the phase.

For a full description of the system, i.e., one which takes into account the anomalous vertices (1.7), in addition to the exact Green's functions for polaritons with $p\sim p_0$ (Fig. 3a):

$$D(p)=\begin{pmatrix} 0 & D^a \\ D^r & S \end{pmatrix}, \tag{1.8}$$

for scattered polaritons ($\alpha=$ pol, Fig. 3b) and for phonons ($\alpha=$ ph, Fig. 3c):

$$G_\alpha(p)=\begin{pmatrix} 0 & G_\alpha^a \\ G_\alpha^r & F_\alpha \end{pmatrix}, \tag{1.9}$$

we need to introduce anomalous propagators. For scattered polaritons and phonons, we call these functions $G_\Diamond(p,-p)$ and $G_\times(-p,p)$ (Figs. 4a and 4b); they are proportional respectively to $\langle T_c ab\rangle$ and $\langle T_c a^+b^+\rangle$, where a^+ and b^+ are creation operators for a polariton and a phonon. For polaritons with $p\sim p_0$, we use the Beliaev functions $D_\Diamond(p,-p)$ and $D_\times(-p,p)$ (Figs. 4c and 4d), corresponding to creation of polariton pairs out of the Bose-condensate and to annihilation of such pairs. We will use a notation analogous

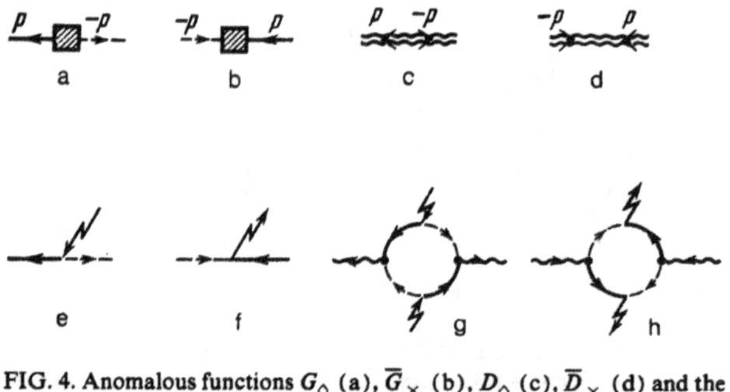

FIG. 4. Anomalous functions $G_\Diamond$ (a), $\overline{G}_\times$ (b), $D_\Diamond$ (c), $\overline{D}_\times$ (d) and the simplest diagrams which correspond to them in (e–h).

to (1.8) and (1.9) for the time components of these functions. The simplest diagrams for the anomalous propagators are shown in Figs. 4e–4h. Further on, we will use the abbreviated notation $G_\alpha \equiv G_\alpha(p)$, $\overline{G}_\alpha \equiv G_\alpha(-p)$, $G_\Diamond \equiv G_\Diamond(p, -p)$, and $\overline{G}_\times \equiv G_\times(-p, p)$, etc., when this will not lead to confusion.

It is convenient to combine these functions, which are all matrices in the temporal indices, into matrices $g_{\alpha\beta}(p)$, $d_{\alpha\beta}(p)$, $\alpha, \beta = 1,2$:

$$g(p) = \begin{bmatrix} G_{pol} & G_\Diamond \\ \overline{G}_\times & \overline{G}^T_{ph} \end{bmatrix}, \quad d(p) = \begin{bmatrix} D_\Diamond & D \\ \overline{D}^T & \overline{D}_\times \end{bmatrix}, \tag{1.10}$$

where "T" denotes the transpose of a matrix. Let us also take

$$g_0(p) = \begin{bmatrix} G_{0,pol} & 0 \\ 0 & \overline{G}^T_{0,ph} \end{bmatrix}, \quad d_0(p) = \begin{bmatrix} 0 & D_0 \\ \overline{D}_0^T & 0 \end{bmatrix}. \tag{1.11}$$

Let us emphasize that the matrices (1.10) and (1.11) are four-rowed matrices, since the quantities G_{pol}, etc., are themselves two-rowed matrices in the temporal indices. From here on we will denote matrices of this kind with lower case letters. The extra indices, i.e., the nontemporal ones, we will refer to as "external" indices and denote by Greek letters. On the graph, we will denote the matrices corresponding to the lines with large black arrows (see Fig. 5).

It is easy to see that any diagram for the Green's functions (1.8), (1.9) etc., can be obtained as a matrix component of the topologically-equivalent graphs for the functions (1.10), in which the propagators D_0 and $G_{0,\alpha}$ are replaced by the matrices (1.11), while the vertex matrix (the triangle in the diagrams in Fig. 5e)

$$i^{1/2}\mu_k m_{\alpha\beta}{}^{\gamma} = i^{1/2}\mu_k \delta_{\alpha,\gamma}(\sigma_x)_{\gamma\beta} \otimes M_{ij}{}^{k} \tag{1.12}$$

has only two non-zero components m^1_{12} and m^2_{21}.

Let us introduce the polarization operators, which consist of sums of all single-particle operators of a given type. We require polarization operators for scattered polaritons (α = pol), phonons (α = ph) and polaritons with $p \sim p_0$:

$$\Sigma_\alpha(p) = \begin{pmatrix} \Omega_\alpha & \Sigma_\alpha{}^r \\ \Sigma_\alpha{}^a & 0 \end{pmatrix}, \quad \Pi_{11}(p) = \begin{pmatrix} \Delta_{11} & \Pi_{11}{}^r \\ \Pi_{11}{}^a & 0 \end{pmatrix}, \tag{1.13}$$

as well as the anomalous operators for creation and annihilation of phonons and scattered polaritons $\Sigma_{02}(p, -p)$ and $\Sigma_{20}(-p, p)$ and also for creation and annihilation of pairs of $p \sim p_0$ polaritons: $\Pi_{02}(p, -p)$ and $\Pi_{20}(-p, p)$.[1)] The simplest diagrams for Σ_{02} and Σ_{20} consist of the vertices

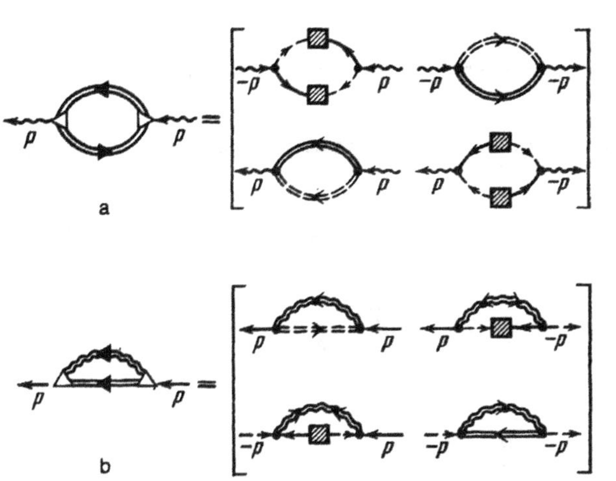

FIG. 6. Diagrams for the polarization operators for polaritons with $p \sim p_0$ (a), and scattered polaritons and phonons (b).

(1.7), i.e., linear in $n_0^{1/2}$ (or Φ). It is convenient to treat them separately, beginning with an expansion of Σ_{02} and Σ_{20} to third order in $n_0^{1/2}$. The simplest diagrams for Π_{02} and Π_{20} appear in the central blocks of Figs. 4g and 4h.

Let us combine the polarization operators into two matrices $\sigma_{\alpha\beta}$ and $\pi_{\alpha\beta}$:

$$\sigma(p) = \begin{bmatrix} \Sigma_{pol} & \Sigma_{02} \\ \Sigma_{20} & \Sigma_{ph}{}^T \end{bmatrix}, \quad \pi(p) = \begin{bmatrix} \Pi_{20} & \Pi_{11}{}^T \\ \Pi_{11} & \Pi_{02} \end{bmatrix}. \tag{1.14}$$

The simplest skeleton diagrams for these operators and their matrix components are shown in Fig. 6.

In addition, let

$$\varphi(p) = \begin{bmatrix} 0 & \Phi(p) \\ \Phi(p) & 0 \end{bmatrix}. \tag{1.15}$$

We obtain the standard equations for the functions (1.10):

$$g = g_0 + g_0(\sigma + \varphi)g, \tag{1.16}$$

$$d = d_0 + d_0 \pi d. \tag{1.17}$$

The structure of these equations for the various matrix components is shown in Fig. 7. It is easy to obtain the solutions to (1.16), (1.17) for the leading-order Green's functions:

$$g^a(p) \equiv \begin{bmatrix} G^a_{pol} & G_\Diamond{}^a \\ \overline{G}_\times{}^a & \overline{G}^T_{ph} \end{bmatrix} = \frac{1}{Z^a(p)} \begin{bmatrix} \overline{G}^{-1}_{0,ph} - \overline{\Sigma}^r_{ph} & \overline{\Sigma}^a_{20} + \Phi \\ \Sigma_{02}{}^a + \Phi & G^{-1}_{0,pol} - \Sigma^a_{pol} \end{bmatrix}, \tag{1.18}$$

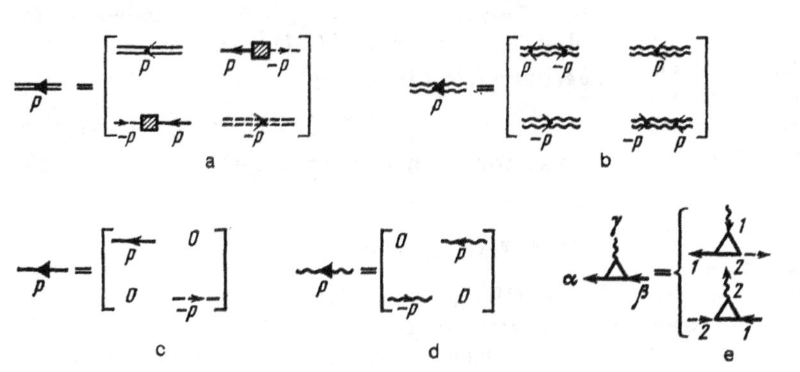

FIG. 5. Graphical notation for the matrices g, d, g_0, d_0, $i^{1/2}\mu m^{\gamma}_{\alpha\beta}$ along with their matrix structure.

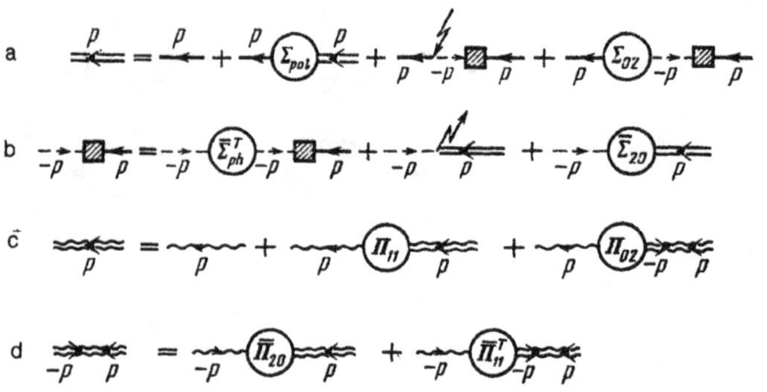

FIG. 7. Equations (1.16) and (1.17) for the components G_{pol} (a), $\bar{G}_\times$ (b), D (c) and $\bar{D}_\times$ (d).

$$d^a(p) \equiv \begin{bmatrix} D_\diamond{}^a & D^a \\ \bar{D}^r & \bar{D}_\times{}^a \end{bmatrix} = \frac{1}{Y^a(p)} \begin{bmatrix} \Pi_{02}{}^a & \bar{D}_0^{-1} - \bar{\Pi}_{11}{}^r \\ D_0^{-1} - \Pi_{11}{}^a & \bar{\Pi}_{20}{}^a \end{bmatrix}, \tag{1.19}$$

where

$$G_{0,pol}^{-1}(p) = \varepsilon - \varepsilon_\pm(\mathbf{p}), \qquad \bar{G}_{0,ph}^{-1} = G_{0,ph}^{-1}(-p) = \pm[\varepsilon + \omega_\pm(-\mathbf{p})],$$

$$D_0{}^{-1} = \varepsilon - \varepsilon_0(\mathbf{p}),$$

$$Z^a(p) = [\bar{G}_{0,ph}^{-1} - \Sigma_{ph}{}^r][G_{0,pol}^{-1} - \Sigma_{pol}^a] - [\Sigma_{20} + \Phi][\Sigma_{02} + \Phi], \tag{1.20}$$

$$Y^a(p) = [\bar{D}_0{}^{-1} - \bar{\Pi}_{11}{}^r][D_0{}^{-1} - \Pi_{11}{}^a] - \Pi_{02}{}^a \bar{\Pi}_{20}{}^a. \tag{1.21}$$

A more cumbersome expression for the static Green's functions can be found using the matrix formulae

$$f = g^r \omega g^a, \qquad s = d^r \delta d^a, \tag{1.22}$$

where

$$f = \begin{bmatrix} F_{pol} & F_\diamond \\ F_\times & \bar{F}_{ph} \end{bmatrix}, \qquad \omega = \begin{bmatrix} \Omega_{pol} & \Omega_{02} \\ \bar{\Omega}_{20} & \bar{\Omega}_{ph} \end{bmatrix},$$

$$s = \begin{bmatrix} S_\diamond, & S \\ \bar{S} & \bar{S}_\times \end{bmatrix}, \qquad \delta = \begin{bmatrix} \bar{\Delta}_{20} & \bar{\Delta}_{11} \\ \Delta_{11} & \Delta_{02} \end{bmatrix}. \tag{1.23}$$

Formula (1.22) follows from (1.16) and (1.17), and is analogous to the formula $F = G^r \Omega G^a$,[3] which is correct in the absence of anomalous Green's functions. For example,

$$F_{pol} = G_{pol}^r [\Omega_{pol} G_{pol}^a + \Omega_{02} \bar{G}_\times{}^a] + G_\diamond{}^r [\bar{\Omega}_{ph} \bar{G}_\times{}^a + \bar{\Omega}_{20} G_{pol}^a]. \tag{1.24}$$

2. THE τ-APPROXIMATION: QUASIPARTICLE (PHONORITON) SPECTRUM

Let us first consider only the lowest-order diagrams in power of $n_0^{1/2}$ (no higher than linear) for the polarization operators. In this approximation, which we call henceforth the τ-approximation, the normal polarization operators do not depend on Φ, while the anomalous ones, whose expansion in powers of Φ begins with quadratic (for Π_{02}, Π_{20}) and cubic (for Σ_{02}, Σ_{20}) terms, equal zero.

Near the corresponding surfaces $\varepsilon = \varepsilon(p)$, in the τ-approximation $\Sigma_\alpha(p)$ and $\Pi_{11}(p)$ can be expressed in terms of the polariton lifetime $\tau_{pol}(\mathbf{p}) = \gamma_{po}^{-1}(\mathbf{p})$ and phonon lifetime $\tau_{ph} = \gamma_{ph}^{-1}$ (see, e.g., Ref. 5):

$$\Sigma_{pol} = i\gamma_{pol} \begin{pmatrix} -2 & -1 \\ 1 & 0 \end{pmatrix}, \qquad \Pi_{11} = i\overset{0}{\gamma}_{pol} \begin{pmatrix} -2 & -1 \\ 1 & 0 \end{pmatrix}, \tag{2.1}$$

$$\Sigma_{ph} = i\gamma_{ph} \begin{pmatrix} -2(1+2N_\pm) & \mp 1 \\ \pm 1 & 0 \end{pmatrix}. \tag{2.2}$$

Here the upper sign, as before, refers to absorption of phonons, the lower one to emission. The solution to (1.17) is trivial:

$$D(p) = \begin{pmatrix} 0 & [\varepsilon - \varepsilon_0(\mathbf{p}) - i\overset{0}{\gamma}_{pol}]^{-1} \\ [\varepsilon - \varepsilon_0(\mathbf{p}) + i\overset{0}{\gamma}_{pol}]^{-1} & \dfrac{-2\pi i \overset{0}{\gamma}_{pol}}{[\varepsilon - \varepsilon_0(\mathbf{p})]^2 + [\overset{0}{\gamma}_{pol}]^2} \end{pmatrix}, \tag{2.3}$$

$$D_\diamond = D_\times = 0, \tag{2.4}$$

and the quantity which replaces the functions $G^a_{pol}, \ldots$ equals

$$Z^a(p) = [\bar{G}_{0,ph}^{-1} - \Sigma_{ph}{}^r][G_{0,pol}^{-1} - \Sigma_{pol}^a] - \Phi_p{}^2. \tag{2.5}$$

Let us investigate the zeroes of the function (2.5). We obtain a quadratic for the excitation spectrum:

$$[\pm\varepsilon \pm \omega_\pm(-\mathbf{p}) \mp i\gamma_{ph}][\varepsilon - \varepsilon_\pm(\mathbf{p}) - i\gamma_{pol}] - \Phi_p{}^2 = 0. \tag{2.6}$$

The behavior of the solutions (2.6) for the anti-Stokes component (the upper sign in 2.6) was analyzed in Ref. 1; we therefore limit ourselves to an analysis of the Stokes component. The two solutions to (2.6), which (following Ref. 1) we will call "phonoritons," take the form

$$\varepsilon_{1,2}(\mathbf{p}) = \tfrac{1}{2}\{\varepsilon_-(\mathbf{p}) - \omega_-(-\mathbf{p}) + i\Gamma$$
$$\pm([\varepsilon_-(\mathbf{p}) + \omega_-(-\mathbf{p}) + i\gamma]^2 - 4\Phi_p{}^2)^{1/2}\}, \tag{2.7}$$

where $\Gamma = \gamma_{ph} + \gamma_{pol}$, $\gamma = \gamma_{pol} - \gamma_{ph}$. Separating out the real and imaginary parts of (2.7),[2)] we obtain

$$\operatorname{Re} \varepsilon_{1,2}(\mathbf{p}) = \tfrac{1}{2}\{\varepsilon_-(\mathbf{p}) - \omega_-(-\mathbf{p})$$
$$\pm 2^{-1/2}[(Q_p{}^2 + R_p{}^2)^{1/2} + Q_p]^{1/2} \operatorname{sgn} R_p\}, \tag{2.8}$$

$$\operatorname{Im} \varepsilon_{1,2}(\mathbf{p}) = \tfrac{1}{2}\{\Gamma \pm 2^{-1/2}[(Q_p{}^2 + R_p{}^2)^{1/2} - Q_p]^{1/2}\}, \tag{2.9}$$

where

$$Q_p = [\varepsilon_-(\mathbf{p}) + \omega_-(-\mathbf{p})]^2 - \gamma^2 - 4\Phi_p{}^2,$$

$$R_p = 2[\varepsilon_-(\mathbf{p}) + \omega_-(-\mathbf{p})]\gamma. \tag{2.10}$$

In certain ranges of momentum the decay constant of one of the phonoritons can change sign (Im $\varepsilon_2(0)$. Over

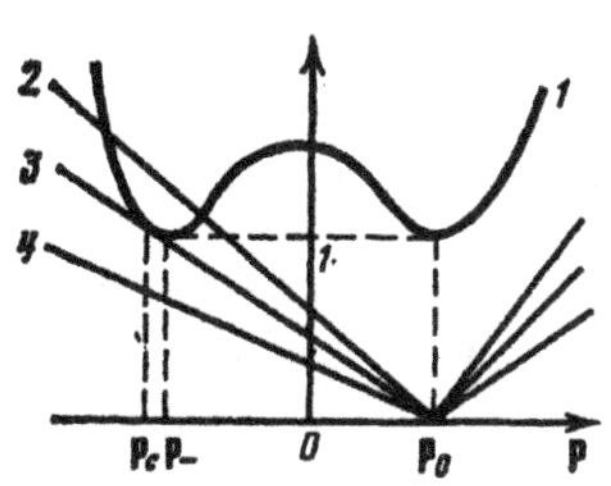

FIG. 8. Graphical solution to equation (2.11); the figure is explained in the text.

these ranges of p, it is possible to generate polaritons and phonons; therefore, the stationary approximation is meaningful only in the region below threshold.

From (2.9), it follows that the boundary surface in momentum space which satisfies the condition $\operatorname{Im}\varepsilon_2 = 0$ is given by the equation

$$1+[\varepsilon_-(\mathbf{p})+\omega_-(-\mathbf{p})]^2/\Gamma^2=\Phi_p^2/\bar{\gamma}^2, \tag{2.11}$$

where $\bar{\gamma} = (\gamma_{\rm pol}\gamma_{\rm ph})^{1/2}$. The properties of equation (2.11) are illustrated in Fig. 8, in which curve 1 shows the dependence of the left side of (2.11) on longitudinal momentum $p_\parallel = \mathbf{p}_0\cdot\mathbf{p}$ for $p_\perp = 0$. Curves 2–4 give the dependence of the right side of (2.11) on $p_\parallel$ for various coherent-mode densities. If the density is smaller than a certain critical value, i.e., $n_0 < n_c$ (curve 4), equation (2.11) has no solution and both phonoriton branches are stable. For $n_0 = n_c$ (curve 3), $\operatorname{Im}\varepsilon_2 > 0$ everywhere, excluding the tangent point $p = p_c$, at which $\operatorname{Im}\varepsilon_2 = 0$. For $n_0 > n_c$ (curve 2) there is a region of momenta around p in which $\operatorname{Im}\varepsilon_2 < 0$ and stimulated polariton scattering can occur. It is not hard to show that to within small terms of order $(\Gamma/\varepsilon_0)^2 \ll 1$

$$\mathbf{p}_c\approx\mathbf{p}_-. \tag{2.12}$$

Therefore

$$n_c\approx 2\bar{\gamma}^2\hbar\rho u/D^2|\mathbf{p}_--\mathbf{p}_0|,\qquad \Phi|_{\mathbf{p}=0,\ n_0=n_c}\approx\bar{\gamma}. \tag{2.13}$$

In Fig. 9 we illustrate schematically the functions $\operatorname{Re}\varepsilon_{1,2}$ (2.18) and $\operatorname{Im}\varepsilon_{1,2}$ (2.9) for $n_0 = n_c$.

3. ABOVE-THRESHOLD REGION: INADEQUACY OF THE τ-APPROXIMATION

Fundamental interest attaches to an investigation of the Green's functions (1.18), (1.22) in the region near threshold—for $n_0 = n_c(1-\lambda)$, $0<\lambda\ll 1$—and in a small region of frequency and momentum around p_-. In this case

$$\Phi_p^2\approx\bar{\gamma}^2(1-\lambda)\,|\mathbf{p}+\mathbf{p}_--\mathbf{p}_0|/|\mathbf{p}_--\mathbf{p}_0|. \tag{3.1}$$

Saving the leading terms in an expansion in $\lambda\ll 1$, we obtain

$$\operatorname{Im}\varepsilon_1(\mathbf{p})\approx\Gamma, \tag{3.2}$$

$$\operatorname{Im}\varepsilon_2(\mathbf{p})\approx(\bar{\gamma}^2/\Gamma)\,(\lambda+ap_\parallel^2+bp_\perp^2), \tag{3.3}$$

$$\operatorname{Re}\varepsilon_1(\mathbf{p})=\varepsilon_-(\mathbf{p})\gamma_{pol}/\Gamma-\omega_-(-\mathbf{p})\gamma_{ph}/\Gamma\approx-u_1p_\parallel+v_1p_\perp^2/2|\mathbf{p}_0|, \tag{3.4}$$

$$\operatorname{Re}\varepsilon_2(\mathbf{p})=\varepsilon_-(\mathbf{p})\gamma_{ph}/\Gamma-\omega_-(-\mathbf{p})\gamma_{pol}/\Gamma\approx-u_2p_\parallel+v_2p_\perp^2/2|\mathbf{p}_0|, \tag{3.5}$$

where

$$a=\left(\frac{c+u}{\Gamma}\right)^2,\qquad b=\frac{(c+u)^2}{8(c-u)c|\mathbf{p}_0|^2},\qquad c=\left|\frac{\partial\varepsilon(\mathbf{p})}{\partial\mathbf{p}}\right|_{\mathbf{p}=\mathbf{p}_-},$$

$$u_{1(2)}=c\gamma_{pol(ph)}/\Gamma-u\gamma_{ph(pol)}/\Gamma, \tag{3.6}$$

$$v_{1(2)}=c\frac{c+u}{c-u}\frac{\gamma_{pol(ph)}}{\Gamma}-u\frac{c+u}{2c}\frac{\gamma_{ph(pol)}}{\Gamma}.$$

In the Green's function (1.18) we can ignore the pole with $\varepsilon = \varepsilon_1(p)$ in view of the large damping of $\operatorname{Im}\varepsilon_1$ compared to $\operatorname{Im}\varepsilon_2$ (see Fig. 9b). Let

$$\varkappa_p=-u_2p_\parallel+v_2p_\perp^2/2|\mathbf{p}_0|+i(\bar{\gamma}^2/\Gamma)\,(\lambda+ap_\parallel^2+bp_\perp^2). \tag{3.7}$$

Then to the same accuracy as (3.2)–(3.5), we obtain

$$g(p)=\begin{bmatrix}(\gamma_{ph}/\Gamma)R(p) & i(\bar{\gamma}/\Gamma)R(p)\sigma_z\\ -i(\bar{\gamma}/\Gamma)\sigma_zR(p) & (\gamma_{pol}/\Gamma)\sigma_zR(p)\sigma_z\end{bmatrix}, \tag{3.8}$$

where

$$R(p)=\begin{pmatrix}0 & (\varepsilon-\varkappa_p)^{-1}\\ (\varepsilon-\varkappa_p^*)^{-1} & -4i\dfrac{\bar{\gamma}^2(1+N_-)}{\Gamma|\varepsilon-\varkappa_p|^2}\end{pmatrix}. \tag{3.9}$$

For small $p^{(3)}$ (for $|\varepsilon| \lesssim \bar{\gamma}^2\lambda/\Gamma$; $|\mathbf{p}| \lesssim \Gamma\lambda^{1/2}/c$), the components $R(p)$ of (3.9) are large: the off-diagonal terms are proportional to λ^{-1}, while $R_{22}\sim\lambda^{-2}$. As a consequence of this, near threshold the τ-approximation is insufficient, since there are diagrams not included in it which diverge for $\lambda\to 0$. For example, a simple estimate of the one-loop diagrams $\pi^{(1)}$ for the polarization operator (Fig. 6a) gives

$$\Pi_{02}^{(1)}\sim\Pi_{11}^{(1)}\propto\begin{pmatrix}\lambda^{-3+n/2} & \lambda^{-2+n/2}\\ \lambda^{-2+n/2} & 0\end{pmatrix}, \tag{3.10}$$

where n is the spatial dimension of the system. In order to improve the τ-approximation it is necessary to extract the leading diagrams in $\lambda^{-\alpha}$ $(\alpha > 0)$. We are therefore up against a problem, which at first glance rules out the possibility of our making such a choice in general. It turns out that if in the skeleton diagrams we substitute the functions d (2.3) and g (3.8) in the τ-approximation, the quantity α increases as we go to higher orders in perturbation theory. Thus, if we introduce the two new vertices (1.5) into a given diagram, there arise diagrams which, although having one extra integration $d\varepsilon d^np$ (giving a small contribution in the anomalous region, whose volume goes as $\lambda^{1+n/2}$), also contain large coefficients of type $fg^{a(r)}$ which go as λ^{-3}. Because in the τ-approximation $d\sim$ constant in the anomalous region, the

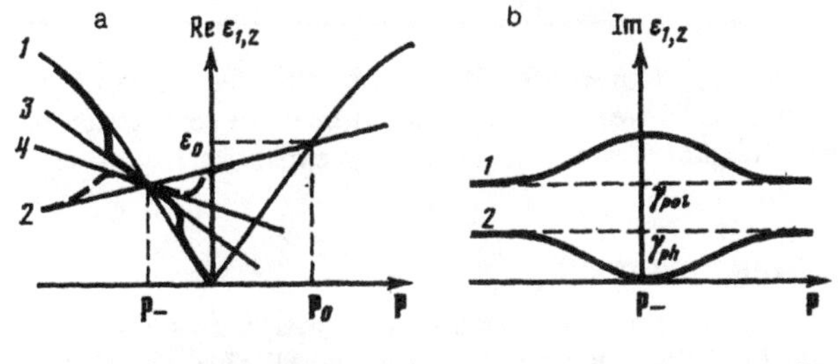

FIG. 9. (a) The dependence of $\operatorname{Re}\varepsilon_1(p)$ (curve 1) and $\operatorname{Re}\varepsilon_2$ (curve 2); (b) the same for $\operatorname{Im}\varepsilon_1(p)$ (curve 1) and $\operatorname{Im}\varepsilon_2(p)$ (curve 2). The straight lines 3 and 4 are projections onto the plane $p = 0$ of the tangent planes to the spectra at the point $\mathbf{p} = \mathbf{p}_-$.

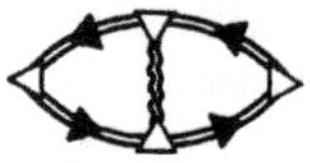

FIG. 10. The "correction" $\pi^{(2)}$ to the diagram $\pi^{(1)}$ in Fig. 6a, which is actually larger than the latter if in place of d and g we substitute their values in the τ-approximation.

magnitude of these diagrams is increased by a large factor $\lambda^{-2+n/2}$. For example, the components of the diagram $\pi^{(2)}$ in Fig. 10

$$\Pi_{02}^{(2)} \sim \Pi_{11}^{(2)} \infty \begin{pmatrix} \lambda^{-5+n} & \lambda^{-4+n} \\ \lambda^{-4+n} & 0 \end{pmatrix} \tag{3.11}$$

can be substantially larger than the components of the diagram for $\pi^{(1)}$ (Fig. 6a).

In this estimate, however, we have ignored the fact that to leading order in λ^{-1} the components of the function g are linear in the single function R. If we include in the polarization operator the large diagram $\pi^{(1)}$ (i.e., depart from the τ-approximation framework), this leads to mutual compensation of the leading terms in λ^{-1} for higher order diagrams, and these diagrams do not grow in size. We will demonstrate this by performing a certain linear transformation: let

$$x^{-1}=\begin{bmatrix} 2^{1/2}(\bar\gamma/\Gamma)I & 2^{1/2}i(\gamma_{pol}/\Gamma)\sigma_z \\ -2^{-1/2}i\sigma_z & -2^{-1/2}(\gamma_{ph}/\bar\gamma)I \end{bmatrix},$$

$$x=\begin{bmatrix} 2^{-1/2}(\gamma_{ph}/\bar\gamma)I & 2^{1/2}i(\gamma_{pol}/\Gamma)\sigma_z \\ -2^{-1/2}i\sigma_z & -2^{1/2}(\bar\gamma/\Gamma)I \end{bmatrix}, \tag{3.12}$$

where

$$x^{-1}x=xx^{-1}=\begin{bmatrix} I & 0 \\ 0 & I \end{bmatrix}, \quad I=\begin{pmatrix} 1 & 0 \\ 0 & 1 \end{pmatrix}.$$

It is easy to verify that for (3.8) the following relation holds

$$\tilde g(p)=x^{-1}g(p)x=\begin{bmatrix} R(p) & 0 \\ 0 & 0 \end{bmatrix}. \tag{3.13}$$

In addition, the vertices which transform according to the rule $\tilde m^{\gamma}_{\alpha\beta}=x^{-1}_{\alpha\alpha'}m^{\gamma}_{\alpha'\beta'}x_{\beta'\beta}$ take the form

$$\tilde m_{11}{}^1=-i(\bar\gamma/\Gamma)M\sigma_z,\qquad \tilde m_{11}{}^2=i(\bar\gamma/\Gamma)\sigma_z M, \tag{3.14}$$

where, for example,

$$\|M\sigma_z\|_{ij}{}^k=M^k_{ij'}(\sigma_z)_{j'j}.$$

(The vertices (3.14) are sufficient for us because only $\tilde g_{11}=R$ is non-zero).

Let us now include the single-loop diagrams $\pi^{(1)}$ in the polarization operator in addition to the term $\Pi^{(0)}$, i.e., (2.1). Then

$$\pi=\pi^{(0)}+\pi^{(1)}=\begin{bmatrix} -\sigma_z\Pi^{(1)} & \sigma_z(\Pi^{(0)}+\Pi^{(1)})\sigma_z \\ \Pi^{(0)}+\Pi^{(1)} & -\Pi^{(1)}\sigma_z \end{bmatrix}, \tag{3.15}$$

where the matrix

$$\|\Pi^{(1)}(p)\|_{ij}=\mu^2\frac{\bar\gamma^2}{\Gamma^2}\int\frac{d^{n+1}q}{(2\pi)^{n+1}}\,\mathrm{Sp}\,\sigma_z M^i R(q)M^j\sigma_z R(q-p), \tag{3.16}$$

and the trace is taken over the omitted indices. Formula (3.15) shows in particular that to leading order in λ^{-1} a formula holds which is analogous to the Hugenholtz-Pines formula for a nonideal Bose gas:

$$\Pi_{11}{}^a\Pi_{11}{}^r-\Pi_{02}{}^a\Pi_{20}{}^a=0. \tag{3.17}$$

The matrix (3.15) is diagonalized (in the "external" indices) with the help of the linear transformation

$$\tilde\pi=y^{-1}\pi y=\begin{bmatrix} \sigma_z\Pi^{(0)} & 0 \\ 0 & -(2\Pi^{(1)}+\Pi^{(0)})\sigma_z \end{bmatrix}, \tag{3.18}$$

where

$$y=2^{-1/2}\begin{bmatrix} I & -\sigma_z \\ \sigma_z & I \end{bmatrix},\quad y^{-1}=y^T. \tag{3.19}$$

If we limit ourselves to terms linear in p for a small-p expansion of $\varepsilon_0(\mathbf{p})$, then $\varepsilon_0(\mathbf{p})=-\varepsilon_0(-\mathbf{p})$ and

$$d_0=\begin{bmatrix} 0 & D_0 \\ \sigma_z D_0\sigma_z & 0 \end{bmatrix}. \tag{3.20}$$

This matrix is diagonalized in the "external" indices by the transformation $\tilde d_0=y^{-1}d_0 y$. Therefore, the solution to (1.17) is also diagonalized and takes the form

$$\tilde d=(\tilde d_0{}^{-1}-\tilde\pi)^{-1}=\begin{bmatrix} D_1\sigma_z & 0 \\ 0 & -\sigma_z D_2 \end{bmatrix}, \tag{3.21}$$

where D_1 coincides with the solution to 2.3 in the τ-approximation, while

$$D_2=\begin{pmatrix} 0 & (D_0{}^{-1}-\Pi^{(0)a}-2\Pi^{(1)a})^{-1} \\ (D_0{}^{-1}-\Pi^{(0)r}-2\Pi^{(1)r})^{-1} & (\Delta^{(0)}+2\Delta^{(1)})\,|D_0{}^{-1}-\Pi^{(0)a}-2\Pi^{(1)a}|^{-2} \end{pmatrix}. \tag{3.22}$$

For $\lambda \ll 1$ the operator $\Pi^{(1)}$ is the same order as (3.10) and $\Pi^{(1)} \gg \Pi^{(0)}$. Thus, in the anomalous region,

$$D_1 \infty \text{ const},\qquad D_2 \infty \begin{pmatrix} 0 & \lambda^{2-n/2} \\ \lambda^{2-n/2} & \lambda^{1-n/2} \end{pmatrix}. \tag{3.23}$$

On the other hand, the transformed vertices $\tilde{\tilde m}_{\alpha\beta}{}^{\gamma}=\tilde m_{\alpha\beta}{}^{\gamma'}y_{\gamma'\gamma}$ take the form

$$\tilde{\tilde m}_{11}{}^1=0,\qquad \tilde{\tilde m}_{11}{}^2=2^{1/2}i(\bar\gamma/\Gamma)\sigma_z M. \tag{3.24}$$

From this it is clear that the large functions R are connected through the vertices not with the finite D_1 but with the functions D_2. It is not difficult to show that the orders of the quantities D_2 are such that the addition of the new vertices to any diagram preserves its order in an expansion in powers of λ^{-1}.

4. MODEL SYSTEM

In the previous paragraph, it was shown that near threshold the τ-approximation is inapplicable, but that in principle it can be improved by a judicious choice of leading diagrams. Below we will construct a self-consistent approximation which takes into account one-loop diagrams in the polarization operators. So as to simplify the problem, we will investigate a model system in which quanta of the coherent mode can decay into two quanta within the same free-particle spectrum.[4)] Apparently, such a simplification is not too critical, since near threshold, as we have seen, even in the three-quanta problem the propagators for phonons and scattered polaritons coincide to within a numerical factor. A further simplification occurs because we are investigating only a one-dimensional system (for example, a polariton wave in a light fiber).

Dropping the subscripts pol and ph, we denote the components of the functions g and g_0 in the following way:

$$g=\begin{bmatrix} G & G_\lozenge \\ G_\times & G^T \end{bmatrix}, \quad g_0=\begin{bmatrix} G_0 & 0 \\ 0 & G_0^T \end{bmatrix}. \tag{4.1}$$

Correspondingly, we transform also the matrix σ (1.14):

$$\sigma=\begin{bmatrix} \Sigma_{11} & \Sigma_{02} \\ \Sigma_{20} & \Sigma_{11}^T \end{bmatrix}. \tag{4.2}$$

Including in the expansions of the spectra around the special points only leading terms linear in p (here p is a one-dimensional vector), we will use the following forms for the free-particle Green's functions:

$$G_0^{-1}(p)=(\varepsilon+w\mathbf{p})\sigma_z, \qquad D_0^{-1}(p)=(\varepsilon-v\mathbf{p})\sigma_z. \tag{4.3}$$

It is easy to show that for each of the functions g, d, σ and π only two of the four matrix components are linearly independent; for these functions, the following relations hold:

$$g=\begin{bmatrix} G & G_\lozenge \\ -\sigma_z G_\lozenge \sigma_z & \sigma_z G\sigma_z \end{bmatrix}, \quad \sigma=\begin{bmatrix} \Sigma_{11} & \Sigma_{02} \\ -\sigma_z\Sigma_{02}\sigma_z & \sigma_z\Sigma_{11}\sigma_z \end{bmatrix}, \tag{4.4}$$

$$d=\begin{bmatrix} D_\lozenge & D \\ \sigma_z D\sigma_z & \sigma_z D_\lozenge\sigma_z \end{bmatrix}, \quad \pi=\begin{bmatrix} \sigma_z\Pi_{02}\sigma_z & \sigma_z\Pi_{11}\sigma_z \\ \Pi_{11} & \Pi_{02} \end{bmatrix}. \tag{4.5}$$

The differences in sign between (4.4) and (4.5) are related to the fact that diagrams for G, $\Sigma_\parallel$, D, $\Pi_\parallel$, D_x and Π_{02} contain an even number of the vertices (1.7), while those for G_x and Σ_{02} contain an odd number.

The matrices (4.4) and (4.5) are diagonalized in the external indices with the help of the transformations (3.12) and (3.19). In the matrix (3.12) we must set $\gamma_{ph}=\gamma_{pol}$:

$$x=x^{-1}=2^{-1/2}\begin{bmatrix} I & i\sigma_z \\ -i\sigma_z & I \end{bmatrix}. \tag{4.6}$$

Then

$$\tilde{g}_0=x^{-1}g_0x=\begin{bmatrix} G_0 & 0 \\ 0 & \sigma_zG_0\sigma_z \end{bmatrix}, \quad \tilde{d}_0=y^{-1}d_0y=\begin{bmatrix} D_0\sigma_z & 0 \\ 0 & -\sigma_zD_0 \end{bmatrix}, \tag{4.7}$$

$$\tilde{g}=x^{-1}gx$$

$$=\begin{bmatrix} G-iG_\lozenge\sigma_z & 0 \\ 0 & \sigma_z(G+iG_\lozenge\sigma_z)\sigma_z \end{bmatrix}=\begin{bmatrix} G_1 & 0 \\ 0 & \sigma_zG_2\sigma_z \end{bmatrix}, \tag{4.8}$$

$$\tilde{d}=y^{-1}dy=\begin{bmatrix} (D+D_\lozenge\sigma_z)\sigma_z & 0 \\ 0 & -\sigma_z(D-D_\lozenge\sigma_z) \end{bmatrix}$$

$$=\begin{bmatrix} D_1\sigma_z & 0 \\ 0 & -\sigma_zD_2 \end{bmatrix}, \tag{4.9}$$

$$\tilde{\sigma}=x^{-1}\sigma x$$

$$=\begin{bmatrix} \Sigma_{11}-i\Sigma_{02} & 0 \\ 0 & \sigma_z(\Sigma_{11}+i\Sigma_{02})\sigma_z \end{bmatrix}=\begin{bmatrix} \Sigma_1 & 0 \\ 0 & \sigma_z\Sigma_2\sigma_z \end{bmatrix}, \tag{4.10}$$

$$\tilde{\pi}=y^{-1}\pi y=\begin{bmatrix} \sigma_z(\Pi_{11}+\Pi_{02}\sigma_z) & 0 \\ 0 & -(\Pi_{11}-\Pi_{02}\sigma_z)\sigma_z \end{bmatrix}$$

$$=\begin{bmatrix} \sigma_z\Pi_1 & 0 \\ 0 & -\sigma_z\Pi_2 \end{bmatrix} \tag{4.11}$$

$$\tilde{\varphi}=x^{-1}\varphi x=\begin{bmatrix} -\Phi\sigma_y & 0 \\ 0 & \Phi\sigma_z\sigma_y\sigma_z \end{bmatrix}. \tag{4.12}$$

The vertices, which transform according to the rule $\tilde{m}^{\alpha}_{\beta\gamma}=x_{\beta\beta'}\,m^{\alpha'}_{\beta'\gamma'}x_{\gamma'\gamma}y_{\alpha'\alpha}$ have four non-zero components

$$\tilde{m}_{11}{}^2=2^{-1/2}i\sigma_zM, \quad \tilde{m}_{22}{}^2=-2^{-1/2}iM\sigma_z,$$

$$\tilde{m}_{12}{}^1=-2^{-1/2}M, \quad \tilde{m}_{21}{}^1=-2^{-1/2}\sigma_zM\sigma_z. \tag{4.13}$$

After the diagonalization equations (1.16) and (1.17) are easily solved. The functions $G_{1,2}$ and $D_{1,2}$ which enter into the solutions [see (4.8), (4.9)] have the form

$$G_{1,2}(p)$$

$$=\begin{pmatrix} 0 & [\varepsilon+w\mathbf{p}-\Sigma^a_{1,2}(p)\pm i\Phi]^{-1} \\ [\varepsilon+w\mathbf{p}-\Sigma^r_{1,2}(p)\mp i\Phi]^{-1} & \Omega_{1,2}|\varepsilon+w\mathbf{p}-\Sigma^a_{1,2}\pm i\Phi|^{-2} \end{pmatrix} \tag{4.14}$$

(the upper sign refers to the function G_1) and

$$D_{1,2}(p)=\begin{pmatrix} 0 & [\varepsilon-v\mathbf{p}-\Pi^a_{1,2}(p)]^{-1} \\ [\varepsilon-v\mathbf{p}-\Pi^r_{1,2}(p)]^{-1} & \Delta_{1,2}|\varepsilon-v\mathbf{p}-\Pi^a_{1,2}|^{-2} \end{pmatrix}. \tag{4.15}$$

5. THE MODEL SYSTEM NEAR THRESHOLD

Let us first investigate the τ-approximation, in which

$$\Sigma_{02}=0, \quad \Sigma_{11}=i\gamma_0\begin{pmatrix} -2 & -1 \\ 1 & 0 \end{pmatrix}, \tag{5.1}$$

where γ_0 is the decay constant for g-particles in zero field. Then

$$G_{1,2}$$

$$=\begin{pmatrix} 0 & [\varepsilon+w\mathbf{p}-i(\gamma_0\mp\Phi)]^{-1} \\ [\varepsilon+w\mathbf{p}+i(\gamma_0\mp\Phi)]^{-1} & -2i\gamma_0[(\varepsilon+w\mathbf{p})^2+(\gamma_0\mp\Phi)^2]^{-1} \end{pmatrix}. \tag{5.2}$$

In the τ-approximation, therefore, the threshold for generation is reached when $\Phi_c=\gamma_0$. Near threshold, for $\Phi=\gamma_0(1-\lambda)$, $\lambda\ll1$, the components of the function $\tilde{g}_{11}=G_1$ are large, whereas the component function G_2 is finite (compare with (3.13), taking into account only the leading terms in λ^{-1}).

We now investigate the possibility of generating g-quanta outside the framework of the τ-approximation, i.e., when Σ_1 and Σ_2 are certain functions of Φ. We denote

$$\operatorname{Im}\Sigma^a_{1,2}(p,\Phi)\equiv\gamma_{1,2}(p,\Phi).$$

From (4.14) it is clear that for $\Phi\sim\Phi_c$, where Φ_c is a root of the equation

$$\gamma_1(0,\Phi)-\Phi=0, \tag{5.3}$$

the imaginary part of the denominator of G_1 is small and changes sign as Φ passes through Φ_c. Therefore, if there is a solution to (5.3), it corresponds to the threshold for generation of g-quanta. Let us assume that such a Φ_c exists (we will verify this *a posteriori*). As in section 3, we will study the system behavior below threshold but close to the threshold region. Let

$$\Phi_c-\Phi=\lambda\Phi_c \tag{5.4}$$

where the quantity $\lambda\ll 1$ is a parameter which determines how we will select our diagrams.

If $\gamma_1(\Phi)$ is differentiable for $\Phi\sim\Phi_c$, then for small p and $\Phi\sim\Phi_c$, analogous to (3.3) we have

$$\gamma_1\approx\Phi_c-(\partial\gamma_1/\partial\Phi)_{\Phi=\Phi_c}\lambda\Phi_c+a\mathbf{p}^2,\quad a\sim w^2/\gamma_0.$$

Therefore

$$(\gamma_1-\Phi)_{\Phi=\Phi_c(1-\lambda)}\approx\lambda\Psi_c+a\mathbf{p}^2, \tag{5.5}$$

where

$$\Psi_c=[1-(\partial\gamma_1/\partial\Phi)_{\Phi=\Phi_c}]\Phi_c. \tag{5.6}$$

For $|\varepsilon|\lesssim\lambda\Psi_c$, $|\mathbf{p}|\lesssim\lambda^{1/2}\Psi_c/w$, the functions $G_{1,2}$ can be cast in the form

$$G_1(p)\approx\begin{pmatrix}0 & [\varepsilon+w\mathbf{p}-i(\lambda\Psi_c+a\mathbf{p}^2)]^{-1}\\ [\varepsilon+w\mathbf{p}+i(\lambda\Psi_c+a\mathbf{p}^2)]^{-1} & -i\alpha_1[(\varepsilon+w\mathbf{p})^2+(\lambda\Psi_c+a\mathbf{p}^2)^2]^{-1}\end{pmatrix}, \tag{5.7}$$

$$G_2(p)\approx\begin{pmatrix}0 & [\varepsilon+w\mathbf{p}-i(\gamma_2+\Phi_c)]^{-1}\\ [\varepsilon+w\mathbf{p}+i(\gamma_2+\Phi_c)]^{-1} & -i\alpha_2[(\varepsilon+w\mathbf{p})^2+(\gamma_2+\Phi_c)^2]^{-1}\end{pmatrix}, \tag{5.8}$$

where

$$-i\alpha_{1,2}=\Omega_{1,2}(0),\qquad i\gamma_2=\Sigma_2^a(0). \tag{5.9}$$

Here it is assumed that the functions $\Omega_{1,2}(p)$ and $\Sigma_{1,2}(p)$ are analytic for $\Phi\sim\Phi_c$, $p\sim 0$. This assumption can be proved within the framework of the self-consistent approximation to be developed below.

The goal of the next investigation is to find self-consistent values for Φ_c, $\alpha_{1,2}$, γ_2, $(\partial\gamma_1/\partial\Phi)_{\Phi=\Phi c}$. The single-loop diagrams for the polarization operators included in the self-consistent approximation are representative in order of magnitude (see section 3). But higher-order diagrams, generally speaking, will also be of the same order in an expansion of λ^{-1}. The situation near threshold, consequently, coincides with the situation near a phase transition, and a full solution requires the use of the renormalization group. We will not solve this problem here, but rather limit ourselves to the mean-field approximation.

6. SELF-CONSISTENT APPROXIMATION

Let us first calculate the polarization operator (Fig. 6a), using the representations (5.7) and (5.8) for the functions $G_{1,2}$ in terms of the parameters Φ_c, $\alpha_{1,2}$, ... In addition, by making use of (4.15), we obtain also the function d. After then calculating $\tilde{\sigma}$ (Fig. 6b), we will compare the expressions we derive with (5.6), (5.9), and thereby obtain the self-consistency equation. The solutions to these equations in terms of the parameters Φ_c, $\alpha_{1,2}$ are presented in the following Table I:

TABLE I.

	$\frac{\Phi_c}{\gamma_0}$	$\frac{\alpha_1}{\gamma_0}$	$\left.\frac{\partial\gamma_1}{\partial\Phi}\right\vert_{\Phi_c}$	$\frac{\alpha_2}{\gamma_0}$	$\frac{\gamma_2}{\gamma_0}$
τ-approximation	1	2	0	2	1
Self-consistent approx.	1	8	−1	4	3

For comparison, we also present the values of these parameters within the τ-approximation (see section 5). In the self-consistent approximation, as is clear from Table I, a linear dependence on the field amplitude appears in the decay constant near threshold, and the occupation numbers are also renormalized. Thus, the functional form of the propagators for scattered polaritons and phonons is not changed. The fact that the value of the amplitude Φ_c is the same for both approximations is related to the neglect of the dependences of g and d on p (within the framework of the self-consistent approximation) for large p. The most significant disagreement between the two approximations is in their description of the particles with $p\sim p_0$ [compare (2.3) with (4.9) and (4.15); the expressions for the operators $\Pi_{1,2}$ in (4.15) are presented below, see (6.3), (6.4), (6.6) and (6.7)]. In the self-consistent approximation, near the coherent wave there appears a low-intensity but broad-spectrum noise amplitude.

We now turn to a description of the self-consistent approximation.[5)]

A. Polarization operator $\tilde{\Pi}$

Let us calculate the operator $\Pi_1(q)=\sigma_z\tilde{\pi}_{11}(q)$, $q=(\omega,\mathbf{q})$ (see (4.11) and Fig. 11a). Using (4.8) and (4.13), we obtain

$$\Delta_1(q)=i\mu^2\int\frac{d\varepsilon\,dp}{(2\pi)^2}[F_1(p)F_2(p-q)-G_1^a(p)G_2^r(p-q) -G_1^r(p)G_2^a(p-q)], \tag{6.1}$$

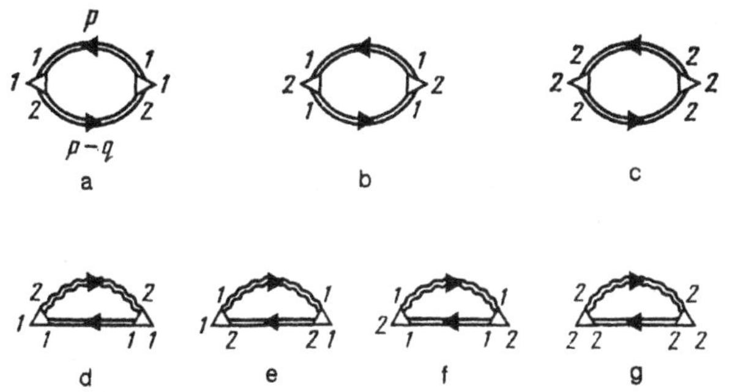

FIG. 11. One-loop diagrams for $\tilde{\pi}_{11}$(a), $\tilde{\pi}_{22}$ (b, c), $\tilde{\sigma}_{11}$ (d, e) and $\tilde{\sigma}_{22}$ (f, g). Here, the numbers are "external" indices.

$$\Pi_1^r(q)=-i\mu^2\int\frac{d\varepsilon\,dp}{(2\pi)^2}[F_1(p)G_2^a(p-q)-G_1^r(p)F_2(p-q)]. \tag{6.2}$$

The leading terms in (6.1) and (6.2) equal

$$\Delta_1(q)\approx\frac{-i\mu^2\alpha_1\alpha_2}{4a^{1/2}[(\omega+w\mathbf{q})^2+(\Phi_c+\gamma_2)^2]}(\lambda\Psi_c)^{-1/2}, \tag{6.3}$$

$$\Pi_1^r(q)\approx\frac{\mu^2\alpha_1}{4a^{1/2}[\omega+w\mathbf{q}+i(\Phi_c+\gamma_2)]}(\lambda\Psi_c)^{-1/2}. \tag{6.4}$$

We note that Π_1 is slowly-varying in the vicinity of $q=0$. With the help of (4.15) we find that over a wide interval of values $|\omega|\lesssim\Phi_c$, $|\mathbf{q}|\lesssim\Phi_c/\omega$

$$D_1(q)\approx D_1(0)$$

$$\approx\frac{4(\lambda\Psi_c a)^{1/2}(\Phi_c+\gamma_2)}{\alpha_1\mu^2}\begin{pmatrix}0 & i\\ -i & -i\alpha_2/(\Phi_c+\gamma_2)\end{pmatrix}. \tag{6.5}$$

We now calculate $\Pi_2(q)=-\tilde{\pi}_{22}(q)\sigma_z$ (Figs. 11b and 11c). The basic contribution to Π_2 comes from the integration over the anomalous region involved in the diagram in Fig. 11b:

$$\Pi_2^r(q)\approx-i\mu^2\int\frac{d\varepsilon\,dp}{(2\pi)^2}F_1(p)G_1^a(p-q)=-iA\lambda^{-1/2}Q^r(\eta,\xi), \tag{6.6}$$

$$\Delta_2(q)\approx\frac{i\mu^2}{2}\int\frac{d\varepsilon\,dp}{(2\pi)^2}[F_1(p)F_1(p-q)-G_1^aG_1^r-G_1^rG_1^a]$$

$$=-iB\lambda^{-1/2}2\,\mathrm{Re}\,P(\eta,\xi), \tag{6.7}$$

where

$$\eta=\frac{\omega+w\mathbf{q}}{\lambda\Psi_c},\quad \xi=\left(\frac{a}{\lambda\Psi_c}\right)^{1/2}\mathbf{q},\quad A=\frac{\mu^2\alpha_1}{8\Psi_c(a\Psi_c)^{1/2}},$$

$$B=\frac{\mu^2\alpha_1^2}{32\Psi_c^2(a\Psi_c)^{1/2}},$$

$$Q^r(\eta,\xi)=i\{(\xi+i\xi^2/2+\eta/2)^{-1}-[\xi(1+\xi^2/4-i\eta/2)^{1/2}+\eta/2]^{-1}(1+\xi^2/4-i\eta/2)^{-1/2}\}, \tag{6.8}$$

$$P(\eta,\xi)=[(\xi^2/2+i\xi)^2+\eta^2/4]^{-1}+[\xi^2(1+\xi^2/4-i\eta/2)-\eta^2/4]^{-1}(1+\xi^2/4-i\eta/2)^{-1/2}. \tag{6.9}$$

In contrast to Π_1, the operator Π_2 is strongly dependent on q in the neighborhood of $q=0$.

B. Polarization operator $\tilde{\sigma}$

Let us calculate the matrix component $\approx -i(\Phi_c+\gamma_2)/2$. (Figs. 11d, 11e). The diagram shown in Fig. 11d makes the following contribution to $\Omega_1(0)$:

$$\frac{i\mu^2}{2}\int\frac{d\varepsilon\,dp}{(2\pi)^2}[F_1(p)S_2(p)+G_1^aD_2^r+G_1^rD_2^a]. \tag{6.10}$$

It can be shown that after integrating over the anomalous region the basic contribution to (6.10), which is finite as $\lambda\to0$ and equal to $-3i\alpha_1/4$, comes from the first term. The integral for large p, where the function under the integral differs only slightly from the $\Phi=0$ function, together with the diagram in Fig. 11e, gives $-2i\gamma_0$. Consequently, the self-consistent value of α_1 is found from the relation

$$-i\alpha_1=\Omega_1(0)\approx-{}^3/_4 i\alpha_1-2i\gamma_0 \tag{6.11}$$

and so $\alpha_1\approx+8\gamma_0$.

Let us now calculate $\Sigma_1^r(0)$. The diagram in Fig. 11d gives

$$-\frac{i\mu^2}{2}\int\frac{d\varepsilon\,dp}{(2\pi)^2}[F_1(p)D_2^a(p)+G_1^r(p)S_2(p)]. \tag{6.12}$$

It can be shown that the integral in (6.12) over the anomalous region gives only a small contribution, proportional to λ and equal to $-i\lambda\Psi_cK$. Numerical calculations give $K\approx0.5$. On the other hand, the integral in (6.12) for large p, together with the diagram in Fig. 11e, gives a finite contribution as $\lambda\to0$ which equals $-i\gamma_0$. Thus, this self-consistency condition takes the form

$$-i\gamma_1|_{\Phi=\Phi_c(1-\lambda)}=\Sigma_1^r(0)=-i\gamma_0-iK\lambda\Psi_c. \tag{6.13}$$

From this, using (5.5) and (5.6), we obtain

$$\Phi_c=\gamma_0,\qquad(\partial\gamma_1/\partial\Phi)_{\Phi=\Phi_c}=K/(K-1)\approx-1. \tag{6.14}$$

It remains to calculate the matrix component $\Sigma_2(0)=\sigma_z\tilde{\sigma}_{22}\sigma_z$ (Figs. 11f, 11g). The diagram in Fig. 11f gives the following contribution to $\Omega_2(0)$:

$$\frac{i\mu^2}{2}\int\frac{d\varepsilon\,dp}{(2\pi)^2}[F_1(p)S_1(p)+G_1^aD_1^r+G_1^rD_1^a]. \tag{6.15}$$

The basic contribution to (6.15), which is finite as $\lambda\to0$ and equal to $-i\alpha_2/2$, is contained in the first term. The large-p integral for both the diagrams 11f and 11g gives $-2i\gamma_0$. Thus, the self-consistency condition for α_2 takes the form

$$-i\alpha_2=\Omega_2(0)\approx-i\alpha_2/2-2i\gamma_0, \tag{6.16}$$

or $\alpha_2\approx4\gamma_0$.

Diagram 11d for $\Sigma_2^r(0)$ gives

$$\frac{i\mu^2}{2}\int\frac{d\varepsilon\,dp}{(2\pi)^2}[F_1(p)D_1^a(p)+G_1^rS_1]. \tag{6.17}$$

The first term in (6.17) integrated over the anomalous region gives a finite contribution as $\lambda\to0$, equal to $\approx -i(\Phi_c+\gamma_2)/2$. Including the large-$p$ contribution to the integral, we obtain the self-consistency condition for γ_2:

$$-i\gamma_2=\Sigma_2(0)\approx-i(\Phi_c+\gamma_2)/2-i\gamma_0, \tag{6.18}$$

i.e., $\gamma_2\approx3\gamma_0$.

CONCLUSION

Let us discuss the applicability of the model we have just investigated to a real polariton system in an optical fiber. In order to use the one-dimensional model, it is necessary that only light waves with $p_\perp = 0$ propagate. Transverse modes will not be excited if the spacing (in frequency) between them satisfies $\Delta\omega \sim W/1$ (where W is the phase velocity of the wave and l is the thickness of the fiber) exceeds the spectral width γ_0. This can be achieved if we take a thin enough fiber, i.e., with $l \ll W/\gamma_0$.

In order to apply the self-consistent approximation, we had to show first that the one-loop polarization operators Π_1 (6.3), (6.4) and Π_2 (6.6)–(6.9) were large compared to the τ-approximation polarization operators

$$\Pi_1 \gg \Pi^{(0)}, \quad \Pi_2 \gg \Pi^{(0)}, \tag{7.1}$$

and second, that the higher-order diagrams in perturbation theory were small compared to $\Pi_{1,2}$. It is easy to show that in order of magnitude

$$\Pi_1 \sim \Lambda\lambda^{-1/2}\Pi^{(0)}, \quad \Pi_2 \sim \gamma_0\Lambda\begin{pmatrix} \lambda^{-3/2} & \lambda^{-1/2} \\ \lambda^{-1/2} & 0 \end{pmatrix}. \tag{7.2}$$

Here, the dimensionless quantity $\Lambda = \mu^2/\omega\gamma_0 < 1$, since the scattering of polaritons by phonons, which gives (as is not difficult to show) a contribution to the polariton width on the order of μ^2/w[6)] is not solely due to the broadening of the polariton levels. There are also in γ_0 contributions from interactions with optical phonons, impurity scattering, and losses at the surfaces of the optical fiber. The scale of nearness to the threshold for stimulated scattering is naturally determined by powers of Λ, which also define the regions I–IV shown in Fig. 12 in which various relations hold between the polarization operators $\Pi_{1,2}$ and $\Pi^{(0)}$. In region I the τ-approximation holds; in region IV the self-consistent approximation holds. In regions II and III, some of the parameters are determined by the τ-approximation while others are determined within the self-consistent scheme.

For the self-consistent approximation, the situation is worse as regards fulfilling the second applicability condition, since as we have already pointed out all the higher-order perturbation diagrams are, generally speaking, of the same order as $\Pi_{1,2}$. The analogous situation arises in the theory of phase transitions; a full solution to the problem requires the use of the renormalization group.

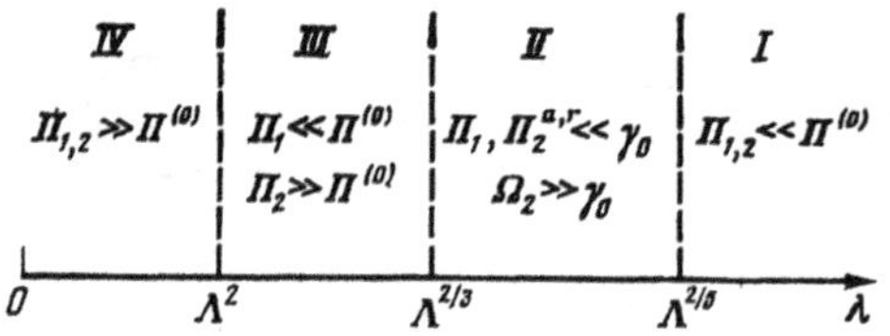

FIG. 12. Ranges of λ over which various inequalities hold between the components Π_1, Π_2, and $\Pi^{(0)}$.

Finally, let us discuss the region of applicability of the undepleted (i.e., prescribed pump approximation. It is not hard to show that the total spatial density of polaritons after scattering near threshold

$$n_- \sim n_c\Lambda\lambda^{-1/2}. \tag{7.3}$$

Consequently, the undepleted pump approximation obtains in regions I–III (Fig. 12) and does not hold near threshold in region IV, where losses to scattering become significant. (The intensity of polaritons before scattering in the framework of the self-consistent approximation is found to be small compared to n_c for all λ.)

1) We use the index notation from Ref. 4.

2) In this case, for calculating $z^{1/2}$ it is convenient to put the cut along the line $\mathrm{Im}\, z = 0$, $\mathrm{Re}\, z > 0$.

3) We will call this small region around $p = 0$ the "anomalous" region.

4) However, we will not assume these two quanta are identical particles in the quantum-mechanical senses. For this reason, the model developed below cannot be directly applied to problems such as the decay of optical phonons into acoustic phonons.

5) A similar calculation was presented in Ref. 7.

6) In the one-dimensional case, in determining the vertex (1.6), in place of a volume density ρ a line density down the optical fiber of ρl^2. Thus, the dimensions of $[\mu^2/w]$ are s^{-1}.

[1] A. L. Ivanov and L. V. Keldysh, Zh. Eksp. Teor. Fiz. **84**, 404 (1983) [Sov. Phys. JETP **57**, 234 (1983)].
[2] Eksitony (Excitons; Eds. E. I. Rashba and M. D. Stredzha). M.: Nauka, Chs. 2,3.
[3] L. V. Keldysh, Zh. Eksp. Teor. Fiz. **47**, 1515 (1964) [Sov. Phys. JETP **20**, 1018 (1964)].
[4] S. T. Beliaev, Zh. Eksp. Teor. Fiz. **34**, 417 (1958) [Sov. Phys. JETP **7**, 289 (1958)]; Zh. Eksp. Teor. Fiz. **34**, 433 (1958) [Sov. Phys. JETP **7**, 299 (1958)].
[5] I. B. Levinson, Zh. Eksp. Teor. Fiz. **65**, 331 (1973) [Sov. Phys. JETP **38**, 162 (1973)].
[6] N. Hugenholtz and D. Pines, Phys. Rev. **116**, 489 (1959).
[7] L. V. Keldysh and S. G. Tikhodeev, FIAN No. 331 (1985) (Preprint).

Translated by F. J. Crowne

Time-dependent Brillouin scattering of an intense polariton wave

L. V. Keldysh

P. N. Lebedev Physics Institute, USSR Academy of Sciences

S. G. Tikhodeev

Institute of General Physics, USSR Academy of Sciences

(Submitted 21 January 1986)
Zh. Eksp. Teor. Fiz. **91**, 78–85 (July 1986)

It is shown that the amplitude of the anti-Stokes line in a Brillouin-scattering spectrum can oscillate with time following the onset of an intense coherent polariton wave. The initial evolution of the correlated noise of scattered polaritons is investigated under conditions such that the intensity of the transmitted polariton wave greatly exceeds the stimulated-scattering threshold.

The analogy between the behavior of stimulated emission and scattering close to threshold, on the one hand, and second-order phase transitions, on the other, has been pointed out and discussed numerous times in the literature.[1-3] The similarity lies in the onset of a macroscopically coherent state superposed on intense fluctuation noise that increases as the threshold is approached. The interaction of the fluctuations with one another and with the coherent mode therefore plays a decisive role in both cases. A fundamental difference between the two is that whereas in phase transitions the fluctuations are in thermodynamic equilibrium, in stimulated emission they are excited by an external source. The statistical properties of the fluctuation ensemble in this second case are therefore, generally speaking, not universal but are determined by the properties of the excitation source and of its interaction with the medium. In particular, they can be described by some additional correlations, such as the correlation of the scattered polaritons and phonons in the specific problem discussed below. In addition, many stimulated processes (particularly scattering) are observed as a rule under essentially nonstationary conditions (pulsed excitation). Establishment of a stationary fluctuation pattern near threshold, on the other hand, is a long-time process. The question of the time-dependent onset of coherence is here much more significant than in the typical formulation of the second-order phase-transition problem. Furthermore, the time available for passage through the regions below and near threshold, is quite possibly too short for the noise level to reach values that can influence substantially the onset of a coherent mode. The evolution of the fluctuations can also be strongly influenced by the size of the samples, whose linear dimensions can be comparable with or even smaller than the photon mean free path.

Our present aim is to investigate certain phenomena of this kind, using as a very simple example the Brillouin scattering of polaritons in semiconductors. The stationary behavior of this system was investigated in Refs. 4 and 5, where the variation of the dispersion of the scattered polaritons under the influence of the exciting radiation and their spectral distribution was studied. We use here the same approach as in Refs. 4 and 5, based on introducing two types of Green's function—retarded and correlational.[6,7] As noted above, a characteristic feature of this system is the onset of the correlation of the scattered polaritons and the emitted phonons, and also of the scattered polaritons with one another. This additional coherence is manifested, in particular, in the intensity oscillations of anti-Stokes scattering following abrupt application of the pump wave. From the formal standpoint, this coherence is accounted for in natural fashion by the appearance of the so-called anomalous Green's functions, similar to the functions introduced by Belyaev[8,9] and by Gor'kov[9,10] in superfluidity and superconductivity theories. This also alters substantially the character of the equations that describe the spectral distribution of the scattered particles. The usual kinetic equation for an incoherent and strongly nonequilibrium many-particle system is replaced by an equation similar to the equation, well know in quantum radiophysics, for the density matrix of a two-level system in a resonant external field, with the off-diagonal matrix elements replaced by the anomalous Green's functions.

Especially noteworthy is also the physical meaning of the self-energy of these equations. In the so-called triangular representation, the off-diagonal elements (retarded and advanced) of the self-energy matrix are related, as in the equilibrium case,[9] to the polarizability of the medium and determine the propagation of the field in it. On the contrary, the diagonal element of the self-energy matrix plays the role of a correlator of the flucutuating field sources. In thermodynamic equilibrium, this term is uniquely related to the off-diagonal elements by the fluctuation-dissipation theorem, but under the conditions far from equilibrium considered here, it becomes a fully autonomous nontrival physical characteristic of the problem.

More specifically, we consider here the evolution of the Stokes and anti-Stokes components in the spectrum of scattered polaritons following an abrupt application to the crystal of an electric field of frequency ε_0 close to that of the polariton resonance. We use the equations obtained[5] for normal and anomalous Greens's functions by a diagram technique for nonequilibrium processes.[6,7] We confine ourselves here to the τ-approximation, i.e., assume that the normal polarization operators are independent of the external-field amplitude, and only terms linear in the field amplitude are included in the anomalous polariton-phonon polarization

 0038-5646/86/070045-05$04.00

operator that takes into account the correlations between the scattered polaritons and phonons. The normal polarization operators, whose imaginary parts specify the widths of the corresponding levels, are integrals over a large range of frequencies and momenta, and vary little in fields that are not too strong, so long as the spectrum restructuring and the change of the occupation numbers are concentrated in regions of frequencies and momenta small compared with ε_0 and $\mathbf{p}_0$ (where $\mathbf{p}_0$ is the quaasimomentum of the polariton wave). An exception, as shown in Ref. 5, is the behavior of the system near the stimulated-scattering threshold. In our problem involving the onset of the field, however, even this case is not dangerous (meaning that the τ-approximation can be used), since the accumulation of weakly damped phonoritons (mixed polariton-phonon modes), which alters the polarization operators, is localized in a narrow spectral range (compared with the combined width of the polariton and phonon levels), and proceeds quite slowly.

We use a mixed momentum-time representation of the Green's functions. In the diagram technique for nonequilibrium processes, the Green's functions are 2×2 matrices. In the "triangular" representation, which we shall use for the most part, the off-diagonal components are retarded and advanced Green's functions, while the nonzero diagonal component is the statistical function. Our procedure is the following. We first solve the equations for the retarded and advanced functions. To obtain a unique solution these equations must be supplemented by boundary conditions. Since the retarded (advanced) functions are proportional to the mean values of the commutators of the corresponding fields, at equal times the normal and anomalous functions are equal to $+i$ and zero, respectively. To find the intensities of the Stokes and anti-Stokes components, we next calculate the statistical functions (proportional to the mean values of the anticommutators), using[6] equations of the type $F=FG^r \Omega G^a$ (Ω is the statistical component of the polarization operator). We shall also find it convenient to combine[4] the normal and anomalous Green's functions into 4×4 matrices whose retarded, advanced, and statistical components are 2×2 matrices similar to (4) (see below). These 4×4 matrices and their temporal 2×2 components will be designated by appropriate lower-case letters.

The equations for the advanced Green's functions take the form

$$\begin{bmatrix} i\frac{\partial}{\partial t}-\varepsilon_{pol}(\mathbf{p}) & -\Phi(\mathbf{p},t)\exp(-i\varepsilon_0 t) \\ -\Phi(\mathbf{p},t)\exp(i\varepsilon_0 t) & \pm\left(i\frac{\partial}{\partial t}-\varepsilon_{ph}(\mathbf{p})\right)\end{bmatrix} \times g^a(t,t',\mathbf{p})=\delta(t-t')I, \qquad (1)$$

$$g^a(t,t',\mathbf{p})\begin{bmatrix} -i\frac{\partial}{\partial t'}-\varepsilon_{pol}(\mathbf{p}) & -\Phi(\mathbf{p},t')\exp(-i\varepsilon_0 t') \\ -\Phi(\mathbf{p},t')\exp(i\varepsilon_0 t') & \pm\left(-i\frac{\partial}{\partial t'}-\varepsilon_{ph}(\mathbf{p})\right)\end{bmatrix} =\delta(t-t')I. \qquad (2)$$

Here

$$I=\begin{bmatrix}1&0\\0&1\end{bmatrix},\quad \varepsilon_{pol}(\mathbf{p})=\varepsilon+i\gamma_{pol},\quad \varepsilon_{ph}(\mathbf{p})=\pm u|\mathbf{p}-\mathbf{p}_0|+i\gamma_{ph},$$

is the polariton dispersion law, u the speed of sound in the semiconductor, $\gamma_{pol(ph)}$ the reciprocal polariton (phonon) lifetime,

$$\Phi(\mathbf{p},t)=\Phi_{\mathbf{p}}\theta(t)=D(|\mathbf{p}-\mathbf{p}_0|n_0/2\hbar\rho u)^{1/2}\theta(t), \qquad (3)$$

D the deformation-potential constant for the polariton, ρ the semiconductor density, n_0 the spatial density of the coherent-mode polaritons,

$$\theta(t)=\begin{cases}0, & t<0\\ 1, & t>0\end{cases},$$

$$g^a(t,t',\mathbf{p})=\begin{bmatrix} G^a_{pol}(t,t',\mathbf{p}) & G^a_{\Diamond}(t,t',\mathbf{p},\mathbf{p}_0-\mathbf{p}) \\ G_\times{}^a(t,t',\mathbf{p}_0-\mathbf{p},\mathbf{p}) & G^r_{ph}(t',t,\mathbf{p}_0-\mathbf{p})\end{bmatrix}, \qquad (4)$$

G^a_{pol} the advanced Green's function of the scattered polariton, G^r_{ph} the retarded one of the photon, and G^a_Δ and $G^a_\times$ the advanced phonon-polariton anomalous Green's functions. Here and elsewhere the superscripts and subscripts refer to the anti-Stokes and Stokes components, respectively. It is assumed that the electromagnetic field is applied at the instant $t=0$. The functions (4) satisfy the boundary conditions

$$g^a(t,t+0,\mathbf{p})=\begin{bmatrix} i & 0\\ 0 & \pm i\end{bmatrix}. \qquad (5)$$

The Stokes and anti-Stokes components in (1), (2), and (5) differ in sign because the positive- and negative-frequency parts of the customarily employed Green's function are chosen for these respective components.

The solutions of Eqs. (1) and (2), satisfying conditions (5), take the following forms: For $t, t'<0$

$$g^a(t,t',\mathbf{p}) =i\begin{bmatrix}\exp\{i\varepsilon_{pol}(t'-t)\} & 0\\ 0 & \pm\exp\{i\varepsilon_{ph}(t'-t)\}\end{bmatrix}\theta(t'-t). \qquad (6)$$

For $t<0<t'$

$$\begin{aligned}
&G^a_{pol}(t,t',\mathbf{p})=[A_1\exp(i\varepsilon_1 t')+A_2\exp(i\varepsilon_2 t')]\exp(-i\varepsilon_{pol}t),\\
&G^a_\Diamond(t,t',\mathbf{p},\mathbf{p}_0-\mathbf{p})\\
&=B\exp(-i\varepsilon_{pol}t)[\exp(i(\varepsilon_1-\varepsilon_0)t')-\exp(i(\varepsilon_2-\varepsilon_0)t')],\\
&G_\times{}^a(t,t',\mathbf{p}_0-\mathbf{p},\mathbf{p})=B\exp(-i\varepsilon_{ph}t)[\exp(i\varepsilon_1 t')-\exp(i\varepsilon_2 t')],\\
&G_{ph}{}^r(t',t,\mathbf{p}_0-\mathbf{p})\\
&=\pm\exp(-i\varepsilon_{ph}t)[A_2\exp(i(\varepsilon_1-\varepsilon_0)t')+A_1\exp(i(\varepsilon_2-\varepsilon_0)t')].
\end{aligned} \qquad (7)$$

Finally, for $t, t'>0$

$$\begin{aligned}
&G^a_{pol}(t,t',\mathbf{p})\\
&=\theta(t'-t)[A_1\exp(i\varepsilon_1(t'-t))+A_2\exp(i\varepsilon_2(t'-t))],\\
&G^a_\Diamond(t,t',\mathbf{p},\mathbf{p}_0-\mathbf{p})=\theta(t'-t)B\exp(-i\varepsilon_0 t')[\exp(i\varepsilon_1(t'-t))\\
&\qquad-\exp(i\varepsilon_2(t'-t))],\\
&G_\times{}^a(t,t',\mathbf{p}_0-\mathbf{p},\mathbf{p})=\theta(t'-t)B\exp(i\varepsilon_0 t)[\exp(i\varepsilon_1(t'-t))\\
&\qquad-\exp(i\varepsilon_2(t'-t))],
\end{aligned} \qquad (8)$$

$$G_{ph}^{r}(t', t, \mathbf{p}_0-\mathbf{p})=\pm\theta(t'-t)[A_2\exp(i(\varepsilon_1-\varepsilon_0)(t'-t))+A_1\exp(i(\varepsilon_2-\varepsilon_0)(t'-t))].$$

Here

$$A_{1,2}=i(\varepsilon_{2,1}-\varepsilon_{pol})/(\varepsilon_{2,1}-\varepsilon_{1,2}),\quad B=\pm i\Phi_p/(\varepsilon_1-\varepsilon_2), \quad (9)$$

and the functions $\varepsilon_{1,2}(\mathbf{p})$ are the split "phonoriton" terms[4,5]:

$$\varepsilon_{1(2)}(\mathbf{p})=\varepsilon_0+{}^1/_2\{\varepsilon_p-\varepsilon_0\pm u|\mathbf{p}-\mathbf{p}_0|+i\Gamma+(-)[(\varepsilon_p-\varepsilon_0\mp u|\mathbf{p}-\mathbf{p}_0|+i\gamma)^2\pm4\Phi_p^2]^{1/2}\},$$
$$\Gamma=\gamma_{pol}+\gamma_{ph},\quad \gamma=\gamma_{pol}-\gamma_{ph}. \quad (10)$$

Since the scattering probability increases with momentum transfer [see (3)], the splitting of the phonoriton terms is greatest for backscattering into modes located near the intersection of the unperturbed polariton spectra and the absorbed and emitted phonons; the characteristic frequencies $\varepsilon_\pm$ and momenta $\mathbf{p}_\pm$ of the latter are determined from the conditions

$$\varepsilon_\pm=\varepsilon_{\mathbf{p}_\pm},\quad \varepsilon_\pm\mp u|\mathbf{p}_\pm-\mathbf{p}_0|=\varepsilon_0,\quad \mathbf{p}_\pm\|\mathbf{p}_0. \quad (11)$$

At $\mathbf{p}\sim\mathbf{p}_\pm$ the occupation numbers of the anti-Stokes and Stokes components of the scattered polaritons are maximal. To find them, we calculate first the statistical normal and anomalous Green's functions, using the equation[5]

$$f(t,t',\mathbf{p})=\begin{bmatrix} F_{pol}(t,t',\mathbf{p}) & F_\Diamond(t,t',\mathbf{p},\mathbf{p}_0-\mathbf{p}) \\ F_\times(t,t',\mathbf{p}_0-\mathbf{p},\mathbf{p}) & F_{ph}(t',t,\mathbf{p}_0-\mathbf{p}) \end{bmatrix}$$
$$=\iint_{-\infty}^{\max(t,t')} dt_1\,dt_2\,g^r(t,t_1,\mathbf{p})\,\omega(t_1,t_2,\mathbf{p})\,g^a(t_2,t',\mathbf{p}), \quad (12)$$

where the statistical components of the polarization operators take in the τ-approximation the form

$$\omega(t_1,t_2,\mathbf{p})=\delta(t_1-t_2)\begin{bmatrix} -2i\gamma_{pol}(1+2N_{0,pol}) & 0 \\ 0 & -2i\gamma_{ph}(1+2N_{0,ph}) \end{bmatrix}, \quad (13)$$

where

$$N_{0,ph}=[\exp(u|\mathbf{p}-\mathbf{p}_0|/k_BT)-1]^{-1},$$
$$N_{0,pol}=[\exp(\varepsilon_p/k_BT)-1]^{-1}$$

are the bare "thermal" occupation numbers of the scattered polaritons and phonons, while the retarded Green's functions are Hermitian adjoints of the advanced ones: $g^r(t,t',\mathbf{p})=[g^a(t',t,\mathbf{p})]^+$.

The equal-time statistical functions are connected with the occupation numbers by the relations (α = pol, ph)[6,7]

$$N_\alpha(\mathbf{p},t)=[iF_\alpha(t,t,\mathbf{p})-1]/2. \quad (14)$$

Simple calculations yield[1)]

$$N_{pol}(\mathbf{p},t)=N_{0,pol}+\left[N_{0,ph}\mp N_{0,pol}+\frac{1\mp1}{2}\right]\frac{\Phi_p^2}{\omega_p^2+(\gamma_1-\gamma_2)^2}$$
$$\times\left[(1-e^{-2\gamma_1 t})\frac{\gamma_{ph}-\gamma_1}{\gamma_1}+(1-e^{-2\gamma_2 t})\frac{\gamma_{ph}-\gamma_2}{\gamma_2}+\frac{2(\omega_p^2+\gamma\Gamma)}{\omega_p^2+\Gamma^2}\right.$$
$$\left.-\frac{2(\omega_p^4+\omega_p^2(\Gamma^2+\gamma^2)+\gamma^2\Gamma^2)^{1/2}}{\omega_p^2+\Gamma^2}e^{-\Gamma t}\sin(\omega_p t+\psi_{pol})\right], \quad (15)$$

where

$$\omega_p=\mathrm{Re}(\varepsilon_1-\varepsilon_2),\quad \gamma_{1,2}=\mathrm{Im}\,\varepsilon_{1,2},$$
$$\mathrm{tg}\,\psi_{pol}=(\omega_p^2+\gamma\Gamma)/\omega_p(\Gamma-\gamma).$$

To obtain the corresponding equation for $N_{ph}(\mathbf{p}_0-\mathbf{p},t)$, we must interchange the subscripts pol and ph in (15) (and, in particular, replace γ by $-\gamma$).

We consider first the anti-Stokes scattering [the superscript in (10) and (15)]. Equation (15) describes the establishment of the stationary occupation numbers of the scattered polaritons. This process is relaxational in weak fields, i.e., at $\Phi_{\mathbf{p}_+}<\Gamma$ and oscillatory in strong fields $\Phi_\mathbf{p}>\Gamma$. The oscillation frequency equals the splitting of the phonoriton terms; oscillations set in when the splitting exceeds the sum of the line widths.

If the electromagnetic field is turned on for a time τ, it can be easily shown that

$$N_\alpha(t>\tau)=N_{0,\alpha}[1-\exp(-2\gamma_\alpha(t-\tau))]+N_\alpha(\tau)\exp(-2\gamma_\alpha(t-\tau)), \quad (16)$$

(α = pol, ph). If, for example, $\tau\sim\omega_{\mathbf{p}_+}\ll\Gamma^{-1}$, the anti-Stokes component modulation depth is large and small changes of the duration of the passing pulse τ or of the intensity of the transmittted wave can give rise to considerable changes, from τ to Γ^{-1}, of the duration of the back-reflected anti-Stokes pulse.

For Stokes scattering, the splitting of the phonoriton terms in the central part of the line near $\mathbf{p}\sim\mathbf{p}_-$ is always small compared with Γ. No line-center oscillations are therefore produced. In sufficiently strong fields, at $\Phi_{\mathbf{p}_-}>(\gamma_{ph}\gamma_{pol})^{1/2}$, the sign of the damping γ_2 is reversed[5] in a certain momentum region near $\mathbf{p}_-$, and stimulated scattering sets in. It follows from (15) that in this case the number of scattered polaritons and of emitted phonons increases exponentially with time. If $\Phi_{\mathbf{p}_-}\gg(\gamma_{ph}\gamma_{pol})^{1/2}$ and there is no thermal source of polaritons ($N_{0,pol}=0$), we have

$$N_{pol}(\mathbf{p}_-,t)\approx{}^1/_4(1+N_{0,ph})\exp(2\Phi_{\mathbf{p}_-}t). \quad (17)$$

Note that $\omega_\mathbf{p}>\Gamma$ at the end points of the Stokes line and oscillations of the occupation numbers are possible.

In strong fields $\Phi_{\mathbf{p}_-}\gg(\gamma_{ph}\gamma_{pol})^{1/2}$ phonoriton backscattering causes a rapid buildup of the transmitted-phonoriton-wave fluctuations, consisting of correlated pairs of polaritons with momenta $\mathbf{p}_0+\mathbf{k}$ and $\mathbf{p}_0-\mathbf{k}$. The approach developed here permits a description of the initial stage of the onset of the noise, for times $t<\Gamma^-$ when the intensity of the scattered polariton wave is still low compared with that of the transmitted wave, and the fixed pump approximation is applicable. In second order of perturbation theory in the polariton-phonon intersection, we obtain[2] at $N_{0,pol}=0$ the following expressions for the density of the "noise" polaritons near the transmitted wave

$$N_{pol}(\mathbf{p}_0+\mathbf{k},t)=\langle a^+(\mathbf{p}_0+\mathbf{k},t)a(\mathbf{p}_0+\mathbf{k},t)\rangle$$

and for the equal-time noise correlator $\langle a(\mathbf{p}_0-\mathbf{k},t)\,a(\mathbf{p}_0+\mathbf{k},t)\rangle$ (where $a(\mathbf{p},t)$ is the polariton annihilation operator in the Heisenberg representation and $|\mathbf{k}|\ll|\mathbf{p}_0|$):

$$N_{pol}(\mathbf{p}_0+\mathbf{k},t)\approx \mu^2(1+N_{0,ph})^2\int \frac{d^3p}{(2\pi)^3}\exp[2\Gamma(\mathbf{p},\mathbf{k})t]$$
$$\times\{[\varepsilon(\mathbf{p}_0+\mathbf{k})-\varepsilon_0-\Omega(\mathbf{p},\mathbf{k})]^2$$
$$+[\Gamma(\mathbf{p},\mathbf{k})+\gamma_{pol}]^2\}^{-1}|B(\mathbf{p})B(\mathbf{p}-\mathbf{k})|^2$$
$$\times[\gamma_2(\mathbf{p}-\mathbf{k})-\gamma_{pol}][\gamma_2(\mathbf{p})-\gamma_{ph}][\gamma_2(\mathbf{p}-\mathbf{k})\gamma_2(\mathbf{p})]^{-1}, \qquad (18)$$

$$\langle a(\mathbf{p}_0-\mathbf{k},t)a(\mathbf{p}_0+\mathbf{k},t)\rangle\approx \mu^2(1+N_{0,ph})^2\exp(-2i\varepsilon_0 t)$$
$$\times\int\frac{d^3p}{(2\pi)^3}\exp[2\Gamma(\mathbf{p},\mathbf{k})t]\{[\varepsilon(\mathbf{p}_0+\mathbf{k})-\varepsilon_0-\Omega(\mathbf{p},\mathbf{k})]^2$$
$$+[\Gamma(\mathbf{p},\mathbf{k})+\gamma_{pol}]^2\}^{-1}B^*(\mathbf{p})B^*(\mathbf{p}-\mathbf{k})A_1(\mathbf{p})A_1(\mathbf{p}-\mathbf{k})$$
$$\times[\gamma_2(\mathbf{p})-\gamma_{ph}]$$
$$\times[\gamma_2(\mathbf{p}-\mathbf{k})-\gamma_{ph}][\gamma_2(\mathbf{p})\gamma_2(\mathbf{p}-\mathbf{k})]^{-1}, \qquad (19)$$

where

$$\mu^2={}^1/_4D^2|\mathbf{p}_--\mathbf{p}_0|/\hbar\rho u,$$

$$\Omega(\mathbf{p},\mathbf{k})=\mathrm{Re}\,[\varepsilon_2(\mathbf{p})-\varepsilon_2^*(\mathbf{p}-\mathbf{k})],$$

$$\Gamma(\mathbf{p},\mathbf{k})=-\mathrm{Im}\,[\varepsilon_2(\mathbf{p})-\varepsilon_2^*(\mathbf{p}-\mathbf{k})]. \qquad (20)$$

Using (9) and (10) we obtain

$$\langle a(\mathbf{p}_0-\mathbf{k},t)a(\mathbf{p}_0+\mathbf{k},t)\rangle\approx\exp(-2i\varepsilon_0 t)N_{pol}(\mathbf{p}_0+\mathbf{k},t) \qquad (21)$$

and

$$N_{pol}(\mathbf{p}_0,t)\approx(1+N_{0,ph})^2\exp(4\Phi_{\mathbf{p}_-}t)\,\Gamma/\Phi_{\mathbf{p}_-}. \qquad (22)$$

We emphasize that Eqs. (18)–(22) are valid in strong fields, when the transmitted-wave intensity exceeds substantially the stimulated-scattering threshold, and during the initial stage of the process, at times t satisfying the condition $\Phi_{\mathbf{p}_-}^{-1}\lesssim t\ll\Gamma^{-1}$.

We note in conclusion that the question of the effects observable outside the crystal calls for additional consideration of the coefficient of polariton passage through the crystal boundary, which is also changed by the change of the polariton spectrum.

APPENDIX

By calculating (12) with the aid of (13) and (6)–(8) we obtain, e.g., at $t>t'$

$$F_\alpha(t,t',\mathbf{p})=\tilde F_\alpha(t,t',\mathbf{p})+(1+2N_{0,\alpha})G_\alpha^r(t,t',\mathbf{p}),$$
$$F_\diamond(t,t',\mathbf{p},\mathbf{p}_0-\mathbf{p})$$
$$=\tilde F_\diamond(t,t',\mathbf{p})\pm(1+2N_{0,ph})G_\diamond^r(t,t',\mathbf{p},\mathbf{p}_0-\mathbf{p}), \qquad (A.1)$$
$$F_\times(t,t',\mathbf{p}_0-\mathbf{p},\mathbf{p})$$
$$=\tilde F_\times(t,t',\mathbf{p})-(1+2N_{0,pol})G_\times^r(t,t',\mathbf{p}_0-\mathbf{p},\mathbf{p}),$$

where $\alpha=$ pol, ph and $f=0$ if $0>t>t'$ and $t>0>t'$. If $t>t'>0$ we have

$$\tilde F_{pol}(t,t',\mathbf{p})=-i|B|^2(E_{11}+E_{22}-E_{12}-E_{21}), \qquad (A.2)$$
$$\tilde F_\diamond(t,t',\mathbf{p})$$
$$=\mp i\exp(-i\varepsilon_0 t')B^*(A_2E_{11}-A_1E_{22}+A_1E_{12}-A_2E_{21}), \qquad (A.3)$$
$$\tilde F_\times(t,t',\mathbf{p})$$
$$=\pm i\exp(i\varepsilon_0 t)B(A_2^*E_{11}-A_1^*E_{22}-A_2^*E_{12}+A_1^*E_{21}), \qquad (A.4)$$

where

$$E_{ij}=(1\mp1+2N_{0,ph}\mp2N_{0,pol})\exp(-i\varepsilon_i^*t+i\varepsilon_j t')$$
$$\times\{1-\exp[i(\varepsilon_i^*-\varepsilon_j)t']\}[i(\varepsilon_i^*-\varepsilon_j)-2\gamma_{ph}]/i(\varepsilon_i^*-\varepsilon_j), \qquad (A.5)$$

$\varepsilon_i=\varepsilon_i(\mathbf{p})$ [see (10)], and $i=1,2$. To obtain the function $\tilde F_{ph}(t,t',\mathbf{p})$, we must interchange the subscripts pol and ph in (A.2) and (A.5)

The correlation properties of the noise near the transmitted wave are described by the corresponding components of the normal and anomalous Green's functions of the direct polaritons[5]

$$D(\mathbf{p}_0+\mathbf{k},t,t'),\quad D_\diamond(\mathbf{p}_0+\mathbf{k},\mathbf{p}_0-\mathbf{k},t,t').$$

It is convenient to continue by transforming from the "triangular" to the "$\pm$" representation of the Green's functions, in which[6,7]

$$G^{+-}={}^1/_2(F-G^r+G^a),\quad G^{-+}={}^1/_2(F+G^r-G^a),$$
$$G^{++}=\theta(t-t')G^{-+}+\theta(t'-t)G^{+-},$$
$$G^{--}=\theta(t'-t)G^{-+}+\theta(t-t')G^{+-}. \qquad (A.6)$$

In the "$\pm$" representation the functions D and $D_\diamond$ are related to the polariton correlation functions by

$$D^{+-}=(\mathbf{p}_0+\mathbf{k},t,t')=-i\langle a^+(\mathbf{p}_0+\mathbf{k},t')a(\mathbf{p}_0+\mathbf{k},t)\rangle,$$
$$D_\diamond^{+-}(\mathbf{p}_0+\mathbf{k},\mathbf{p}_0-\mathbf{k},t,t')=-i\langle a(\mathbf{p}_0-\mathbf{k},t')a(\mathbf{p}_0+\mathbf{k},t)\rangle. \qquad (A.7)$$

In the τ approximation the backscattering of the phonoritons into modes close to the transmitted wave is disregarded, and in the absence of a thermal polariton source ($N_{0,\mathrm{pol}}=0$) we have[5]

$$D_0^{+-}=D_{0,\diamond}=0,\quad D_0^{-+}(\mathbf{p},t>t')=-i\exp[i\varepsilon_{pol}^*(\mathbf{p})\cdot(t'-t)]. \qquad (A.8)$$

In second order of perturbation theory in the polariton-phonon interaction, taking diagrams a and b of Fig. 1 into account, we obtain for the equal-time functions the following expressions

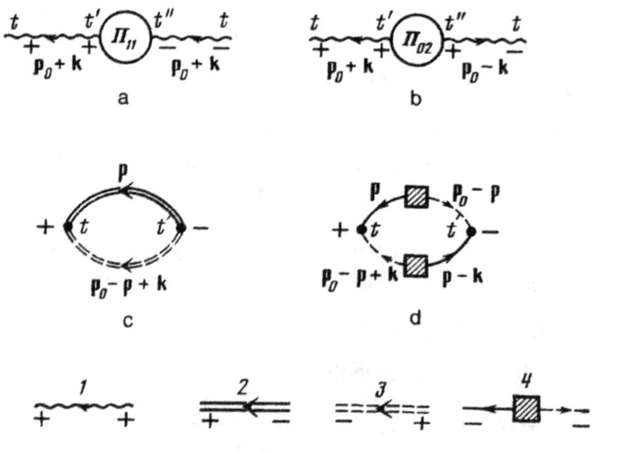

FIG. 1. Diagrams of second-order perturbation theory in the polariton-phonon interaction for the Green's function $D^{+-}(\mathbf{p}_0+\mathbf{k},t,t)$ (a), $D_\diamond^{+-}(\mathbf{p}_0+\mathbf{k},\mathbf{p}_0-\mathbf{k},t,t)$ (b), and for the polarization operators $\Pi_{11}^{+-}(\mathbf{p}_0+\mathbf{k},t,t')$ (c), $\Pi_{02}^{+-}(\mathbf{p}_0+\mathbf{k},\mathbf{p}_0-\mathbf{k},t,t')$ (d). Notation: 1 — D_0^{++}, 2 — G_{pol}^{+-}, 3 — G_{ph}^{-+}, 4 — $G_\diamond^{--}$

$$D^{+-}(\mathbf{p}_0+\mathbf{k},t,t)=\iint_{-\infty}^{t} dt'\,dt''D_0^{-+}(\mathbf{p}_0+\mathbf{k},t,t')$$
$$\times\ \Pi_{11}^{+-}(\mathbf{p}_0+\mathbf{k},t',t'')D_0^{-+}(\mathbf{p}_0+\mathbf{k},t'',t),\qquad (A.9)$$
$$D_0^{+-}(\mathbf{p}_0+\mathbf{k},\mathbf{p}_0-\mathbf{k},t,t)=\iint_{-\infty}^{t} dt'\,dt''D_0^{-+}(\mathbf{p}_0+\mathbf{k},t,t')$$
$$\times\ \Pi_{02}^{++}(\mathbf{p}_0+\mathbf{k},\mathbf{p}_0-\mathbf{k},t',t'')D_0^{-+}(\mathbf{p}_0-\mathbf{k},t,t''),\qquad (A.10)$$

where the normal and anomalous polarization operators (diagrams c and d of Fig. 1) are equal to

$$\Pi_{11}^{+-}(\mathbf{p}_0+\mathbf{k},t,t')$$
$$=-2i\mu^2\int\frac{d^3p}{(2\pi)^3}G_{pol}(\mathbf{p},t,t')G_{ph}^{+-}(\mathbf{p}_0-\mathbf{p}+\mathbf{k},t,t'),\qquad (A.11)$$
$$\Pi_{02}^{+-}(\mathbf{p}_0+\mathbf{k},\mathbf{p}_0-\mathbf{k},t,t')=-2i\mu^2\int\frac{d^3p}{(2\pi)^3}$$
$$\times G_0^{+-}(\mathbf{p},\mathbf{p}_0-\mathbf{p},t,t')G_0^{-+}(\mathbf{p}-\mathbf{k},\mathbf{p}_0+\mathbf{k}-\mathbf{p},t',t),\qquad (A.12)$$
$$\Pi^{++}(t,t')=-\theta(t-t')\Pi^{-+}(t,t')-\theta(t'-t)\Pi^{+-}(t',t).\qquad (A.13)$$

In the derivation of (A.9) and (A.10) we used the fact that in the "$\pm$" representation the Green's functions are independent of the time index at the maximum time. Calculating (A.9) and (A.10) with the aid of (A.1)–(A.5) for Stokes scattering far above the stimulated-scattering threshold, we obtain Eqs. (18) and (19).

[1)]The final expressions for the statistical Green's functions (12) are given in the Appendix [Eqs. (A.1)–(A.5)].
[2)]The calculations are given in the Appendix [Eqs. (A.6)–(A.13)]

[1]P. C. Martin, in: Low Temperature Physics, J. G. Daunt, D. D. Edwards, F. J. Milford, and M. Yagub, eds., Plenum, New York, 1965, p. 9; V. de Giorgio and M. O. Scully, Phys. Rev. **A2**, 1170 (1970). P. Graham and H. Haken, Zs. Phys. **231**, 31 (1970).
[2]G. Haken, *Synergetics*, Springer, 1978.
[3]S. A. Akhmanov, Yu. E. D'yakov, and A. S. Chirkin, Introduction to Statistical Radiophysics and Optics [in Russian], Nauka, 1981, Chap. 7.
[4]A. L. Ivanov and L. V. Keldysh, Zh. Eksp. Teor. Fiz. **84**, 404 (1983) [Sov. Phys. JETP **57**, 234 (1983)].
[5]L. V. Keldysh and S. G. Tikhodeev, *ibid.* **90**, 1852 (1986) [63, 1086 (1986)]; FIAN Preprint No. 331, 1985.
[6]L. V. Keldysh, Zh. Eksp. Teor. Fiz. **47**, 1515 (1964) [Sov. Phys. JETP 20, 1018 (1965)].
[7]E. M. Lifshitz and L. P. Pitaevskii, *Physical Kinetics*, Pergamon, 1981, Chap. 10.
[8]S. T. Belyaev, Zh. Eksp. Teor. Fiz., **34**, 417, 433 (1958) [Sov. Phys. JETP 7, 289, 299 (1958)].
[9]A. A. Abrikosov, L. P. Gor'kov, and I. E. Dzyaloshinskii, *Quantum Field Theoretical Methods in Statistical Physics*, Pergamon, 1965.
[10]L. P. Gor'kov, Zh. Eksp. Teor Fiz. **36**, 1918 (1959) [Sov. Phys. JETP **9**, 1364 (1959)].

Translated by J. G. Adashko

Brillouin scattering of a noisy polariton wave

N. A. Gippius, L. V. Keldysh, and S. G. Tikhodeev

P. N. Lebedev Physical Institute, Academy of Sciences of the USSR

Institute of General Physics, Academy of Sciences of the USSR

(Submitted 25 June 1986)
Zh. Eksp. Teor. Fiz. **91**, 2263–2275 (December 1986)

We consider the Brillouin scattering of a strong polariton wave, assumed to be narrow-band Gaussian noise, in a direct-gap semiconductor. We show that notwithstanding the incoherence of the pumping the effects caused by the coherence of the scattered polaritons and the phonons which are in resonance with them—the mixing of polariton and phonon states, the renormalization of the spectra, the formation of a gap in the density of states and in the spectral density of scattered anti-Stokes polaritons, and oscillations in the intensity of the anti-Stokes waves after the pumping is switched on—are conserved (although they are appreciably weakened when the intensity and the spectral width of the pump are increased). (These effects were considered for the case of coherent pumping by Ivanov, Keldysh, and Tikhodeev, [Sov. Phys. JETP **57**, 234 (1983); **63**, 1086 (1986); **64**, 45 (1986)].)

§1. INTRODUCTION

Characteristic for the behavior of systems which are removed from equilibrium by a strong external field is the occurrence of correlations which are additional to the ones occurring in a state of thermodynamic equilibrium. As a rule, such correlations are not universal (in contrast to the equilibrium case) and are determined by the external field and the specifics of the system; they turn out to affect appreciably the behavior of the system. Phenomena typical of the situation described here take place in Brillouin scattering of a strong coherent polariton wave in a semiconductor. The coherence between scattered polaritons and resonance phonons which arises in this case leads[1] to the formation of mixed polariton-phonon (phonoriton) modes, the restructuring of the spectrum and of the occupation numbers of both the phonons and the polaritons.[1)] There is experimental evidence in favor of the phonoriton restructuring of the spectrum during anti-Stokes scattering by optical phonons in CdS.[3] Various effects caused by this restructuring (near the threshold of induced scattering when the transient wave is abruptly switched on) are considered theoretically in Refs. 4 and 5. In a formal description of these effects the coherence of a strong polariton wave (pump) was used in an essential way. It is therefore of interest to analyze which of the effects considered "survive" when one uses an incoherent pump.

In the present paper we consider the anti-Stokes[2)] Brillouin scattering of a strong noisy polariton wave. In this paper we describe such an electromagnetic wave with a frequency ε_0 close to the polariton resonance frequency as Gaussian noise with a vanishing average field amplitude, $\langle E \rangle = 0$, and a pair correlation function

$$\langle E(0,t)E^*(\mathbf{r},t')\rangle \propto n_0 \exp\{-\delta|t'-t| + i\varepsilon_0(t'-t) - i\mathbf{p}_0\mathbf{r}\}. \quad (1.1)$$

Here $\varepsilon_0 = \varepsilon(\mathbf{p}_0)$, $\mathbf{p}_0$ is the quasimomentum of the wave, $\varepsilon(\mathbf{p})$ the polariton dispersion law, n_0 the spatial polariton density which is connected with the intensity of the passing wave through the relation

$$I = \hbar\varepsilon_0 n_0 c_0, \quad c_0 = |\partial\varepsilon(\mathbf{p})/\partial\mathbf{p}|_{\mathbf{p}=\mathbf{p}_0}, \quad (1.2)$$

and δ is the spectral width of the noise (the reciprocal of the correlation time). In other words, such a wave is a macro-occupied polariton mode (the number of particles n_0V in it is proportional to the volume) with a quasimomentum $\mathbf{p}_0$ and random phase. The averaging in (1.1) is performed over the appropriate density matrix which from an experimental point of view is equivalent to averaging over an ensemble of realizations. However, when we study stationary phenomena this averaging is equivalent, by virtue of the ergodicity of stationary random processes,[6] to averaging over the observation time $t \gg \delta^{-1}$.

The simplest to describe is the scattering in the case of narrow-band noise $\delta \ll \Gamma$ where $\Gamma = \gamma_{pol} + \gamma_{ph}, \gamma_{pol(ph)}$ is the reciprocal of the polariton (phonon) life time. From a formal point of view one finds the solution for $\delta = 0$. The general prescription for finding any final answer reduces [see §2, Eq. (2.5)] to averaging the appropriate expression evaluated for coherent pumping over a Rayleigh intensity distribution. Such a result is natural for Gaussian noise (the so-called adiabatic approximation, see Ref. 6). Many of the results obtained are therefore intuitively obvious. We analyze in §3 the behavior of the polariton density of states (the imaginary part of the retarded Green function) and the spectral density of the scattered polaritons in the frequency and momentum range close to the anti-Stokes resonance for stationary backward scattering. We show that when the pumping intensity increases there occurs a pseudo-gap (as in disordered systems[7]) in the density of states which is considerably more smeared out than the gap for coherent pumping. In particular, the density of states in the center of the gap decreases with increasing I proportional to $I^{-1}\ln I$ and not to I^{-1} as for coherent pumping.[1]

We consider in §3 also non-stationary scattering when the pumping is switched on suddenly. In the coherent case there occur after the switching on of the pumping oscillations in the intensity of the anti-Stokes line[5] similar to the

 0038-5646/86/121344-08$04.00

nutations of a two-level system. In the case of noisy pumping the oscillations of the anti-Stokes line are damped. This result corresponds to the suppression of the nutations of a two-level system in a noisy field.[6]

At the end of §3 we consider Stokes scattering. In the approximation for the given noisy pumping used in the present paper there do not exist stationary solutions for the Stokes lines in contrast to the scattering of a coherent wave when there are stationary solutions right up to the threshold I_c for induced scattering (stochastic instability[6]). The growth in the amplitude of the Stokes waves starts without having a threshold; in the time range $t > \tilde{t}$, where $\tilde{t} \propto I^{-1}$ it proceeds faster than exponentially, proportional to $\exp[t^2\Gamma/\tilde{t}]$.

In the case of Gaussian noise with a finite spectral width, considered in §4, there is no general rule for calculating any quantities such as there is when $\delta = 0$. However, in the framework of the τ-approximation (see Ref. 4) one can for the case of stationary scattering completely sum the perturbation theory series in the external field for the retarded Green function. The summation method (reduction to an infinite continued fraction) was, as far as we know, first applied in Ref. 8 to calculate the linear polarizability of a three-level system. This method was used also in the theory of disordered systems[9] to calculate the electron density of states. It was shown in Ref. 9 that writing the solution as a continued fraction is convenient for a numerical analysis. As to the statistical Green function, even in the τ-approximation and the stationary case one can only carry out an exact summation under the condition that the phonon life time is considerably longer than the polariton life time, or vice versa. (From an experimental point of view this case is, of course, the most common one.) Taking into account that δ is finite leads (as in Ref. 9) to a yet larger smearing out of the pseudo-gap in the density of states; as $\delta/\Gamma \to 0$ the solution goes over into the one obtained for $\delta = 0$ by averaging over the Rayleigh distribution.

On the whole we can conclude that when a noisy polariton wave is scattered in a semiconductor the effects connected with the additional coherence of the scattered polaritons and phonons do not disappear although they are considerably weakened.

Concluding this section we consider how these effects must manifest themselves experimentally. We asusme that we use as a pump a narrow-band noisy source with $\delta \ll \Gamma$. Stationary effects (renormalization of the spectrum and of the populations) can be studied using a single realization under the condition that the observation time $t \gg \delta^{-1}$. Transient processes (oscillations of the anti-Stokes and growth of the Stokes components) develop over times $t \lesssim \Gamma^{-1}$. Noise effects must thus manifest themselves when one averages over a series of pulses (of length $t_{\text{pulse}} \ll \delta^{-1}$) in each of which one observes the scattering of a coherent pump.

§2. PERTURBATION THEORY FOR SCATTERING OF AN INCOHERENT WAVE

We use as in Refs. 4 and 5 a diagram technique for non-equilibrium processes.[10,11] We consider first stationary scattering. The rules for constructing a diagram perturbation theory in terms of the polariton-phonon interaction which was formulated in Ref. 4 remain valid, except for the rule for describing the external field. As the amplitude of the field is zero on average the anomalous vertices[3)] of (I.1.7) for the creation and annihilation of a polariton with $p = p_0$ are also zero. The anomalous Green functions (I.1.8) and (I.1.9) also vanish. All quantities with diagrams which in the case of coherent scattering contain different numbers of anomalous creation and annihilation vertices are, in general, also zero due to averaging over the phase.

The statistical component of the free polariton Green function which is proportional to the correlator (1.1) depends on the intensity of the external field. We isolate it and take it into account separately. After Fourier transforming with respect to the frequencies and momenta it has (in the triangular representation) the form

$$\begin{pmatrix} 0 & 0 \\ 0 & -4in_0(2\pi)^3\delta(\mathbf{p}-\mathbf{p}_0)\,\delta/[(\varepsilon-\varepsilon_0)^2+\delta^2] \end{pmatrix}. \tag{2.1}$$

As we assume the field to be Gaussian, the higher correlators vanish. The action of an incoherent external field is in the resonance approximation thus completely described by a diagram perturbation theory containing the lines (2.1) besides the free propagators of scattered polaritons and phonons (I.1.2) to (I.1.4) (Fig. 1). It is convenient for what follows to combine the latter with the vertices (I.1.5) of the polariton-phonon interaction and to write them in the form (see Fig. 1)

$$\begin{aligned}\Xi_{i'j'i''j''}(p', p'') = {} & \Phi_{\mathbf{p}'}\Phi_{\mathbf{p}''}(\sigma_x)_{i'j'}(\sigma_x)_{i''j''}(2\pi)^4 \\ & \times\delta(k'+p-p')(2\pi)^4\delta(k''+p-p'') \\ & \times(2\pi)^3\delta(\mathbf{p}-\mathbf{p}_0)\,2\delta/[(\varepsilon-\varepsilon_0)^2+\delta^2],\end{aligned} \tag{2.2}$$

where

$$\Phi_{\mathbf{p}} = D(n_0|\mathbf{p}-\mathbf{p}_0|/2\hbar\rho u)^{1/2}, \quad k=(\mathbf{k}, \omega), \quad p=(\mathbf{p}, \varepsilon),$$

$i', j', i'', j'' = 1,2$ are time indexes.[4)]

Comparing the perturbation-theory series constructed thus with the series for coherent pumping one can easily formulate the following correspondence rule illustrated by Fig. 2. To obtain all diagrams for any quantity A in the incoherent case one must construct all diagrams for A with the same number of anomalous creation and annihilation vertices in the coherent case. After that one must join by lines (2.2) the creation vertices with annihilation vertices in all possible ways. We note that from a diagram for the coherent case with m anomalous vertices of each kind we obtain $m!$ diagrams for the incoherent pumping. For finite δ these diagrams are, in general not equal to one another. However, in the limiting case of narrow-band noise, $\delta = 0$, the relation

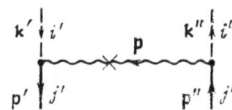

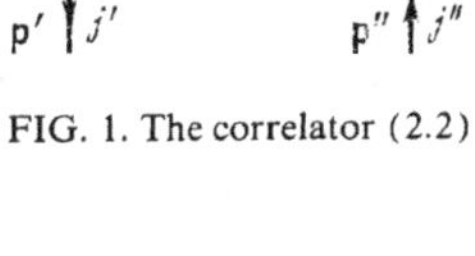

FIG. 1. The correlator (2.2) of the external field, Ξ.

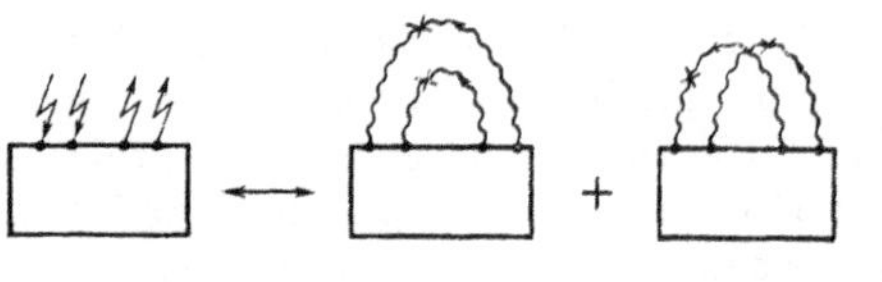

FIG. 2. Rule for the correspondence of diagrams for coherent and noisy polariton waves.

$$\Xi(p', p'') = \Phi(p') \otimes \Phi(p''), \tag{2.3}$$

where

$$\Phi_{ij}(p') = \Phi_{\mathbf{p}'}(2\pi)^4 \delta(p-p_0)(2\pi)^4 \delta(k'+p-p')(\sigma_x)_{ij} \tag{2.4}$$

is the anomalous vertex (I.1.7) for coherent pumping, is satisfied. Hence, in that case all $m!$ diagrams are equal to one another and to the original diagram for coherent pumping.

This statement allows us to obtain a general rule for the summation of diagrams for any quantity $A(n_0)$ describing the scattering of a narrow-band incoherent polariton wave with density n_0 [or intensity I of (1.2)] if we know the corresponding function $A_{\text{coh}}(n_0)$ for the coherent case:

$$A(n_0) = \int_0^\infty e^{-\zeta} A_{coh}(\zeta n_0)\, d\zeta = \int_0^\infty P_{n_0}\{n\} A_{coh}(n)\, dn, \tag{2.5}$$

where

$$P_{n_0}\{n\} = n_0^{-1} \exp\{-n/n_0\} \tag{2.6}$$

is an exponential distribution corresponding to the Rayleigh distribution of the amplitude.

To prove Eq. (2.5) we must expand $A_{\text{coh}}(n_0)$ in a perturbation theory series in powers of n_0:

$$A_{coh}(n_0) = \sum_{m=0}^{\infty} A_{m,coh} n_0^m. \tag{2.7}$$

We then get in the incoherent case (when $\delta = 0$)

$$A(n_0) = \sum_{m=0}^{\infty} m! A_{m,coh} n_0^m. \tag{2.8}$$

Using the representation

$$m! = \int_0^\infty \zeta^m e^{-\zeta}\, d\zeta \tag{2.9}$$

and interchanging summation and integration in (2.8) we prove (2.5).

In concluding this section we consider the non-stationary scattering of an incoherent wave when it is suddenly switched on. This problem was solved for coherent pumping in the τ-approximation in Ref. 5. We shall assume that the noisy wave, switched on at time $t = 0$, is Gaussian noise with a correlation function differing from (1.1) by additional factors $\theta(t)\theta(t')$, where $\theta(t)$ is the step function:

$$\langle E(0, t) E^*(\mathbf{r}, t') \rangle \propto n_0 \theta(t) \theta(t')$$
$$\times \exp\{-\delta|t-t'| + i\varepsilon_0(t'-t) - i\mathbf{p}_0\mathbf{r}\}. \tag{2.10}$$

As in the case of stationary scattering the anomalous Green functions [the off-diagonal components of the matrix (II.4)] vanish. One sees easily that the above formulated rule of correspondence between diagrams for the coherent and incoherent cases remains valid. In the limit as $\delta \to 0$ the dependence on t and t' of the corresponding function $\Xi(p't, p''t')$ can be factorized and an equation such as (2.3) is satisfied. The summation rule (2.5) is thus also valid for non-stationary scattering.

§3. SCATTERING OF NARROW-BAND NOISE, $\delta \to 0$

We use Eqs. (I.1.18) and (I.2.5) and calculate in the τ-approximation the polariton density of states $|\text{Im} G^r_{\text{pol}}(p)|$ and the spectral density of the backward scattered polaritons $N_{\text{pol}}(p)$ for coherent pumping:[5)]

$$|\text{Im}\, G^r_{pol}(p)|_{coh} = [\gamma_{pol}|b(p)|^2 + \Phi_{\mathbf{p}}^2 \gamma_{ph}]\,|Z^r(p)|^{-2}, \tag{3.1}$$

$$[N_{pol}(p) = {}^1/_2 i(F_{pol} - G^r_{pol} + G^a_{pol})]_{coh} = \gamma_{ph} N_+ \Phi_{\mathbf{p}}^2 |Z^r(p)|^{-2}, \tag{3.2}$$

where

$$Z^r(p) = a(p) b(p) - \Phi_{\mathbf{p}}^2,$$
$$a(p) = \varepsilon - \varepsilon(\mathbf{p}) + i\gamma_{pol}, \quad b(p) = \varepsilon - \varepsilon_0 - u|\mathbf{p} - \mathbf{p}_0| + i\gamma_{ph}, \tag{3.3}$$
$$N_+ = [\exp(\hbar u|\mathbf{p}_0 - \mathbf{p}_+|/k_B T) - 1]^{-1}$$

is the equilibrium number of resonance phonons with momentum $\mathbf{p}_0 - \mathbf{p}_+$ which is in resonance for the anti-Stokes backward scattering, see (I.1.1); u is the sound speed. In (3.2) and henceforth we neglect the thermal source of polaritons $N_{0,\text{pol}} = 0$.

To obtain the corresponding functions $|\text{Im}\, G^r_{\text{pol}}|, N_{pol}$ for the incoherent case as $\delta \to 0$ we use Eq. (2.5). For a qualitative comparison of the behavior in the coherent and the incoherent cases we performed numerical calculations the results of which are given in Figs. 3 and 4 (for $|\text{Im} G^r_{\text{pol}}|$) and Figs. 5 and 6 (for N_{pol}). Figures *a* refer to the coherent and figures *b* to the incoherent case. The quantities $|\text{Im} G^r_{\text{pol}}|$, N_{pol}/N_+ are shown as functions of the frequency and of the longitudinal momentum. The central point in all figures [with coordinates $p_+ = (\varepsilon_+, \mathbf{p}_+)$, see (I.1.1)] is the region of anti-Stokes resonance for backward scattering. In that point the polariton and absorbed phonon terms intersect. Quantities with the dimensions of frequency $(\varepsilon, (G^r_{\text{pol}})^{-1}, \text{N}^{-1}_{\text{pol}})$ are measured in units γ_{pol}, and momenta in units $2\gamma_{\text{pol}}/c_0$. In the calculations we used the following parameter values: $u = 1/3$, $c_0 = 2$, $\gamma_{\text{pol}} = 1$, $\gamma_{\text{ph}} = \frac{1}{2}$. The pumping strength in dimensionless units $\Phi^2_+/\gamma^2_{\text{pol}}$ $(\Phi_+ \equiv \Phi_{\mathbf{p}_+})$ is equal to 1 (Figs. 3 and 5) and 9 (Figs. 4 and 6).

It is very clear from Figs. 3(a) and 4(a) how the phonon and polariton modes mix when the intensity of the coherent pumping increases, tails occur in the polariton density of states and extend along the phonon term, and a gap is formed in the density of states. In the incoherent case (Figs. 3b and 4b) there also occurs a mixing and a trough in the density of states is formed but much less well pronounced. The spectral density of the backward scattered polariton wave (Figs. 5 and 6) behaves similarly.

The speed at which the density of states diminishes at the center of the gap when the pumping strength increases

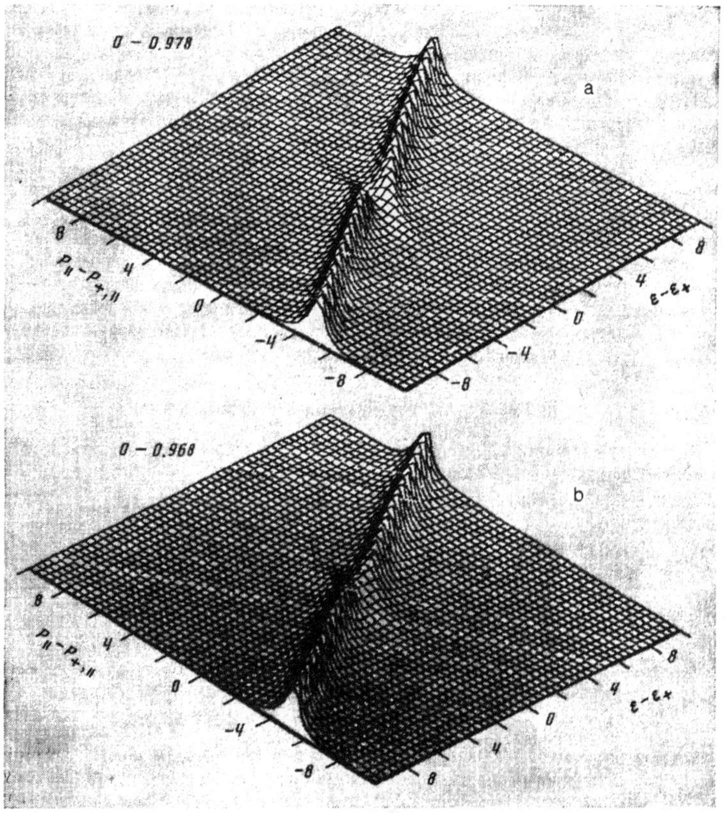

FIG. 3. The polariton density of states $|\mathrm{Im}\, G^r_{\mathrm{pol}}|\gamma_{\mathrm{pol}}$ as function of frequency and longitudinal momentum for $\Phi_+/\gamma_{\mathrm{pol}} = 1$ (explanation in the text): a: coherent case, b: incoherent case (the numbers in the upper left-hand corner of each figure are the minimum and maximum values of the function shown, $p_\perp = p_{+,\perp}$).

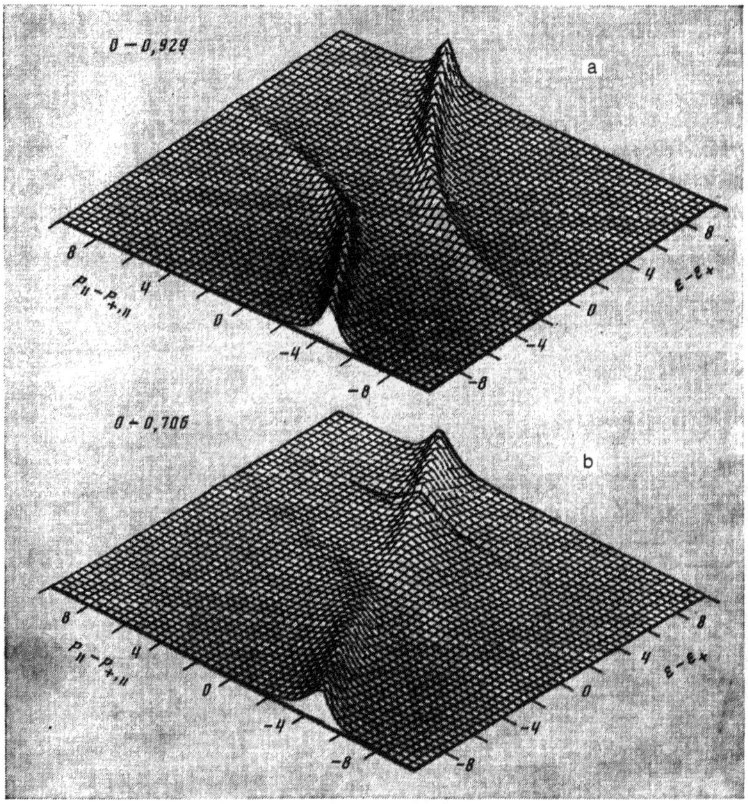

FIG. 4. The same as in Fig. 3, for $\Phi_+/\gamma_{\mathrm{pol}} = 3$.

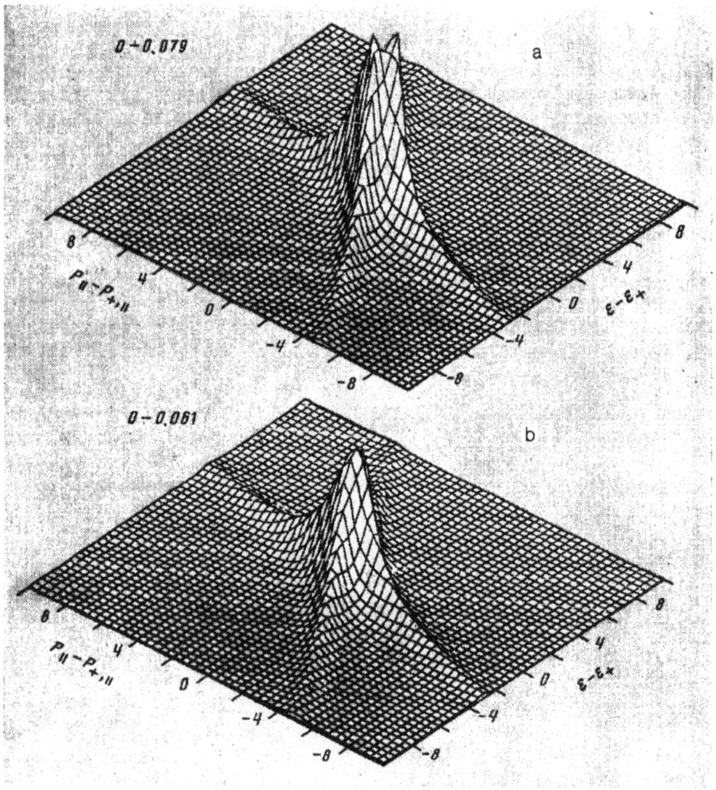

FIG. 5. The spectral density of scattered polaritons $N_{pol}\gamma_{pol}/N_+$ for $\Phi_+/\gamma_{pol} = 1$: a: coherent case, b: incoherent case.

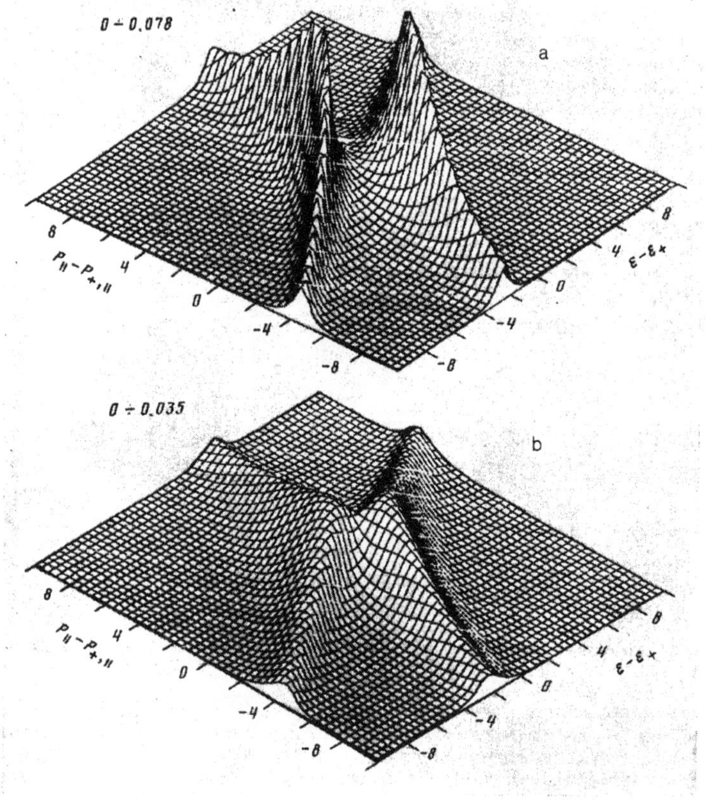

FIG. 6. The same as in Fig. 5 for $\Phi_+/\gamma_{pol} = 3$.

can be estimated quantitatively. It follows from (3.1) that

$$|\operatorname{Im} G^r_{pol}(p_+)|_{coh}=\gamma^{-1}_{pol}\xi/(1+\xi), \tag{3.4}$$

where $\xi = \gamma_{\rm pol}\gamma_{\rm ph}\Phi_+^{-2}$. Using (2.5) we get for incoherent scattering

$$|\operatorname{Im} G^r_{pol}(p_+)|=\gamma^{-1}_{pol}\int_0^\infty e^{-\zeta}(\zeta+\xi)^{-1}\,d\zeta=\gamma^{-1}_{pol}\xi e^{\xi}E_1(\xi). \tag{3.5}$$

In strong fields when $\xi \ll 1$ the exponential integral $E_1(\xi)\sim \exp(-\xi)|\ln\xi|$. Hence, the density of states in the center of the gap decreases in the incoherent case when the pumping strength increases as $\xi\,|\ln\xi|\propto I^{-1}\ln I$ which is appreciably more slowly than in the coherent case (porportional to $\xi\propto I^{-1}$).

The intensity of the scattered polaritons is for non-stationary scattering of a coherent pump described by Eq. (II.5). In the simplest case $\gamma_{\rm pol}=\gamma_{\rm ph}=\Gamma/2$ this formula has for the anti-Stokes component the form

$$[N_{pol}(\mathbf{p},t)]_{coh}=2N_+\left[\frac{\Omega^2}{\Omega^2+\Gamma^2}-\frac{\Omega e^{-\Gamma t}}{(\Omega^2+\Gamma^2)^{1/2}}\sin(\Omega t+\varphi)\right], \tag{3.6}$$

where

$$\Omega^2=\Delta^2+4\Phi_p^2,\quad \Delta=\varepsilon(\mathbf{p})-\varepsilon_0-u|\mathbf{p}-\mathbf{p}_0|,\quad \operatorname{tg}\varphi=\Omega/\Gamma.$$

The time-dependence of (II.15) together with the corresponding dependence in the incoherent case calculated using (2.5) for the same parameter values as before is for $\Phi_+/\gamma_{\rm pol}=3$ shown in Fig. 7. As earlier, *a* refers to the coherent and *b* to the incoherent case. It is clear that the averaging of (2.5) leads to a suppression of the oscillations in the intensity of the scattered polariton wave except for the first period.

This result corresponds to the damping of the nutations of a two-level system in a noise field.[6]

In concluding this section we give some results referring to the Stokes scattering of a narrow-band noisy wave. The considerations given in §2 are formally independent of whether we consider anti-Stokes or Stokes scattering. One can thus expect that Eq. (2.5) remains valid also in the Stokes case. However, in our statement of the problem when the pumping intensity is assumed to be given by an external source there is no stationary solution in arbitrarily weak fields (stochastic instability[6]) in contrast to the scattering of a coherent pump when there are stationary solutions right up to the threshold of induced scattering. This is clear from (2.5) in which the integration is performed over all intensities, among which there are also those which exceed the threshold in the coherent case. We consider therefore the non-stationary problem. We write down that part of (II.5) which gives the exponential increase of the Stokes wave $[N_{\rm pol}]_{\rm coh}$ when $I>I_c$ for $\gamma_{\rm ph}=\gamma_{\rm pol}=\Gamma/2$ and $p=p_-$ [see (I.1.1)]:

$$[N_{pol}(\mathbf{p}_-,t)]_{coh}\sim-\frac{1+N_-}{4}\frac{\Phi_-}{\gamma_2(\mathbf{p}_-)}\exp[-2\gamma_2(\mathbf{p}_-)t], \tag{3.7}$$

where

$$\gamma_2(\mathbf{p}_-)=-\Phi_-+\Gamma/2,\quad \Phi_-=\Phi_{\mathbf{p}_-},$$

$$N_-=[\exp(\hbar u|\mathbf{p}_--\mathbf{p}_0|/k_BT)-1]^{-1}.$$

For a noisy pump we have

$$N_{pol}(\mathbf{p}_-,t)\sim\frac{1+N_-}{2}\int_0^\infty d\zeta\,\frac{\zeta^2\Phi_-}{\zeta\Phi_--\Gamma/2}\exp(-\zeta^2+2\zeta\Phi_-t-\Gamma t). \tag{3.8}$$

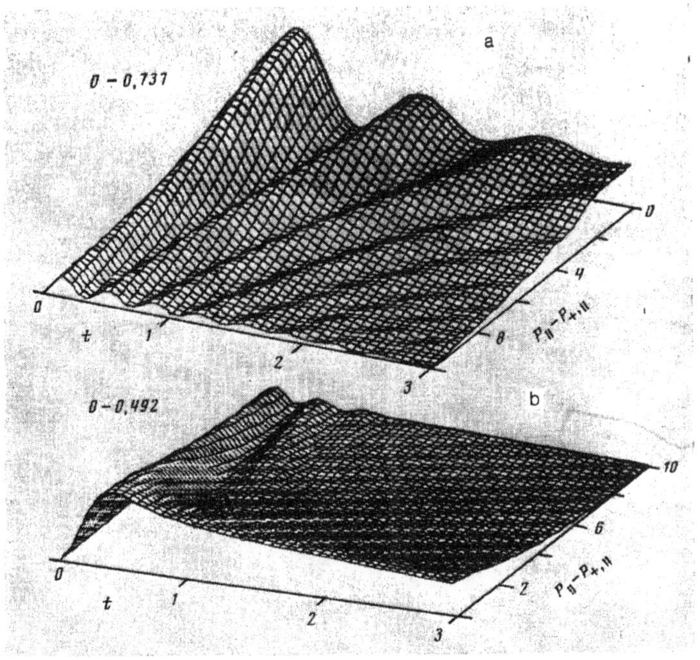

FIG. 7. The function $N_{\rm pol}(t,p_\parallel)/N_+$ for $\Phi_+/\gamma_{pol}=3$. The time is measured in units $\gamma_{\rm pol}^{-1}$ and the momentum in $2\gamma_{\rm pol}/c_0$.

The integrand in (3.8) contains a simple pole on the integration contour. This non-integrable singularity arises in the threshold region for induced scattering in the framework of the τ-approximation which is, as was shown in Ref. 4, not applicable in that region. We assume that taking the divergent diagrams near the threshold completely into account leads to the singularity in (3.8) becoming integrable. Using the Laplace method to estimate the integrals[12] and noting that the singularity of the factor of the exponent does not fall for large t in the important region of integration we find that for $t \gg \tilde{t} = \Gamma \Phi_-^{-2}$

$$N_{pol}(\mathbf{p}_-, t) \sim {}^1/_2 \pi^{1/2} (1+N_-) \Phi_- t \exp [(\Phi_- t)^2 - \Gamma t]. \quad (3.9)$$

For the scattering of a noisy wave of arbitrary intensity the Stokes waves must thus grow faster than exponentially. Leaving this growth regime occurs with a delay which is inversely proportional to the pumping strength (thanks to this there does not arise a paradox when we are considering pumping with $I \to 0$). For clarity we recall that this effect must occur after averaging over a large number of realizations.

Under actual conditions, of course, the pumping strength in each point of the semiconductor is not fixed by an external source (as in our idealized statement of the problem). The growth process is limited by particles leaving the passing wave and the amplitude of the Stokes waves emerges at a stationary value. It follows from our considerations that the establishment of a stationary picture proceeds completely differently for coherent and for noisy pumps.

§4. SCATTERING OF NOISE WITH A FINITE SPECTRAL WIDTH

If $\delta \neq 0$ the rule (2.3) for factorization is not satisfied and there does not exist a general rule like (2.5) for summing any diagrams. However, in the framework of the τ-approximation one can for the stationary case solve the problem for the retarded and advanced Green functions by a method proposed by Elyutin.[8] This method breaks down already for the statistical Green functions and allows us to solve the problem only when $\gamma_{pol} \ll \gamma_{ph}$ or vice versa (and arbitrary δ).

We start with the calculation of G^r_{pol}. In the framework of the τ-approximation any diagram for G^r_{pol} consists (see Fig. 8) of a "spine" containing a product of alternating functions

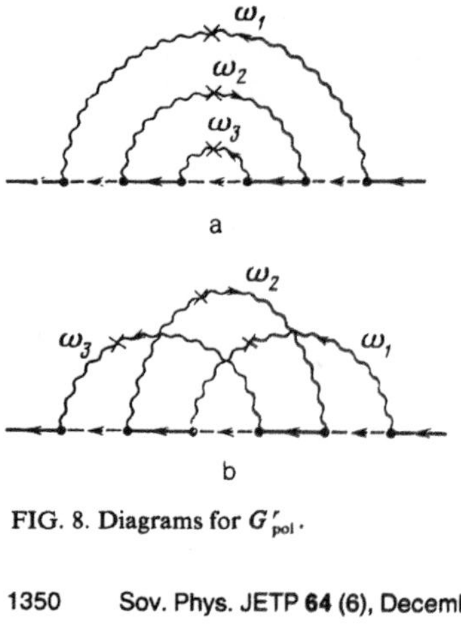

FIG. 8. Diagrams for G^r_{pol}.

$$G_{\tau,pol}(\varepsilon-\omega_1+\omega_2-\ldots) = [a(\varepsilon-\omega_1+\omega_2-\ldots)]^{-1}, \quad (4.1)$$

$$G^r_{\tau,ph}(\varepsilon-\omega_1+\omega_2-\ldots) = [b(\varepsilon-\omega_1+\omega_2-\ldots)]^{-1} \quad (4.2)$$

and "ribs"—arbitrarily entangled lines Ξ, (2.2), which after integration over momenta are reduced to the form

$$\Xi(\omega_j) = \Phi_p^2 2\delta[(\omega_j-\varepsilon_0)^2+\delta^2]^{-1}. \quad (4.3)$$

The functions $a(\varepsilon)$ and $b(\varepsilon)$ occurring in (4.1), (4.2) are defined in (3.3). Here and henceforth we shall not write down the momentum arguments: $\mathbf{p}$ for the $G^r_{\tau,pol}$ lines and $\mathbf{p}-\mathbf{p}_0$ for $G^r_{\tau,ph}$.

The method for summing such diagrams[8] is based upon the fact that their magnitude depends only on the number of ribs passing over each line of the spine and does not depend on the entanglement of the ribs. For instance, the diagrams of Fig. 8 have the same magnitude and are equal to

$$(\Phi_p^2)^3 a(\varepsilon)^{-2}[b(\varepsilon)+i\delta]^{-2}[a(\varepsilon)+2i\delta]^{-2}[b(\varepsilon)+3i\delta]^{-1}.$$

This property follows from the analyticity of the functions $G^r_{\tau,\alpha}$ in the upper ε-halfplane. It enables us to calculate G^r_{pol} as a sum of "simple" diagrams (such as Fig. 8a) for which the ribs are not entangled and for which, hence, the vertices are not renormalized. One needs only correctly take into account the number of diagrams which are equal in magnitude to the given simple diagram. As a result (see Refs. 8 and 9) finding the function G^r_{pol} reduces to solving an infinite set of coupled equations

$$\begin{aligned}
G^r_{pol} &= a^{-1}[1+\Phi_p^2 G^r_{1,ph} G^r_{pol}],\\
G^r_{1,ph} &= [b+i\delta]^{-1}[1+\Phi_p^2 G^r_{1,pol} G^r_{1,ph}],\\
G^r_{1,pol} &= [a+2i\delta]^{-1}[1+2\Phi_p^2 G^r_{2,ph} G^r_{1,pol}],\\
&\vdots\\
G^r_{m,pol} &= [b+(2m-1)i\delta]^{-1}[1+m\Phi_p^2 G^r_{m,pol} G^r_{m,ph}],\\
G^r_{m,ph} &= [a+2mi\delta]^{-1}[1+(m+1)\Phi_p^2 G^r_{m+1,ph} G^r_{m,pol}].\\
&\vdots
\end{aligned} \quad (4.4)$$

We can write the solution of the set (4.4) in the form of an infinite continued fraction

$$G^r_{pol}(\varepsilon) = \frac{1}{a(\varepsilon)-}\,\frac{\Phi_p^2}{b(\varepsilon)+i\delta-}\,\frac{\Phi_p^2}{a(\varepsilon)+2i\delta-}\,\frac{2\Phi_p^2}{b(\varepsilon)+3i\delta-}\,. \quad (4.5)$$

To find G^r_{ph} one must in (4.5) interchange the functions a and b.

We now turn to the calculation of $F_{pol}(p)$. In the τ-approximation any diagram for F_{pol} also consists of ribs and vertebra which in contrast to the diagrams for G^r_{pol} contain (at an arbitrary place) one function such as

$$F_{\tau,pol} = 2i \operatorname{Im} G^r_{\tau,pol} \quad (N_{0,pol}=0) \quad (4.6)$$

or

$$F_{\tau,ph}=2i(1+2N_+)\,\mathrm{Im}\,G^r_{\tau,ph}, \quad (4.7)$$

to the left of which stand the $G^r_{\tau,\alpha}$ and to the right the $G^a_{\tau,\alpha}$, α = pol, ph. As the functions (4.6), (4.7) are not analytical either in the vertex or in the lower ε-half-plane diagrams with entangled ribs are not equal to the corresponding simple diagram. We were not able to obtain a general rule of summation similar to (4.5) for any relation between δ, γ_{ph}, and γ_{pol}. However, in the case when one of the dampings is appreciably less than the other ($\gamma_{pol} \ll \gamma_{ph}$ or vice versa) one easily finds a solution. For instance, when $\gamma_{pol} \ll \gamma_{ph}$

$$N_{pol}(p)=2N_+|\,\mathrm{Im}\,G^r_{pol}|, \quad (4.8)$$

where G^r_{pol} is the infinite continued fraction (4.5).

To prove (4.8) we evaluate the imaginary part of any diagram for G^r_{pol}. It is proportional to the imaginary part of the vertebra

$$\mathrm{Im}(G^r_{\tau,pol}G^r_{\tau,ph}G^r_{\tau,pol}\ldots).$$

To evaluate the imaginary part we use an identity which is valid for any complex numbers $\alpha_1, \alpha_2, \ldots, \alpha_n$:

$$\mathrm{Im}\,(\alpha_1\alpha_2\ldots\alpha_n)=\mathrm{Im}\,\alpha_1\,\mathrm{Re}\,(\alpha_2^*\alpha_3^*\ldots\alpha_n^*) +\mathrm{Im}\,\alpha_2\,\mathrm{Re}(\alpha_1\alpha_3^*\ldots\alpha_n^*)+\ldots\mathrm{Im}\,\alpha_n\,\mathrm{Re}(\alpha_1\alpha_2\ldots\alpha_{n-1}),$$

and also Eqs. (4.6), (4.7). We get

$$\mathrm{Im}(G^r_{\tau,pol}G^r_{\tau,ph}G^r_{\tau,pol}\ldots)=\frac{1}{2i}F_{\tau,pol}\,\mathrm{Re}\,(G^a_{\tau,ph}G^a_{\tau,pol}\ldots)$$
$$+\frac{F_{\tau,ph}}{2i(1+2N_+)}\mathrm{Re}\,(G^r_{\tau,pol}G^a_{\tau,pol}\ldots)+\ldots=\frac{1}{2i(1+2N_+)}$$
$$+\ [F_{\tau,pol}G^a_{\tau,ph}G^a_{\tau,pol}\ldots+G^r_{\tau,pol}F_{\tau,ph}G^a_{\tau,pol}\ldots+\ldots]$$
$$+\frac{iN_+}{1+2N_+}[F_{\tau,pol}G^a_{\tau,ph}G^a_{\tau,pol}\ldots+G^r_{\tau,pol}G^r_{\tau,ph}F_{\tau,pol}\ldots+\ldots]. \quad (4.9)$$

The first term on the right-hand side of (4.9) is the sum of all vertebra diagrams for F_{pol} with the topological structure considered.[6)] The second term (proportional, as should be the case, to the number of thermal phonons) is small provided the polariton damping is small, as it is proportional to γ_{pol}. When $\gamma_{pol} \ll \gamma_{ph}$ and for arbitrary δ for each diagram of a given topological structure therefore the relation

$$F_{pol}=2i(1+2N_+)\,\mathrm{Im}\,G^r_{pol}. \quad (4.10)$$

holds. This proves Eq. (4.8). It follows from the proof that a similar relation holds for F_{ph}:

$$F_{ph}=2i(1+2N_+)\,\mathrm{Im}\,G_{ph}^r. \quad (4.11)$$

Using (4.5), (4.8) and the results of the numerical analysis of similar expressions given in Ref. 9 we may conclude that taking into account a finite δ leads to an even larger (as compared to the case when $\delta = 0$) smearing out of the pseudo-gap and of the singularities of the spectral density of the scattered polaritons. One can also verify that the small parameter which leads, when it tends to zero, to the solutions obtained here going over into the solution (2.5) for $\delta = 0$ is, indeed, $\delta/\Gamma \ll 1$.

[1)] The effect of the exciton-photon interaction on the phonoriton restructuring of the spectrum was analyzed in Ref. 2.

[2)] Some results regarding Stokes scattering are given at the end of §3.

[3)] Here and henceforth formulae from Ref. 4 are indicated by the Roman number I and those from Ref. 5 by a II.

[4)] In contrast to the representation chosen in Ref. 4 we shall in the present paper use the positive frequency part of the phonon Green function for the anti-Stokes scattering. The directions of the lines of the scattered polaritons and of the phonons in Fig. 1 are thus the same in contrast to the directions chosen in Ref. 4.

[5)] An error slipped into Eq. (I.2.2). The off-diagonal components of the matrix Σ_{ph} must change place.

[6)] We have here not written the Re sign in the right-hand side of (4.9) as either it is real (for mirror-symmetric diagrams) or it becomes real when we add to (4.9) the mirror-image diagram.

[1] A. L. Ivanov and L. V. Keldysh, Zh. Eksp. Teor. Fiz. **84**, 404 (1983) [Sov. Phys. JETP **57**, 234 (1983)].

[2] A. L. Ivanov, Zh. Eksp. Teor. Fiz. **90**, 158 (1986) [Sov. Phys. JETP **63**, 90 (1986)].

[3] G. S. Vygovskiĭ, G. P. Golubev, E. A. Zhykov, *et al.* Pis'ma Zh. Eksp. Teor. Fiz. **42**, 134 (1985) [JETP Lett. **42**, 164 (1985)].

[4] L. V. Keldysh and S. G. Tikhodeev, Zh. Eksp. Teor. Fiz. **90**, 1852 (1986) [Sov. Phys. JETP **63**, 1086 (1986)].

[5] L. V. Keldysh and S. G. Tikhodeev, Zh. Eksp. Teor. Fiz. **91**, 78 (1986) [Sov. Phys. JETP **64**, 45 (1986)].

[6] S. A. Akhmanov, Yu. E. D'yakov, and A. S. Chirkin, Vvedenie v Statisticheskuyu Radiofiziku i Optiku [in Russian] Nauka, Moscow, 1981.

[7] M. A. Sadovskiĭ, Thesis, Physical Institute Acad. Sc. USSR, Moscow, 1974.

[8] P. V. Elyutin, Opt. Spektrosk. **43**, 542 (1977) [Opt. Spectrosc. (USSR) **43**, 318 (1977)].

[9] M. A. Sadovskiĭ, Zh. Eksp. Teor. Fiz. **77**, 2070 (1979) [Sov. Phys. JETP **50**, 989 (1979)].

[10] L. V. Keldysh, Zh. Eksp. Teor. Fiz. **47**, 1515 (1964) [Sov. Phys. JETP **20**, 1018 (1965)].

[11] E. M. Lifshitz and L. P. Pitaevskiĭ, (Physical Kinetics Nauka, Moscow, 1979, Ch. 10 [English translation published by Pergamon Press, Oxford].

[12] N. G. de Bruijn, Asymptotic Methods in Analysis North-Holland, 1958.

Translated by D. ter Haar

CONTEMP. PHYS., 1986, VOL. 27, NO. 5, 395–428

The electron–hole liquid in semiconductors†

L. V. Keldysh, P.N. Lebedev Physical Institute of the Academy of Sciences, Moscow 117924, U.S.S.R.

ABSTRACT. In highly excited semiconductors at low enough temperatures nonequilibrium electrons, holes and excitons condense into droplets of a metallic degenerate Fermi liquid, the so called electron–hole liquid. General properties of this new quantum liquid are reviewed including possible types of its phase diagram; the strong dependence of the phase diagram on the band and crystalline structure of the semiconductor, magnetic field etc.; the kinetics of electron–hole drop nucleation, growth and decay. Electron–hole drops can easily be accelerated by some external forces up to velocities close to that of sound. Intense movement of drops also occurs because of the so called phonon wind-drag by intense flows of nonequilibrium phonons, arising in recombination processes inside the drops themselves or in the thermalization of excited carriers.

1. Introduction

In quantum many-body theory the usual way of describing a macroscopic system is in terms of elementary excitations such as phonons, magnons, electrons and holes, etc., above some ground state. The convenience and usefulness of the elementary excitation concept itself are obvious if the excitation interaction is negligible or small enough compared to its energy, and the system of elementary excitations can be treated as a gas. It fails, however, in the vicinity of second-order phase transitions, where for some group of elementary excitations the interaction becomes strong, resulting in sharp changes in the nature of the excited states and of the corresponding physical properties. Similar conditions can be achieved far from equilibrium if some excitation modes are intensely pumped and the number of elementary excitations in these modes becomes large enough. Condensation of nonequilibrium charge carriers in semiconductors into the metallic degenerate Fermi liquid of electrons and holes is today one of the best-studied examples of this phase transition far from equilibrium. Arising in highly excited semiconductors, the electron–hole liquid (EHL) has been the subject of numerous theoretical and experimental investigations for the last fifteen years and the results of this work are summarized in several reviews [1–9]. In this article the presentation partially follows that in [9].

2. Interaction and bound states in the system of nonequilibrium charge carriers

In the case of semiconductors the so-called free charge carriers, electrons and holes, are the most important type of elementary excitations. At distances much larger than that between atoms they interact via the usual Coulomb forces, reduced by dielectric screening,

$$V(r_{12})=\frac{e_1e_2}{\varepsilon r_{12}}. \tag{1}$$

Here V is the potential energy of two point charges, e_1 and e_2, and r_{12} is the distance between them. The overwhelming majority of the phenomena in the physics of

---

† This article is based on the paper Electron–Hole Droplets in Semiconductors (from *Modern Problems of Condensed Matter Sciences*, Vol. 6, 1983, pp. xi-xxxvii, published by Elsevier North-Holland) and includes material from a lecture presented at the Enrico Fermi Summer School, 1983, which subsequently appeared in *Nuovo cimento* (Italian Physical Society). Permission to use this material is gratefully acknowledged.

semiconductors can be understood in terms of free charge carriers, neglecting their interaction, because usually this interaction is small compared to the thermal kinetic energy $k_B T$ if the temperature T is not too low and the concentration of electrons and holes n is not too large:

$$\frac{e^2}{\varepsilon|\mathbf{r}|} \sim \frac{e^2}{\varepsilon} n^{1/3} \ll k_B T. \tag{2}$$

The values of the dielectric constant ε for semiconductors are typically large, $\varepsilon \gtrsim 10$, and it is one of the most important properties determining the usefulness of the free-charge-carrier concept itself. For $T \sim 100$ K the inequality (2) is fulfilled at all concentrations up to $n \sim 10^{18}$ cm^{-3}. The problem to be considered here is what happens to the system of free charge carriers if the temperature is low enough and the charge carrier density is high enough, so that the above-mentioned inequality is violated and the interaction becomes strong.

The conditions of high concentration and low temperature are incompatible in intrinsic (pure enough) semiconductors at thermodynamic equilibrium. In this case

$$n \sim \left(\frac{mk_B T}{\hbar^2}\right)^{3/2} \exp\left[-\frac{E_g}{2k_B T}\right] \tag{3}$$

and, as the temperature decreases, decreases much faster than the temperature itself (E_g is the energy gap, i.e. the minimum energy necessary to produce one electron–hole pair). But concentration can be made arbitrarily large at any temperature, under non-equilibrium conditions, when additional electron hole pairs are artificially produced by some external source (illumination, injection through contacts and so on). In this sense the subject of this article is a system of nonequilibrium charge carriers at low temperatures. However, the nonequilibrium nature of this system will be understood here in some restricted sense. The reason is that there exist two very different time scales in the problem under consideration: the thermalization time τ and the recombination time (nonequilibrium charge carrier lifetime) $\tau_0 \gg \tau$. This means that charge carriers, initially produced with kinetic energies much larger than $k_B T$, thermalize very fast, i.e. acquire a nearly equilibrium energy distribution corresponding to the crystal lattice temperature. But after that they exist for much longer time intervals in such a quasi-equilibrium state, and the only nonequilibrium parameter is the total number of carriers, which is fixed at an arbitrary value by the external excitation source and does not correspond to formula (3). In this sense the nonequilibrium charge carrier system will be treated approximately in what follows as an equilibrium system of electrons and holes, but with an arbitrarily fixed total number of particles. Some of its unusual properties arising only from the nonequilibrium nature of the system will be described in the last section of this paper. One of the best known manifestations of the Coulomb interaction between electrons and holes is the existence of Wannier–Mott excitons, bound states of the electron and hole, similar to the hydrogen atom and positronium. The resemblance of the Wannier–Mott exciton to the positronium is closer than that to the hydrogen atom for two reasons: unlike the electron and the proton, the effective mass values of the electron and hole in semiconductors differ usually by no more than one order of magnitude, and the lifetime of the exciton is finite owing to the possible recombination of the electron and the hole. But quantitatively excitons differ drastically from both the hydrogen atom and positronium. Evaluated by the well-known Bohr formulae the exciton binding energy E_{ex} and the effective radius a_{ex} are

$$E_{ex}=\frac{1}{2}\frac{e^4m}{\varepsilon^2\hbar^2}\sim(10^{-1}-10^{-3})\,\text{eV}, \tag{4a}$$

$$a_{ex}=\frac{\varepsilon\hbar^2}{me^2}\sim(10^{-6}-10^{-7})\,\text{cm}, \tag{4b}$$

where e, $\hbar$ and m are respectively the electronic charge, Planck's constant and the reduced effective mass of the electron and hole. They differ from the Rydberg and the Bohr radius by several orders of magnitude because of the above-mentioned large values of the dielectric constant and the relatively small values of m, which are usually smaller by an order of magnitude than the free electron mass. One important consequence of these numerical estimates, which is fundamental to the whole of the following discussion of the properties of the interacting charge carrier system and to the validity of equations (4) themselves, should now be explained. Values of a_{ex} which are large compared to interatomic distances in the host crystal justify not only the applicability to the problem of excitons of the interaction law in the Coulomb form (1), but also the possibility of treating this problem as that of two particles interacting in a spatially homogenous effective medium, i.e. ignoring details of the real crystal structure and crystal potential, which manifest themselves only indirectly in the values of effective mass and dielectric constant. Because the values of E_{ex} are small compared to the binding energies of atoms and valence electrons, even at such high concentrations as will be discussed later, the interaction of free charge carriers, including exciton formation, does not noticeably influence the host crystal structure, its bound electron and phonon spectra etc. In other words, within some approximation, free charge carriers and excitons constitute an autonomous subsystem for which the host crystal represents a neutral spatially uniform background, some kind of vacuum determining the energy spectrum of carriers (effective masses) and their interaction (dielectric constant). The evolution of the properties of this subsystem, depending on the temperature T and the concentration n, may be understood from simple qualitative considerations, based on the following observations:

(1) It is, as a whole, an electroneutral assembly of many oppositely charged particles, interacting according to the Coulomb law (1) and, therefore, analogous to a system of electrons and nuclei (protons, for example). The values E_{ex} and a_{ex} (4) for this system are the natural quantum scales of energy and length in the same way as the Rydberg and the Bohr radius are the natural scales of energy and length in atoms, molecules and solids, because no other quantities of energy and length dimension can be constructed from e^2/ε, $\hbar$ and m—the only intrinsic dimensional parameters entering the dynamics of the system under consideration.

(2) The most fundamental qualitative difference of the electron–hole system from that of the electron–proton system is the absence of heavy particles in the former, since the effective masses of electrons m_e and holes m_h are usually of the same order of magnitude. Therefore, the adiabatic approximation—one of the most fruitful concepts in molecular and solid-state physics—is not valid in our problem because its accuracy is $(m_e/m_h)^{1/4}$.

At high temperatures and low enough concentrations, when

$$n \ll \left(\frac{mk_B T}{\hbar^2}\right)^{3/2} \exp\left[-\frac{E_{ex}}{2k_B T}\right], \tag{5}$$

the free-charge-carrier system is nearly a perfect gas or, more correctly, nearly a perfect completely ionized plasma. Exciton concentration is negligibly small, owing to the thermal dissociation, and interaction is weak as inequality (2) is automatically fulfilled. The system also remains nearly perfect even at arbitrarily low temperature if the concentration is large enough, $na_{ex}^3 \gg 1$, but in this case the system is a degenerate plasma, and the interaction is weak compared to its Fermi energy. In the intermediate concentration range

$$\left(\frac{mk_B T}{\hbar^2}\right)^{3/2} \exp\left[-\frac{E_{ex}}{2k_B T}\right] \lesssim n \lesssim a_{ex}^{-3}, \tag{6}$$

which exists apparently only at low temperatures where $k_B T \lesssim E_{ex}$, the interaction becomes strong and its influence on the system properties dominating. This is the range of nontrivial states of the system and the main subject of the following discussion.

If the concentration is not large, $na_{ex}^3 \ll 1$, then as the temperature decreases, violating inequality (5), the majority of electrons and holes combine into excitons. Recalling the above-mentioned analogy of excitons to atoms, we can treat this state of the nonequilibrium charge carrier system as an atomic gas or a weakly ionized plasma. Proceeding with this analogy, one can expect at still lower temperatures the formation of excitonic molecules, biexcitons, proposed initially by Moskalenko [10] and Lampert [11]. The existence of biexcitons is established with certainty now, both theoretically and experimentally [12, 13], but, unlike excitons, which differ from hydrogen atoms only quantitatively, they differ qualitatively from hydrogen molecules in some respects. The reason for this difference, very important for the following analysis of the strongly interacting electron–hole system, is the above-mentioned absence of the adiabaticity in the electron–hole problem. For this reason in biexcitons, unlike the hydrogen molecule, not only the electrons but also the holes are strongly delocalized, which results, as will be shown now, in a sharp reduction of the molecular binding (dissociation) energy E_D.

In figure 1 the usual interaction potential of two hydrogen atoms in the singlet state is shown schematically. It would be essentially the same for excitons if $m_e \ll m_h$ and the natural quantum scales a_{ex} and E_{ex} are used instead of the usual atomic Bohr radius and Rydberg. The dissociation energy differs from the potential well depth U_0 by a small amount, equal to the zero-vibration energy $\frac{1}{2}\hbar\omega_0 \sim (m/M)^{1/2} U_0$. Here m and M are electron and proton masses. If it were possible to reduce smoothly the heavy particle mass M, the ground-state energy level would shift upwards, as shown in figure 1, due to the increase of zero-point vibration amplitude and energy. For this reason the relative contribution of zero-point vibrations to the total biexciton energy is greater by more than an order of magnitude than in the molecule H_2 and almost completely compensates the adiabatic attractive potential. For a difference of electron and hole masses within one order of magnitude, such qualitative estimates give $E_D \lesssim 0{\cdot}1\ E_{ex}$ for the biexciton dissociation energy. This value is confirmed both by more accurate variational estimates and by the available experimental data [12, 13]. For hydrogen the ratio of the molecular dissociation energy to the atomic binding energy is 0·35. Since $(m_e/m_h)^{1/4}$ is

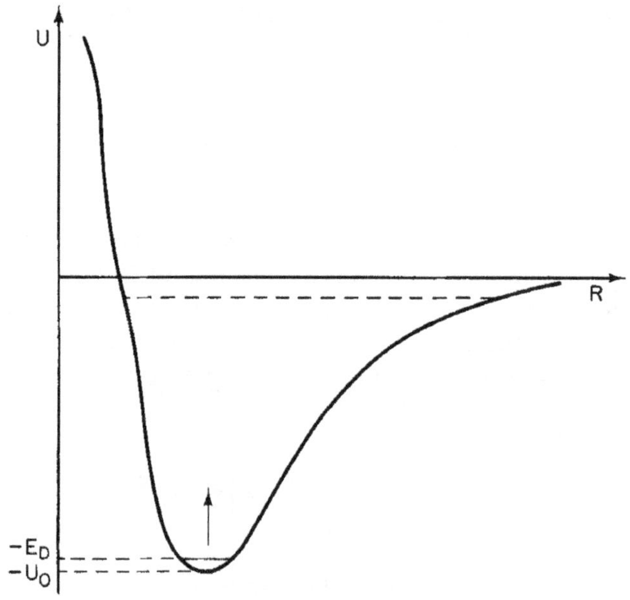

Figure 1. Interaction potential of two excitons in reduced 'excitonic' units (E_{ex}, a_{ex}). The arrow indicates the direction of the ground-state energy level shift as the electron to hole mass ratio changes from 0 to 1.

always of the order of unity, the zero-point vibration amplitude in a biexciton is of the order of a_{ex} and probably even larger since the dissociation energy is so small. Thus holes are indeed completely delocalized inside the excitonic molecule and this molecule is a very loosely bound system.

So, at low enough temperatures and concentration, fulfilling the condition (6), the nonequilibrium electron–hole system exists in the form of a gas of excitons, biexcitons and free charge carriers. According to the same analogy of electron–hole and electron–nuclei systems one would expect that, at still lower temperature or higher concentration (gas pressure), the nonequilibrium electron–hole system would undergo something like a gas–liquid phase transition, accompanied by the formation of some condensed phase (liquid) of nonequilibrium charge carriers [14]. The word 'liquid' is used here in its usual meaning: a system of a macroscopically large number of particles occupying a macroscopically large volume and bound together by internal interaction forces. It is characterized by a definite equilibrium density (concentration of electron–hole pairs) n_l, binding energy (work function) per electron–hole pair E_l and surface tension, i.e. the capacity to form a sharp, stable boundary, separating it from the gas phase. Contrary to the usual behaviour of the electron–hole plasma or the exciton gas, it does not tend to spread over the whole crystal volume, but occupies only a definite part of it, $V_l = N_l/n_l$, where N_l is the total number of particles in this phase. But sharing these general features with any other known liquid, the electron–hole liquid (EHL) is unique in many other properties, which will be described below.

In this article the modern status of EHL study and understanding will be briefly presented, illustrated by a few experimental and theoretical results. Comprehensive review of the subject, including full lists of references, may be found in the monographs [7–9].

3. Thermodynamics of the electron–hole liquid

3.1. *The electron–hole liquid: qualitative description*

In this section, the main properties and manifestations and the existence region of the EHL in different types of semiconductor will be discussed under the assumption of thermodynamic equilibrium within the charge carrier system, i.e. neglecting the electron–hole recombination process, which, as explained above, is a good starting approximation. In this case the considerations in the preceding section may be conveniently illustrated and essentially supplemented by a schematic phase diagram (figure 2) on the plane of variables $(T, \bar{n})$. Here T is the temperature, and $\bar{n}$ is the mean concentration of electron–hole pairs: $\bar{n} = N/V$. N is the total number of pairs and V the volume of the excited region, which for simplicity is considered to be uniformly excited. In figure 2 as a scale of temperature and concentration their values are used at the spatially uniform gaseous and liquid phases of the nonequilibrium carrier system. The shaded region G+L is that of the parameter values where the spatially uniform distribution appears to be unstable and there occurs separation into liquid-phase droplets with equilibrium density $n_l(T)$, surrounded by exciton, biexciton and free-carrier gas with equilibrium density $n_g(T)$. Here $n_l(T)$ and $n_g(T)$ are correspondingly the right and left branches of the curve limiting the region of phase coexistence in figure 2. In figure 2 as a scale of temperature and concentration their values are used at the so-called critical point (T_c, n_c), i.e. at the point at which the difference between gas and liquid disappears. At $T/T_c > 1$ there is no density at which the phase transition occurs, i.e. the nonequilibrium carrier concentration increases continuously with increase of excitation level. The value T_c is likely to be determined by the particle binding energy in the condensed phase E_l. There exists the well-known empirical correlation $k_B T_c \approx 0{\cdot}1\, E_l$ [15] valid for many liquids as well as for the nonequilibrium carrier liquid phase in semiconductors. According to the above consideration of energy and length scales in the nonequilibrium carrier system the orders of magnitude of the main parameters of the condensed-phase and its existence region may be estimated: $n_c \sim n_l \sim a_{ex}^{-3}$, $10 k_B T_c \sim E_l \sim E_{ex}$, i.e. the mean interparticle distance in the condensed phase should be of the order of a_{ex} and the binding energy per electron–hole pair of the order of E_{ex}.

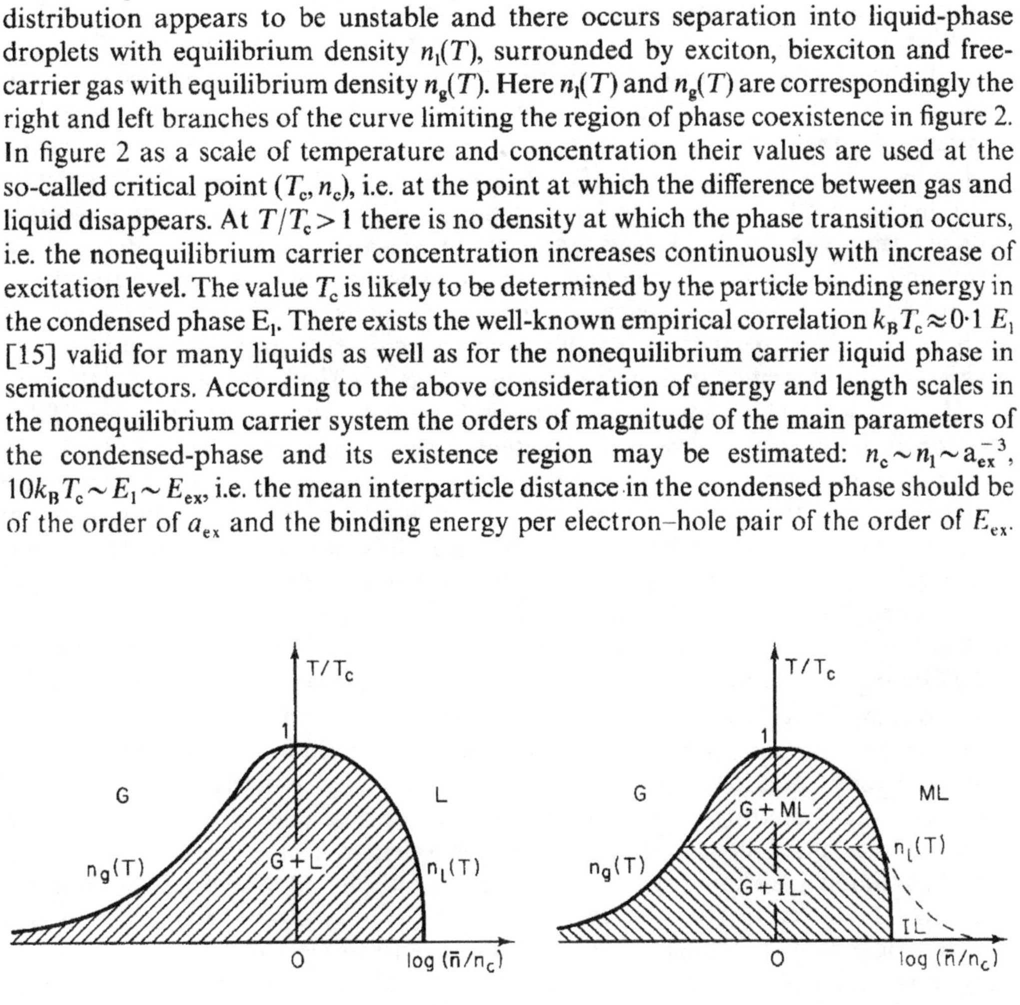

(*a*) (*b*)

Figure 2. Possible types of nonequilibrium charge carrier system phase diagrams. The letters G and L denote the existence region of spatially uniform gas and liquid states; ML and IL metallic liquid and insulating liquid. The coexistence region of electron–hole droplets and gas is hatched.

Let us now discuss the structure and the main physical properties of the condensed phase. At this stage the above-mentioned fundamental difference between the nonequilibrium carrier system and that of the electron and nuclei system, that is, the absence of heavy particles, acquires decisive significance. Because of this there is no temperature at which crystallization, i.e. the formation of a 'solid' phase in the nonequilibrium carrier system, is possible. Essentially, because $(m_e/m_h)^{1/4}$ is not a small parameter, there is no length dimension scale, except a_{ex}. Therefore, if crystallization occurred, the particle zero-vibration amplitude around the equilibrium positions would be of the order of a_{ex}, that is of the order of interparticle distances, and this, according to existing melting criteria, should already result in melting at zero temperature. Hence the non-equilibrium carrier condensed phase is a liquid with extreme quantum properties. However, there exist liquids very different in structure and properties: molecular, metallic, ionic (electrolytes), and so on. Proceeding from the exciton analogy to hydrogen atoms, one might suppose that the nonequilibrium carrier condensed phase is a molecular liquid of biexcitons, weakly bound with each other. However, this is not correct, and the analogy itself at this point is not unambiguous. Indeed, the closest analogues to hydrogen, the alkali metals, condense not into molecular but into metallic liquids. The reason for this qualitative difference is a considerable difference in binding (dissociation) energy E_D of the corresponding molecules. The strongly bound ($E_D \approx 0{\cdot}35$ Ryd) molecule H_2 with a completely saturated valence bond and, therefore, interacting with other surrounding molecules only via weak Van der Waals forces, appears to be stable and energetically the most favourable structural unit in the liquid phase. Intermolecular distances here exceed essentially interatomic distances within a molecule, and the interaction of each atom with its partner within the same molecule exceeds by several orders of magnitude its interaction with neighbouring molecules. The molecules Na_2, K_2, Cs_2, etc. ($E_D < 0{\cdot}1$ Ryd) which are much more loosely bound owing to the presence of filled electron shells, do not survive in the condensed phase. Intramolecular binding is not strong enough to provide an energy advantage for a liquid whose components are molecular rather than atomic, where each atom would interact strongly with a few nearest, approximately equidistant, neighbours. In this latter case an intense electron exchange among all nearest neighbours results in complete electron delocalization, i.e. in the formation of a metallic liquid.

The exciton binding in the excitonic molecule is also rather weak, as explained above, owing to the absence of heavy nuclei and the correspondingly large zero-point vibration amplitude. Moreover, large zero-point vibration amplitudes not only weaken the intramolecular bond but, in addition, if molecules existed in the liquid phase, these large zero point vibrations would greatly increase the overlap of the wavefunctions of excitons in neighbouring molecules and the electron exchange between them. As a result, the condensed phase in the charge carrier system in semiconductors cannot be a molecular liquid. Like liquid alkali metals it is a metallic liquid, where neither excitonic molecules nor excitons themselves are present [14]. Electrons and holes in this electron–hole liquid (EHL) are 'free' in the same sense as electrons are free in metals: they move freely and more or less independently of each other (without violating macroscopic electroneutrality) within the liquid volume, but they cannot leave it, unless they are supplied with additional energy exceeding the so-called work function. The main qualitative difference of the EHL from ordinary metals is the complete quantum delocalization of both electrons and holes. From the above-mentioned estimates of the main condensed-phase parameters, the temperature of Fermi degeneracy T_F of the carriers in it is

$$k_B T_F \sim \frac{\hbar^2}{m} n_l^{2/3} \sim \frac{\hbar^2}{m a_{ex}^2} = 2E_{ex} \gtrsim k_B T.$$

so throughout the whole domain of its existence the EHL is a degenerate two-component Fermi liquid.

Within this general picture the EHL parameters and properties in different semiconductors and under different experimental conditions can be rather diverse. They are extremely sensitive to the peculiarities of the electron spectrum of the semiconductor and to any external influence. In those semiconductors where the electron and hole masses differ by an order of magnitude or more the heavy carriers may appear to be nondegenerate at temperatures of the order of the critical one. In these semiconductors the spatial correlation (short-range order) in the heavy-carrier arrangement already resembles the short-range order in the ion arrangement in molten metals, but still shows pronounced quantum effects, especially at $T \ll T_c$. To produce a crystal with long-range order, even at $T=0$, a difference in carrier effective mass of more than two orders of magnitude is necessary [16], which is unlikely in intrinsic semiconductors. Still more essential is band degeneracy and, especially, the so-called multivalley band structure, i.e. the presence (due to crystal symmetry) of several equivalent electron or hole groups, or both. Such are the band structures of many well-known semiconductors: Ge, Si, C, GaP, $A^{IV}B^{VI}$ group compounds, etc. It appears that in this case, n_l, E_l and T_c are much greater than they would be in semiconductors with the same effective-mass values and dielectric constant, but with the simple single-valley spectrum both for electrons and holes. The origin of this phenomenon is not difficult to explain qualitatively [17]. The particle energy in the EHL is composed of kinetic (Fermi) energy, which is positive, and potential energy from the Coulomb interaction, which is essentially negative, because due to the correlation in particle movement each particle is surrounded mainly by oppositely charged particles. The equilibrium density is determined by the minimum condition for total energy, i.e. by some balance of these two contributions, as shown in figure 3. A transition from the single-valley to the multivalley case at a fixed concentration would disturb this balance, as the Fermi energy, determined by the number of particles in each valley, would be essentially

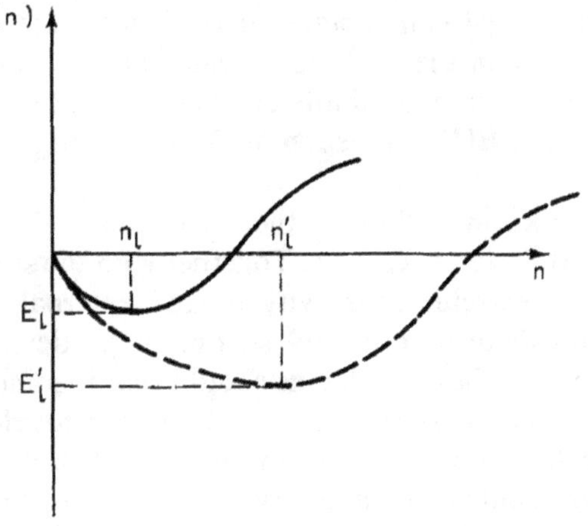

Figure 3. Concentration dependence of the EHL energy E_l in single valley (full line) and multi-valley (dashed line) cases.

reduced, while the potential energy dependent in a first approximation only on the mean distance between the particles, i.e. on their total concentration, would be virtually unchanged. The disturbed balance would result in a spontaneous contraction of the system to such a concentration that the growth in the Fermi energy would compensate any further increase in potential energy. Thus both concentration and binding energy are much greater in the new equilibrium position than in the initial one, corresponding to the single-valley band structure (figure 3), i.e. the multivalley band structure considerably increases the EHL stability and the range of its existence in the plane of the variables $(\bar{n}, T)$. The exciton binding energy does not depend on the number of valleys, since the electron and hole which comprise the exciton each belong to only one of the corresponding valleys. In the same direction and for similar reasons the EHL stability is also enhanced if the effective masses are strongly anisotropic. Indeed, the Fermi energy depends on the density of states, i.e. on the so-called density-of-states effective mass $m_d = (m_1 m_1 m_3)^{1/3}$, where m_i are the principal values of the effective mass tensor: the relationship $E_F = (\pi^2\hbar^2/2m_d)(3/\pi n)^{2/3}$. Thus, if, starting from the isotropic case $m_{1,2,3} = m$, one of the masses, e.g. m_1, increased, at a fixed concentration n, then E_F would evidently decrease $\sim (m/m_1)^{1/3}$. In the same way as in the multivalley case, the decrease in E_F results in an increase in equilibrium density and EHL binding energy. At the same time, the mean distance between particles in an exciton is mainly determined by the smallest of the masses, since, in the case of anisotropic masses, the exciton is elongated in the direction in which the quantum delocalization effect is most strongly pronounced, i.e. in the smallest mass direction. Even at $m_1 \to \infty$, a_{ex} diminishes only by a factor of two, E_{ex} correspondingly increases four-fold, while E_1 grows as $(m_1/m)^{1/5}$.

Such external actions as uniaxial stress or magnetic field, lowering the crystal symmetry, destroy the valley equivalence, so electrons (holes) remain only in some of them if the action is strong enough. Choosing a different stress or field direction, one can produce a different number of equivalent valleys in the same semiconductor and so it is possible to change the EHL parameters within wide limits.

Intervalley electron transitions are comparatively rare. Therefore, in discussing the EHL structure, electrons (holes) from different valleys may be approximately considered to be different types of particles. From this standpoint, the EHL in multivalley semiconductors is a multicomponent Fermi liquid, and external actions make it possible to change the number of its components or their relative concentrations arbitrarily. In certain cases these changes occur via a first-order phase transition [18], which is easy to understand on the basis of the above discussion of the increase in binding-energy in the multivalley case. The possibility of varying all the main EHL parameters within wide limits enables us to observe many new physical phenomena and makes the EHL an ideal model for studying collective phenomena in multielectron systems.

Compared to all known liquid, the EHL has the least mass density, $(m_e + m_h)n_1 \sim m a_{ex}^{-3} \sim (10^{-6}\text{–}10^{-8})\,\text{kg m}^{-3}$. Together with the small binding energy this is also the cause of its extreme sensitivity to any external influence: electric and magnetic fields, crystal deformation, and so on. In particular, the liquid is easily accelerated and flows inside the crystal. One should bear in mind, however, that, owing to electric neutrality, this flow is accompanied neither by an electrical current nor by any transfer of matter. As a hole represents the absence of an electron, an electron–hole pair has an effective mass, but not a real one (with a precision of $E_g/c^2 \sim 10^{-36}$ kg, where E_g is the energy gap width in the semiconductor in question and c is the light velocity). The above-estimated mass density is the *effective* mass density, which determines the

EHL inertial properties and its response to external forces, but does not describe the matter content. Like the exciton, which is essentially the excitation energy quantum in a crystal, the EHL is spatially condensed excitation energy with density $n_1 E_g \sim 10^6\ \mathrm{J\,m^{-3}}$, and its flow is first of all a transfer of excitation energy. This energy drastically changes the crystal properties, transforming it into a new phase state. In the region where it is concentrated, n_1 of the interatomic bonds per unit volume are broken, and the n_1 electron–hole pairs thus formed are in a metallic Fermi liquid state as described above.

These transformations are especially spectacular at excitation levels and temperatures corresponding to the G+L region in the phase diagram of figure 2. In this parameter region the EHL exists as macroscopic droplets—the so-called electron–hole drops (EHD).

According to the above considerations, EHD are mobile semimetallic phase regions inside insulating crystals, while from another point of view they are stable bunches of excitation energy. EHD motion is transfer of excitation energy as well as that of metallic conductivity and all other EHL properties. The ability to move freely inside a crystal without damaging it is one of the most remarkable features of EHD, distinguishing it from other macroscopic objects and emphasizing its quantum nature.

3.2. *Theory of the electron–hole Fermi liquid*

The central problem of EHL theory is finding the dependence of the EHL phase diagram parameters on electron and hole spectra and other semiconductor characteristics. Two main parameters—the dielectric constant and some mean electron and hole mass—determine the effective Bohr radius and Rydberg, i.e. the scales along the n and T axes in the phase diagram. In addition, the shape of the phase diagram may depend on the electron and hole mass ratio, the anisotropy, the number of equivalent valleys in the valence and conduction bands, the frequency dependence of the dielectric constant, etc. A priori three qualitatively different situations are possible:

(1) The binding energy per electron–hole pair in the EHL $|E_l|$ exceeds $E_{ex}+\frac{1}{2}E_D$, the energy per single exciton in a molecule. In this case the EHL is energetically the lowest state of the non-equilibrium carrier system and at a sufficiently low temperature condensation occurs at any $n<n_l$. In this case the phase diagram has the qualitative appearance shown in figure 2(*a*).

(2) $E_1<E_{ex}+\frac{1}{2}E_D$. In this case the biexciton gas of low density ($n\to 0$), existing in a Bose-condensed state at low enough temperatures is the ground state of the nonequilibrium charge carrier system. However according to Brinkman and Rice [19] the effective interaction of biexcitons at low temperatures is strongly repulsive because of the absence of heavy particles and, consequently, the dominating role of quantum effects [19]. Thus with an increase in the level of excitation, the energy per excitonic molecule increases (decreases in absolute value) and at some $\bar{n}=n_{g0}$ becomes equal to $2E_1$; after that condensation occurs. The qualitative appearance of the phase diagram is illustrated in figure 4(*a*) for this case. Letters BG mark the area of gas degeneracy of the excitonic molecules. In the same way as the phase diagram in figure 2(*a*) was compared with the alkali-metal phase diagram, the phase diagram in figure 4(*a*) may be called hydrogen-like. However, the essential difference from the phase diagram of hydrogen is that the region in the hydrogen phase diagram corresponding to the classical molecular liquid and crystal corresponds to the quantum molecular gas region in figure 4.

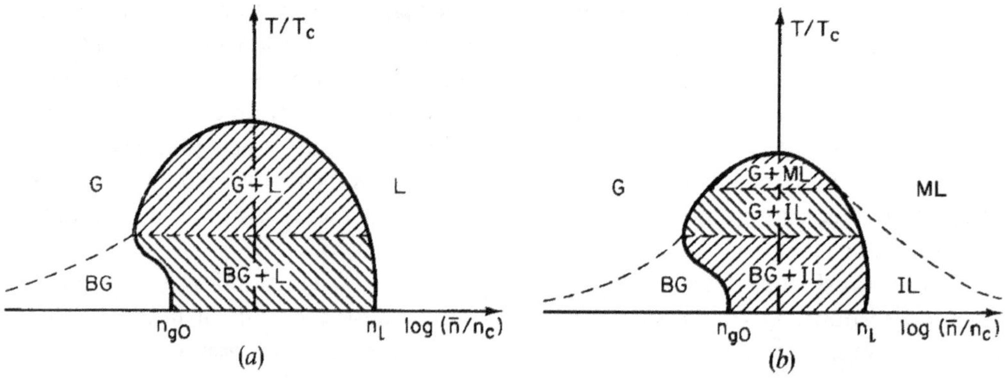

Figure 4. Phase diagram types, corresponding to a gas of Bose condensed excitonic molecules, BG being the ground state of the electron–hole system.

(3) So far the complete electron and hole delocalization in the EHL has been assumed to correspond to the usual metallic or, more exactly, semi-metallic nature of their energy spectrum. However, it is known [20–22] that in certain special cases, e.g. at single almost isotropic valleys for both electrons and holes, the semimetallic spectrum is unstable at low temperatures. Collective hole and electron interaction results in gap formation at the Fermi level and the spectrum becomes of insulator type. With increase in temperature or density the gap decreases and disappears; a semi-metallic state is restored. The phase diagram corresponding to this possibility is shown in figures 2(*b*) and 4(*b*), where the letters IL denote insulating liquid of the above type and ML a semi-metallic liquid. The region of separation into liquid and gaseous phases is hatched in figures 2 and 4. In different temperature intervals different phases coexist: one gaseous (G or BG) with one liquid (ML or IL). The B–BG and IL–ML transitions are likely to be second-order phase transitions and phase coexistence is impossible for them.

3.2.1. *Ground-state energy: dependence on band structure*

The question of which type of phase diagram corresponds to which type of semiconductor may be solved either by a quantitative theory, accounting for all the band structure peculiarities and other semiconductor parameters mentioned above, or experimentally. Quantitative theories of the EHL were initiated by Brinkman *et al.* [23], Combescot and Nozières [24], Brinkman and Rice [19], Vashishta *et al.* [25, 26]. The results of these and many other papers are described in reviews [7, 9]. Here I summarize only the most general results and then illustrate the EHL theory by a more complete examination of some simplified but correctly solvable models.

The most important factors for the determination of the type of phase diagram are the effective mass anisotropy and, especially, the multivalley structure of the electron and hole spectra. For the simplest case, when both the valence and conduction bands have one extremum with isotropic masses, all the existing calculations indicate $|E_1| < E_{ex} + \frac{1}{2}E_D$, i.e. in favour of one of the phase diagrams of figure 4. The largest calculated values for E_1 are $0{\cdot}99E_{ex}$ for the ML phase [25] and $1{\cdot}08E_{ex}$ for IL. The corresponding value of $n_1(T \to 0)$ is $0{\cdot}03a_{ex}^{-3}$ for the ML case. However, the energy difference $|E_1 - E_{ex} - \frac{1}{2}E_D|$ does not exceed a few per cent and, possibly, does not exceed the calculation accuracy. As explained above, the multivalley band structure essentially enhances E_l without noticeably changing E_{ex} and E_D. For semiconductors with an extremely anisotropic electron spectrum, particularly the so-called quasi-one-dimensional (polymer) and quasi-two-dimensional (layered) systems, or those with a large number of equivalent valleys, the problem of the EHL and its phase diagram turn out to be solvable exactly and analytically [27–29]. For such model system, as we shall now demonstrate, in agreement with the qualitative reasoning given above, the binding energy per particle pair in the EHL appears to be much greater than the exciton and excitonic molecule binding energies, and its equilibrium density is $n_l \gg a_{ex}^{-3}$.

In the usual approach to the calculation of the ground-state energy of many interacting fermion systems this energy is expressed as a sum of three contributions

$$E(n) = E_0(n) + E_{exch}(n) + E_{corr}(n). \tag{7}$$

Here E_0 is the kinetic energy of the noninteracting particles, E_{exch} is the Hartree–Fock exchange energy and E_{corr} the correlation energy due to dynamic spatial correlation of the particles. The calculation of E_0 and E_{exch} is straightforward for any system and does not present any difficulties. It is the last term in (7) which depends on the complicated dynamics of the many-particle system and presents the main problem in the $E(n)$ calculation. In the case of the Coulomb interaction there exists a well-known general formula, expressing E_{corr} in terms of the polarizability of the system $\chi(\mathbf{k}, \omega)$, dependent on wave-vector $\mathbf{k}$ and frequency ω:

$$E_{corr}(n) = \frac{1}{2n}\int_0^1 \frac{d\lambda}{\lambda}\int \frac{d^3k\,d\omega}{(2\pi)^4}\left[\frac{4\pi\chi(\mathbf{k}, i\omega; \lambda)}{1+4\pi\chi(\mathbf{k}, i\omega; \lambda)} - 4\pi\chi^{(0)}(\mathbf{k}, i\omega; \lambda)\right]. \tag{8}$$

Here $\chi(\mathbf{k}, i\omega; \lambda)$ is the polarizability for an imaginary frequency $i\omega$, of the system of particles with charges $\pm\sqrt{\lambda}e$ (+ for holes) at concentration (density) n; $\chi^{(0)}$ is the first approximation of χ. In what follows it is natural to use 'excitonic' scales, i.e. to put $e^2/\varepsilon = \hbar = m = 1$. Then E is measured in units of E_{ex} and the only parameter in (8) is the dimensionless concentration n (na_{ex}^3). The only limiting case permitting a strict calculation of (8) is that of high density $n \gg 1$. In this case $\chi \approx \chi^{(0)}$ and higher-order corrections are of the order $p_F^{-1} \sim n^{-1/3} \ll 1$. This is the well-known random-phase approximation (RPA), described in any textbook on quantum many-body theory. Generally speaking, it is not appropriate for the EHL theory any more than for the theory of common metals, because the equilibrium densities corresponding to the minimum energy are of the order of unity. However, we shall now present a few model systems where the EHL equilibrium densities are $n_1 \gg 1$ and the RPA treatment appears to be adequate. The general feature of these models is that, owing to peculiarities of their band structure (multivalley, anisotropic), the Fermi momentum p_F and energy are anomalously small, and there exists a concentration range where

$$1 \ll p_F \ll n^{1/4} \tag{9}$$

holds. These model systems are:

(1) *Quasi-one-dimensional*, i.e. the system of conducting filaments with neglible transfer of electrons and holes from one filament to another and the density of filaments N (number per unit of surface area perpendicular to their direction) satisfies $Na_{ex}^2 \gg 1$ (i.e. $N \gg 1$ in dimensionless units). a_{ex} here is given by (4) with m the longitudinal mass. In this model p_F is determined by a linear concentration of carriers in every filament,

$$p_F = \frac{\pi}{2}\frac{n}{N}, \quad (N \ll n \ll N^{4/3}). \tag{10}$$

In the brackets here and below the range of concentrations satisfying inequalities (9) is indicated.

(2) *Quasi-two-dimensional*, i.e. a system of parallel conducting planes (layers) with interplane distance $c \ll 1$ (in units of a_{ex})

$$p_F = (2\pi nc)^{1/2}, \quad (c^{-1} \ll n \ll c^{-2}) \tag{11}$$

(3) *Multivalley or with large effective mass anisotropy*; if the number of valleys ν or the effective-mass ratio M/m can be considered as a large enough parameter for both electrons and holes, M and m—principle values of the mass tensor

$$p_{F\mathrm{e,h}} = \pi \left(\frac{3}{\pi} \frac{n}{\tilde{\nu}_{\mathrm{e,h}}} \right)^{1/3}, \quad (\tilde{\nu}_{\mathrm{e,h}} \ll n \ll \tilde{\nu}_{\mathrm{e,h}}^{8/5}) \tag{12}$$

Here $\tilde{\nu} = \nu(M/m)^{1/2}$, if one of the effective masses M is much larger than the other two, and $\tilde{\nu} = \nu(M/m)$, if one of the effective masses m is much smaller than the other two. In both cases it is the smallest mass m which enters the exciton binding energy and the effective radius and, therefore, the definition of length and energy scales.

(4) *An electron–hole system in a strong magnetic field* $\mathscr{H} \gg 1$ (in units $\mathscr{H}_0 = e^3 m^2 c/\varepsilon^2 \hbar^3$). In this case, which is similar to the quasi-one-dimensional case, all the carriers are confined to the lowest Landau level and move freely in narrow 'Landau tubes' along the magnetic-field direction. The number of tubes per unit area of the plane orthogonal to the magnetic field is $\mathscr{H}/2\pi$:

$$p_F = 2\pi^2 \frac{n}{\mathscr{H}}, \quad (H \ll n \ll H^{4/3}). \tag{13}$$

In all these systems, owing to the respective large parameter $\Lambda = N, c^{-1}, \nu, M/m$ or $\mathscr{H}/2$, a concentration range exists which satisfies the inequalities (9) (as indicated in formulae (10)–(13) in brackets).

In this range, owing to the first inequality (9), the RPA is valid and, therefore

$$\chi(\mathbf{k}, i\omega; \lambda) \approx \chi^{(0)}(\mathbf{k}, i\omega; \lambda) = \frac{\lambda}{4\pi} V_{\mathbf{k}} \sum_{i=\mathrm{e,h}} 2\Lambda_i \int \frac{\mathrm{d}^d p}{(2\pi)^d} f_i(\mathbf{p}) \frac{\varepsilon_i(\mathbf{p}+\mathbf{k}) - \varepsilon_i(\mathbf{p})}{[\varepsilon_i(\mathbf{p}+\mathbf{k}) - \varepsilon_i(\mathbf{p})]^2 + \omega^2}. \tag{14}$$

In (14) $f_{\mathrm{e,h}}(\mathbf{p})$ are Fermi distribution functions of electrons and holes, and $\varepsilon_i(p)$ their dispersion laws; d is the dimensionality of the system under consideration, which is unity for the cases of filaments and the strong magnetic field, two for the layered system and three for the multivalley system. $V_{\mathbf{k}}$ is the matrix element of the bare Coulomb interaction, accounting generally for structural peculiarities of the system (electron and hole confinement to filaments, layers, 'Landau tubes', etc.). But for wave-vectors which are small compared to the inverse structural unit length ($N^{1/2}, \mathscr{H}^{1/2}, c^{-1}$), which are only essential for the following considerations, this matrix element is always nearly equal to the Coulomb interaction Fourier transform, i.e. $V_{\mathbf{k}} \approx 4\pi k^{-2}$. The dominating contribution to the integral over wave-vectors in (8), as one can easily check below, comes from $|\mathbf{k}| \sim n^{1/4}$ and $\omega \sim n^{1/2}$, which owing to the second inequality (9) are large compared to p_F and E_F respectively. Therefore, only the values of $\chi^{(0)}$ corresponding to such momenta are needed, and in (14) $|\mathbf{k}| \gg p_F \gtrsim |\mathbf{p}|$ and $\varepsilon_i(\mathbf{p}+\mathbf{k}) \approx \varepsilon_i(\mathbf{k}) \gg \varepsilon(\mathbf{p})$. Then formula (14) reduces to

$$\chi^{(0)}(\mathbf{k}, i\omega; \lambda) \approx 2\lambda \sum_{i=\mathrm{e,h}} \Lambda_i \frac{\varepsilon_i(\mathbf{k})}{|\mathbf{k}|^2[\varepsilon_i^2(\mathbf{k}) + \omega^2]} \frac{n}{\Lambda_i} = 2\lambda n \sum_{i=\mathrm{e,h}} \frac{\varepsilon_i(\mathbf{k})}{|\mathbf{k}|^2[\varepsilon_i^2(\mathbf{k}) + \omega^2]}. \tag{15}$$

It should be remembered that $\varepsilon_i(\mathbf{k})$ (but not $|\mathbf{k}|^2$) depend here only on the d components of the wave-vector $\mathbf{k}$. In the multivalley case contributions to (15) of both electrons and holes should be averaged over all equivalent valleys, which may be differently oriented in momentum space. After substitution of (15) into (8) and the introduction of new integration variables $\mathbf{k}=(64\pi\lambda n)^{1/4}\,\xi$ and $\omega=(4\pi\lambda n)^{1/2}\zeta$,

$$E_{\text{corr}}(n)=\frac{8}{5\pi^3}(4\pi n)^{1/4}\int \mathrm{d}^d\xi\,\mathrm{d}\zeta\, f(\xi,\zeta).$$

Here $f(\xi,\zeta)$ is a rational function depending on no other parameters than m_e/m_h. The evaluation of $E_{\text{corr}}(n)$ now becomes straightforward and results in

$$E_{\text{corr}}(n)=-\frac{32\pi}{5[\Gamma(\frac{1}{4})]^2}A_d\left(\frac{4}{\pi}n\right)^{1/4}\approx -1{\cdot}625A_d n^{1/4}. \tag{16}$$

The constant A_d depends on the dimensionality d of the system and very slowly on the electron and hole effective-mass ratio, but does not depend on the concentration n or the parameter Λ. In the case $m_e=m_h$, $A_3=1$, $A_2\approx 1,2$ and $A_1=2$.

The most remarkable features of (16) are universal for all models, that is, the dependence of E_{corr} on concentration as $n^{1/4}$ and its independence of Λ, whereas the values of $E_0(n)$ and $E_{\text{exch}}(n)$ are strongly reduced in these models, owing to the large values of Λ.

The exchange energy in all these models appears to be small compared to the correlation energy, because, like $E_0(n)$, it is determined only by the concentration in one filament, layer, valley, etc. Therefore, (7) transforms to

$$E(n)=E_0(n)-An^{1/4}, \tag{17}$$

where $A\approx 1{\cdot}625\,A_d$. It is easy to check that for all cases under investigation $E(n)$ reaches its minimum in the concentration interval where inequalities (9) are valid. Let us demonstrate this claim for a quasi-two-dimensional system, where $E_0(n)=\frac{1}{2}\pi nc$. Then $A\approx 1{\cdot}95$,

$$n_{\min}\equiv n_1(T=0)=\left(\frac{A}{2\pi c}\right)^{4/3}\approx 0{\cdot}21c^{-4/3}, \tag{18}$$

$$E_{\min}\equiv E_1(T=0)=-\frac{3\pi}{2}\left(\frac{A}{2\pi}\right)^{4/3}c^{-1/3}\approx -0{\cdot}99c^{-1/3}. \tag{19}$$

Numerical results here and below correspond to the case $m_e=m_h$. Thus both the equilibrium concentration n_1 and the binding energy of the EHL are large (in units of a_{ex}^{-3} and $2E_{ex}$ respectively) if c is small enough compared to a_{ex}. From (18) $p_F\sim c^{-1/6}$ at the equilibrium density and both inequalities (9) are satisfied.

Not only the ground-state energy and the equilibrium density at zero temperature, but the whole thermodynamics of the EHL can be calculated for these models in the same way. For all temperatures $T\lesssim T_c$ the correlational contribution to the thermodynamic potential is the same as that given by (16), because the corresponding general formula differs from (8) only by changing the integration over ω to a sum over $\omega_j=2\pi jT$. But, as was seen, the principal contribution to this integral comes from $\omega\sim n^{1/2}\gg n^{1/4}\sim E_1\gtrsim T_c$. So for nonzero temperatures $E_0(n)$ in (17) has only to be replaced by the chemical potential of the noninteracting Fermi gas $\mu_0(n,T)$ in order to obtain the equation of state of the electron–hole system: $\mu(n,T)=\mu_0(n,T)-\frac{5}{4}An^{1/4}$.

Here $\mu(n, T)$ is the chemical potential of an electron–hole pair. For a quasi-two-dimensional system this looks like

$$\mu(n, T) = 2T \ln\left(\exp\left[\frac{\pi nc}{2T}\right] - 1\right) - \tfrac{5}{4}An^{1/4}. \tag{20}$$

The dependence (20) of μ on n at different temperatures is shown schematically in figure 5. The family of curves $\mu(n, T)$ has a typical van der Waals appearance and describes the coexistence of two phases. All the parameters of the phase diagram can be found from (20), including the equilibrium concentrations of both phases $n_l(T)$ and $n_g(T)$. The critical point is $n_c \approx 0{\cdot}036c^{-4/3} \approx 0{\cdot}17n_l(T=0)$, $T_c \approx 0{\cdot}102c^{-1/3} \approx 0{\cdot}103E_l(T=0)$. Note the excellent agreement with the empirical rule $kT_c \approx 0{\cdot}1E_l$, mentioned above. Three more qualitative results obtained from the analysis of equation (20) are worthy of special notice. The critical temperature appears to be very close to the Fermi degeneracy temperature, corresponding to $n = n_c$. This means that the electron–hole liquid actually exists only as a degenerate Fermi liquid. In the vicinity of the critical point the gas phase is a completely ionized electron–hole plasma and the transition is a gas–liquid type transition from a nondegenerate to a degenerate plasma. And, finally, within the model under consideration no other phase transitions occur except that of gas–liquid type with EHL formation.

Results for other models are similar and I will mention only the case of the strong magnetic field. In this case E_1 and T_c increase $\sim \mathcal{H}^{2/7}$, and n_1 and $n_c \sim \mathcal{H}^{8/7}$: $E_1(T=0) \approx -0{\cdot}42\mathcal{H}^{2/7}$, $n_1 \approx 0{\cdot}03\mathcal{H}^{8/7}$, or $E_1 \sim H^{2/5}$, $n_1 \sim \mathcal{H}^{6/5}$, if $m_h \gg m_e$ and electrons only are in the ultraquantum limit. As is known, the binding energy of the exciton increases in this limit as $\ln^2 \mathcal{H}$, and therefore, in a strong enough field the EHL becomes energetically favourable for arbitrary band structures. The most interesting property of

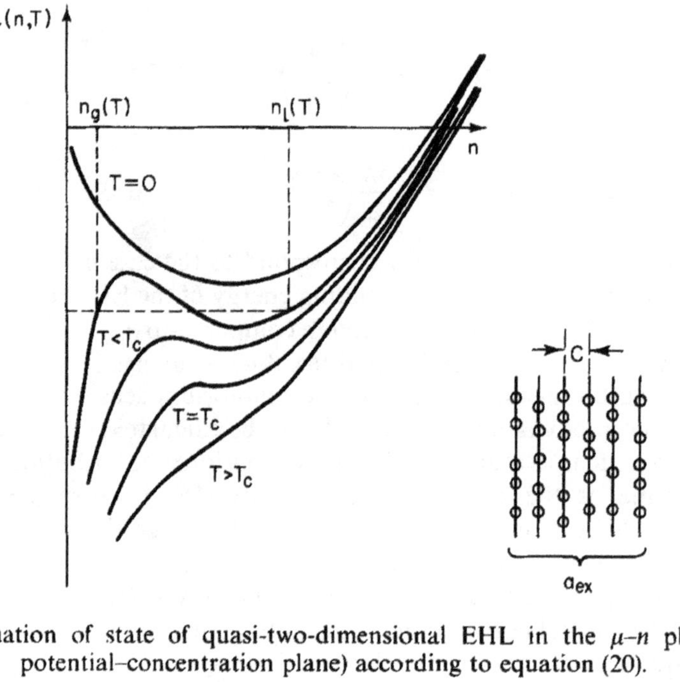

Figure 5. Equation of state of quasi-two-dimensional EHL in the μ–n plane (chemical potential–concentration plane) according to equation (20).

the EHL in a strong magnetic field is that it is insulating at low temperatures, as, owing to the one-dimensional character of the carrier movement, the appearance of the gap at the Fermi level becomes inevitable [30, 31]. This gap suppresses the scattering of carriers by phonons and thus greatly increases the EHL mobility. The coherent nature of the gap gives one reason to expect properties similar to superfluidity. The EHL is the only condensed matter which can be studied experimentally (in laboratories) in strong fields, corresponding to the ultraquantum limit. For usual matter the critical field for this regime is of the order 10^{10} G (10^6 T) and is inaccessible on Earth. However, such high fields are postulated under some astrophysical conditions (neutron stars, etc.).

Thus for semiconductors with multivalley band structure or sufficiently pronounced effective mass anisotropy the phase diagram of the nonequilibrium carrier system is certainly of the simplest shape, as depicted in figure 2 (*a*). It is only in the case of single valley and not too anisotropic spectra of electrons and holes that the phase diagram of figures 4 (*a*) and (*b*) might be realized: 4 (*b*) for almost isotropic and not very different electron and hole masses and 4 (*a*) for extremely different, moderately anisotropic masses. It should be emphasized once again that the accuracy of the existing calculations for single-valley, weakly anisotropic spectra does not enable us to eliminate completely the possibility that the phase diagram type of figure 2 is perfectly general.

The gas–liquid transition in the nonequilibrium carrier system is closely connected with the metal–insulator transition: from a weakly ionized poorly conducting exciton gas to a metallic EHL. The problem of the coexistence of these two transitions was first discussed (for mercury vapour condensation) in the well-known paper by Landau and Zel'dovitch [32]. The phase diagrams of figures 3 and 4 illustrate two different possibilities: complete coincidence of these transitions (figures 2 (*a*) and 4 (*a*)) and the case where the metal–insulator transition in some temperature interval occurs in the phase which is already condensed (figures 2 (*b*) and 4 (*b*)). Many authors (Insepov and Norman [33], Rice [7, 34], Ebeling *et al.* [35]) proposed still another possibility: that the metal–insulator transition should still occur in the gaseous phase owing to the screening of the Coulomb interaction, resulting in exciton destruction (the so-called Mott transition). However, well-known qualitative reasoning (Mott) [36] as well as the subsequent calculations show the Mott transition to be that of the first order and, consequently, to be accompanied by a specific volume change and a phase separation. Thus the dense phase formed in this transition possesses all the typical features of the EHL and there is no reason for the occurrence of still another phase transition at a further density increase.

3.2.2. *The influence of crystal ionicity*

Not only may the structure of the electron spectrum be different in different semiconductors, but also the nature of the electron–hole interaction may also be different. All the preceding considerations, as well as the theoretical papers mentioned above, are based on the assumption that the interaction is purely Coulombic and its specific character in a particular crystal is determined by only one constant—the dielectric constant. This assumption is well established for covalent semiconductors such as germanium and silicon. However, compounds such as gallium arsenide or cadmium sulphide are partially ionic and a noticeable contribution to their polarizability is due to the ion displacement, i.e. their lattice deformation. In such semiconductors the inertia of the heavy ions results in the Coulomb interaction becoming retarded in time intervals of the order of the period of the lattice vibrations or, in other words, the dielectric constant becomes frequency dependent in the optical-phonon frequency region. In the random phase approximation (RPA) the introduction of the interaction with optical phonons into the calculation of the EHL energy corresponds [37] to the replacement of ε in (1) by the frequency-dependent dielectric constant $\varepsilon(\omega)$,

$$\varepsilon(\omega)=\varepsilon_\infty\varepsilon_0\frac{\omega_1^2-\omega^2}{\varepsilon_\infty\omega_1^2-\varepsilon_0\omega^2}. \qquad (21)$$

Here ε_0 and ε_∞ are the static and high-frequency (electronic) values of the dielectric constant; ω_1 is the frequency of the longitudinal optical phonon. Such interaction naturally changes the exciton, biexciton and EHL binding energies and their ratios.

Direct calculations [7, 38–40] using (21) show that, in many polar semiconductors even with a small degree of crystal ionicity, the EHL becomes energetically more favourable than the biexcitonic gas, even with a single-valley spectrum and complete effective mass isotropy.

Finally let us make one more remark concerning the general status of EHL theory. The absence of heavy particles of nuclear type and the crystallization connected with them, the possibility of changing all the main parameters within a wide range, the comparatively easy accessibility of the so-called extreme conditions ($\bar{n}\sim n_c$, $T\sim T_c$ or $n\gg n_1$, superstrong magnetic fields), the possibility of observing directly binding energy and particle energy distribution in recombination radiation spectra, all these properties make the EHL an ideal model for the experimental and theoretical study of the quantum electron liquid—one of the fundamental objects of condensed-matter physics. Therefore, it is no wonder that practically all the methods invented for a theoretical description of the electron liquid in metals have also been tested in the EHL computations. As a matter of fact, all these methods are one or other modification of the RPA. Their general aim is to supplement the RPA treatment, which is asymptotically exact at very high densities $na_0^3\gg 1$ (a_0 is the Bohr radius), in order to obtain a reasonable description at electron densities $na_0^3\sim 10^{-1}$–10^{-3}, usual for metals. Taking into account certain factors (usually short-range electron–electron and electron–hole correlations, i.e. local-field corrections) and ignoring others, they are semi-empirical in the sense that there is no rigorous proof of the approximations used and no estimates of their accuracy within the theory itself for the density interval mentioned above. It is only the agreement with experiment that enables us to evaluate their real effectiveness. The same problem confronts EHL theory by changing only the scales of length a_0 and Rydberg by a_{ex} and E_{ex}. From this standpoint the success of the theory in application to the EHL would demonstrate the adequacy of the quantum many-body theory methods referred to above for the treatment of multielectron systems with densities typical for atoms, molecules and condensed media. However, the excellent agreement between the EHL theory and experimental data in Ge and Si described below should not be overestimated. Owing to the multivalley band structure and the strong effective mass anisotropy, the EHL equilibrium density in these semiconductors is abnormally great and really close to the limit of applicability of RPA. From this point of view, the absence of quantitative agreement between theory and experimental data in semiconductors with a simpler electronic spectrum might not be accidental, but may reflect the real limitations of the methods used in the domain of the usual 'metallic' densities $na_0^3\sim 10^{-3}$–10^{-1}. Further experimental data on the EHL in different semiconductors and its systematic analysis will provide a final evaluation of these methods and probably indicate how to perfect them.

2.3. *Some experimental results*

Many excellent experimental studies of the EHL in semiconductors have been performed during the last fifteen years and are reviewed in [8, 9]. Only very few of them will be presented here in order to illustrate the most general and important EHL manifestations.

EHL experimental studies were started by Pokrovski and Svistunova [41] in 1969. In the spectrum of low-temperature luminescence of germanium they found the radiation due to the recombination of electron–hole pairs in the EHL.

The appearance of this radiation had a threshold character at decreasing temperature or increasing excitation level in accordance with the phase transition picture; its spectrum reflected the Fermi distribution of recombining carrier energies. Hence the equilibrium concentration in the EHL in germanium was determined to be $n_l \approx 2 \times 10^{23}\,\mathrm{m}^{-3}$. Figure 6 [42] shows the luminescence spectrum of Ge consisting of the luminescence line of the free excitons (FE) and the recombination radiation line of the EHL for a very special choice of temperature at which the intensities of both lines are comparable. At the same excitation level but at a few tenths of a degree higher in temperature the EHL line completely disappears, but at a few tenths of a degree lower in temperature it becomes dominating and the exciton line decreases sharply. The line shape of the EHL luminescence corresponds to the calculated recombination radiation of a degenerate electron–hole plasma (solid line). Its full width equals the sum of electron and hole Fermi energies, and therefore only slightly depends on temperature; it is much larger than the line width of exciton luminescence, which is $k_B T$. The energy difference between the long-wavelength edge of the FE line and the short-wavelength edge of the EHL line is just the work function $\phi = |E_1 - E_{ex}|$ of the EHL. Thus luminescence data contain information concerning the main EHL parameters n_l, E_l and also the electron and hole energy distribution in the EHL, n_g, and so on. Their study at different temperatures makes possible the reconstruction of the whole phase diagram of the nonequilibrium charge carrier system, as was done first for germanium [43] and silicon [44]. Up to now luminescence spectra have been the main source of information about the EHL and the only source for many direct-gap semiconductors with short lifetimes of non-equilibrium charge carriers.

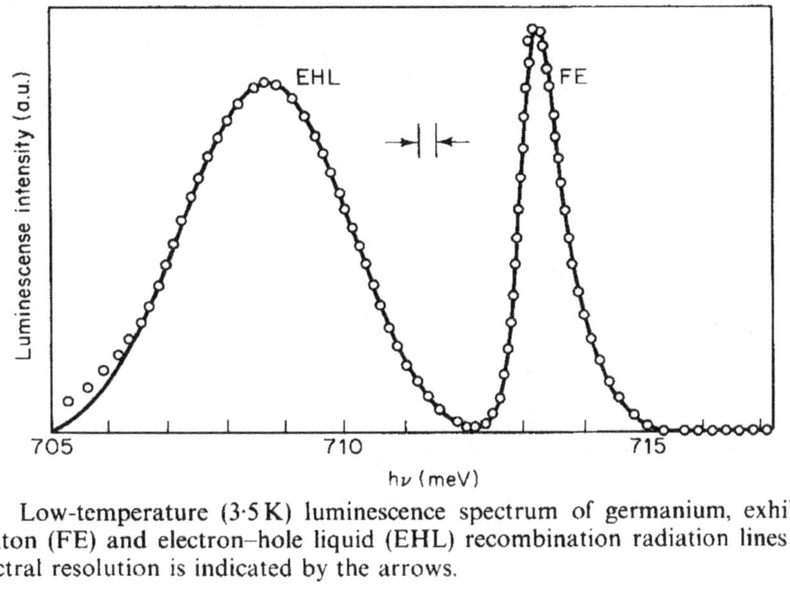

Figure 6. Low-temperature (3·5 K) luminescence spectrum of germanium, exhibiting free exciton (FE) and electron–hole liquid (EHL) recombination radiation lines [42]. The spectral resolution is indicated by the arrows.

The phase diagrams of nonequilibrium charge carrier systems that have been most completely studied are in silicon and, more especially, germanium. The former is presented in figure 7. Both of them are undoubtedly of the simplest type of figure 2 (*a*) owing to the multivalley band structure and the large effective mass anisotropy in both germanium and silicon. In both cases very good quantitative agreement of theoretically calculated values and experimental data (2–4% difference) is now achieved:

$$E_l(T=0)=6\,\mathrm{meV},\ n_l(T=0)=2{\cdot}3\times10^{23}\,\mathrm{m}^{-3},\ T_c=6{\cdot}7\,\mathrm{K},\ n_c=0{\cdot}6\times10^{23}\,\mathrm{m}^{-3}$$

in germanium and

$$E_l(T=0)=23\,\mathrm{meV},\ n_l(T=0)=3{\cdot}5\times10^{24}\,\mathrm{m}^{-3},\ T_c=28\,\mathrm{K},\ n_c=1{\cdot}2\times10^{24}\,\mathrm{m}^{-3}$$

in silicon. Note again the very accurate fulfilment of the rule $k_B T_c \approx 0{\cdot}1\,E_l(T=0)$. In addition, another relation of the same type, namely $n_c = \mathrm{const}\, n_l(T=0)$ has been proposed empirically [8] and theoretically justified by Vashishta *et al.* in [9] with const ≈ 0·2, but its accuracy is much poorer.

E_l values in germanium and silicon are relatively large, approximately 1·5 E_{ex}. There exists direct experimental evidence that this is due primarily to the multivalley band structure: the uniaxial stress of crystals, reducing the crystal symmetry and, therefore, the number of equivalent valleys, significantly reduces the EHL binding energy, critical temperature and equilibrium density [17]. In figure 8 the dependence of the energy maxima of free excitons and the EHL luminescence lines on the uniaxial stress P is depicted. The shift of the exciton position is approximately linear and follows that of the band gap E_g. But the EHL lines at small stresses shift in the opposite direction so that their distance from the exciton line, reflecting the EHL binding energy, decreases. And only for $P > P_{cr} \approx 2{\cdot}7\times10^6\,\mathrm{kg\,m}^{-2}$, when all the electrons occupy only the two lower valleys instead of four in unstressed germanium (these two remain equivalent for this stress direction) the further shift of the EHL line becomes the same as that of the exciton line, i.e. the EHL binding energy becomes constant but smaller than

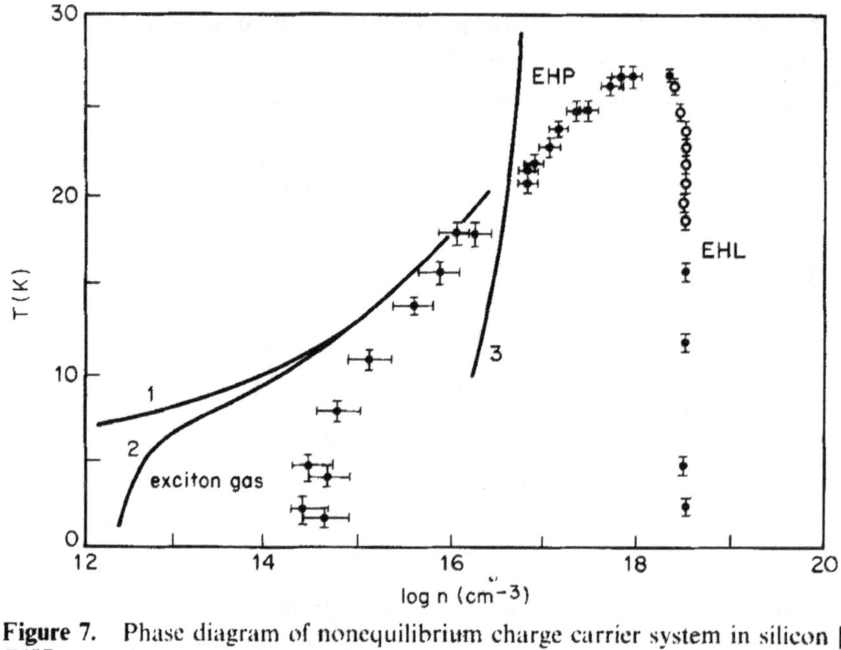

Figure 7. Phase diagram of nonequilibrium charge carrier system in silicon [44]. EHP nondegenerate electron hole plasma. 3 theoretical Mott transition curve.

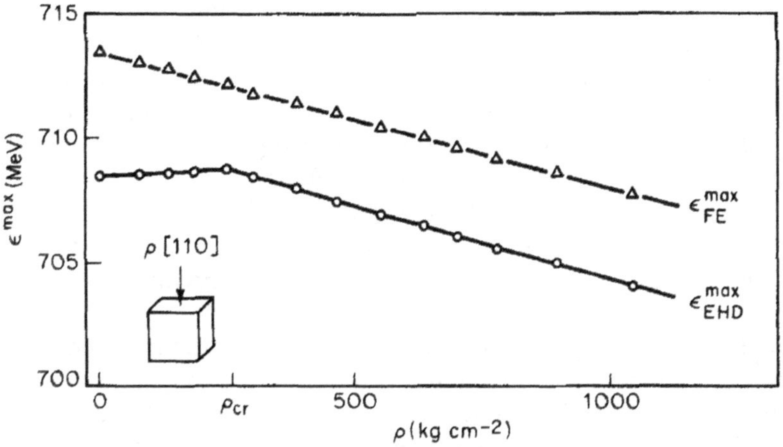

Figure 8. Stress dependence of positions of the maxima of FE and EHL lines, shown in figure 6 for $T=3$ K [17].

at zero stress. The same experiments [17] demonstrated that non-uniformity of the stress results in the fast movement of EHDs in the direction of increasing stress. At moderate stresses $P \gtrsim P_{cr}$ they travel for a distance of the order of 1 cm with velocities close to that of sound. This phenomenon provided makes possible the aggregation of a cloud of small EHDs into one 'large drop' [45]—a macroscopic volume of the EHL, which is very convenient for the investigation of the properties and the many new fascinating phenomena in the EHL, including its direct visualization, as described by Wolfe and Jeffries [9].

There exist many manifestations of two-phase coexistence in the non-equilibrium carrier system at $T<T_c$. One of the most spectacular is the time dependence of the exciton concentration in the gas phase n_g after excitation of a specimen by a short intense illumination pulse at $t=0$ (figure 9) [46]. It differs drastically from the usual exponential decay with lifetimes of 4–8 μs which one observes in this type of experiment but at higher temperatures or lower excitation levels. Instead, the concentration of

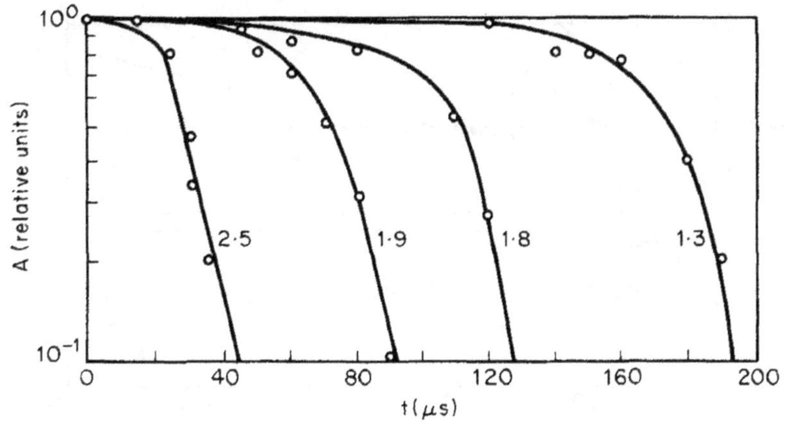

Figure 9. Time decay of exciton concentration under the conditions of existence of EHL in a germanium sample, excited by an intense short pulse at $t=0$ [46]. The values 2·5, 1·9, 1·8 and 1·3 show preliminary attenuation of the exciting pulse compared to some maximum value.

excitons coexisting with the EHL remains virtually constant during time intervals exceeding 20–30 times their lifetime, because evaporation of new excitons from the EHDs just compensates their recombination in the gas phase, maintaining the equilibrium value $n_g \approx n_g(T)$ (saturated vapour pressure). And only after the disappearance of all EHDs due to recombination and evaporation, does the remaining exciton concentration decrease fast with the usual lifetime. Obviously, the time interval of constant (quasi-equilibrium) exciton concentration increases as the initial total volume of the EHL increases. This also agrees with the data given in figure 9, where the numbers on different curves designate the attenuation of the preliminary exciting-pulse relative to some maximum value.

The Fermi liquid nature of the EHL was first demonstrated by the observation of the oscillatory dependence of its luminescence intensity on magnetic field [47]. This dependence is shown in figure 10 together with the Landau levels of electrons and holes, as a function of the magnetic-field strength $\mathscr{H}$, and the crossing of these levels with the Fermi levels E_{Fe} and E_{Fh}. The reason for these oscillations is the same as for the de Haas–van Alphen effect. Like any other thermodynamic quantity, the equilibrium concentration of the EHL oscillates as a function of the quantizing magnetic field, and, therefore, the probability of radiative recombination also oscillates [48]. At still larger fields all electrons are in the lowest Landau level, i.e. the ultraquantum limit is reached.

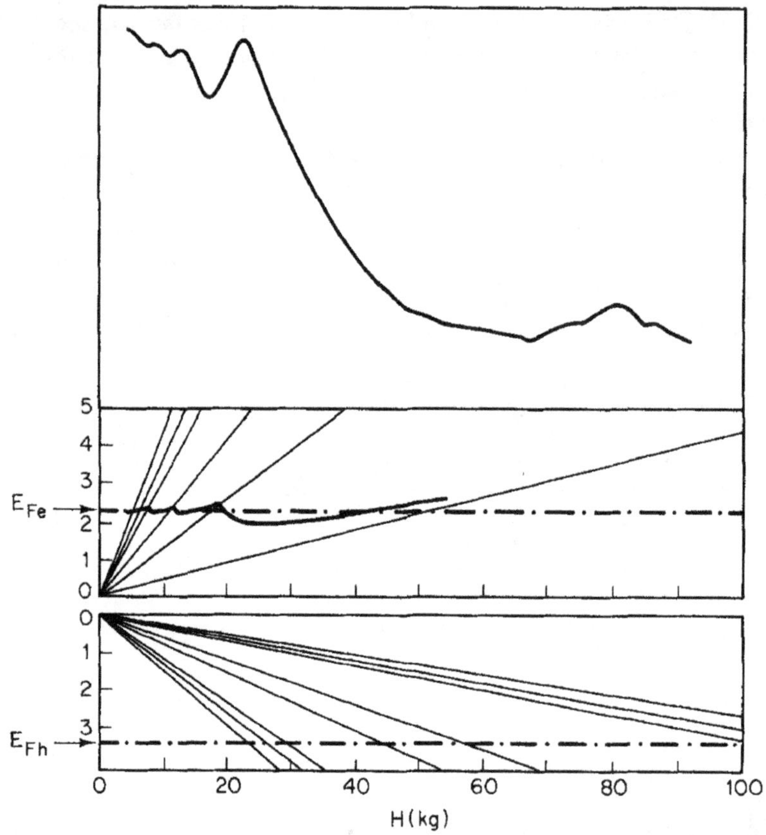

Figure 10. Magnetic-field dependence of the intensity of EHL luminescence in germanium, $H \| [100]$, $T = 1{\cdot}5$ K. Also shown are Landau levels in conduction and valence bands, crossing electron and hole Fermi levels [47].

An approximately linear increase of the equilibrium EHL density with magnetic field strength was reported in this regime [49], closely corresponding to the above theoretical prediction. But any experimental evidence for the existence of the insulating liquid phase is still lacking, and it remains one of the most intriguing possibilities of future EHL research.

Other manifestations of the Fermi liquid state of the EHL are the observation of plasma resonance [50], Alven waves [51] and the direct measurement of conductivity [52], which appeared to be rather large, corresponding to a carrier momentum relaxation time 10^{-10}–10^{-11} s, limited probably be electron–hole scattering. It should be noted that the relaxation time of the momentum of the EHD as a whole, measured in experiments with EHD movement, appears to be two orders of magnitude larger, because obviously interparticle collisions inside the EHD do not contribute to this relaxation, and it is determined only by electron–phonon scattering. All the data described above, like many other, refer to germanium and silicon.

There are numerous reports on the observation of the EHL and the establishment of its phase diagram in many other semiconductors, most of them reviewed by Kulakovskii and Timofeev in [9]. However, as a rule, these data are less complete and conclusive as compared with those on germanium and silicon and obtained only from spontaneous- or stimulated-luminescence data. In connection with the above discussion of possible types of phase diagram, special interest attaches to materials with one valley and a more or less isotropic spectrum, such as gallium arsenide or cadmium sulphide. Though it is too early to draw any final conclusions, it seems that, in spite of the above-mentioned theoretical estimates, experimental data on these semiconductors give evidence in favour of a phase diagram of the type of figure 2 (*a*). The cause of this discrepancy (perhaps not the only one) may be that all these semiconductors are partially ionic, which, as explained above, changes the stability conditions of the EHL. Accounting for the frequency dependence of the dielectric constant in some cases eliminates the qualitative contradiction of theory and experiment concerning the type of phase diagram, but fails to produce quantitative agreement similar to that achieved for germanium and silicon: calculated values of the EHL binding energy remain smaller and equilibrium densities essentially larger than those reported in experimental studies. Also the results of calculations by different authors are noticeably different. Perhaps the reason is the inadequacy of the RPA and other methods closely related to it because of the relatively small ($\sim 10^{-2}$) equilibrium density in the single-valley semiconductors.

4. Electron–hole drops

4.1. *Kinetics of electron–hole drop nucleation, growth and decay*

The concepts of phase diagram, phase transition, phase coexistence, etc., strictly speaking, refer to systems in thermodynamic equilibrium. For a description of the nonequilibrium charge carrier system in semiconductors they are applicable only approximately, according to the smallness of the thermalization time, i.e. the relaxation time of the carrier kinetic energy, compared with their lifetime, determined by recombination processes. This time ratio reaches 10^4–10^5 in the case of Ge and Si and 10–10^2 in the so-called direct-band semiconductors of groups $A^{III}B^{V}$ and $A^{II}B^{VI}$, of which gallium arsenide and cadmium sulphide are typical examples. Thus with a greater or lesser degree of accuracy thermalization occurs if the lattice temperature is not too low. But even if carrier thermalization may be considered complete, the finite carrier lifetime qualitatively changes some details of the phase diagram and produces quite a number of phenomena and properties that distinguish the EHL from any other liquid. The reason is that the condensation process itself, i.e. the formation of liquid-phase nuclei and their further growth to the macroscopic EHD, is much slower than thermalization, as it requires the participation of a macroscopically large number of particles. At this stage the finite carrier lifetime plays a decisive role. It limits the EHD growth, complicates the nucleation process, etc. However, it does not limit the existence time of the EHD if some excitation source continuously produces new electron–hole pairs whose condensation compensates the recombination in the EHL volume.

From this point of view, all experiments with the EHL can be divided into two essentially different groups: stationary and pulsed. In the latter the specimen is excited by a short intense pulse with duration less than the carrier lifetime, and subsequent EHD formation and decay are observed. The growth (or decay) kinetics for every individual EHD is described by a simple balance equation [3, 53] for the total number of particles in it:

$$\frac{\mathrm{d}}{\mathrm{d}t}\left(\frac{4\pi}{3}R^3 n_1\right)=4\pi R^2\gamma v_{\mathrm{T}}[n-n_{\mathrm{g}}(T,R)]-\frac{4\pi}{3\tau_1}R^3 n_1, \tag{22}$$

where R is the spherical EHD radius, n is the exciton concentration in the gaseous phase, v_{T} is their thermal velocity, γ the so-called accommodation coefficient, i.e. the probability of an exciton approaching the EHD being absorbed, τ_1 and n_1 the carrier lifetime and their equilibrium concentration in the EHL. The first term on the right-hand side is the difference between the number of excitons captured by the EHD surface from the surrounding gas and the number evaporated from this surface. The flow of evaporated excitons from the detailed-balance condition is expressed in terms of the gas concentration $n_{\mathrm{g}}(T,R)=n_{\mathrm{g}}(T)\exp[2\sigma/n_1 RT]$ (σ is the EHL surface tension), which would be in equilibrium with EHD of radius R at a given temperature in the absence of recombination. The second term on the right-hand side of equation (22) is the recombination rate in the EHD volume. A complete quantitative description of the condensation kinetics in the nonequilibrium charge carrier system is based on the nucleation analysis and the time evolution of the EHD size distribution function [53–59]. But the qualitative picture of the phenomena and the significance of the recombination process in them can already be understood from equation (22).

Figure 11 presents schematically three curves corresponding to temperatures $T_1>T_2>T_3$ and describing the connection of $\Delta n=n-n_{\mathrm{g}}(T)$ (gaseous-phase supersaturation) with the EHD radius R in a stationary state, obtained by equating the righthand side of equation (22) to zero. The domain of parameter values $(\Delta n, R)$ above such a curve corresponds at a given temperature to growing EHD, i.e. $\mathrm{d}R/\mathrm{d}t>0$, and the domain under this curve to decaying EHD, $\mathrm{d}R/\mathrm{d}t<0$. The typical U-shaped form of the curves for T_1 and T_2 is due to the combined action of two factors. At small R surface tension diminishes the work function of the droplet and in this way increases the pressure of the saturated vapour above it. At larger R the supersaturation, necessary to support the drop in a steady state, increases because the exciton flux must compensate for the recombination inside the drop, and this flux is proportional to supersaturation and surface area, while the number of recombining particles grows with the volume of the drop. The first of these effects is common to any liquid. The second is due to the finite lifetime of nonequilibrium carriers and is specific to the EHL. It produces in the curves $\Delta n(R,T)$ of figure 11 a characteristic minimum at $R=R_{\min}(T)$,

$$R_{\min}(T)=\frac{1}{n_1}\left[\frac{6\sigma\tau_1\gamma v_{\mathrm{T}}n_{\mathrm{g}}(T)}{k_{\mathrm{B}}T}\right]^{1/2}. \tag{23}$$

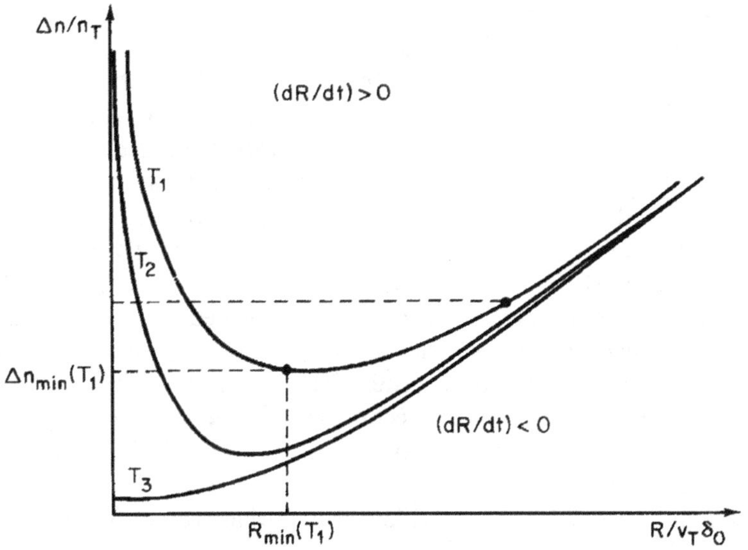

Figure 11. Super-saturation of the exciton gas against steady-state electron hole drop radius for different temperatures $T > T_2 > T_3$. At a temperature T, R_{min} (T) is the minimum stable EHD radius and $\Delta n_{min}(T)$ the minimum saturation necessary for EHD existence.

The supersaturation corresponding to this point,

$$\Delta n_{min}(T) = 2\left[n_g(T)\frac{2\sigma}{3v_T\gamma\tau_1 k_B T}\right]^{1/2}, \tag{24}$$

is at a given temperature the minimum supersaturation; only above this can the EHD exist. Therefore, because of kinetic factors, the threshold nonequilibrium carrier concentration for condensation is $n_g(T) + \Delta n_{min}(T)$, and not simply $n_g(T)$, determined from the thermodynamic equilibrium conditions of the EHL and the carrier and exciton gas.

There are two stationary values of R for every $\Delta n > \Delta n_{min}$. However, it is easy to understand that only one value of R, that on the ascending $(R > R_{min})$ branch of the curve $\Delta n(R)$, is stable, because the signs of dR/dt at the (n, R) plane sections close to this branch, are such that, even after an accidental small deviation of R from its stationary value at given Δn, R returns to its original value. By contrast the stationary values of R at the descending $(R < R_{min})$ branch are unstable: after a small deviation they do not return, but tend either to $R \to 0$ or to the ascending branch. They correspond to the so-called critical nuclei, well known in condensation theory. The average picture considered above does not enable us to describe their formation process, as according to (22) droplets with sizes less than that of a critical nucleus at given Δn do not grow, but evaporate. Critical nucleus formation requires a big fluctuation to overcome the thermodynamic barrier, separating the exciton gas (in figure 11 it corresponds conventionally to $R \to 0$) from the EHL, consisting of macroscopic EHD with radii on the ascending branch. The reverse process is also possible; then the barrier is overcome in the opposite direction from the liquid phase side: a fluctuational decrease of the stationary existing EHD to the critical embryo size with its subsequent evaporation. In a more elaborate description, accounting for fluctuations and EHD size distribution [54–59], stable R values correspond to sharp maxima of a distribution function and critical embryo size to a deep minimum. The greater the embryo size and the smaller Δn, the less is the probability of a fluctuation necessary for this embryo formation. Therefore, when Δn rising reaches $\Delta n_{\min}$, the EHD existence becomes possible; virtually, however, they do not appear as an enormous time is required for their nucleation. The gaseous phase remains metastable. Only at considerably larger Δn, corresponding to a smaller size of the critical embryo, does intense formation of EHD start. If Δn is then lowered again, all EHD diminish in size according to the $\Delta n(R)$ curve of figure 11 but continue to survive down to Δn values only slightly exceeding $\Delta n_{\min}$, when the probability of their evaporation due to fluctuations already becomes overwhelming. Thus, at concentrations not greatly exceeding the condensation threshold value, the nonequilibrium carrier system possesses pronounced hysteresis [60], as the EHD number and the total volume of the liquid phase depend not only on the present excitation level, but also on its prehistory. This behaviour is typical of first-order phase transitions and is not in the least peculiar to the EHL. But the EHL is unique inasmuch as the memory of initial conditions is preserved in it during times many orders of magnitude greater than the lifetime of the electrons and holes of which it is composed [61, 62].

With a decrease in temperature the minimum value $R_{\min}(T)$ of the stable radius of the EHD quickly diminishes and at sufficiently low temperatures becomes $\sim n_l^{-1/3}$. This means that droplets containing only a few pairs of particles are involved. Essentially these are not yet EHL droplets but the so-called multiexciton complexes. It is clear that continuing the curves in figure 11 into the range of still smaller R is meaningless. Therefore, at such and still lower temperatures the $\Delta n(R)$ dependence assumes the appearance depicted in figure 11 by the curve T_3 with no minimum and no descending branch corresponding to critical embryos. Therefore, there exists no thermodynamic barrier to EHD formation. And this changes the condensation picture drastically [54–59]. Hysteresis phenomena are completely absent. The system responds to an increase in excitation level, first of all by increasing the number of complexes being formed. The size of each of them remains very small and they grow only slowly with excitation level. Close to threshold these are complexes of a few excitons and it is only at essentially larger excitation levels that they acquire all the characteristic features of EHL drops, including definite and constant values of n_l and E_l. Just such a picture was observed at low temperatures in Si [63] and later in other semiconductors. Under such conditions nonequilibrium charge carrier condensation loses all the typical features of a first-order phase transition. A difference between the gaseous and the condensed phase gradually develops as the excitation level increases, corresponding rather to a second-order phase transition picture [64, 65]. Perhaps it would be more correct to speak of the absence of a well-defined phase transition, because the formation of the multiexciton complexes has no definite threshold. The process only weakly depends on the temperature and becomes noticeable at carrier concentrations determined by the competition between the elastic capture of excitons and their recombination.

Figure 12 shows schematically the EHD existence region under stationary excitation conditions, taking into account the above discussion on the role of the charge carrier finite lifetime and the absence of equilibrium associated with it. For comparison the branch $n_g(T)$ of the phase diagram of figure 2 (*a*) corresponding to the system in the thermodynamic equilibrium is also shown.

In the case of pulse excitation it is the mutual influence of EHL and exciton gas that most noticeably manifests itself. Generally the decay kinetics of both phases appear to

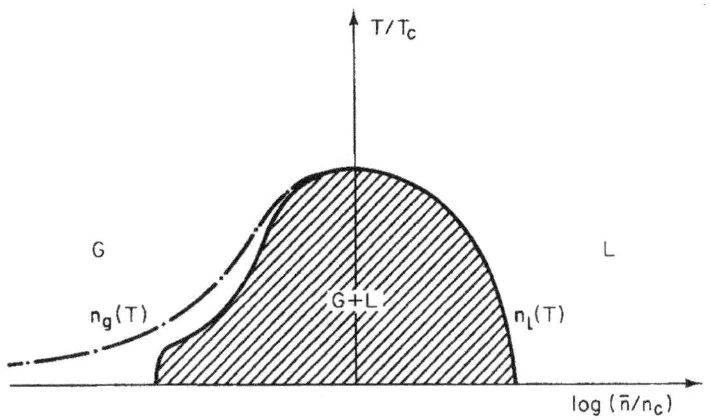

Figure 12. Electron–hole drop existence region accounting for the effect of charge carrier finite lifetime. The broken line is the $n_g(T)$ low-density branch of the phase separation curve of figure 2(*a*).

be essentially nonexponential. At sufficiently low temperatures EHD evaporation maintains the exciton concentration at a nearly constant level (saturated vapour pressure) for a time longer by more than an order of magnitude than the exciton lifetime and several times greater than the charge carrier lifetime in EHDs themselves, as shown in figure 9 and explained above. At higher temperatures it is evaporation and not recombination that is the dominant decay mechanism for sufficiently small droplets. In this case the reduction of the EHD radius occurs not exponentially with time but linearly [8]. EHDs completely disappear in a time interval which at small initial supersaturation may be considerably less than the carrier lifetime, recombination occurring mainly in the gaseous phase.

The most powerful method of direct observation and investigation of EHDs is that of light scattering [66–71]. In this type of experiment both the EHD radii and their number can be measured. In figures 13 and 14 the temperature dependences of these quantities are shown at a fixed excitation level. Typical values of EHD radii in germanium are 1–10 μm and increase as the temperature increases. The number of drops per unit volume N decreases rapidly at higher temperatures, typical values being 10^9–10^{14} m^{-3}. Both temperature dependences can be easily understood in terms of the above considerations. The number of drops is determined by the nucleation process which increases very rapidly with supersaturation. At a given excitation level immediately after the excitation source is switched on, nucleation still starts and the exciton concentration is the same for any temperature. This means that the relative supersaturation is much higher at low temperatures than at higher temperatures, because the threshold concentration decreases rapidly with temperature. Therefore, at low temperatures the nucleation process is very fast. While growing, these EHDs absorb excitons, supersaturation reduces and the birth of new nuclei stops. The total number of nonequilibrium charge carriers is fixed by the excitation source, and it relates the radii of the EHD with their number. So at low temperatures many small drops arise, and at higher temperatures a smaller number of larger drops. It is interesting that under stationary conditions all drops have the same radius, since it is

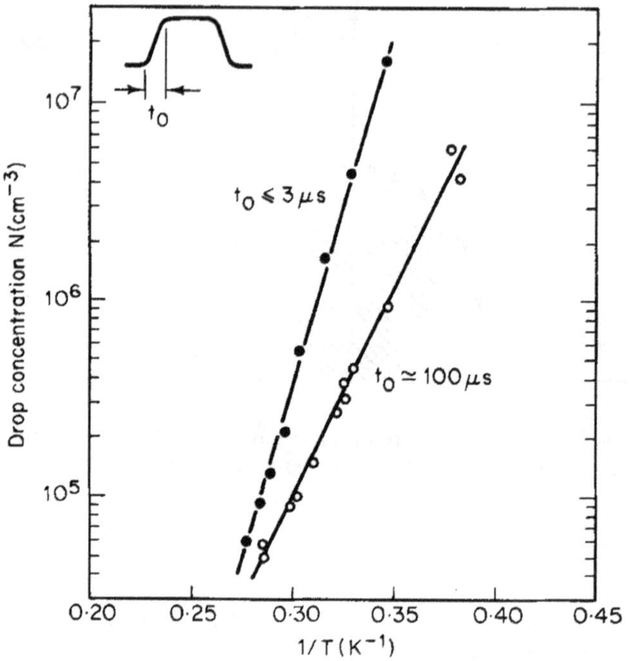

Figure 13. Temperature dependence of the number of electron–hole drops per unit volume. Different curves correspond to the same excitation levels, but with different rise times t_0 [70].

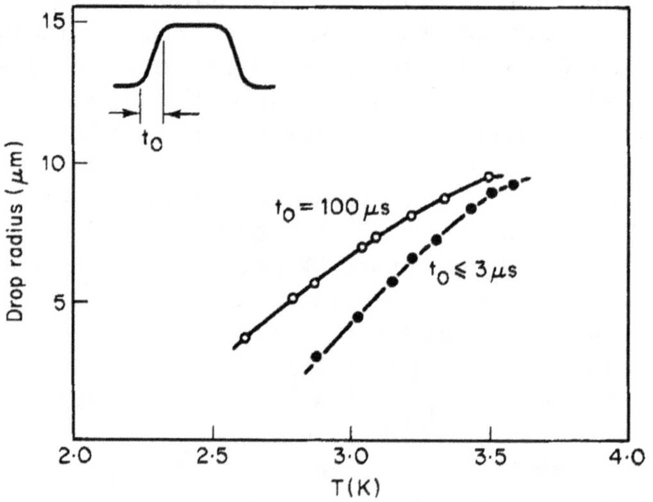

Figure 14. Electron–hole drop radii under the same experimental conditions as in figure 13 [70].

determined, as figure 11 shows, only by the exciton concentration in the gas phase. Strictly speaking, this claim refers only to drops in a volume whose linear dimensions do not exceed the exciton diffusion length, which is ~1 mm in germanium. Different curves in figures 13 and 14 correspond to different rise times of the same excitation level and thus demonstrate the dependence of the drop number on the initial supersaturation and the subsequent memory of this initial period of their formation.

Figure 14 shows also that, corresponding to different experimental conditions, both curves tend to saturation at the same limiting value $R=10\,\mu$m. Indeed detailed investigation [70] shows that neither a further increase in excitation level nor in temperature results in EHD radii exceeding this value. Reaching it, all the curves become constant. The only exceptions are the 'large drops' artificially produced by a nonuniform strain of the crystals mentioned above. The reason for this sharp limitation on the EHD radii in unstrained crystals is explained below. The unusual quantum nature of the accommodation coefficient γ is also worthy of notice. There exists experimental evidence that γ is relatively small, $\sim 0{\cdot}1$, but increases with temperature. The probable reason seems to be the absence of heavy particles and the pronounced quantum effects for excitons. In some approximation the EHD may be treated as a potential well for excitons with a sharp boundary. The width of the boundary is $\sim n_l^{-1/3} \sim a_{ex}$, much smaller than the exciton de Broglie wavelength $\hbar(mk_BT)^{-1/2}$, if $k_BT \ll E_{ex}$. However, it is a well-known quantum-mechanical effect that the reflection coefficient for slow particles from a potential well tends to unity as the particle energy tends to zero. In other words, the probability of exciton penetration into the EHD tends to zero as $(k_BT/E_{ex})^{1/2}$. This effect is absent in ordinary gas–liquid systems, because owing to the presence of heavy nuclei the de Broglie wavelength of the atoms remains small in atomic units at all temperatures and the interface boundary cannot be treated as sharp.

4.2. *Moving drops*

One of the most remarkable features of EHDs is their high mobility. They can be accelerated by spatially nonuniform magnetic and electric fields, deformations, and so on. Among these possibilities nonuniform deformation is the most effective. As a matter of fact, the forbidden gap E_g, i.e. the electron–hole pair energy at rest, is known to depend on stress. Therefore, in nonuniformly strained crystals the energy of every electron–hole pair and of the EHD as a whole is different at different points, being equivalent to some potential energy. This means the existence of forces proportional to the local deformation,

$$\left.\begin{aligned} \mathbf{f} &= -n_l \operatorname{grad}(D_{ik}\varepsilon_{kl}), \\ D_{ik} &= D_{ik}^{(e)} + D_{ik}^{(h)}. \end{aligned}\right\} \tag{25}$$

Here $\mathbf{f}$ is the volume force density, $D_{ik}^{(e,h)}$ the deformation potentials for electrons and holes, and ε_{ik} are the deformations. The movement of the drop is damped owing to the scattering of carriers by thermal phonons. But at low temperature the damping is reduced by the Fermi degeneracy of the carriers and decreases as T^5. At liquid helium temperature the value of the damping coefficient is of the order of $10^9\,\mathrm{s}^{-1}$ in germanium. That means that at stresses of the order $10^7\,\mathrm{kg\,m}^{-2}$ EHDs can be accelerated to a speed close to that of sound, as was experimentally demonstrated in [17]. Up to now there exists no experimental evidence of EHD movement with a velocity exceeding that of sound, perhaps because of additional damping or even the disintegration of the drop by coherent radiation of Mach waves. But even with smaller velocities they can travel during their lifetime macroscopic distances as large as 0·1–1 cm.

4.3. *The phonon wind*

At excitation levels reasonably exceeding the threshold value another set of phenomena related to the so-called 'phonon wind' becomes crucial for condensation kinetics

[72, 73]. The point is that most of the excitation energy eventually dissipates into heat, i.e. into phonons. Phonons are emitted both during the thermalization process of nonequilibrium carriers immediately after their creation and as a result of their subsequent recombination. Therefore, the areas where carrier generation takes place and EHDs exist and where excitation energy is concentrated are sources of strong nonequilibrium phonon fluxes. Interacting with carriers, free or bound in excitons, biexcitons and EHDs, phonons are partially reabsorbed and transfer their energy and momentum (or, more accurately, quasimomentum) to the carriers. The average momentum transferred to the carriers per time unit is equivalent to an effective force acting on the carriers. Its value is proportional to the phonon flux density, the direction being determined by that of their propagation. In a certain approximation EHL volume forces produced by the phonon wind are similar to the electrostatic forces in a uniformly charged liquid. In fact, every element of the EHL volume is a source of radially propagating phonon flux, whose density decreases in inverse proportion to the square of the distance. Consequently any two elements of the EHD, whether they are within the same EHD or belong to different EHDs, repel each other with a force inversely proportional to the square of the distance between them, i.e. as if in agreement with the Coulomb law. Thus the EHL may be characterized by some effective charge density ρ which has nothing to do with any real electric charge, but can be expressed in terms of the concentration of electron–hole pairs, their recombination rate, the emitted phonon spectrum, the effective cross-section of their absorption, etc.

$$\rho^2 = C\frac{n_1^{4/3}E_g}{\tau_1}\frac{D^2m^2}{\hbar^2 ds^2}. \qquad (27)$$

Here D is the deformation potential, d the crystal density and s the sound velocity. Only long-wave phonons with momenta $|\mathbf{k}| \leqslant 2p_F$ interact with carriers. Therefore, the coefficient C accounts for the part of dissipated energy released in these phonons. Crude estimates give $C \sim 10^{-2}$, $\rho \sim 10^2$ CGSE in Ge and $\rho \sim 10^4$ CGSE in Si.

One of the most striking manifestations of the phonon wind is the instability of large EHL volumes [73]. The phonon wind strength grows in proportion to the EHD linear size and, as the droplet radius exceeds some critical value,

$$R_c = \left(\frac{15}{2\pi}\frac{\sigma}{\rho^2}\right)^{1/3} \qquad (28)$$

(σ is the EHL surface tension), the force produced by the phonon wind exceeds that of the surface tension. The EHD becomes unstable to quadrupole deformation and divides into two drops with radii less than R_c, as shown in figure 15.

This process is quite similar to the fission of a large atomic nucleus. Thus the EHD cannot grow to sizes larger than R_c. This explains the limitation on the radii of EHDs discussed above and in figure 11 the curves must be limited both at small and large values of R: $n_l^{-1/3} \lesssim R \leqslant R_c$. According to theoretical estimates and some experimental data in germanium $R_c \simeq 10\ \mu$m and in silicon $R_c \sim 1\ \mu$m. This claim does not contradict the observation mentioned above of much larger drops in germanium [45], because these drops are strain confined. There exist experimental indications that in silicon such large drops do not exist, owing to the much more intense phonon wind. Instead, under the conditions of a strain-produced potential well, a cloud of small droplets arises, hanging in the field of the phonon wind produced by themselves [74]. The formation of an EHL volume with size considerably exceeding R_c is also possible as a result of a

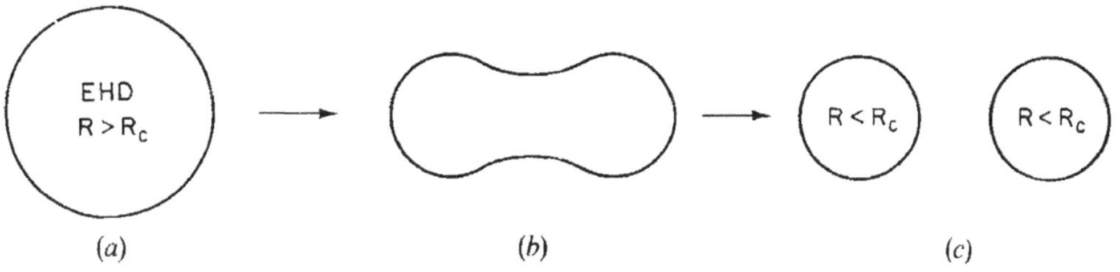

Figure 15. Deformation (*b*) and division (*c*) of electron–hole drops induced by phonon wind.

short intense exciting pulse, immediately producing an electron–hole pair concentration $n \geqslant n_l$ in a given volume. The subsequent events, shown in figure 16, remind one of a microexplosion. Under the action of the phonon wind the EHL surface becomes unstable: the so-called capillary wave amplitudes increase, so that EHDs with sizes of the order R_c begin to break off from it. This process continues until the whole EHL volume transforms into a cloud of EHDs, flying away due to mutual repulsion: the bigger the initial EHL volume, the greater the velocities of the drops. A similar flying-apart also occurs in the case on an exciting pulse producing not a single large EHL volume, but a dense cloud of small EHDs. This EHD cloud flying off in different directions at near-sonic velocities was observed experimentally for the first time by Damen and Worlock [71]. The dynamics of this process, shown in figure 17, has been well studied experimentally [75, 76]. It is extremely anisotropic owing to the anisotropy of the sound velocity and the deformation potential [77]. Impressive photographic images of the cloud of EHDs ejected from the excitation region under stationary and pulsed conditions are shown in figures 18 and 19 [77, 78]. The ejection of EHDs from the initial excitation region was shown to occur not only because of their mutual repulsion. It is clearly visible in figures 17 and 19 that the EHD clouds move away as a whole from the point where they were born. Evidently the phonon flux produced in the carrier thermalization process is of considerable importance. However, it is quite unexpected that this flux should exist after the end of the excitation pulse. During a time some orders greater than the carrier thermalization time the phonon wind source, the so-called 'hot spot', exists in the initial excitation region [79]. It is likely that the optical and short-wave acoustic phonons with a small free-path length, generated during the thermalization process, cannot leave the generation region, but

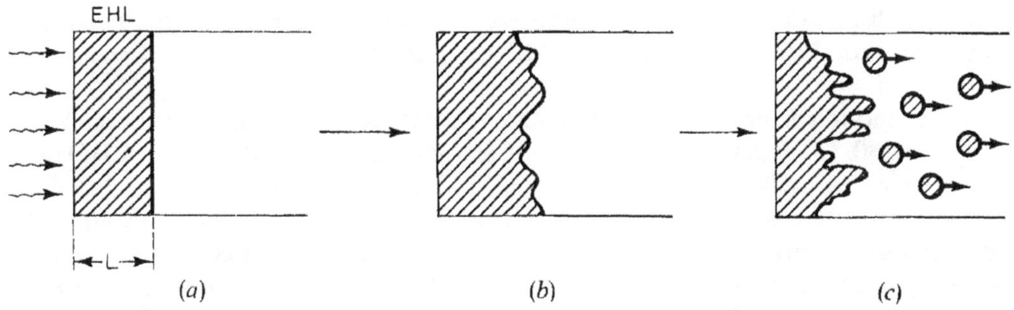

Figure 16. (*a*) Formation of a large EHL volume, (*b*) growing capillary wave instability of surface and (*c*) decay of an initial volume into a cloud of flying-away electron–hole drops.

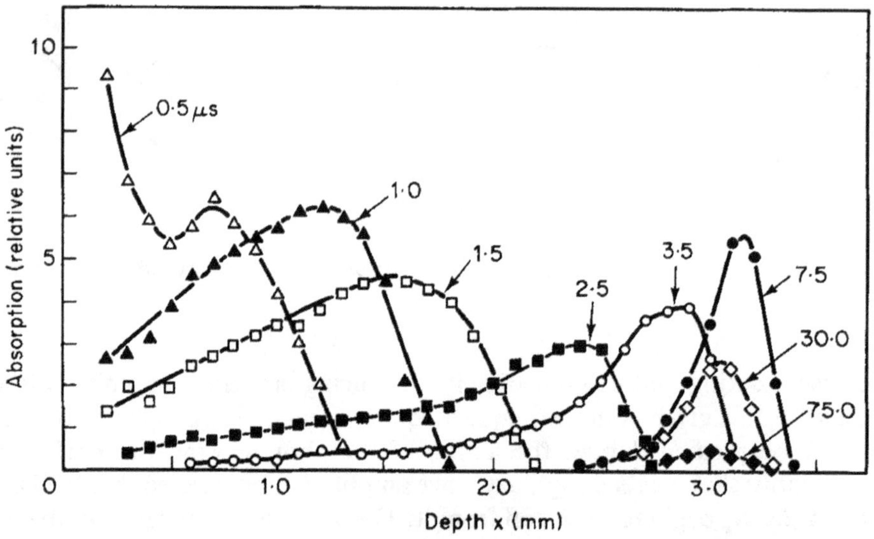

Figure 17. Positions and shapes of electron–hole drop cloud, measured by absorption of $\lambda = 3{\cdot}39\,\mu$m radiation at different delay times after a short exciting pulse [75].

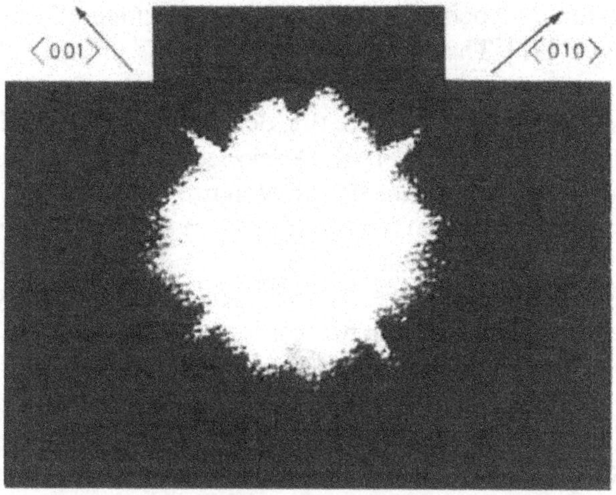

Figure 18. Anisotropic electron–hole drop cloud under conditions of stationary excitation, sharply focused at the central point of the image [77].

gradually decay and produce a phonon wind of long-wave acoustic phonons with macroscopically large mean free-paths. Such phonons interact strongly with the charge carriers.

EHD motion from the excitation region also occurs under intense stationary excitation [80, 81]. It determines the size and shape of the sample area where EHDs and the surrounding excitonic gas are dispersed, and, therefore, defines the conditions of EHD generation, growth and decay. Together with such large-scale movements in the nonequilibrium carrier system, the phonon wind changes its small-scale structure: phonon fluxes emanating from each individual EHD induce correlation in their relative position and movement, blow out excitons from the EHD neighbourhood, decrease its growth rate even at $R < R_c$, etc.

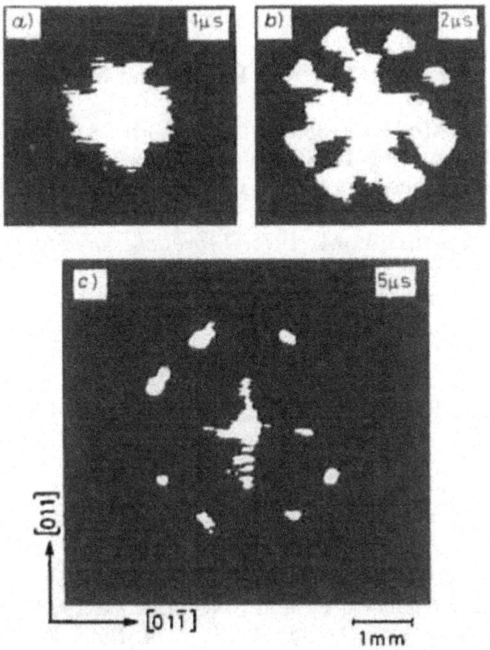

Figure 19. Time-resolved images of electron–hole droplet could after pulse excition. Delay times are shown in the right upper corners [78].

The phonon wind is only one of the manifestations of the interaction of EHD with different types of crystal deformations, static or dynamic (ultrasonic, phonons, and so on). Many other electrical, magnetic and optical phenomena, connected with the EHL, have been observed and described in reviews [1–9]. To date the EHL has been studied and understood most completely in germanium and silicon owing to the high mobility and long lifetimes of the carriers in these semiconductors. In many other semiconductors the experimental data are much scarcer. Undoubtedly the investigation of these materials will greatly increase the number of EHL manifestations and transformations in semiconductors under high excitation conditions.

References

[1] KELDYSH, L. V., 1970, *Usp. fiz. Nauk*, **10**, 514.
[2] KELDYSH, L. V., 1971, *Sb: Eksitony v poluprovodnikah*, edited by B. M. Vul (Moscow: Nauka), p. 5.
[3] POKROVSKII, YA. E., 1972, *Phys. Stat. Sol.* **A**, **11**, 385.
[4] VOOS, M., and BENOIT À LA GUILLAUME, C., 1975, *Optical Properties of Solids: New Developments*, edited by B. O. Serafin (Amsterdam: North-Holland), p. 143.
[5] BAGAEV, V. S., 1975, *Springer Tracts Modern Physics Vol. 73, Excitons at High Density*, edited by H. Haken and S. Nikitine (Berlin, New York: Springer-Verlag), p. 72.
[6] JEFFRIES, C. D., 1975, *Science*, **89**, 955.
[7] RICE, T. M., 1977, *Solid State Physics, Vol. 32*, edited by H. Ehrenreich, F. Seitz and D. Turnball (New York, San Francisco, London: Academic Press), p. 1.
[8] HENSEL, J. C., PHILLIPS, T. G., and THOMAS, G. A., 1977, *Solid State Physics, Vol. 32*, edited by H. Ehrenreich, F. Seitz and D. Turnball (New York, San Francisco, London: Academic Press), p. 88.
[9] JEFFRIES, C. D., and KELDYSH, L. V., 1983, *Modern Problems of Condensed Matter Science Vol. 6: Electron-Hole Liquid in Semiconductors* (Amsterdam: North-Holland).
[10] MOSKALENKO, S. A., 1958, *Optika Spektrosk.*, **5**, 147.

[11] LAMPERT, M. A., 1958, *Phys. Rev. Lett.*, **1,** 450.
[12] GRUN, J. B., HÖNERLAGE, B., and LEVY, R., 1982, *Modern Problems in Condensed Matter Sciences, Vol. 2: Excitons,* edited by R. I. Rashba and M. D. Sturge (Amsterdam: North-Holland), Chap. 11, p. 459.
[13] TIMOFEEV, V. B., 1982, *Modern Problems in Condensed Matter Sciences, Vol. 2: Excitons,* edited by E. I. Rashba and M. D. Sturge (Amsterdam: North-Holland), Chap. 9, p. 349.
[14] KELDYSH, L. V., 1968, *Proceedings of the Ninth International Conference on Semiconductors* (Moscow: Nauka), p. 1303.
[15] LANDAU, L. D., and LIFSHITS, E. M., 1975, *Teoreticheskaya fizika, Vol. 5, Statisticheskaya fizika,* Chap. 1 (Moscow: Nauka).
[16] CARE, C. M., and MARCH, M. H., 1975, *Adv. Phys.*, **24,** 101.
[17] BAGAEV, V. E., GALKINA, T. I., GOGOLIN, O. V., and KELDYSH, L. V., 1969, *Pis'ma ZhETF,* **10,** 309.
[18] KIRCZENOV, G., and SINGWI, K. S., 1979, *Phys. Rev.* **B, 20,** 4171.
[19] BRINKMAN, W. F., and RICE, T. M., 1973, *Phys. Rev.* **B, 7,** 1508.
[20] KELDYSH, L. V., and KOPAEV, YU. V., 1964, *Fizika tverd. Tela,* **6,** 2791.
[21] DES CLOIZEAUX, J., 1965, *J. Phys. Chem. Solids,* **26,** 259.
[22] HALPERIN, B. I., and RICE, T. M., 1968, *Solid St. Phys.*, **21,** 115.
[23] BRINKMAN, W. F., RICE, T. M., ANDERSON, P. W., and CHUI, S.-T., 1972, *Phys. Rev. Lett.*, **28,** 961.
[24] COMBESCOT, M., and NOZIERES, P., 1972, *J. Phys. C,* **5,** 2369.
[25] VASHISHTA, P., BHATTACHARYYA, P., and SINGWI, K. S., 1973, *Phys. Rev. Lett.*, **30,** 1284.
[26] VASHISHTA, P., DAS, S. G., and SINGWI, K. S., 1974, *Phys. Rev. Lett.*, **33,** 911.
[27] KELDYSH, L. V., and ONISHCHENKO, T. A., 1976, *Pis'ma ZhETF,* **24,** 70.
[28] ANDRYUSHYN, E. A., BABICHENKO, V. S., KELDYSH, L. V., ONISHCHENKO, T. A., and SILIN, A. P., 1976, *Pis'ma ZhETF,* **24,** 210.
[29] ANDRYUSHYN, E. A., KELDYSH, L. V., and SILIN, A. P., 1977, *Zh. éksp. téor. Fiz.*, **73,** 1163.
[30] FENTON, E. W., 1968, *Phys. Rev.*, **170,** 816.
[31] BRAZOVSKII, S. A., 1972, *Zh. éksp. téor. Fiz.*, **62,** 820.
[32] LANDAU, L. D., and ZEL'DOVICH, YA. B., 1943, *Acta Phys. Chem. U.S.S.R.*, **18,** 194.
[33] INSEPOV, S. A., and NORMAN, G. E., 1972, *Zh. éksp. téor. Fiz.*, **62,** 2290.
[34] RICE, T. W., 1974, *Proceedings of the Twelfth International Conference on the Physics of Semiconductors, Stuttgart,* edited by M. H. Pilkuhn (Stuttgart: Teubner), p. 23.
[35] EBELING, W., KRAIEFT, W. D., and KREMP, D., 1976, *Theory Bound States and Ionisation Equilibrium in Plasma and Solids. Vol. 5,* edited by R. Rompe and M. Steenback (Berlin: Academie-Verlag).
[36] MOTT, N. F., 1961, *Phil. Mag.*, **62,** 287.
[37] KELDYSH, L. V., and SILIN, A. P., 1975, *Zh. éksp. téor. Fiz.*, **69,** 1053.
[38] BENI, G., and RICE, T. M., 1976, *Phys. Rev. Lett.*, **37,** 874.
[39] ROSSLER, M., and ZIMMERMAN, R., 1977, *Phys. Stat. Sol.* (b), **83,** 85.
[40] MULLER, G. O., and ZIMMERMAN, R., 1979, *Physics of Semiconductors: Proceedings of the Fourteenth International Conference on the Physics of Semiconductors* (Edinburgh: SUSSP).
[41] POKROVSKII, YA. E., and SVISTUNOVA, K. I., 1969, *Pis'ma ZhETF,* **9,** 435; Pokrovskii, YA. E., and SVISTUNOVA, K. I., 1970, *FTP,* **4,** 491.
[42] THOMAS, G. A., FROVA, A., HENSEL, J. C., MILLER, R. E., and LEE, P. A., 1969, *Phys. Rev. B,* **13,** 1692.
[43] THOMAS, G. A., RICE, T. W., and HENSEL, J. C., 1974, *Phys. Rev. Lett.*, **33,** 219.
[44] DITE, A. F., KULAKOVSKY, V. D., and TIMOFEEV, V. B., 1977, *Zh. éksp. téor. Fiz.*, **72,** 1156.
[45] WOLFE, J. P., HANSON, W. L., HALLER, E. E., MARKIEWICZ, R. S., KITTEL, C., and JEFFRIES, C. D., 1975, *Phys. Rev. Lett.*, **34,** 1291.
[46] MANENKOV, A. A., MILJAEV, V. A., MIHILOVA, G. N., SANINA, V. A., and SEFEROV, A. S., 1976, *Z. éksp. téor. Fiz.*, **70,** 695.
[47] BAGAEV, V. S., GALKINA, T. I., PENIN, N. A., STOPACHINSKY, V. B., and CHURAEVA, M. A., 1972, *Pis'ma ZhETF,* **16,** 120.
[48] KELDYSH, L. V., and SILIN, A. P., 1973, *Fizika tverd. Tela,* **15,** 1532.
[49] STORMER, H. L., and MARTIN, R. W., 1979, *Phys. Rev.* **B, 20,** 4213.
[50] VAVILOV, V. S., ZAYATS, V. A., and MURZIN, V. N., 1969, *Pis'ma ZhETF,* **10,** 304.
[51] MARKIEVICH, R. S., WOLFE, J. P., and JEFFRIES, C. D., 1974, *Phys. Rev. Lett.*, **32,** 1357.

[52] Kamiinskii, A. S., Pokrovskii, Y. E., and Zhidkov, A. E., 1977, *Z. éksp. terr. Fiz.*, **72,** 1960.
[53] Bagaev, V. S., Zamkovets, N. V., Keldysh, L. V., Sibeldin, N. N., and Tsvetkov, V. A., 1976, *Z. éksp. téor. Fiz.*, **70,** 1501.
[54] Silver, R. N., 1975, *Phys. Rev.* B, **11,** 1569.
[55] Silver, R. N., 1975, *Phys. Rev.* B, **12,** 5689.
[56] Westervelt, R. M., 1976, *Phys. Stat. Sol.* (*b*), **74,** 727.
[57] Westervelt, R. M., 1976, *Phys. Stat. Sol.* (*b*), **76,** 31.
[58] Etienne, B., Benoit à La Guillaume, C., and Voos, M., 1976, *Phys. Rev.* B, **14,** 712.
[59] Staehli, J. L., 1976, *Phys. Stat. Sol.* (*b*), **75,** 451.
[60] Lo, T. K., Feldman, B. J., and Jeffries, C. D., 1973, *Phys. Rev. Lett.*, **31,** 224.
[61] Westervelt, R. M., Staehli, J. L., and Haller, E. E., 1978, *Phys. Stat. Sol.* (*b*), **90,** 557.
[62] Westervelt, R. M., Culbertson, J. C., and Black, B. S., 1979, *Phys. Rev. Lett.*, **42,** 267.
[63] Kaminskii, A. S., and Pokrovskii, Ya E., 1970, *Pis'ma ZhETF*, **11,** 381.
[64] Combescot, M., and Combescot, R., 1976, *Phys. Lett.* A, **56,** 228.
[65] Koch, S., and Haug, H., 1979, *Physics Lett.* A, **74,** 250.
[66] Pokrovskii, Ya. E., and Svistunova, K. I., 1971, *Pis'ma ZhETF*, **13,** 297.
[67] Sibeldin, N. N., Bagaev, V. S., Tsvetkov, V. A., and Penin, N. A., 1973, *Fizika tverd. Tela* (*Leningrad*), **15,** 177.
[68] Worlock, J. M., Damen, T. C., Shaklee, K. L., and Gordon, J. P., 1974, *Phys. Rev. Lett.*, **33,** 771.
[69] Voos, M., Shaklee, L. K., and Worlock, J. M., 1974, *Phys. Rev. Lett.*, **33,** 1161.
[70] Bagaev, V. S., Zamkovets, N. V., Keldysh, L. V., Sibeldin, N. N., and Tsevtkov, V. A., 1976, *Z. éksp. téor. Fiz.*, **70,** 1501.
[71] Damen, T. C., and Worlock, J. M., 1976, *Proceedings of the Third International Conference on Light Scattering in Solids, Campinas, Brazil* (Paris: Flammarion), p. 183.
[72] Bagaev, V. S., Keldysh, L. V., Sibeldin, N. N., and Tsevtkov, V. A., 1976, *Zh. éksp. téor. Fiz.*, **70,** 702.
[73] Keldysh, L. V., 1976, *Pis'ma ZhETF*, **23,** 100.
[74] Gourley, P. L., and Wolfe, J. P., 1971, *Phys. Rev.* B, **14,** 5970.
[75] Zamkovets, N. V., Sibeldin, N. N., Stopachinsky, V. B., and Tsvetkov, V. A., 1978, *Zh. éksp. téor. Fiz.*, **74,** 1147.
[76] Kavetskaya, I. V., Sibeldin, N. N., Stopachinsky, V. B., and Tsvetkov, V. A., 1978, *Fizika tverd. Tela*, **20,** 3608.
[77] Greenstein, M., and Wolfe, J. P., 1978, *Phys. Rev. Lett.*, **41,** 715.
[78] Greenstein, M., Tamor, M. A., and Wolfe, J. P., 1983, *Solid St. Commun.*, **45,** 355.
[79] Hensel, J. C., and Dynes, R. C., 1977, *Phys. Rev. Lett.*, **39,** 969.
[80] Doehler, J. Mattos, J. C. V., and Worlock, J. M., 1977, *Phys. Rev. Lett.*, **38,** 726.
[81] Greenstein, M., and Wolfe, J. P., 1981, *Phys. Rev.* B, **24,** 3318.

12

Macroscopic Coherent States of Excitons in Semiconductors

L. V. Keldysh

P. N. Lebedev Physics Institute
Leninsky Prospect 53
117924 GSP, Moscow B-333
Russia

Abstract

Initially put forward by Moskalenko [1] *and Blatt et al.* [2], *the idea of a possible Bose–Einstein condensation* (BEC) *of excitons in semiconductors has attracted the attention of both experimentalists and theoreticians for more than three decades. At different stages of this long history, the results of their efforts have been described and discussed in review articles* [3–10]. *A brief introduction and summary of the main qualitative conclusions of this older work is presented here* (*Sections* 1 *and* 2), *followed by a more detailed discussion of some more recent developments* (*Sections* 3 *and* 4).

1 Electronic Excitations in Semiconductors

Schematically presented in Fig. 1 is the typical electronic spectrum of a semiconductor: two bands (or two groups of bands) of continuous spectrum – conduction (c) and valence (v) – separated by the energy gap $E_g = E_{c,\min} - E_{v,\max}$. In the ground state, all of the states in the valence band(s) are occupied by valence electrons of the semiconductor, and all states in the conduction band are empty.

The lowest single-particle electronic excitations are an additional electron (e) in the conduction band or a single empty state – a hole (h) – in the valence band. Both of these excitation types are mobile fermions (spin $= 1/2$) characterized by effective masses, m_e and m_h and effective charges $e_e = e$ and $e_h = -e$, respectively. Here e is the usual (negative) elementary charge. Effective charges of electrons and holes are universal, unlike the effective masses, which are different not only in different semiconductors but also in different bands of the same semiconductor and for different directions in the same band (anisotropy).

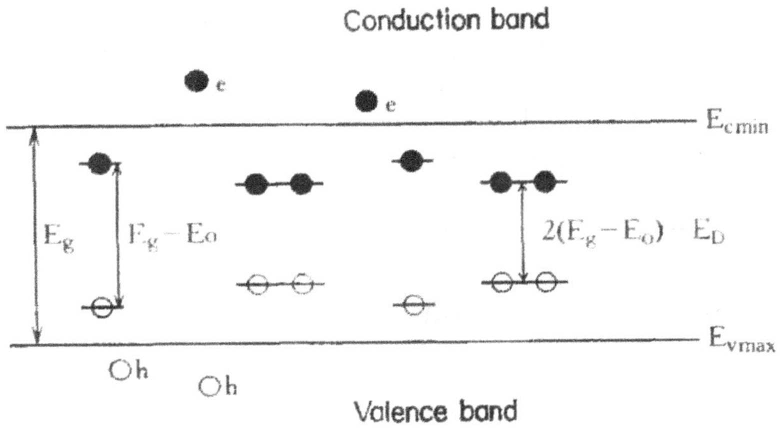

Fig. 1. Schematic picture of the main components of the nonequilibrium electron–hole plasma – free electrons and holes, excitons and excitonic molecules – with their characteristic energies. The top and bottom represent the continuous spectrum of single-particle excitations – electrons and holes. The energy levels of electrons and holes bound in excitons or excitonic molecules have no exact meaning and are depicted only for the sake of illustration. Only the total energy of a complex as a whole is well defined.

Being oppositely charged (quasi)free particles, the electron and hole attract one another by Coulomb forces just as the electron and proton do. And like an electron and proton which can be bound into a hydrogen atom, the electron and hole can be bound via this interaction into the two-particle excitation called the Wannier–Mott exciton [11, 12]. Being the combination of two fermions, an exciton is expected to be a boson, and so the problem of the possibility of BEC of excitons arises. But to discuss this problem, we will need more information about the possible states and transformations in the system of many excitons, or, more correctly, in the many electron–hole system. The above-mentioned similarity to the electron–proton system can be extended to the many-particle system and so suggests the existence of the excitonic molecule (biexciton), the bound state of two excitons similar to a hydrogen molecule [13, 14] and, moreover, the gas–liquid-type phase transition of exciton and biexciton "gas" into an electron–hole Fermi liquid (EHL) – the bound state of a macroscopically large number of electrons and holes. [15, 6, 9, 10]

But this similarity is not complete. The electron–hole system possesses some peculiar properties distinguishing it from any other. The most important are

(1) Greatly reduced Coulomb interaction due to dielectric screening in the host crystal. Typical values of dielectric constant ϵ in semiconductors are of the order of 10. Therefore, by the well-known Bohr formulas, the binding energy of a hydrogen-like exciton

$$E_0 = \frac{e^4 m}{2\epsilon^2\hbar^2} \sim 10^{-1} - 10^{-3}\,\mathrm{eV}, \tag{1}$$

is a few orders of magnitude smaller than the atomic binding energy, and the exciton effective radius

$$a_0 = \frac{\epsilon\hbar^2}{me^2} \sim 10^{-7} - 10^{-6}\mathrm{cm} \tag{2}$$

is macroscopically large compared to interatomic distances in the host crystal. Formulas (1) and (2) represent the natural units of quantum scales of energy and length in the interacting electron–hole system, in just the same sense as the Rydberg and Bohr radius are the natural scales in atomic, molecular and solid state physics. Therefore in what follows we will use these units. Instead of the temperature, a dimensionless ratio kT/E_0, designated by T, will be used, and instead of particle concentration the product na_0^3, designated by n. In this notation, the usual formula for the critical temperature of an excitonic gas BEC, neglecting any interaction corrections, is given by

$$T_c = 6.62\,\frac{m}{M}\,(\frac{n}{g})^{2/3}. \tag{3}$$

Here $M = m_e + m_h$ is the effective mass of the exciton, equal to the sum of electron m_e and hole m_h effective masses; $m = m_e \cdot m_h/M$ is their reduced mass, entering (1) and (2); and g is the exciton ground state statistical weight (degeneracy).

(2) The effective masses m_e and m_h in any particular semiconductor, though different, are usually of the same order of magnitude (or differ at most by one order of magnitude). The absence in the electron–hole system of really heavy particles such as nuclei results in a dominant role of quantum effects at all temperatures $T \leq 1$. In particular as formula (3) shows, $T_c \sim 1$ for $n \sim 1$, unlike, for example, in ^{4}He where $T_c \sim 10^{-5}$ in corresponding units. Other important manifestations of the absence of heavy particles are very large zero vibrations of excitons in the excitonic molecule. The amplitude of this vibration is of the order of, or even larger than, the excitonic radius a_0 itself. So biexcitons appear to be

very loosely bound complexes with binding energy E_D not exeeding 0.1, compared to 0.35 in the hydrogen molecule [16, 17].

For the same reason, nothing like an "excitonic crystal" can exist. The weak van der Waals attraction dominating at large intermolecular distances is not able to confine light particles such as excitons or biexcitons. According to Ref. [18], the s-wave scattering length of two excitonic molecules is large ($\simeq 7$) and positive (repulsive). Therefore a condensed phase of excitonic molecules – a molecular "liquid" similar to liquid hydrogen – also cannot exist. But an electron–hole liquid (EHL) similar to metallic hydrogen or alkali metals does exist [15, 6, 9, 10]. Unlike common metals, in the EHL not only electrons but also holes ("nuclei") are Fermi degenerate. If the effective masses of both electrons and holes are more or less isotropic, the EHL at low temperatures transforms to an "excitonic insulator" phase [19–21]. In this two-Fermi-liquid state the collective pairing of electrons and holes in the vicinity of Fermi surfaces arises, quite similar to BCS pairing in superconductors. This pairing manifests itself in the appearance of the energy gaps Δ around the Fermi surfaces, as shown in Fig. 2. These gaps may be considered a remnant of the binding energy of a single exciton, transformed by collective many-body interactions in the condensed phase. The gap diminishes fast with increasing particle density or m_e–m_h difference, and it does not exist at all if the anisotropy is large. In that sense, the excitonic insulator state in the nonequilibrium electron–hole (e–h) system is a coherent BEC state of high density ($n \geq 1$) excitons, as the superconducting state is a coherent BEC state of Cooper pairs. Like Cooper pairs, e–h pairs in an excitonic insulator have very large radii $\sim \Delta^{-1/2}$, much larger than interparticle distance. So they are true collective phenomenon. But, unlike Cooper pairs in superconductors which are charged ($-2e$), excitons are electrically neutral.

The above-described phenomena can be schematically represented by the phase diagrams depicted in Figs. 3-4. They show possible states of the e–h system in the plane ($\bar{n}$, T). Here $\bar{n}$ is the average particle density.

The phase diagram type of any particular semiconductor is defined by the relationship of the binding energy per e–h pair in the biexciton gas, given by $(E_0 + E_D/2)$, and in the EHL, indicated by $-\mu_l$. Strictly speaking, μ_l is the difference between the chemical potential per e–h pair in the EHL and E_g, where E_g is the minimum energy of the free e–h pair. Different phases presented in Figs. 3 and 4 are: G, the gas (plasma) of weakly interacting electrons, holes, excitons and excitonic molecules; BG, the degenerate gas of excitons and(or) excitonic molecules; ML, the

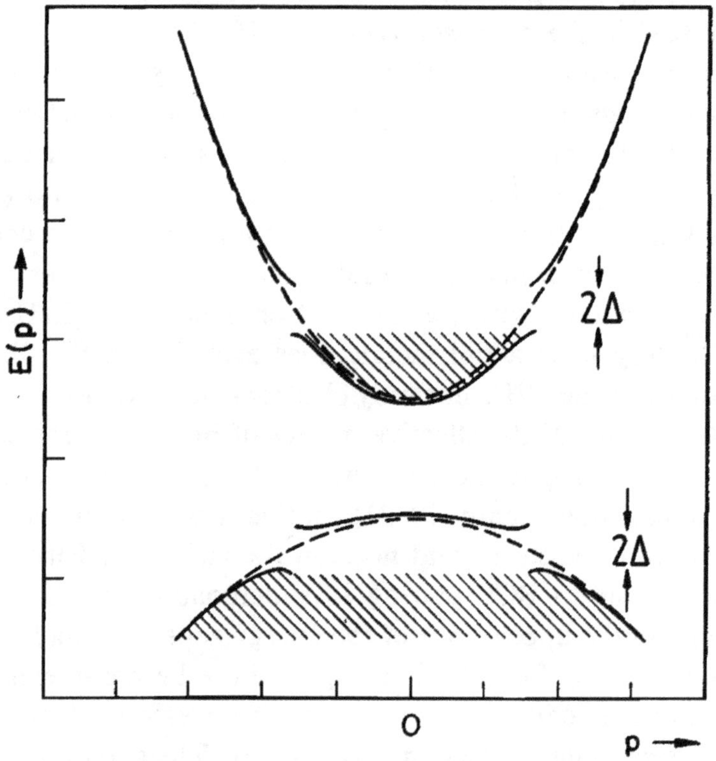

Fig. 2. Energy spectrum in the nonequilibrium "excitonic insulator" phase. The striped regions are energy ranges occupied by electrons.

metallic liquid (EHL); and IL, the insulating liquid (excitonic insulator). The striped area is the region of phase coexistance: droplets of liquid phase surrounded by gas, degenerate or nondegenerate. Coherent states BG or IL are present in the phase diagrams of the types shown in Figs. 3(b) and 4(a),(b). But up to now only the simplest type, shown in Fig. 3(a), is firmly established and studied experimentally, in germanium and silicon.

Both phases in Fig. 3 correspond to the case $|\mu_l| > (E_0 + E_D/2)$; that is, the EHL is more tightly bound than excitonic molecules. It is theoretically proven and experimentally confirmed that the fulfilment of this condition is strongly favored by multivalley band structure, i.e. the presence in the electronic spectrum of a semiconductor of a few equivalent groups of electrons or holes (or both). Just such a band structure is characteristic of both germanium and silicon. No BEC exists in the gas phase in this case, exactly as in the case of helium. It does exist in the phase diagrams depicted in Fig. 4, corresponding to the fulfillment

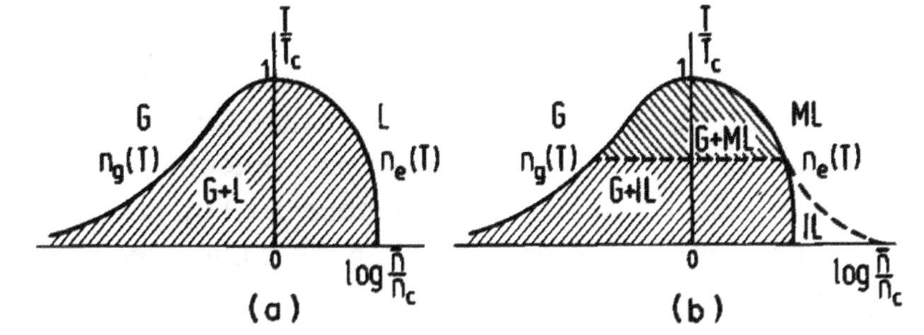

Fig. 3. Possible types of nonequilibrium electron–hole system phase diagrams in semiconductors with electron–hole liquid more tightly bound than the excitonic molecule. The striped regions are the coexistence domains of the liquid and gas phases.

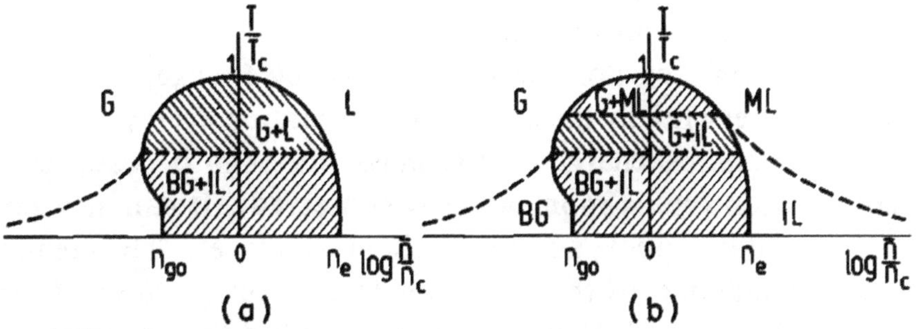

Fig. 4. The same as in Fig. 3, but for semiconductors with excitonic molecules more tightly bound than electron–hole liquid.

of the opposite condition,

$$|\mu_l| < E_0 + 0.5E_D\,. \tag{4}$$

Theoretically, such a case may be expected in semiconductors with single-valley band structure for both electrons and holes, Fig. 4(a) corresponding to semiconductors with a very large difference of the electron and hole effective mass and moderate anisotropy, and Fig. 4(b) to comparable effective masses and small anisotropy. Experimentally, the fulfillment of (4) is not firmly established in any semiconductor. And it should be noted also that the difference of binding energies per *e–h* pair in the EHL and the biexciton in the case corresponding to Fig. 4(b) is very small and comparable to the accuracy of theoretical computation. So even the theoretical possibility of the phase diagram shown in Fig. 4(b) is, strictly speaking, not proven.

(3) Another important difference of the *e–h* system from any ordinary

matter is its essentially nonequilibrium nature. For sure, it also exists under equilibrium conditions, but without any condensation phenomena, because the number of *e–h* pairs itself is then determined by the equilibrium conditions, i.e. by the balance of the thermal creation and recombination processes. Therefore the chemical potential for pairs must be equal to zero, while the above-described condensation phenomena can take place only if it is equal to or larger than the minimal pair excitation energy. It implies

$$\min(E_g - (E_0 + 0.5E_D), E_g + \mu_l) \leq 0, \tag{5}$$

which means that the binding energy per pair in the excitonic molecule or in the EHL becomes larger than the minimal energy for a free *e–h* pair creation, E_g. Such a possibility was also considered theoretically [19–22], and it was recognized that it implies the reconstruction of the electronic structure of the host semiconductor.

Indeed, the term " exciton" was introduced by Frenkel for the quasiparticle which is essentially the quantum of excitation energy. The exciton does not carry electric charge or real mass, although it possesses effective mass (exciton *crystal* momentum is conserved). It can also possess angular momentum, dipole and magnetic moments, etc. The creation of an exciton actually means the transition of an electron to some excited level. The fulfillment of (5) in that sense signifies the existence of another ground state electronic configuration than is supposed in Fig. 1, with lower total energy. The equilibrium condensation of excitons with zero chemical potential is, then, another formal language to describe an electronic phase transition, i.e. reconstruction of electronic states in the valence and conduction bands, accompanied by the formation of a charge or spin density wave, magnetic ordering, or something like that. Nothing like superfluidity can arise because the Hamiltonian does not conserve the total number of excitons (*e–h* pairs). This inevitably results in pinning of the "condensate" by interband scattering matrix elements of the Coulomb interaction [23], which in this language appears as the source of the spontaneous creation of two *e–h* pairs lifting the gauge invariance of the Hamiltonian to yield a phase transformation, different in different bands, of the kind

$$\psi_j \to \psi_j \exp(i\phi_j). \tag{6}$$

Here $j = (c, v)$ is the band index. Also, interband scattering of electrons by static lattice distortions – uniform or periodic – can serve as a single pair creation source and thus a pinning mechanism.

Quite different is the problem of the collective properties of the nonequilibrium dense *e–h* system produced in a semiconductor by some external action, usually illumination. Then the total number of *e–h* pairs N (or as used in Figs. 3 and 4, their average density $\bar{n} = N/V$, where V is the volume of the specimen) really becomes an independent variable with a value controlled by an external source. The conservation of N is broken by recombination processes. But in some semiconductors the direct radiative recombination is forbidden by a parity selection rule (as in Cu_2O) or by quasimomentum conservation, as in germanium and silicon. The lifetime of the nonequilibrium *e–h* system in such a case may be much longer than the thermalization time and the system appears to be in quasiequilibrium, the only nonequilibrium parameter being the total number of particles N itself. Under this condition, the phase diagrams of Figs. 3 and 4 have an exact meaning. On the other hand, in semiconductors with a dipole-allowed excitonic radiative transition, the lifetimes are of the order $10^{-9} - 10^{-10}$ s, and a dense enough *e–h* system is always far from equilibrium, so that these phase diagrams can be considered only as qualitative indications of trends in its evolution.

One more question of importance is whether excitons are really bosons [24]. The reason for asking this is that the exciton is a compound system composed of two fermions, and increasing the density of excitons and their overlap also means increasing the local density of fermions, perhaps leading to problems with the Pauli principle. Also, from a formal point of view, the exciton annihilation operator $\hat{A}_{\mathbf{P}}$ is composed of electron and hole annihilation operators $\hat{a}_{e\mathbf{p}}$ and $\hat{a}_{h\mathbf{p}}$ ($\mathbf{P}, \mathbf{p}$ are momenta)

$$\hat{A}_{\mathbf{P}} = \sum_{\mathbf{p}} \varphi_0(\mathbf{p}) \hat{a}_{e\frac{\mathbf{P}}{2}+\mathbf{p}} \hat{a}_{h\frac{\mathbf{P}}{2}-\mathbf{p}}\,, \tag{7}$$

where φ_0 is the wave function describing the relative *e–h* motion in the exciton. The commutator of the two exciton operators, $\hat{A}^{\dagger}_{\mathbf{P}}$ and $\hat{A}_{\mathbf{P}'}$, is then

$$[\hat{A}_{\mathbf{P}}, \hat{A}^{\dagger}_{\mathbf{P}'}] = \delta_{\mathbf{P},\mathbf{P}'} + \sum_{\mathbf{q}} \left\{ \varphi_0^*(\mathbf{q} - \frac{\mathbf{P}'}{2}) \varphi_0(\mathbf{q} - \frac{\mathbf{P}}{2}) \hat{a}^{\dagger}_{e\mathbf{q}-\mathbf{Q}} \hat{a}_{e\mathbf{q}+\mathbf{Q}} \right.$$
$$\left. + \varphi_0^*(\frac{\mathbf{P}'}{2} - \mathbf{q}) \varphi_0(\frac{\mathbf{P}}{2} - \mathbf{q}) \hat{a}^{\dagger}_{h\mathbf{q}-\mathbf{Q}} \hat{a}_{h\mathbf{q}+\mathbf{Q}} \right\}, \tag{8}$$

where $\mathbf{Q} = (\mathbf{P}' - \mathbf{P})/2$. The righthand side of (8) differs from the usual one for Bose operators $\delta_{\mathbf{P},\mathbf{P}'}$ by operators that are of the order of n, the exciton density. So for $n \sim 1$, the operator (7) has nothing to do with usual

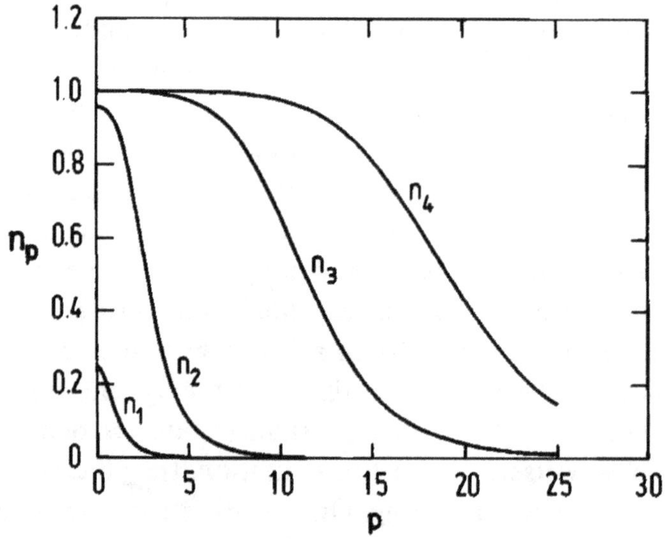

Fig. 5. Momentum distribution function for electrons (or holes) in a system of Bose-condensed excitons. Different curves correspond to different exciton densities ($n_1 < n_2 < n_3 < n_4$).

Bose operators. What it really means and what happens to excitons at high densities is illustrated by Fig. 5, in which the momentum distribution function of electrons (or holes) for the system of Bose-condensed excitons of different densities is presented. For a single exciton, it is $|\varphi_0(\mathbf{p})|^2$, the probability of finding an electron with momentum $\mathbf{p}$ in the exciton. For low concentration it increases proportional to density, since all excitons are identical: $f(\mathbf{p}) = n\,|\varphi_0(\mathbf{p})|^2$. But as $f(\mathbf{p} = 0)$ approaches 1 with increasing n, it cannot continue to increase because of the Pauli principle – $f(\mathbf{p})$ is the fermion distribution function. So at larger densities, the electrons and holes bound into excitons are forced to occupy states with larger momenta until their $f(\mathbf{p})$ also aproaches 1, and so the distribution smoothly transforms to the Fermi distribution function. The binding energy of the exciton transforms to the narrow gap around the Fermi level which diminishes fast as the kinetic Fermi energy becomes large compared to the Coulomb interaction. Thus Bose-condensed excitons continuously transform to the "excitonic insulator" described above, and finally to a degenerate EHL.

Such transformations are not unique for excitons. They must hold also for ^{4}He or any other atomic or molecular system. But it is very difficult experimentally to compress, for example, helium to densities corresponding to $n \sim 1$. The peculiarities of excitons are their large radii and the possibility of creating them by external excitation, so that high compression can be easily realized.

2 Polaritons

A very peculiar, but undoubtedly also the most important, case is presented by BEC of dipole-active excitons. Not only is it always essentially a nonequilibrium problem, as explained above, but the nature of the quasiparticles themselves is very special. They are the well-known polaritons – the quantum superposition of photons and electronic excitation (excitons) [25]. In other words, a polariton is a real photon in a medium: an exciton representing a polarization cloud accompanying propagation of electromagnetic field through the medium [26]. These polarization effects are especially strong under resonant conditions, that is, for photon energies $\hbar\omega$ close to the energy of exciton creation $\hbar\omega_0 = (E_g - E_0)$. Schematically depicted in the Fig. 6 is the well-known polariton dispersion law. Strong mixing arises close to the intersection point of the undisturbed photon and exciton branches $(\mathbf{k}_0, \omega_0)$, and the typical picture of level repulsion appears. So the problem of BEC of dipole-active excitons is essentially that of photons, or, in other words, the problem of a coherent electromagnetic wave of finite amplitude in the polariton frequency range. It is a very familiar problem but is still far from trivial. It includes all the many-body phenomena typical of BEC, but they acquire the meaning of nonlinear optics phenomena. Two different problems are usually considered in this context.

The accumulation of a macroscopic number of initially incoherent excitation quanta in a single-photon mode is lasing. It was recognized long ago [27–31] that the appearance of a coherent mode from multimode intense noise in a pumped system is a very typical example of a nonequilibrium phase transition. The approaches used for its general theoretical treatment are based on mean-field approximations, the Langevin equation, the Fokker–Planck equation, etc. Their detailed review is beyond the scope of this article. As applied to the particular case of dipole-active excitons in semiconductors, the qualitative picture of the lasing process should look like the following [32]. Initially excited electrons and holes combine into excitons, which dissipate their kinetic energy by emission of phonons – quanta of the crystal lattice vibrations – and so move downwards in energy along the lower polariton branch (LPB) in Fig. 6. Approaching the exciton–photon intersection point, which corresponds to very small momenta $\sim E_g/c$ (c the light velocity), they smoothly

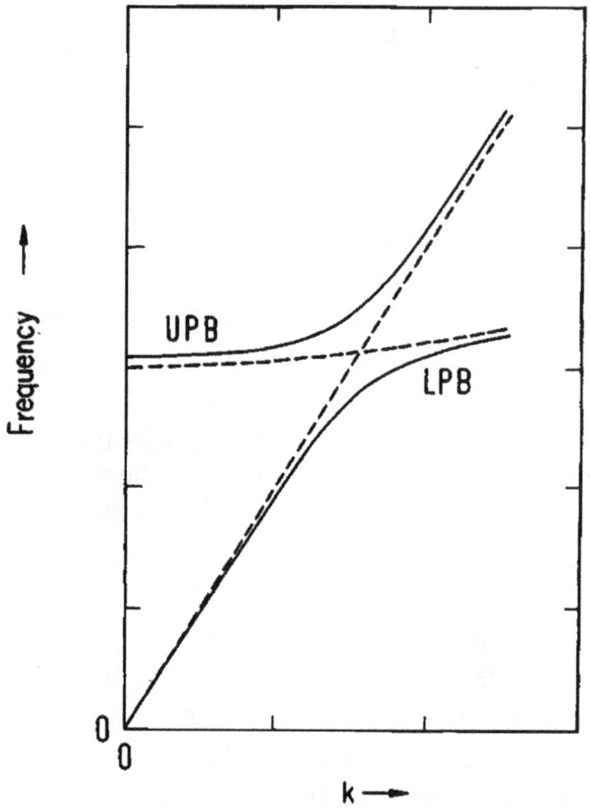

Fig. 6. Polariton dispersion law.

transform into polaritons, with the photonic component increasing at the expense of excitonic. But this transformation is accompanied by a decrease of energy dissipation rate, since just the excitonic component interacts with lattice vibrations, and also because of the dramatic decrease in the density of states due to increase of the polariton velocity below the intersection point. Thus a "bottleneck" arises in this momentum range where polaritons accumulate. Because of the small density of states in the bottleneck region, polariton occupation numbers there easily reach values of the order of unity even if the total exciton density is small, $n \ll 1$. Then the mechanism, common in laser physics, of line narrowing by stimulated emission starts. As applied to the polariton kinetics, it looks like stimulation of transitions to the bottleneck region. Under the condition of sufficiently strong stationary pumping, this process results in the imaginary part of the optical susceptibility becoming negative for one or a few polariton modes. This denotes an instability of the stationary incoherent radiation regime and the appearance of a finite-amplitude coherent polariton mode, i.e. BEC of polaritons. In the transient regime, as a peak in polariton frequency distribution becomes large enough and narrow, the exciton interaction synchronizes all of the modes inside this peak. So in a finite time interval, the coherent state of a macroscopically large number of polaritons arises. It should be noted that existing semiconductor lasers usually operate either at excitation densities $n \gg 1$ or with participation of optical phonons, impurities or lattice defects, and do not correspond to this idealized picture.

Another direction of research is the nonlinear optics proper, i.e. propagation of the intense electromagnetic wave or pulse and its interaction with the medium under conditions of polariton formation [31, 33]. In such a case the coherence is induced by some external source of radiation. In what follows, mainly this type of problem will be discussed. The main difference in the formal description of these phenomena corresponding to the above physical picture, compared to common "excitonic insulator" theory, is the presence in the Hamiltonian of the source of e–h pairs – the electromagnetic field f. The additional term is

$$H_{eh} = \int [\psi_e(\mathbf{r}) f^*(\mathbf{r}, t)\psi_h(\mathbf{r}) + \psi_h^\dagger(\mathbf{r}) f(\mathbf{r}, t)\psi_e^\dagger(\mathbf{r})] d^3r, \tag{9}$$

where

$$f = ie\hbar\mathscr{E}\frac{\mathbf{p}_{cv}}{mE_g E_0} \tag{10}$$

is the matrix element of the electronic dipole transition between the conduction and valence bands induced by the electric field $\mathscr{E}$; $\mathbf{p}_0$ is the momentum of the photons; and $\mathbf{p}_{cv}$ is the interband momentum matrix element. In its physical interpretation, f is the interband Rabi frequency in terms of $E_0/\hbar$.

It was indicated in Ref. [34] that for photon energies above the absorption threshold $\hbar\omega > E_g$, a coherent field produces a collective excited state of semiconductors quite similar to the excitonic insulator state, even neglecting the Coulomb interaction. Creating electrons and holes coherently in correlated pairs, the field itself acts as the pairing force. Also, an energy gap arises in the electronic spectrum at the Fermi level with a magnitude $|f|$ (Fig. 2). The obvious condition for the existence of such a state is $f\tau \gg 1$, where τ is electron (hole) relaxation time. The complete treatment of this problem needs inclusion of many-body effects. This was done in Refs. [35–36] (see also the reviews and monographs [31, 33], [38–40] and especially [37], where the problem has received its most complete formulation, in terms of nonequilibrium Green functions [41]. (The details of this technique are presented in Refs. [42–51], [31] and many other reviews and monographs.) Here we summarize only a few general definitions and basic relations and also some features specific to the nonequilibrium *e–h* system.

The nonequilibrium Green's function is a 2×2 matrix, which in the so-called "triangular representation" looks as follows:

$$\hat{G} = \begin{pmatrix} 0 & G^{(a)} \\ G^{(r)} & F \end{pmatrix}. \tag{11}$$

$G^{(a,r)}$ are advanced and retarded functions, respectively, defined in the usual way

$$G^{(r)}(x,x') = -i\Theta(t-t') < [\psi(x), \psi^{\dagger}(x')]_{\pm} >, \tag{12}$$

$$G^{(a)}(x,x') = [G^{(r)}(x',x)]^{*}, \tag{13}$$

$$F(x,x') = -i < [\psi(x), \psi^{\dagger}(x')]_{\mp} > . \tag{14}$$

Here $x = (\mathbf{r}, t)$; indices $+(-)$ in (12) and $-(+)$ in (14) refer to the case of Fermi (Bose) statistics. In the electron–hole system all of these functions are 2×2 matrices in band indices (c, v). In this notation, the interband (electron–hole) Green's function

$$\hat{G}_{vc}(x,x') \equiv \hat{G}_{he}(x,x') = \begin{pmatrix} 0 & G_{he}^{(a)} \\ G_{he}^{(r)} & F_{he} \end{pmatrix} \tag{15}$$

describes the exciton (electron–hole pair) condensate just as a Gor'kov function in the theory of superconductivity describes a condensate of Cooper pairs:

$$\begin{aligned} G_{he}^{(r)}(x,x') &= -i\Theta(t-t') < [\psi_v(x), \psi_c^{\dagger}(x')]_+ > \\ &= -i\Theta(t-t') < [\psi_h^{\dagger}(x), \psi_e^{\dagger}(x')]_+ >, \end{aligned} \tag{16}$$

$$\begin{aligned} F_{he}(x,x') &= -i < [\psi_v(x), \psi_c^{\dagger}(x')]_- > \\ &= -i < [\psi_h^{\dagger}(x), \psi_e^{\dagger}(x')]_- > . \end{aligned} \tag{17}$$

Diagonal in both types of matrix indices, the functions F_{ee} and F_{hh} at coincident times $t = t'$ are closely related to the single-particle density matrices and in the momentum representation to the distribution functions (occupation numbers)

$$F_{ee}(\mathbf{p}; t, t) = -i[1 - 2n_{e\mathbf{p}}(t)]. \tag{18}$$

The interband function $F_{he}(\mathbf{p};t,t)$ can be considered as an effective exciton wave function. This interpretation follows from the comparison of definitions (17) and (16) and is exact, at least for $n \ll 1$. In a dense many-body system it should not be treated too literally.

In general, Green's functions can be found from the Dyson equation

$$\left(i\frac{\partial}{\partial t} - \frac{\hbar^2}{2m_i}\nabla^2\right)\hat{G}_{ij}(x,x') - \int \hat{\Sigma}_{ik}(x,x'')\hat{G}_{kj}(x'',x')d^4x'' = \delta_{ij}\delta(x-x'), \tag{19}$$

where "self-energy" matrices $\hat{\Sigma}_{ij}$ also have "triangular" form

$$\hat{\Sigma}_{ij} = \begin{pmatrix} \Omega_{ij} & \Sigma_{ij}^{(r)} \\ \Sigma_{ij}^{(a)} & 0 \end{pmatrix} \tag{20}$$

and can be expressed in terms of Green's functions and the interaction potential in any approximation by a set of Feynman graphs. The specific feature of the system under consideration is that, apart from the infinite set of diagrams accounting for different order interaction processes, the interband self-energy Σ_{eh} contains, according to (9), the driving force contribution

$$\hat{\Sigma}_{ij}^{(0)}(x,x') = \begin{pmatrix} 0 & f^*(x) \\ f(x) & 0 \end{pmatrix} \cdot \delta(x-x'). \tag{21}$$

The electric field $\mathscr{E}$ entering into Eq. (10) has to be found self-consistently from Maxwell's equation,

$$\nabla \times \nabla \mathscr{E} + \frac{1}{c^2}\frac{\partial^2(\epsilon\mathscr{E})}{\partial t^2} = -\frac{4\pi}{c^2}\frac{\partial^2 \mathscr{P}_{ex}}{\partial t^2}, \tag{22}$$

where the resonant excitonic contribution to the polarization [4] is

$$\mathscr{P}_{ex} = e < \psi_e(x)\mathbf{r}\psi_h(x) > = -\frac{e\hbar}{2mE_g}\left[\mathbf{p}_{cv} \cdot F_{he}(x,x)\right]^* \tag{23}$$

and ϵ in (22) accounts for all the nonresonant contributions.

Equations (19)–(23) are the complete set describing formally all nonlinear polarization problems. The interband matrix function $\hat{G}_{eh}$ represents the macroscopic coherent state of electron–hole pairs (excitons). In general, the theory under consideration is that of a driven two-component Fermi liquid with all its many-body aspects, with the additional complications due to being far from equilibrium. Therefore, the real problem in any particular case is the choice of an appropriate approximation.

If the retardation effects in the self-energies may be neglected, i.e. $\hat{\Sigma}_{ij} \sim \delta(t-t')$, which is the case at least in the mean-field approximation,

(19) can be reduced to the equation for the single-time ($t = t'$) density matrix [31, 33, 36, 37, 51]

$$f_{ij}(\mathbf{r},\mathbf{r}';t) = F_{ij}(\mathbf{r},\mathbf{r}';t=t'). \tag{24}$$

Symbolically written, this equation is similar to the well-known Bloch equation for two-level systems,

$$i\frac{\partial \hat{f}}{\partial t} = \left(\hat{h}_0 + \hat{\Sigma}^{(r)}\right)\hat{f} - \hat{f}\left(\hat{h}_0 + \hat{\Sigma}^{(a)}\right). \tag{25}$$

Here the two-band Hamiltonian matrix is introduced by

$$(h_0)_{ij} = \left(E_{0i} - \frac{\hbar^2}{2m_i}\nabla^2\right)\cdot\delta_{ij}. \tag{26}$$

Matrices $(\hat{\Sigma}^{(r)})_{ij}$ and $(\hat{\Sigma}^{(a)})_{ij}$ are defined as $(\hat{\Sigma}^{(r)})_{ij} = \Sigma_{ij}^{(r)}$, $(\hat{\Sigma}^{(a)})_{ij} = \Sigma_{ij}^{(a)}$; E_{0i} and m_i are the corresponding band-edge energy and the effective mass (negative in the valence band).

In the simplest case of stationary and spatially homogeneous pumping, the solution of (25) is

$$f_{ee,hh} = f_{e,h}^{(0)} - \frac{\gamma_{e,h}}{\gamma_e+\gamma_h}\frac{\lambda^2|\,\Sigma_{eh}\,|^2}{|\,E(\mathbf{p})+\Sigma^{(r)}(\mathbf{p})\,|^2+\lambda^2|\,\Sigma_{eh}\,|^2}(f_e^{(0)}+f_h^{(0)}), \tag{27}$$

$$f_{eh}(\mathbf{p}) = \Sigma_{eh}(\mathbf{p})^2\frac{E(\mathbf{p})+\Sigma^{(r)}(\mathbf{p})}{|\,E(\mathbf{p})+\Sigma^{(r)}(\mathbf{p})\,|^2+\lambda^2|\,\Sigma_{eh}\,|^2}(f_e^{(0)}+f_h^{(0)}), \tag{28}$$

where

$$\lambda^2 = \frac{(\gamma_e+\gamma_h)^2}{\gamma_e\gamma_h}, \qquad \Sigma^{(r)}(\mathbf{p}) = \Sigma_e^{(r)}(\mathbf{p}) + \Sigma_h^{(r)}(\mathbf{p}),$$

$$E(\mathbf{p}) = E_g - \hbar\omega_0 + \frac{p^2}{2m}. \tag{29}$$

Here, $\mathbf{p}_0$ and ω_0 are the momentum and frequency of the pumping wave, respectively, and

$$\gamma_{e,h} = Im\Sigma_{e,h}^{(a)}, \quad f_{e,h}^{(0)} \equiv -\frac{\Omega_{ee,hh}}{2\gamma_{e,h}}. \tag{30}$$

In the mean-field approximation, the self-energies themselves can be expressed in terms of f_{ij},

$$\Sigma_{ij}^{(r)}(x,x';t) = \frac{1}{2}V(\mathbf{r}-\mathbf{r}')\left[if_{ij}(\mathbf{r},\mathbf{r}';t) - \delta_{ij}\cdot\delta(\mathbf{r}-\mathbf{r}')\right] + \Sigma_{ij}^{(0)}\cdot\delta(\mathbf{r}-\mathbf{r}') \tag{31}$$

and so the set of equations (25)–(30) is complete. Under stationary conditions they reduce to two equations for the self-energies,

$$\frac{1}{2}\int V_{\mathbf{p},\mathbf{p}'}\frac{\left(\mid f_e^{(0)}\mid+\mid f_h^{(0)}\mid\right)\Sigma_{eh}(\mathbf{p}')\cdot\left(E(\mathbf{p}')+\Sigma^{(r)}(\mathbf{p}')\right)}{\mid E(\mathbf{p}')+\Sigma^{(r)}(\mathbf{p}')\mid^2+\lambda^2\mid\Sigma_{eh}(\mathbf{p}')\mid^2}\frac{d^3p'}{(2\pi)^3}$$
$$+\ \Sigma_{eh}(\mathbf{p})=f, \tag{32}$$

$$\Sigma^{(r)}(\mathbf{p}) = \frac{1}{2}\int V_{\mathbf{p},\mathbf{p}'}\frac{\lambda^2\left(\mid f_e^{(0)}\mid+\mid f_h^{(0)}\mid\right)\mid\Sigma_{eh}(\mathbf{p}')\mid^2}{\mid E(\mathbf{p}')+\Sigma^{(r)}(\mathbf{p}')\mid^2+\lambda^2\mid\Sigma_{eh}(\mathbf{p}')\mid^2}\frac{d^3p'}{(2\pi)^3}. \tag{33}$$

The most important equation here is (32). It defines the gap in the renormalized excitation spectrum,

$$\epsilon(\mathbf{p}) = \sqrt{\mid E(\mathbf{p})+\Sigma^{(r)}(\mathbf{p})\mid^2+\mid\Sigma_{eh}(\mathbf{p})\mid^2}+\frac{p^2}{2}\left(\frac{1}{m_e}-\frac{1}{m_h}\right). \tag{34}$$

For $\hbar\omega_0 > \mid E_g+\Sigma^{(r)}(0)\mid$ it is just equal to $\mid\Sigma_{eh}\mid$. So (32) is similar to a BCS equation, but inhomogeneous and with a slightly different integral kernel. The last difference is due to the nonequilibrium nature of our problem. At moderate field $\mid f\mid\leq 1$, and not too close to the resonance $\mid\Sigma_{eh}\mid \ll E_g-\hbar\omega_0$, Eq. (32) can be transformed to an inhomogeneous Schrodinger equation for the exciton wave function. At very large intensities $\mid f\mid\gg 1$, it gives the gap $\mid\Sigma_{eh}\mid\approx f$, as was obtained in Ref. [34]. The functions $f_{e,h}^{(0)}$ entering (32)–(33), defined by Eq. (30), are connected to spontaneous fluctuations of occupation numbers. In thermodynamic equilibrium, the ratios on the right-hand side of (31) are found to be exactly

$$-\frac{\Omega_{e,h}(\mathbf{p})}{2\gamma_{e,h}(\mathbf{p})} = i\left[1-2n_{\mathbf{p}}(T)\right] \equiv i\tanh\frac{\epsilon_{e,h}(\mathbf{p})-\mu}{2kT}, \tag{35}$$

where $n_{\mathbf{p}}(T)$ is the equilibrium Fermi distribution function, μ is the chemical potential and $\epsilon_{e,h}(\mathbf{p})$ are the energies of corresponding single-particle excitations. Correct calculation of both the diagonal self-energy matrix element $\Omega(\mathbf{p})$ and the damping $\gamma(\mathbf{p})$ demands consideration of relaxation processes at least to lowest order. But this is out of the scope of our mean-field approximation. For our present considerations, it does not seem to be crucial, since in an intrinsic semiconductor, $n_{\mathbf{p}}(T)$ is very small and it seems reasonable to suppose that the difference of $\mid f_{e,h}^{(0)}\mid$ from 1 may also be neglected in a pumped system, at least for fast processes.

Nonlinearites in (31) account for the exchange interaction of electrons and holes and so-called phase-space filling, which means reduction of the interaction due to the blocking of some virtual scattering processes because of occupation of corresponding final states.

The only shortcoming of this approach is the difficulty of including the correlation effects, the most important of them being the formation of excitonic molecules. Because of the polaritonic nature of excitons, the existence of biexcitons implies the resonant mutual scattering of photons. In the following sections, we will show that at $|f| \sim 1$ it can completely dominate the optical properties in the vicinity of the exciton resonance.

3 Excitonic Molecules in a Coherent Cloud of Virtual Excitons

The existence of excitonic molecules – biexcitons (EM) – in semiconductors is well-established theoretically [13, 14, 16, 17] and experimentally [52–62]. It is also well recognized [62–67] that, as the excitons themselves are the essential characteristic feature of semiconductor optics, the biexciton contribution may be very important in the nonlinear optical phenomena in the frequency range close to the intrinsic absorption threshold. As a rule, two different approaches have been used in the theoretical description of EMs and their contribution to physical phenomena: the direct microscopic treatment [68, 50], considering the EM as a four-particle bound state – two electrons and two holes – and a phenomenological one [63–66], introducing the EM as an independent boson excitation capable of decaying into an exciton and a photon. Both of these approaches, in many cases very successful, have essential disadvantages for application to the problem under consideration. It seems that the first approach is too complicated to be used as a starting point for a true many-body problem. The second one is inadequate to account for the changes of EM properties and parameters, similar to those described above for excitons, due to interactions in a dense exciton–biexciton system. Nevertheless, according to modern ideas [24, 37] about many-body exciton physics, just these transformations, including an exciton's and other complexes' continuous destruction, constitute the essence of the high-excitation problem. Moreover, the assignment of definite statistics (e.g. Bose–Einstein) to compound particles at such high densities becomes inadequate, as explained in Section 2. At moderate exciton densities both of these effects contribute corrections of the order n, where n is the exciton dimensionless density. For the EM it will evidently be proportional to na_{bx}^3. Here, a_{bx} is the effective EM radius (in terms of the exciton radius). As EM binding energies E_{bx} are typically an order of magnitude smaller than the corresponding exciton Rydberg $E_0 = 1$, usually a_{bx} must be a few times larger than unity. So there is a relatively wide interval of densities in which excitons are virtually unchanged, whereas the EM may transform drastically. Nonlinear optical phenomena due to excitons coherently driven by an intense electromagnetic pump, with the densities in the interval described above, are the main subjects of this section.

For this purpose the following model seems to be adequate: excitons will be considered as true boson particles, described in terms of the Bose field operator $\psi(r,t)$ and the dispersion law $\epsilon_p^{(0)} = E_g - 1 + p^2/2m$. So defined, $\psi(x)$ is essentially $F_{eh}(x,x)$ of Section 2. In contrast with that section, an EM will be treated as composite particles – the bound state of two excitons arising due to the interaction potential $V(\mathbf{r},\mathbf{r}')$. Obviously such an interaction has to be attractive. But then the well-known problem in quantum liquid theory arises: at low temperatures, a system of Bose particles with an attractive interaction is unstable against spontaneous contraction. In order to avoid this difficulty, two equivalent types of excitons $\psi_\alpha(\alpha = 1,2)$ with the same parameters may be introduced. The only distinction is that the excitons of different types attract each other ($U_{12} = U_{21} = -V < 0$), whereas particles of the same type have a repulsive interaction ($U_{11} = U_{22} \equiv U > V > 0$). Such an appoach leads simultaneously to the existence of bound complexes of two excitons of different types and to the stability of the many-particle system as a whole, i.e. the stability of the "molecular gas" of excitons. Moreover, the proposed model is not too artificial, since two types of exciton may be considered as corresponding to singlet excitons with mutually opposite directions of electron (and also hole) spins. The main difference of this picture from reality is the absence of triplet excitons. Nevertheless, it seems that such an assumption is not crucial because a triplet exciton in the dipole approximation does not interact with photons (though it must in a real system, due to electron exchange.) A similar model has been used in Ref. [50]. Thus the exciton–exciton interaction is introduced as a 2×2 matrix:

$$\hat{U} = U\hat{I} - V\hat{\tau}_1, \tag{36}$$

where

$$\hat{I} = \begin{pmatrix} 1 & 0 \\ 0 & 1 \end{pmatrix}, \tag{37}$$

$$\hat{\tau} = \begin{pmatrix} 0 & 1 \\ 1 & 0 \end{pmatrix}.$$

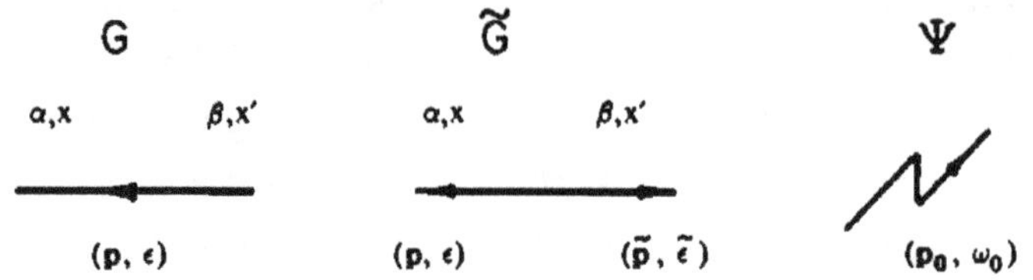

Fig. 7. Schematic illustration of elements in the Feynman graphs: Green's functions and coherent exciton field amplitude.

Qualitatively this corresponds to the exchange interaction of two hydrogen-like atoms (excitons).

From the theoretical point of view, the system under consideration is a "driven Bose liquid". It arises from a macroscopically occupied state (mode) $\psi_\alpha(\mathbf{r}, t)$ of the excitons coherently created by an external electromagnetic field $\mathscr{E}(\mathbf{r}, t)$. As the result of mutual scattering of excitons of this mode, the other modes become populated and also the EM may arise. A consistent description of this system may be given using the nonequilibrium Green's functions technique [41], [46]–[49] and [31]. The following functions are involved in such a treatment:

(1) Normal Green's functions $\hat{g}_{\alpha\beta}(x, x')$, which describe the propagation of a particle from space-time point $x = (\mathbf{r}, t)$ to x' including the possible process of the change of the particle type $(\alpha \to \beta)$.
(2) Anomalous Green's functions [46]–[49] $\hat{\tilde{g}}_{\alpha\beta}\ (x, x')$ which characterize the macroscopically coherent states and describe the correlated appearance (or disappearance – $\hat{\tilde{g}}^{+}_{\alpha\beta}(x, x')$) of two particles α and β at points x and x', respectively. Being the manifestation of the presence of a macroscopically large number of coherently correlated pairs, these functions contain complete information about the EM. In essence, $\tilde{g}_{12}$ at equal times $t = t'$ is the effective EM wave function. Throughout the following discussion we will denote matrices (operators) by the 'hat' $(\hat{o})$ symbol and anomalous Green's functions and self energies by a 'tilde' $(\tilde{o})$.

Graphical representation of all these functions (in real space the same as in momentum representation) is shown in the Fig. 7. Here, $\mathbf{p}_0$ and ω_0 are the momentum and frequency (energy) of the pump wave quanta, respectively, $\tilde{\mathbf{p}} = 2\mathbf{p}_0 - \mathbf{p}, \tilde{\epsilon} = 2\omega_0 - \epsilon$. It is easy to understand the origin of the correlation of pairs of particles with momenta $\mathbf{p}$ and $\tilde{\mathbf{p}}$. Such a pair correlation arises as the result of the mutual scattering of two initial excitons with the momenta $\mathbf{p}_0$. Similar correlations are well-known in the phenomenological description of the nonlinear optical processes such as four-wave mixing, self-phase modulation, etc. In what follows, in order to consider the resonant phenomena in the vicinity of the exciton and biexciton (two-photon) resonances, the "rotating coordinate frame" method will be used. In this case, the frequency ω_0 of the pump wave becomes the origin of the frequency (energy) axis, so that $\omega \to \omega - \omega^0, \epsilon^0_{\mathbf{p}} = E_g - 1 + p^2/M - \omega^0$ and $\tilde{\omega} = -\omega$. As the photon wave vector $\mathbf{p}_0$ is very small in comparison to typical momenta of scattered excitons, one can neglect it in the process of biexciton creation and put $\tilde{\mathbf{p}} \simeq -\mathbf{p}$. For example, $\epsilon^{(0)}_{\tilde{\mathbf{p}}} \simeq \epsilon^{(0)}_{-\mathbf{p}} = \epsilon^{(0)}_{\mathbf{p}}$. Only for an analysis of the optical manifestations of the renormalization phenomena in the exciton–EM system is it necessary to treat the difference between $\tilde{\mathbf{p}}$ and $-\mathbf{p}$ explicitly. But in what follows we will assume the special arrangement permitting realization of BEC of excitonic molecules with exactly zero momentum: pumping by two counterpropagating, circularly polarized waves [69, 70], which are in our model independent sources of excitons of different species α. Then obviously $\tilde{\mathbf{p}} = -\mathbf{p}$.

As explained above, according to the nonequilibrium Green's function formalism, each matrix element of $\hat{g}_{\alpha\beta}$ or $\hat{\hat{g}}_{\alpha\beta}$ matrices is itself a 2×2 matrix,

$$\hat{g}_{\alpha\beta} = \begin{pmatrix} 0 & g^{(a)}_{\alpha\beta} \\ g^{(r)}_{\alpha\beta} & f^{ex}_{\alpha\beta} \end{pmatrix}. \tag{38}$$

Here again, $g^{(r,a)}_{\alpha\beta}$ denote retarded and advanced Green's functions, and the diagonal element $f^{ex}_{\alpha\beta}$ is closely related to the distribution function. But unlike the treatment of excitons in the preceding section, here we are dealing with bosonic Green's functions. Therefore for $t = t'$,

$$f^{ex}_{\alpha\beta}(\mathbf{p}; t, t) = -i[1 + 2n_{\mathbf{p}}(t)]. \tag{39}$$

The + sign in the above formula corresponds to Bose particles, and $n_{\mathbf{p}}$ are the occupation numbers of the excitons.

The Green's functions have to be found from the Dyson equations, which are in this case similar to the corresponding equations for the Bose liquid [46, 51]:

$$\left(i\frac{d}{dt} - \epsilon^{(0)}_{\mathbf{p}}\right) \hat{g}_{\alpha\beta}(\mathbf{p}; t, t') - \int \hat{\sigma}_{\alpha\gamma}(\mathbf{p}; t, t_1)\hat{g}_{\gamma\beta}(\mathbf{p}; t_1, t')dt_1$$

$$\int \hat{\hat{\sigma}}^{+}_{\alpha\gamma}(\mathbf{p}; t, t_1)\hat{\hat{g}}_{\gamma\beta}(\mathbf{p}; t_1, t')dt_1 = \delta_{\alpha\beta}\delta(t - t'), \tag{40}$$

$$\left(-i\frac{d}{dt} - \epsilon^{(0)}_{\bar{\mathbf{p}}}\right) \hat{\tilde{g}}_{\alpha\beta}(\bar{\mathbf{p}};t,t') - \int \hat{\sigma}_{\alpha\gamma}(\bar{\mathbf{p}};t,t_1)\hat{\tilde{g}}_{\gamma\beta}(\bar{\mathbf{p}};t_1,t')dt_1$$

$$\int \hat{\tilde{\sigma}}_{\alpha\gamma}(\bar{\mathbf{p}};t,t_1)\hat{g}_{\gamma\beta}(\bar{\mathbf{p}};t_1,t')dt_1 = 0\,. \tag{41}$$

Here, the normal and anomalous self-energies $\hat{\sigma}_{\alpha\beta}$ and $\hat{\tilde{\sigma}}_{\alpha\beta}$ include all interaction processes in the system and are represented as usual by an infinite series of Feynman graphs. The simplest class, which corresponds to the self-consistent field approximation (SCF), is shown in Fig. 8. The dashed lines in these graphs denote the interaction matrix $\hat{U}$. Nonequilibrium self-energies are also 2×2 matrices of the type

$$\sigma_{\alpha\beta} = \begin{pmatrix} \Omega^{ex}_{\alpha\beta} & \sigma^{(r)}_{\alpha\beta} \\ \sigma^{(a)}_{\alpha\beta} & 0 \end{pmatrix}. \tag{42}$$

Here, $\Omega^{ex}_{\alpha\beta}$ is the noise correlator in the system, closely related to the probability of incoherent creation and annihilation of excitons. The macroscopic wave function $\psi_\alpha(\mathbf{p}_0,t)$ of coherent excitons satisfies the following field equation:

$$\left(i\frac{d}{dt} - \epsilon^{(0)}_{\mathbf{p}_0}\right) \psi_\alpha(\mathbf{p}_0,t) - \int \sigma^{(r)}_{\alpha\beta}(\mathbf{p}_0;t,t')\psi_\beta(\mathbf{p}_0;t')dt'$$

$$- \int \tilde{\sigma}^{+}_{\alpha\beta}(\mathbf{p}_0;t,t')\psi^*_\beta(\mathbf{p}_0;t')dt' = f_\alpha(t), \tag{43}$$

or, in a general spatially inhomogeneous case,

$$\left(i\hbar\frac{\partial}{\partial t} + \frac{\hbar^2}{2m}\nabla^2 - \epsilon_0\right) \psi_\alpha(\mathbf{r},t) - \int \sigma^{(r)}_{\alpha\beta}(\mathbf{r},\mathbf{r}';t,t')\psi_\beta(\mathbf{r}';t')dr'dt'$$

$$- \int \tilde{\sigma}^{+}_{\alpha\beta}(\mathbf{r},\mathbf{r}';t,t')\psi^*_\beta(\mathbf{r}';t')dr'dt' = f_\alpha(\mathbf{r},t), \tag{44}$$

where $\epsilon_0 = \epsilon^{(0)}_{\mathbf{p}=0}$ is the detuning of the pump frequency from the exciton resonance, and the slowly varying electromagnetic field amplitude $f^{ex}_\alpha(\mathbf{r},t)$ is the source of the dipole-active coherent excitons. This function is determined by the dipole matrix element $\mathbf{m}_\alpha$ of the α-exciton–photon interaction:

$$f^{ex}_\alpha(t) = m_{\alpha i}\mathscr{E}_i(t) = \frac{ie\hbar}{m_0 E_g}(p_{cv})_{\alpha i}\phi(0)\mathscr{E}_i(t). \tag{45}$$

Here, $(p_{cv})_{\alpha i}$ is the matrix element of the interband transition, m_0 is the electron mass and $\phi(0)$ is the value of the exciton ground state wave function for the coincident electron and hole coordinates.

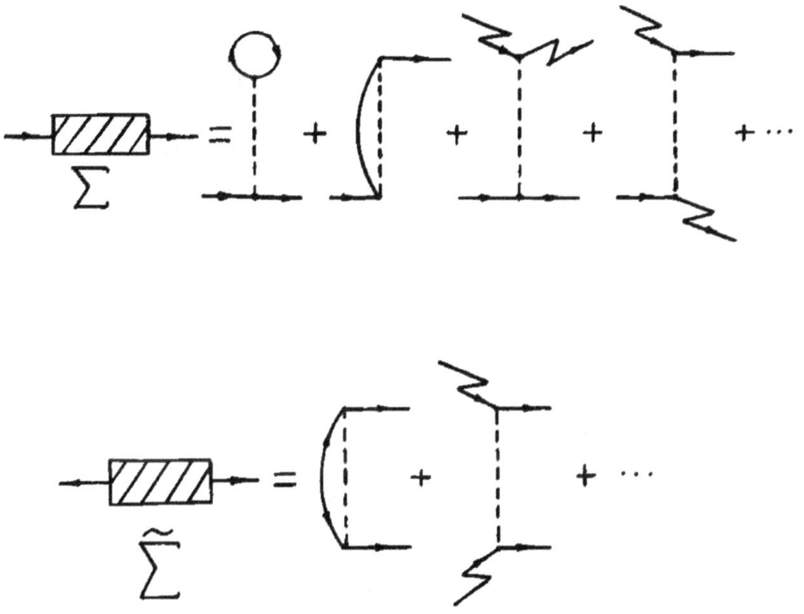

Fig. 8. Diagonal (Σ) and anormalous ($\tilde{\Sigma}$) self-energy graphs in the mean-field approximation.

The equations (40)–(41) are the matrix equations in the sense of the definitions (38) and (42). In fact, the equations for $g^{(r)}$ and $\tilde{g}^{(r)}$ look exactly like (40) and (41) with the substitution of the proper self-energy functions. The same statement holds for the advanced Green's functions. As to the function $f^{ex}_{\alpha\beta}(\mathbf{p};t,t')$, it can be expressed in terms of the self-energies $\Omega^{ex}_{\alpha\beta}$, and the retarded and advanced functions [41]:

$$f^{ex}_{\alpha\beta}(\mathbf{p};t,t') = \int g^{(r)}_{\alpha\gamma_1}(t,t_1)\Omega^{ex}_{\gamma_1\gamma_2}(t_1,t_2)g^{(a)}_{\gamma_2,\beta}(t_2,t')dt_1dt_2\,. \qquad (46)$$

In what follows, the assumption that the self-energies $\Omega^{ex}_{\alpha\beta}$ do not considerably deviate from their quasiequilibrium values will again be used. Also, all the retardation effects in the self-energies will be neglected. Some results of such a treatment are presented in Section 4.

If the duration of the pumping pulse exceeds the relaxation time of excitons, the quasi-stationary state arises. In this case, all self-energies become time-independent functions and the following formulas can be used:

$$\hat{g} = \frac{1}{2}(g^+ + g^-)\cdot\hat{I} + \frac{1}{2}(g^+ - g^-)\cdot\hat{\tau}_1\,, \qquad (47)$$

$$\hat{g} = \frac{1}{2}(\tilde{g}^+ + \tilde{g}^-) \cdot \hat{I} + \frac{1}{2}(\tilde{g}^+ - \tilde{g}^-) \cdot \hat{\tau}_1, \tag{48}$$

$$\sigma^\pm = \sigma_{11} \pm \sigma_{12}, \qquad \tilde{\sigma}^\pm = \tilde{\sigma}_{11} \pm \tilde{\sigma}_{12}. \tag{49}$$

Here, the symbols + and − correspond to the symmetric $\psi_+ = 1/\sqrt{2}(\psi_1 + \psi_2)$ and antisymmetric $\psi_- = 1/\sqrt{2}(\psi_1 - \psi_2)$ superpositions of different types of excitons,

$$g^\pm(p) = \frac{\epsilon + \epsilon_{\mathbf{p}}^{(0)} + \sigma^\pm}{\epsilon^2 - (\epsilon_{\mathbf{p}}^\pm)^2}; \quad \tilde{g}^\pm = -\frac{\tilde{\sigma}^\pm}{\epsilon^2 - (\epsilon_{\mathbf{p}}^\pm)^2}; \tag{50}$$

$$\epsilon_{\mathbf{p}}^\pm = \sqrt{(\epsilon_{\mathbf{p}}^{(0)} + \sigma^\pm)^2 - |\,\tilde{\sigma}^\pm\,|^2}\,; \tag{51}$$

$$\psi_+ = \frac{f}{\epsilon_0 + \sigma^+ + \tilde{\sigma}^+}. \tag{52}$$

The values $\epsilon_{\mathbf{p}}^\pm$ represent the energy difference between the excitonic levels and the carrier frequency of the pump. In a sense, the last formula (52) is an odd result. Instead of the usual resonance relation of the exciton polarization ψ to the field f, the denominator in Eq. (52) is quite different from $\epsilon_{\mathbf{p}}^\pm$. This result is one of the manifestations of the presence of pair coherence in the system under consideration.

The polarizability $\chi(\mathbf{k}, \omega)$ which describes the propagation of a weak probe signal $(\mathbf{k}, \omega)$ in a semiconductor in the presence of a pump is given by the following expression:

$$\chi(\mathbf{k}, \omega) = \left|\mathbf{m}_+ \frac{c}{\omega}\right|^2 \left(g(k) + \frac{4\pi\,|\,\mathbf{m}_+\,|^2|\,\tilde{g}(k)\,|^2}{\tilde{\omega}^2 - c^2\tilde{k}^2 - 4\pi\,|\,\mathbf{m}_+\,|^2\,\tilde{g}(\tilde{k})}\right). \tag{53}$$

This formula directly manifests the interconnection of the polariton branches $\mathbf{k}$ and $\tilde{\mathbf{k}}$. The proper Green's function $D(k)$ of the probe electromagnetic field is given by

$$D(k) = \frac{(\omega - \epsilon_{\mathbf{k}}^+)A^+ + \nu_{\mathbf{p}}\,|\,\mathbf{m}_+\,|^2\,\epsilon_{\mathbf{k}}^+}{A^-A^+ - 2\nu_{\mathbf{k}}\,|\,\mathbf{m}_+\,|^2\,\omega_{\mathbf{k}}\epsilon_{\mathbf{k}}^+}, \tag{54}$$

with

$$A^- = (\omega - \omega_{\mathbf{k}})(\omega - \epsilon_k^+) - |\,\mathbf{m}_+\,|^2, \tag{55}$$

$$A^+ = (\omega + \omega_{\mathbf{k}})(\omega + \epsilon_k^+) - |\,\mathbf{m}_+\,|^2, \tag{56}$$

and

$$\nu_{\mathbf{p}} = \frac{\epsilon_{\mathbf{p}}^{(0)} + \sigma^+}{\epsilon_{\mathbf{p}}^+} - 1 = \sqrt{1 + 2n_{\mathbf{p}}} - 1. \tag{57}$$

The poles of $D(k)$ describe the four new polariton-like branches. Each of these branches reflects the joint propagation of the four particles – two excitons and two photons with wave vectors $\mathbf{k}$ and $\tilde{\mathbf{k}}$. Their dispersion laws are

$$(\omega_{\mathbf{p}}^{\pm})^2 = \frac{1}{2}\Bigg(2\mid \mathbf{m}\mid^2 + (\epsilon_{\mathbf{p}}^{+})^2 + (\omega_{\mathbf{p}}^{(0)})^2$$

$$\pm\sqrt{\left[(\epsilon_{\mathbf{p}}^{+})^2 - (\omega_{\mathbf{p}}^{(0)})^2\right]^2 + 4\mid \mathbf{m}\mid^2 \left[(\epsilon_{\mathbf{p}}^{+} + \omega_{\mathbf{p}}^{(0)})^2 + 2\nu_{\mathbf{p}}\epsilon_{\mathbf{p}}^{+}\omega_{\mathbf{p}}^{(0)}\right]}\Bigg), \quad (58)$$

Here $\omega_{\mathbf{p}}^{(0)} = c|\mathbf{p}|$ and $n_{\mathbf{p}} = -0.5[1 + if^{ex}(\mathbf{p}; t, t)]$ are the exciton occupation numbers. It is easy to see from (54), and especially (58), that the coupling of two polaritons $\mathbf{p}$ and $\tilde{\mathbf{p}}$ is determined by $\nu_{\mathbf{p}}$, which in its turn is governed essentially by the $\mid \tilde{\sigma}^{+} \mid^2$, that is, by the EM density.

4 Results

In the framework of the model introduced, the considerations in the previous section are quite general. In order to obtain a more detailed description accounting for explicit dependences of self-energies on pumping frequency and intensity, the mean-field approximation will be used below. In this case a matrix equation similar to (25) can be used for the density matrix $\hat{f}^{ex}_{\alpha\beta}(\mathbf{r}, \mathbf{r}'; t, t)$:

$$i\hbar\frac{\partial}{\partial t}\hat{f}^{ex} = \left(\hat{h}_0 + \hat{\sigma}^{(r)}\right)\hat{f}^{ex} - \hat{f}^{ex}\left(\hat{h}_0 + \hat{\sigma}^{(a)}\right) \quad (59)$$

and retarded self-energies can be calculated in terms of the density matrix

$$\sigma_{\alpha\beta}(\mathbf{r}, \mathbf{r}'; t) = \frac{1}{2}U_{\alpha\beta}(\mathbf{r}, \mathbf{r}')\left[if^{ex}_{\alpha\beta}(\mathbf{r}, \mathbf{r}'; t) + 2\psi_\alpha(\mathbf{r}, t)\psi_\beta(\mathbf{r}', t)\right.$$

$$\left. -\delta_{\alpha\beta}\cdot\delta(\mathbf{r} - \mathbf{r}')\right] \quad (60)$$

and

$$\hat{h} = \hat{I}\cdot\left(E_g - E_0 - \hbar\omega_0 - \frac{\hbar^2}{2M}\nabla^2\right). \quad (61)$$

It should be noted that, in its physical content, the mean-field approximation used here is essentially different from that used in Section 2. Having introduced the exciton as a basic "elementary particle", we neglect the possibility of treating its internal structure but include in the mean-field treatment the exciton–exciton iteraction and even biexciton formation. As explained above, the accuracy of this approach is of the order of n or

$a_{bx}^{-1} \ll 1$. Within the same accuracy, we can neglect the effective radius of the exciton–exciton iteraction potential, which is obviously of the order of 1. Then considering Fourier components of matrix elements of $\hat{U}$ as two different constants, we can adjust their values using the two most important experimental data – the dissociation energy of the excitonic molecule E_d and the low-energy mutual scattering length of two biexcitons. To be specific in what follows, the results for a two-dimensional problem will be presented, i.e. excitons in a quantum well (QW) structure and a pump wave propagating normal to the QW plane direction will be treated. Many current experiments are done in such a geometry. Its obvious advantages are that the exciting field amplitude is independent of coordinates and undistorted by propagation phenomena, the excitons are coherently created with zero momentum, etc. The well-known problem of the destruction of the long-range order in two-dimensional systems by fluctuations seems in our case to be unimportant because the coherence is induced by an external source and so no phase degeneracy exists, and also because the finite lifetime and excitation pulse duration essentially restrict the logarithmic divergence of long-wave fluctuations.

Being no longer interested in the exciton internal structure, we will now introduce new ("biexcitonic") scales of variables more convenient for the following discussion, by putting the biexciton dissociation energy E_d and its effective radius

$$a_{bx} = \frac{\hbar}{\sqrt{ME_d}} \tag{62}$$

equal to unity. As a result, time intervals in what follows will be measured in terms of $\hbar/E_d$ and two-dimensional exciton densities in units of a_{bx}^{-2}. Dimensionless coupling constants will be introduced as

$$\alpha = \frac{mU}{4\pi\hbar^2}; \quad \lambda = \frac{mV}{4\pi\hbar^2}; \quad \gamma = \frac{\alpha\lambda}{\alpha+\lambda}. \tag{63}$$

In order to avoid explicit treatment of relaxation processes only two limiting regimes will be considered here: (1) stationary and (2) transient, for time intervals short compared to any relaxation process.

4.1 Stationary Regime

This can be achieved if the excitation-pulse duration is large compared to either the exciton phase-relaxation time or the coherence interval for the pump field itself. Then formulas (47)–(51) within the above-described approximations give the possibility of expressing all the interesting variables in terms of two of them – the anomalous self-energies $\tilde{\sigma}^{\pm}$:

$$\sigma^{\pm} = 4\pi\left[\left(\alpha - \frac{1}{2}\lambda\right)\frac{n_{+}+n_{-}}{2} - \frac{1}{2}\lambda n^{\pm} + \left(\alpha - \frac{1\pm 1}{2}\lambda\right)n_0\right], \quad (64)$$

where

$$n^{\pm} = \frac{1}{4\pi}\cdot\tilde{\sigma}^{\pm}\cdot \mathrm{Arsh}\left(\frac{\tilde{\sigma}^{\pm}}{\epsilon^{\pm}_{\mathrm{min}}}\right), \quad (65)$$

and

$$\epsilon^{\pm}_{\mathrm{min}} = \sqrt{(\epsilon_0 + \sigma^{\pm})^2 - |\,\tilde{\sigma}^{\pm}\,|^2}. \quad (66)$$

Here, $n_0 = |\,\psi_{+}\,|^2$ is the density of coherent excitons in the single-particle condensate (polarization) and $\epsilon_0 = \epsilon^{(0)}_{\mathrm{p}=0}$ is the detuning of the pumping wave from the unperturbed excitonic level. Two more equations define the anomalous self-energies themselves:

$$\tilde{\sigma}^{+}\cdot \ln\epsilon^{+}_{\mathrm{min}} - \tilde{\sigma}^{-}\cdot\ln\epsilon^{-}_{\mathrm{min}} = -2n_0\,; \quad (67)$$

$$\tilde{\sigma}^{+}\left(1 - \gamma\ln\epsilon^{+}_{\mathrm{min}}\right) + \tilde{\sigma}^{-}\left(1 - \gamma\ln\epsilon^{-}_{\mathrm{min}}\right) = 2\gamma n_0\,. \quad (68)$$

Together with formula (52), equations (64)–(68) are a complete set, describing in the mean-field approximation the stationary states in the model under consideration. They can be further simplified if $\lambda \ll \alpha$. Then it follows immediately from (68) that $\tilde{\sigma}^{-} \approx -\tilde{\sigma}^{+}$ and then from (64)–(66) it follows that $\sigma^{-} \approx \sigma^{+}$ and $\epsilon^{-}_{\mathrm{min}} \approx \epsilon^{+}_{\mathrm{min}}$. The equations now reduce to

$$\tilde{\sigma}\cdot\ln\epsilon_{\mathrm{min}} = -n_0; \quad (69)$$

$$\sigma = 4\pi\,(n+n_0) = \alpha\left[\tilde{\sigma}\mathrm{Arsh}\left(\frac{\tilde{\sigma}}{\epsilon_{\mathrm{min}}}\right) + n_0\right]. \quad (70)$$

Some results of the numerical solution of these equations are depicted in Figs. 9–12. The first three of them show "resonant curves", i.e. dependences of response – ψ_{+} and $\tilde{\sigma}$ – on the detuning of pumping frequency from the exciton resonance at a fixed value of a pumping field. The zero of the frequency axis corresponds in these plots to the position of the exciton level, and −1 to the biexciton resonance.

In these diagrams, the anomalous self-energy $\tilde{\sigma}$ is represented by full lines and polarization ψ_{+}, designated as c_0, by broken lines. Strictly speaking, $\tilde{\sigma}$ cannot be considered as the "biexciton amplitude," in analogy with the exciton amplitude ψ_{+}. But at low densities this interpretation is exact, as can readily be seen from (65). It can be seen from these plots that in the frequency range $-1 < \omega_0 < 1$ and for fields ≥ 0.2 the response is dominated by excitonic molecules. No resonant structure

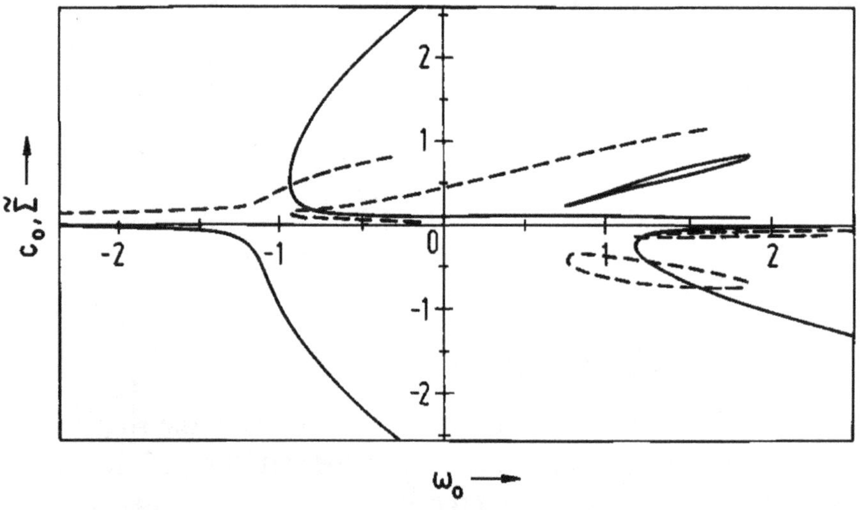

Fig. 9. Coherent exciton amplitude c_0 (broken line) and anomalous self-energy $\tilde{\Sigma}$ (full lines) dependences on the pumping field frequency ω_0 in a stationary regime for pumping field strength $f = 0.2$ and effective repulsion coupling constant $\alpha = 0.6$.

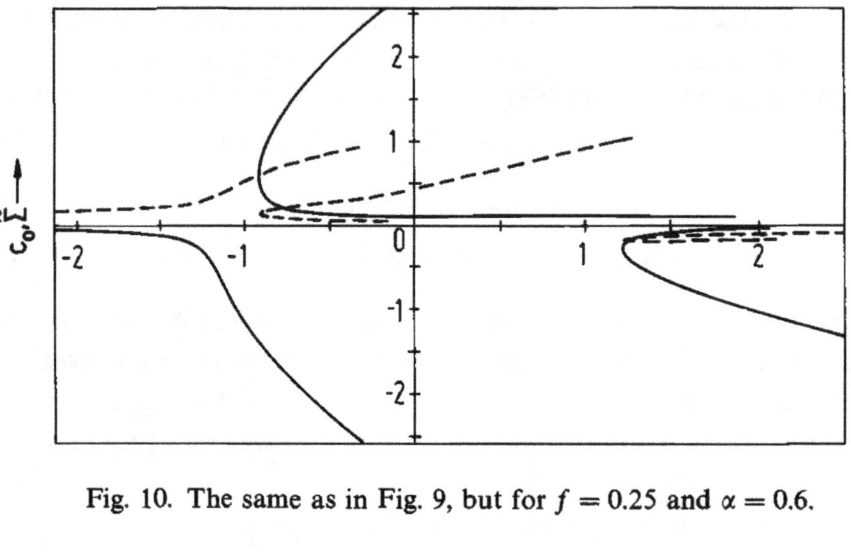

Fig. 10. The same as in Fig. 9, but for $f = 0.25$ and $\alpha = 0.6$.

exists around $\omega_0 = 0$. Only a loop in the interval $0.5 < \omega_0 < 2$ in Fig. 9, also disappearing at higher field, is a small remnant of a "pure" excitonic resonance observed at low fields. Instead of that, a typical resonant structure for excitonic molecules is observed, accompanied by a strongly nonlinear resonance of the excitonic polarization in the vicinity of $\omega_0 = -1$. For frequencies above this stationary resonance the response

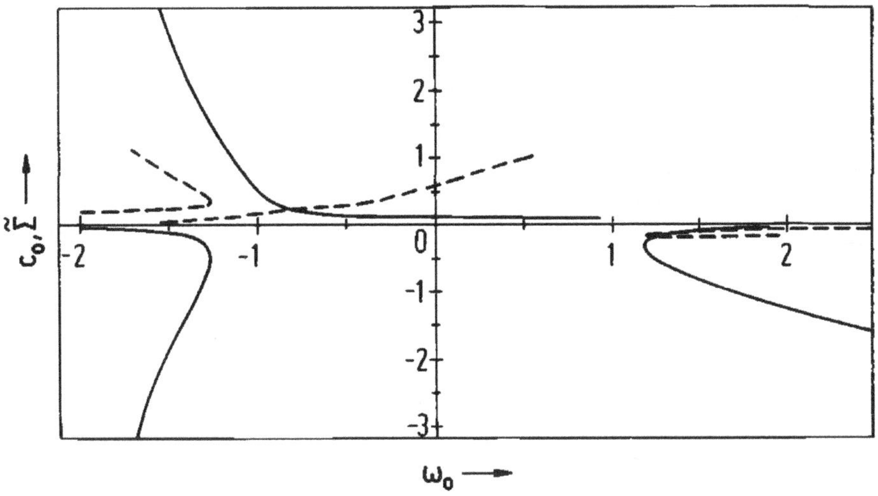

Fig. 11. The same as in Fig. 9, but for $f = 0.25$ and $\alpha = 0.3$.

becomes multistable. Two new states arise. The one with the smaller value of excitonic amplitude corresponds to the one with larger biexciton amplitude, and vice versa. Strictly speaking, solutions with $\omega_0 > 0, c_0 < 0$ are unstable because of stimulated scattering into some states in the lower polariton branch. Therefore, they will not be discussed in what follows. Among the three stable ($c_0 > 0$) solutions, one is "exciton dominated", i.e. beyond the immediate vicinity of the resonance, the majority of excitons are in the single-particle condensate-coherent polarization cloud. Two other solutions are "biexciton dominated" – the majority of excitons are bound into molecules at the expense of coherent polarization. But nonlinear processes become greatly enhanced, such as four-wave mixing (FWM) which increases with $\tilde{\sigma}$, as can readily be seen from (54)–(57) and the following formula, relating the exciton distribution function to the anomalous self-energy

$$n_{\mathbf{p}} = \frac{1}{2} \frac{| \tilde{\sigma} |^2}{| \epsilon_{\mathbf{p}}^{(0)} + \sigma |^2 - | \tilde{\sigma} |^2} . \tag{71}$$

As $\tilde{\sigma}$ increases, this distribution transforms to a narrow peak with the maximum value increasing as $\tilde{\sigma}^2$ and the halfwidth decreasing inversely proportional to $| \tilde{\sigma} |^{1/2}$. This means an increase of the effective biexciton radius proportional to $| \tilde{\sigma} |^{1/2}$.

As the pumping field strength increases, the resonance broadens, different branches shifting in opposite directions from the resonant frequency. So for any frequency $\omega_0 > -1$, at some critical field value dependent

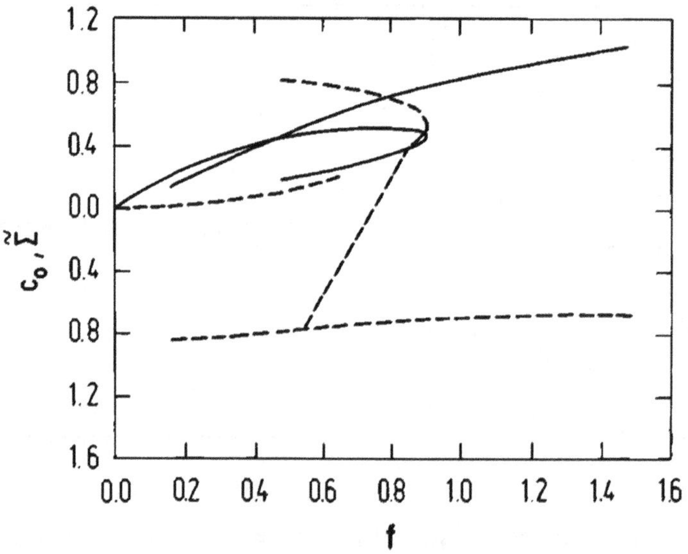

Fig. 12. Field dependence of the coherent exciton amplitude c_0 (full lines; note the difference with Figs. 9–11!) and anomalous self-energy $\tilde{\Sigma}$ (broken lines) for pumping frequency $\omega_0 = -0.75$.

on frequency, two solutions (both with $\tilde{\sigma}$ positive) – the "exciton dominated" one and one "biexciton dominated"– converge and at still larger fields disappear. Only the other "biexciton dominated" state– that with $\tilde{\sigma}$ negative – persists. Such behavior is clearly seen in Fig. 12 where field dependences of response ψ_+ (full lines) and $\tilde{\sigma}$ (broken lines) for $\omega_0 = -0.75$ and $\alpha = 0.75$ are depicted. The common low-field "exciton dominated" solution (the line coming from zero in Fig. 12) corresponding to relatively small biexciton density ($\tilde{\sigma} \sim f^2$) disappears at the field value $f \sim 1$. The system must jump to the only remaining stationary state, which will manifest itself in an abrupt increase of polarizability and large increase of the FWM signal.

Both Figs. 9 and 10 correspond to a relatively large value of the effective-repulsion coupling constant $\alpha = 0.6$. At $\alpha < 0.5$ the characteristic appearance of the biexciton resonance is essentially different [51], as if the effective nonlinearity changes its sign. It is shown in Fig. 11 for $\alpha = 0.3$. Now the multistable regime arises for frequencies below the unperturbed biexciton resonance, starting at

$$\omega_{\min} = -\left[1 - \frac{\alpha^2 z^2}{1-\alpha}\right] \cosh z \tag{72}$$

with a finite value of biexciton density $|\,\tilde{\sigma}_0\,| \approx \sinh z$. Here z is the root

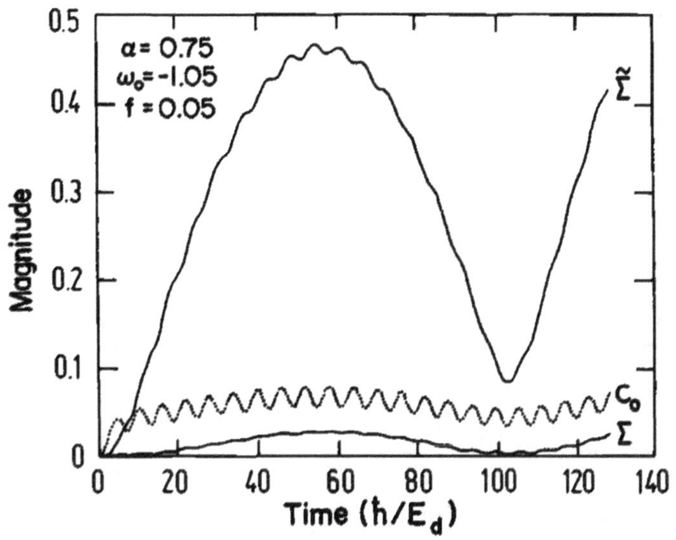

Fig. 13. Time dependences of the coherent exciton amplitude c_0 and normal (Σ) and anomalous ($\tilde{\Sigma}$) self-energies after smooth switching on of a pump field (at $t = 0$) for the pumping field strength $f = 0.05$ and pumping frequency $\omega_0 = -1.05$. Time is in units of $\hbar/E_d$.

of the equation

$$\tanh z = \frac{\alpha}{1-\alpha} z. \qquad (73)$$

In such a case, two more branches of stationary states exist, not shown in Fig. 11, with very large biexciton densities $|\tilde{\sigma}| > |\tilde{\sigma}_0|$. It seems that if α is not too close to 0.5, these densities appear to be too large for excitonic molecules to really survive.

4.2 Transient Regime

This regime is shown in Figs. 13–16. No relaxation processes are taken into account. The pump field onset is supposed to have the form $f(t) = [1 - \exp(-t/\tau)]$ with $\tau = 4$. This results in a partial smearing of the Rabi oscillations for excitons. In these diagrams, one can see even more dramatically than in the static regime that, in the vicinity of the biexciton resonance, excitonic molecules completely dominate the system response. Especially spectacular is the time dependence of the normal self-energy, which, according to (70), contains the contributions of both bound and free excitons proportional to their densities. The striking similarity of the time evolution of both self-energies confirms

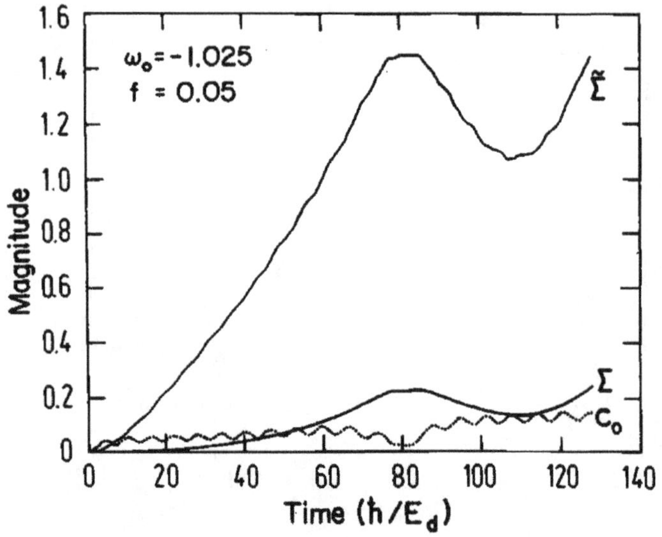

Fig. 14. The same as in Fig. 13, but $f = 0.05, \omega_0 = -1.025$.

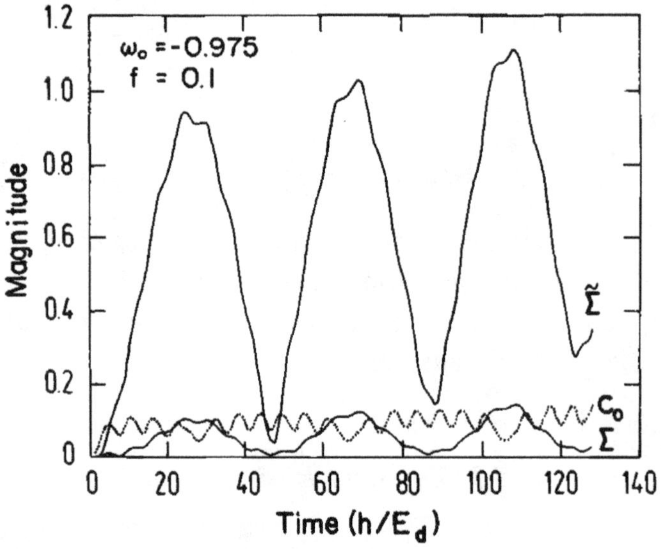

Fig. 15. The same as in Fig. 13, but $f = 0.1, \omega_0 = -1.05$.

the overwhelming contribution of biexcitons. It is evident for both frequencies – 0.05 detuning from the biexciton resonance in Figs. 13 and 15 and 0.025 in Figs. 14 and 16 – and for both (relatively small) pumping fields – 0.05 in Figs. 13–14 and 0.1 in Figs. 15–16.

The Rabi oscillations for excitonic molecules are very slow and extremely nonlinear. At higher exitation and smaller detuning (Fig. 16)

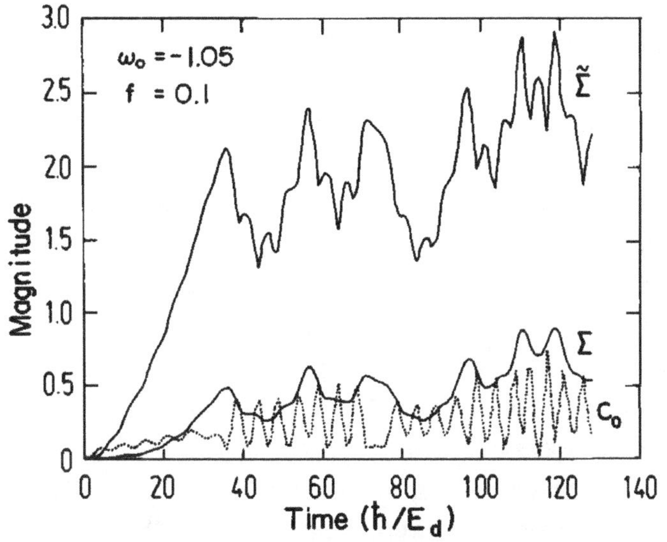

Fig. 16. The same as in Fig. 13, but $f = 0.1, \omega_0 = -1.025$.

they even seem to become chaotic, which does not seem unexpected for such a highly nonlinear system.

While small in reduced units, the field values supposed in Figs. 9–16 are really very large, corresponding to pumping intensities of the order of tens and hundreds of GW/cm^2. These phenomena can be observed only in experiments with ultrashort pulses. Therefore, the details of the relaxation mechanism become both important and interesting, as the standard Boltzmann equation approach becomes inadequate. Different theoretical approaches to this problem are presented in Refs. [71–73]. Being, in a sense, closely related to the problem of the coherence persistance, it remains one of most important topics in the nonlinear optics of semiconductors.

Acknowledgements. This work was partially supported by NATO Collaborative Research Grant CRG 930084. It is also a pleasure to acknowledge here many illuminative and stimulating discussions with H. Haug, A. Mysyrowicz, S. Moskalenko, N. Nagasawa, J. Wolfe, S. Tikhodeev and especially A. Ivanov, whose assistance was very important. I am also grateful to the Institute of Theoretical Physics of Frankfurt University for hospitality while completing this work.

References

[1] S.A. Moskalenko, Fiz. Tverd. Tela **4**, 276 (1962).
[2] I.M. Blatt, K.W. Boer and W. Brandt, Phys. Rev. **126**, 1691 (1962).
[3] S.A. Moskalenko, *Bose–Einstein Condensation of Excitons and Biexcitons*, (Kishinev, RIO, 1970).
[4] L.V. Keldysh, in *Problems of Theoretical Physics* (Nauka, Moskow, 1972), p.433.
[5] E. Hanamura and H. Haug, Phys. Rep. **3**, 209 (1977).
[6] T.M. Rice, Solid State Phys. **32**, 1 (1977).
[7] A. Mysyrowicz, J. Phys. (Paris) **41** Suppl. 7, 281 (1980).
[8] C. Comte and P. Nozieres, J. Phys. (Paris) **43**, 1069, 1083 (1982).
[9] L.V. Keldysh, Electron–hole liquid in semiconductors, in *Modern Problems of Condensed Matter Science* **6**, C.D. Jeffries and L.V. Keldysh, eds. (North-Holland, Amsterdam, 1987).
[10] L.V. Keldysh, Contemp. Phys. **27**, 395 (1986).
[11] G.H. Wannier, Phys. Rev. **52**, 191 (1937).
[12] N.F. Mott, Trans. Farad. Soc. **34**, 500 (1938).
[13] S.A. Moskalenko, Opt. Spectrosk. **5**, 147 (1958).
[14] M.A. Lampert. Phys. Rev. Lett. **1**, 450 (1958).
[15] L.V. Keldysh, *Proc. 9th Int. Conf. on Physics of Semicond.* (Moscow, 1968), p.1303.
[16] O. Akimoto and B. Hanamura, J. Phys. Soc. Jap. **33**, 1537 (1972).
[17] W.F. Brinkman, T.M. Rice, and B. Bell, Phys. Rev. B **8**, 1570 (1973).
[18] W.F. Brinkman and T.M. Rice, Phys. Rev. B **7**, 1508 (1973).
[19] L.V. Keldysh and Yu.V. Kopaev, Sov. Phys. Solid State **6**, 2219 (1965).
[20] J. des Cloizeaux, J. Phys. Chem. Solids **26**, 259 (1965).
[21] B.I. Halperin and T.M. Rice, Solid State Phys. **21**, 115 (1968).
[22] E.A. Andryushin, L. V. Keldysh and A.P. Silin, Sov. Phys. JETP **46**, 616 (1977).
[23] R.R.Guseinov and L.V.Keldysh, Sov. Phys. JETP **36**, 1193, (1973).
[24] L.V. Keldysh and A.N. Kozlov, Sov. Phys. JETP **27**, 521 (1968).
[25] J.J. Hopfield, Phys. Rev. **112**, 1555 (1958)
[26] V.M. Agranovich and V.L. Ginzburg, *Crystal Optics with Spatial Dispersion and Excitons* (Springer, Berlin, 1984).
[27] H.Haug, Z. Phys. **200**, 57 (1967), Phys. Rev. **184**, 338 (1969).
[28] R. Graham and H. Haken, Z. Phys. **237**, 31 (1970).
[29] V. Degiorgio and M.O. Scully, Phys. Rev. A **2**, 1170 (1970).
[30] H. Haken, Rev. Mod. Phys. **47**, 67 (1975).

[31] H. Haug and S.W. Koch, *Quantum Theory of Optical and Electronic Properties of Semiconductors* (World Scientific, London, 1993).

[32] Y. Toyozawa, Progr. Theor. Phys., Supplem.**12**, 11 (1959)

[33] A. Stahl and I. Balslev, *Electrodynamics of the Semiconductor Band Edge*, Springer Tracts in Modern Physics **110** (Springer, Berlin, 1987).

[34] V.M. Galitskii *et al.*, Zh. Eksp. Teor. Phys. **57**, 207 (1969).

[35] V.F. Elesin and Yu.V. Kopaev, Zh. Eksp. Teor. Phys. **63**, 1447 (1972).

[36] S. Schmitt-Rink and D.S. Chemla, Phys. Rev. Lett. **57**, 2752 (1986).

[37] S. Schmitt-Rink, D.S. Chemla and H. Haug, Phys. Rev. B **37**, 941 (1988).

[38] S. Schmitt-Rink, D.S. Chemla and D.A.B. Miller, Adv. Phys. **38**, 89 (1989).

[39] H. Haug and S. Schmitt-Rink, Progress in Quant. Electron. **9**,3 (1984).

[40] R. Zimmerman, *Many Particle Theory of Highly Excited Semiconductors*, (Teubner, Leipzig, 1988).

[41] L.V. Keldysh, Sov. Phys. JETP **20**, 1018 (1965).

[42] E.M. Lifshitz and L.P. Pitaevski, *Statistical Physics*, Part 2 (Pergamon, Oxford, 1980), chapters 10–11.

[43] J. Rammer and H. Smith, Rev. Mod. Phys. **58**, 323 (1986).

[44] G.D. Mahan, Phys. Rev. **145**, 251 (1987).

[45] K. Henneberger, Physica **A150**, 419 (1988).

[46] H. Haug, in *Optical Nonlinearities and Instabilities in Semiconductors*, H. Haug, ed. (Academic, New York, 1988) p. 53.

[47] A.L. Ivanov and L.V. Keldysh, Sov. Phys. JETP **57**, 234 (1983).

[48] L.V. Keldysh and S.G. Tikhodeev, Sov. Phys. JETP **63**, 1086 (1986).

[49] N.A. Gippius, L.V. Keldysh and S.G. Tikhodeev, Sov. Phys. JETP **64**, 1344 (1986).

[50] A.L. Ivanov, L.V. Keldysh and V.V. Panashchenko, Sov. Phys. JETP **72**, 359 (1991).

[51] L.V. Keldysh, Solid State Comm. **84**, 37 (1992)

[52] A. Mysyrowicz, J.B. Grun, R. Levy, A. Bivas, and S. Nikitine, Phys. Lett. A **26**, 615 (1968).

[53] H. Suoma, T. Goto, T. Ohta, and M. Ueta, J. Phys. Soc. Japan, **29**, 697 (1970).

[54] N. Nagasawa, S. Koizumi, T. Mita, and M. Ueta, J. Lumin. **12/13**, 587 (1976).

[55] F. Henneberger, K. Henneberger, and J. Voight, Phys. Stat. Sol. (b) **83**, 439 (1977).
[56] V.D. Phach, A. Bivas, B. Hönerlage, and J.B. Grun, Phys. Status Solidi (b) **84**, 731 (1977).
[57] J.L. Oudar, A. Maruani, E. Batifol, and D.S. Chemla, J. Opt. Soc. Am. **68**, 1638 (1978).
[58] M. Ueta, T. Mito and T. Itoh, Solid State Comm. **32**, 43 (1979).
[59] N. Nagasawa, T. Mito, and M. Ueta, J. Phys. Soc. Japan **41**, 929 (1981).
[60] B. Hönerlage, R. Levy, J.B. Grun, C. Klingshirn, and K. Bohnert, Phys. Rep. **124**, 161 (1985).
[61] V.D. Kulakovski, V.G. Lysenko and V.B. Timofeev, Sov. Phys. Usp. **28**, 735 (1985).
[62] M. Ueta, H. Kanazaki, K. Kobajashi, Y. Tojozava, and E. Hanamura, *Excitonic Processes in Solids* (Springer, Berlin, 1986).
[63] F. Henneberger and J. Voigt, Phys. Status Solidi (b) **76**, 313 (1976).
[64] V. May, K. Henneberger, and F. Henneberger, Phys. Status Solidi (b) **94**, 611 (1979).
[65] P.I. Khadzhi, S.A. Moskalenko, and S.N. Belkin, JETP Lett. **29**, 200 (1979).
[66] H. Haug, R. März, and S. Schmitt-Rink, Phys. Rev. A **77**, 287 (1980).
[67] R. Levy, B. Hönerlage, and J. B. Grun, in *Optical Nonlinearities and Instabilities in Semiconductors*, H. Haug, ed. (Academic, London, 1986), p. 181.
[68] M. Combescot and R. Combescot, Phys. Rev. Lett. **61**, 117 (1988).
[69] M. Hasuo, N. Nagasawa, and A. Mysyrowicz, Phys. Stat. Sol. (b) **173**, 255 (1992).
[70] M. Hasuo, N. Nagasawa, T. Itoh and A. Mysyrowicz, Phys. Rev. Lett. **70**, 1303 (1993).
[71] R. Zimmerman, Phys. Stat. Sol.(b) **173**, 129 (1992).
[72] H. Haug, Phys. Stat. Sol.(b) **173**, 139 (1992).
[73] M. Hartmann and W. Schäfer, Phys. Stat. Sol.(b) **173**, 165 (1992).

Physics – Uspekhi **60** (11) 1187 – 1193 (2017)

FROM THE ARCHIVE

PACS numbers: 32.80.Rm, 42.50.Hz, 42.65.Re

IN MEMORY OF LEONID VENIAMINOVICH KELDYSH

Multiphoton ionization by a very short pulse

L V Keldysh

DOI: https://doi.org/10.3367/UFNe.2017.10.038229

Abstract. The detachment of a bound electron by an electric field pulse whose duration ranges from a fraction of to a few times the optical cycle but is long compared to h/I (h — Planck constant, I — binding energy) is studied theoretically, simulating the ionization of atoms by extremely short laser pulses. Because of the strong nonlinearity, the solution to the problem does not reduce to the sum of monochromatic harmonic contributions and depends significantly on the pulse shape features. A general analysis is carried out for an analytical pulse shape, and exact formulas are given for standard pulse shapes such as solitonlike, gaussian, lorenzian, etc., one or a half optical cycle in duration. The intensity and pulse length dependences of the ionization probability are of a near-universal tunneling type at high intensities. However, at moderate intensities in the multiphoton regime, these dependences differ widely for different pulse shapes, with ionization probabilities always a few orders of magnitude higher than for ionization by a monochromatic wave of the same intensity and mean frequency.

Keywords: multiphoton processes, tunnel and multiphoton ionization, relativistic ionization, very short laser pulses, intense laser radiation

Remarkable progress in generating very high intensity laser fields was closely related to a corresponding reduction in pulse duration [1, 2]. Therefore, actual multiphoton processes, i.e., those requiring the simultanious participation of many ($\gg 1$) photons, are observed typically in experiments with ultrashort pulses (USPs). With this reduction continuing, pulselengths become comparable to the optical field cycle duration [3–8]. Under such conditions, the usual concept of transition (ionization) probability per unit time makes no sense. The only meaningful quantity remains the total — after the whole pulse — transition probability. Moreover, the frequency spectrum of the pulses under consideration is very broad and, because of the extreme nonlinearity of the process, its probability does not reduce to the sum of independent harmonics contributions.

The physical essence of ionization process in the high intensity USP case may be thought of as an interaction of and competition among the contributions of many harmonies, depending not only on the spectrum, but also on the phase relations of different harmonies, i.e., higher order field correlations. In other words, this means that the result is very sensitive to the exact pulse shape. In this article, some extreme particular cases are theoretically studied, corresponding to USPs a few or even half an optical cycle long.

Recently, a few groups have investigated both experimentally and theoretically an even more extreme limiting case, in a sense — the ionization of atoms by a pulse much shorter than characteristic electron times (the inverse optical transition frequency between neighbouring energy levels). They realized these conditions experimentally with alkali atoms, excited to very high Rydberg states, corresponding to quasiclassical electron motion and small interlevel distances. A field pulse acted in such a case as an (quasi)instantaneous kick, moving electron from one — bounded — Kepler orbit to another — unbounded.

In contrast to this, the problem discussed below is essentially a quantum one: ionization from the tightly bound state, e.g., the ground state, by a pulse one or one-half optical cycle long, but much longer than the 'atomic cycle' — h/I, h being the Planck constant and I the ionization energy. This means that the average energy of a photon in the pulse is small compared to the ionization energy. For an atomic electron, this is slowly varying perturbation, and therefore the same adiabatic treatment can be applied to this problem which was exploited earlier [9] for ionization by intense monochromatic waves. The basis for this is the observation that the final — free — state of an electron in the process under consideration is much more sensitive to such type of perturbations than the initial one — strongly bound and localized. So, the transition probability is calculated as that of a first order transition from the unperturbed initial atomic state to the final 'exact' state of the free electron in the strong time-dependent electric field. The latter of these states accounts for the field action nonperturbatively and contains the main contribution to the transition amplitude. For the fields below atomic, i.e., intensities up to the PW/cm^2 range, the most (and only) important defect in this approach is neglecting electron–ion Coulomb interaction in the final state, i.e., a Born-type approximation. The significance of such approximate solutions may seem questionable now. This kind of quantum problem — a single electron in an external field, including both atomic and electromagnetic — is certainly within the limits of modern computing abilities. During the last decade, several algorithms have been proposed and successfully applied to the problems of multiphoton ionization and some other related ones, such as UV higher harmonics generation. Still, analytic solutions, even semiquantitatively correct, also have their advantages, not being restricted by any definite set of parameter values. They may be useful in representing an

L V Keldysh Lebedev Physical Institute, Russian Academy of Sciences, Leninskii prosp. 53, 119991 Moscow, Russian Federation

The manuscript was written not earlier than the fall of 1997 and not later than the fall of 2000
Uspekhi Fizicheskikh Nauk **187** (11) 1280 – 1287 (2017)
DOI: https://doi.org/10.3367/UFNr.2017.10.038229
Written in English by the author

overall view of the process and trends due to variation of parameters, or as a starting point in analyzing more complicated, e.g., multielectron, systems.

Let the spatially uniform time-dependent electric field $\mathbf{F}(t)$ be given by

$$\mathbf{F}(t) = \mathbf{F} f'(\omega t)\,. \tag{1}$$

Here, $f'(x)$ is the derivative of the function f over its argument x, and ω is the inverse characteristic timescale of the pulse. In such a field, the wave function of the free electron is

$$\psi_{\mathbf{p}}(\mathbf{r}, t) = \exp\left[\frac{\mathrm{i}}{h}\left(\mathbf{p}(t)\,\mathbf{r} \quad \int_0^t \frac{p^2(t')}{2m}\,\mathrm{d}t'\right)\right] \tag{2}$$

with

$$\mathbf{p}(t) = \mathbf{p} + \frac{e\mathbf{F}}{\omega} f(\omega t)\,. \tag{3}$$

Following the usual first-order perturbation theory, the transition probability from the initial state $\psi_0(\mathbf{r}) \exp[(\mathrm{i}/h)\,It]$ to the final state $\psi_{\mathbf{p}}(\mathbf{r}, t)$ can be calculated:

$$w_{i\mathbf{p}} = \frac{e^2\mathbf{F}^2}{h^2\omega^2} \times \left|\int_{\infty}^{\infty} \mathrm{d}x\, R_{\parallel}\left(\mathbf{p} + \frac{e\mathbf{F}}{\omega} f(x)\right) \exp\left(\mathrm{i}\frac{I}{h\omega}\,\Phi(x)\right)\right|^2, \tag{4}$$

with phase function $\Phi(x)$ defined as

$$\Phi(x) = \frac{1}{I}\int_0^x \left(I + \frac{p^2(x')}{2m}\right)\mathrm{d}x' \quad \mathrm{i}\frac{h\omega}{I}\ln f'(x)\,. \tag{5}$$

Here, $R_{\parallel}(\mathbf{p})$ is the transition matrix element of the coordinate component parallel to the field $\mathbf{F}$,

$$R_{\parallel}(\mathbf{p}) = \int \exp\left[\quad\left(\frac{\mathrm{i}}{h}\right)\mathbf{pr}\right] \mathbf{nr}\,\psi_0(\mathbf{r})\,d^3r\,,$$

and $\mathbf{n}$ is the unit vector in the field direction.

In order to make the following analysis more vivid, it is convenient to use the representation of all quantities involved in the natural 'atomic' scale, i.e., to define

$$\Omega = \frac{h\omega}{I}\,, \quad \mathbf{q} = \frac{\mathbf{p}}{\sqrt{2mI}}\,, \quad \mathcal{E} = \frac{eh\mathbf{F}}{\sqrt{2mI^3}}\,. \tag{6}$$

Certainly, this means the corresponding transformation of coordinate and time scales. Then, the dimensionless matrix element should be defined as

$$M(\mathbf{q}) = \left(\frac{\sqrt{2mI}}{h}\right)^{5/2} R_{\parallel}(\mathbf{p})\,. \tag{7}$$

According to the above claim, the whole consideration in this article is for $\Omega \ll 1$.

The crucial parameter of the theory is then the ratio of dimensionless field to the frequency:

$$\lambda = \frac{|\mathcal{E}|}{\Omega}\,, \tag{8}$$

which is exactly inverse to the parameter γ introduced in [9] (if ω is considered a characteristic frequency of the process).

The factor $1/\Omega$ being the large parameter of the theory, the integral in (4) can be calculated by the stationary phase method. The stationary phase point(s) in the complex variable x plane is found from the equation

$$\left.\frac{\partial\Phi(x, \mathbf{q})}{\partial x}\right|_{x_s} = 1 + \big(\mathbf{q} + \mathbf{n}\lambda f(x_s)\big)^2 \quad \mathrm{i}\Omega\,\frac{f''(x_s)}{f'(x_s)} = 0\,. \tag{9}$$

Then, the transition probability

$$w_{i\mathbf{p}} = 2\pi\Omega\lambda^2 \left|\sum_s \frac{M\big(\mathbf{q} + \mathbf{n}\lambda f(x_s)\big)}{\sqrt{|\Phi''(x_s, \mathbf{q})|}} \exp\left(\quad\frac{\mathrm{i}}{\Omega}\,\Phi(x_s, \mathbf{q})\right)\right|^2, \tag{10}$$

with summation over all saddlepoints x_s. Contributions of different saddlepoints are exponentially different and only the dominating one must be kept in (10). Generally, there is one such dominating saddle point — that corresponding to the lowest value of the positive imaginary part of $\Phi(x_s, \mathbf{q})$. However, in many cases, due to some symmetry of the pulse function $f(x)$, pairs or groups of equivalent saddle points exist with equal values of $\operatorname{Im}\Phi(x_s, \mathbf{q})$ but different phase factors $\operatorname{Re}\Phi(x_s, \mathbf{q})$. Interference of their contribution results in oscillations of the ionization probability as a function of pulse parameters λ and Ω.

Considered as a function of its argument $\mathbf{q}$, this probability is the momentum distribution function of emitted electrons. Note, however, that $\mathbf{q}$ in this formulae is momentum at the time instant when $f(x) = 0$. So, if $f(\infty) \neq 0$, as it is, e.g., in examples 1 and 4 below, the momentum distribution of ejected electrons is distribution (10) but shifted by $\delta\mathbf{q} = \mathbf{n}\lambda f(\infty)$, as is done in formulae (17) and (43) for the above mentioned examples 1 and 4.

Typically, distribution is Gaussian around some average momentum $\mathbf{q}_{\mathrm{m}}$, to be defined from the condition of minimum of $\operatorname{Im}\Phi(x_s, \mathbf{q})$, which, accounting for (9), reduces to

$$q_{\parallel\mathrm{m}} = \quad\frac{\lambda}{x''_{s\mathrm{m}}}\operatorname{Im}\left[\int_0^{x_{s\mathrm{m}}} \mathrm{d}x\, f(x)\right] \tag{11}$$

and $\mathbf{q}_{\perp} = 0$, with $q_{\parallel}$ and $\mathbf{q}_{\perp}$ being the momentum components parallel and perpendicular to the field direction, $x_{s\mathrm{m}} \equiv x_s(\mathbf{q}_{\mathrm{m}})$, and being the x''_s — imaginary part of x_s.

In the vicinity of this sharp maximum, taking into account (9) and (11), the imaginary part of $\Phi(x_s, \mathbf{q})$ can be transformed into

$$\operatorname{Im}\Phi(x_s, \mathbf{q}) = x''_{s\mathrm{m}} + \operatorname{Im}\left\{\int_0^{x_{s\mathrm{m}}} \left[\lambda^2 f^2(x) \quad q^2_{\|\mathrm{m}}\right]\mathrm{d}x + x_{s\mathrm{m}} q_{\perp}^2 + \left[x_{s\mathrm{m}} \quad \mathrm{i}\big(\lambda f'(x_{s\mathrm{m}})\big)^{\;1}\right]\big(q_{\parallel} \quad q_{\|\mathrm{m}}\big)^2\right\}, \tag{12}$$

with halfwidths defined by the second derivatives of the exponent in (10) over components of the momentum.

A comment should be made about the pre-exponential factor in (10). In deriving this formula, the matrix element $M\big(\mathbf{q} + \mathbf{n}\lambda f(x)\big)$ was treated as a regular function, slowly varying in the vicinity of x_s: $M(\mathbf{q}) \simeq M_0 \equiv M(0)$. However, typically $M(\mathbf{q})$ contains a pole — a singularity $M(\mathbf{q}) = M_0/(1 + q^2)$, in the momentum complex plane [9]. In the whole range of nonlinear absorbtion $\lambda \gg \lambda_{\mathrm{c}}$, with λ_{c} defined below by (46), terms in $\Phi(x_s)$ proportional to λ or λ^2 are much larger than the last term $\sim \Omega$. This pole then comes very close to the position of the saddle point. This modifies slightly the evaluation of the integral in (4): instead of the saddle point contribution, we have half of the residue at that point, which enhances the pre-exponent in (10) by the factor $\pi/(4\mathcal{E}|f'(x_s)|)$. If two (or a few) equivalent saddle points (and poles of M) are present in (10), each contribution to the transition amplitude must be multiplied by $\operatorname{sign}\operatorname{Im}[f(x_s)]\,[\pi/(4\mathcal{E}|f'(x_s)|)]^{1/2}$. However, strictly speaking, these corrections to the preexponential factor (also like the one discussed below and due to violation of the standard stationary phase method in the vicinity of the singularity in the pulseshape function itself) must be ignored: the pre-exponential factor in (10) and some following formulae should be considered correct only by the order of magnitude, because of the abovementioned Born-type approximation.

Results for a few particular but quite representative examples are shown below.

1. Solitonlike half-cycle pulse (HCP)

$$f(x) = \tanh x \tag{13}$$

which means for electric field strength

$$\mathcal{E}(t) = \frac{\mathcal{E}}{(\cosh \omega t)^2}. \tag{14}$$

Momentum-resolved ionization probability

$$w_i(\Omega, \lambda, \mathbf{q}) = \frac{\pi\Omega}{\sqrt{\Omega^2+\lambda^2}}(\lambda^2+\zeta^2)\,|M_0|^2 \times \exp\left\{\;\frac{2}{\Omega}\left[(1+\lambda^2)\arctan\frac{\zeta}{\lambda}\;\lambda\zeta + \arctan\frac{\zeta}{\lambda}(\mathbf{q}\;\mathbf{n}\lambda)^2\right]\right\} \tag{15}$$

with parameter

$$\zeta = \frac{1}{\lambda}\left[\sqrt{\Omega^2+\lambda^2}\;\Omega\right]. \tag{16}$$

Accounting for the above-mentioned pole in the transition matrix element, formula (15) should be modified to

$$w_i(\Omega, \lambda, \mathbf{q}) = |\pi M_0|^2 \times \exp\left[\;\frac{1}{\Omega}\left((1+\lambda^2)\pi + 2\arctan\frac{\zeta}{\lambda}(\mathbf{q}\;\mathbf{n}\lambda)^2\right)\right] \times \sinh^2\left[\frac{1}{\Omega}\left((1+\lambda^2)\arctan(\lambda)+\lambda\right)\right]. \tag{17}$$

Moreover, formulae (15) and (17) are derived by the stationary phase method applied to evaluate the integral in (4). However, for $f(x) = \tanh x$ with field decreasing, the saddle point approaches $i\pi/2$—the singularity of $f(x)$ itself (not that of the matrix element). In the linear absorption regime, $\lambda < \Omega \ll 1$, this violates conditions of the stationary phase method applicability, also modifying the numerically pre-exponential factor. In the framework of the general analysis below, the exact (in the Born approximation) formulae will be derived that are also valid for a weak field limit. Coinciding with (15) and (17) in the nonlinear field range, for weak fields they contain correction factors S^{reg} for (15) and S^{sing} for (17), represented in formulae (52) and (54).

2. Solitonlike one-cycle pulse (OCP)

$$f(x) = \frac{3\sqrt{3}}{4\cosh^2 x}. \tag{18}$$

The numerical factor is introduced to normalize $|f'(x_m)|$ to unity at both extrema of the field strength.

Momentum-resolved ionization probability

$$w_i(\Omega, \lambda, \mathbf{q}) = 8|2\pi M_0|^2 \exp\left[\;\frac{\pi}{\Omega}(1+q^2)\right] \times \left[1\;\cos\left(\frac{2}{\Omega}\operatorname{Re}\Phi(x_s,\mathbf{q})\right)\right]\sinh^2\left(\;\frac{1}{\Omega}\operatorname{Im}\tilde{\Phi}(x_s,\mathbf{q})\right), \tag{19}$$

for which

$$\operatorname{Im}\tilde{\Phi}(x_s,\mathbf{q}) \equiv \operatorname{Im}\Phi(x_s,\mathbf{q})\;\frac{\pi}{2}(1+q^2) = (1+q^2)\left(x_s''\;\frac{\pi}{2}\right)\;\frac{1}{6}\left[(5q_\parallel\;2\tilde{\lambda})\,\eta + \sqrt{1+q_\perp^2}\,\xi\right], \tag{20}$$

$$\operatorname{Re}\Phi(x_s,\mathbf{q}) = (1+q^2)\,x_s'\;\frac{1}{6}\left[(5q_\parallel\;2\tilde{\lambda})\,\xi\;\sqrt{1+q_\perp^2}\,\eta\right]. \tag{21}$$

Saddle point x_s defined by

$$x_s'' \equiv \operatorname{Im} x_s = \frac{\pi}{2}\;\frac{1}{2}\arccos\frac{\sqrt{1+q_\perp^2+(\tilde{\lambda}\;q_\parallel)^2}\;\tilde{\lambda}}{\sqrt{1+q^2}}, \tag{22}$$

$$x_s' \equiv \operatorname{Re} x_s = \frac{1}{2}\tanh^{\;1}\left[\frac{\xi}{\tilde{\lambda}+\sqrt{1+q_\perp^2+(\tilde{\lambda}\;q_\parallel)^2}}\right], \tag{23}$$

field parameter $\tilde{\lambda} = (3\sqrt{3}/4)\,\lambda$ and

$$\xi = \sqrt{2\tilde{\lambda}\left[\sqrt{1+q_\perp^2+(\tilde{\lambda}\;q_\parallel)^2}+\tilde{\lambda}\;q_\parallel\right]}, \tag{24}$$

$$\eta = \sqrt{2\tilde{\lambda}\left[\sqrt{1+q_\perp^2+(\tilde{\lambda}\;q_\parallel)^2}\;\tilde{\lambda}+q_\parallel\right]}. \tag{25}$$

Formula (19) is presented in the form corresponding to the singular matrix element $M(\mathbf{q})$ as described above. The function sinh(...) in (19) accounts for contributions of two pairs of poles (saddle points): one pair with $x_s'' < \pi/2$ and another symmetrically above $\pi/2$. The contribution of the latter pair is significant only at the weakest fields $\lambda \ll \Omega^2$. Thanks to this, formula (19) describes correctly (up to numerical factor ~ 1) the linear absorption. In the whole nonlinear range $\lambda \gg \Omega^2$, this contribution is negligible and sinh does not differ from half of the exponential function of the same argument.

Momentum $q_{\parallel m}$, corresponding to the distribution function maximum, should be found from the equation

$$q_{\parallel m}\, x_{sm}''\Big|_{q_\perp=0} = \eta \tag{26}$$

and substituted into (19)–(25). For small $\lambda \ll 1$, it is approximately $q_{\parallel m} \approx (2\tilde{\lambda})^{1/2}/\pi$. For large fields $\lambda \gg 1$, its value approaches $2\tilde{\lambda}/3$.

Oscillations in the field and momentum dependences in (19) arise because of interference contributions due to the pair of saddle points, symmetrical relative to an imaginary axis. In the total — momentum integrated — ionization probability, their amplitude decreases with a field increase as a result of destructive interference of different momenta contributions:

$$W_{\mathrm{i}}(\Omega, \lambda) = \sqrt{\frac{2\pi\Omega}{u}} \frac{\Omega}{x''_{sm}} |M_0|^2 \exp\left[\quad \frac{\pi}{\Omega}(1 + q_{\mathrm{m}}^2)\right]$$
$$\times \left[1 \quad \exp\left(\quad \frac{2\tilde{\lambda}}{\pi\Omega}\right) \cos\left(\frac{4\sqrt{2\tilde{\lambda}}}{3\Omega}\right)\right]$$
$$\times \sinh^2\left(\quad \frac{1}{\Omega} \operatorname{Im} \tilde{\Phi}(x_{sm}, \mathbf{q}_{\mathrm{m}})\right), \qquad (27)$$

with

$$u = \left(x''_s + \frac{1}{4} \frac{q\xi + \eta}{(1 + q^2)\sqrt{1 + (\tilde{\lambda} \quad q)^2}}\right)_{\mathbf{q}=\mathbf{q}_{\mathrm{m}}}.$$

The oscillating term is written here in a form valid only for $\lambda \ll 1$, as for larger fields this term becomes negligible.

3. Gaussian one-cycle pulse (OCP)

$$f(x) = \exp\left(\frac{1 \quad x^2}{2}\right). \qquad (28)$$

Corresponding field pulse shape

$$\mathcal{E}(t) = \quad \mathcal{E}\omega t \exp\left(\frac{1 \quad (\omega t)^2}{2}\right). \qquad (29)$$

Then,

$$\Phi(x_s, \mathbf{q}) = (1 + q^2)\, x_s + 2\tilde{\lambda} q_{\|} \operatorname{Erf}\left(\frac{x_s}{\sqrt{2}}\right) + \tilde{\lambda}^2 \operatorname{Erf}(x_s). \qquad (30)$$

Here, $\tilde{\lambda} \equiv \sqrt{e}\,\lambda$, $\operatorname{Erf}(x)$ is the error integral

$$\operatorname{Erf}(x) = \int_0^x \exp(\quad y^2)\, dy,$$

$$x_s(\mathbf{q}) = \sqrt{\ln \frac{\tilde{\lambda}^2}{1 + q^2} \mp 2\mathrm{i} \arccos \frac{q_{\|}}{1 + q^2}}. \qquad (31)$$

Only saddlepoints in the upper halfplane of x are relevant. Thus, the signs of the roots must be chosen with positive imaginary parts. Therefore, the signs of the real part are different for two saddlepoints. This is just an example of two equivalent saddlepoint interferences.

$$x''_s \equiv \operatorname{Im} x_s = \frac{1}{\sqrt{2}}$$
$$\times \sqrt{\sqrt{\left(\ln \frac{\tilde{\lambda}^2}{1+q^2}\right)^2 + 4\left[\arccos\left(\quad \frac{q_{\|}}{\sqrt{1+q^2}}\right)\right]^2} \quad \ln \frac{\tilde{\lambda}^2}{1+q^2}}, \qquad (32)$$

$$x'_s \equiv \operatorname{Re} x_s = \mp \frac{1}{\sqrt{2}}$$
$$\times \sqrt{\sqrt{\left(\ln \frac{\tilde{\lambda}^2}{1 + q^2}\right)^2 + 4\left[\arccos\left(\quad \frac{q_{\|}}{\sqrt{1 + q^2}}\right)\right]^2} + \ln \frac{\tilde{\lambda}^2}{1 + q^2}}. \qquad (33)$$

The equation for $\mathbf{q}_{\mathrm{m}}$ in this case looks like

$$\mathbf{q}_{\mathrm{m}} \int_0^{x''_{sm}} \left[1 \quad \exp\left(\quad x''_{sm} u + \frac{u^2}{2}\right) \cos(x'_{sm} u)\right] du$$
$$= \quad \int_0^{x''_{sm}} \exp\left(\quad x''_{sm} u + \frac{u^2}{2}\right) \sin(x'_{sm} u)\, du. \qquad (34)$$

As x_{sm} itself is a function of $\mathbf{q}_{\mathrm{m}}$, this equation, along with (31), a system of two coupled equations defining both x_{sm} and $\mathbf{q}_{\mathrm{m}}$.

The imaginary parts of the Erf functions in (30) can also be represented by integrals similar to those in (34). Taking into account (31),

$$\tilde{\lambda}^2 \operatorname{Im} \operatorname{Erf}(x_s) = \quad \int_0^{x''_s} \exp(\quad 2x''_{sm} u + u^2)$$
$$\times \left[(1 \quad q_{\|\mathrm{m}}^2) \cos(2x'_{sm} u) + 2q_{\|\mathrm{m}} \sin(2|x'_{sm}|u)\right] du. \qquad (35)$$

Unlike the general form of (31)–(33), equations (34) and (35) are written for $\mathbf{q} = \mathbf{q}_{\mathrm{m}}$.

All these formulae become substantially simplified and more transparent in the 'multiphoton' ($\lambda \ll 1$) and 'tunneling' ($\lambda \gg 1$) parameter ranges. For moderate intensities ($\lambda \ll 1$), momentum-resolved ionization probability

$$w_{i\mathbf{q}} \approx 4\pi\Omega \frac{\lambda^2}{\lambda^2 + \lambda_{\mathrm{c}}^2} x''_{sm} \; M(0) \;^2 \left[1 + \cos \frac{\pi \quad 4q_{\|}}{x''_{sm}\Omega}\right]$$
$$\times \exp\left[\quad \frac{2}{\Omega}\left((1 + q^2)\, x''_{sm} \quad \frac{1}{2x''_{sm}}\right)\right], \qquad (36)$$

with x''_{sm} given by

$$x''_{sm} = \sqrt{\ln \frac{1}{\tilde{\lambda}^2 + \lambda_{\mathrm{c}}^2}} \gg 1, \qquad (37)$$

$$q_{\|\mathrm{m}} \approx \quad \frac{\pi}{2x''^2_{sm}(x''^2_{sm} \quad 1)} \ll 1. \qquad (38)$$

Strictly speaking, formula (36) is correct for low fields ($\lambda \ll \lambda_{\mathrm{c}} \equiv \exp[\quad 1/(2\Omega^2)]$, linear absorption) and moderate fields ($\lambda_{\mathrm{c}} \ll \lambda \ll 1$). In the intermediate range ($\lambda \sim \lambda_{\mathrm{c}}$), it seems to be a reasonable interpolation. Oscillations of transition probability to any particular momentum due to the interference of two saddlepoint contributions are very strong — up to complete cancellation. However in the total (momentum integrated) ionization probability, they are gradually damped with a field increase because of the momentum dependence of their phases,

$$W_{\mathrm{i}} = 8 \frac{\lambda^2 x''^2_{sm}}{\lambda^2 + \lambda_{\mathrm{c}}^2} \left(\frac{\pi\Omega}{2x''_{sm}}\right)^{5/2} |M(0)|^2$$
$$\times \left[1 + \exp\left(\quad \frac{2}{\Omega x''^3_{sm}}\right) \cos \frac{\pi}{x''_{sm}\Omega}\right]$$
$$\times \exp\left[\quad \frac{2}{\Omega}\left(x''_{sm} \quad \frac{1}{2x''_{sm}}\right)\right]. \qquad (39)$$

As to the strong field tunneling regime ($\lambda \gg 1$), formulae (48)–(50) are universal for any pulse shape, the only difference being in the particular value of the parameter a — the curvature at the pulse top. For a Gaussian pulse, $a = 2$.

4. Lorenzian half-cycle pulse (HCP)

$$f(x) = \arctan x \,. \tag{40}$$

Corresponding field pulse shape

$$\mathcal{E}(t) = \frac{\mathcal{E}}{1 + (\omega t)^2} \,. \tag{41}$$

The saddle point is then

$$x_s = \mathrm{i} \tanh\left(\frac{\sqrt{1+q_\perp^2} + \mathrm{i} q_\parallel}{\lambda}\right), \tag{42}$$

and momentum distribution of ionization probability

$$w_\mathrm{i}(\Omega, \lambda, \mathbf{q}) = |\pi M_0|^2 \exp\left[\ \ \frac{2}{\Omega}\left(|x_{sm}| \ \ \lambda^2 \varphi(|x_{sm}|)\right)\right] \times \exp\left\{\ \ \frac{2}{\Omega}\left[q_\perp^2 |x_{sm}| + \left(q_\parallel \ \ \frac{\pi\lambda}{2}\right)^2 \times \left(|x_{sm}| + \frac{2}{\lambda}(1 \ \ |x_{sm}|^2)\right)\right]\right\}, \tag{43}$$

with

$$|x_{sm}| = \tanh \frac{1}{\lambda} \,, \tag{44}$$

$$\varphi(x) = \frac{1}{4}\int_0^x \ln^2 \frac{1+y}{1 \ \ y} \mathrm{d}y \,. \tag{45}$$

The numerical solution of (9) and (11) is strightforward for any reasonable pulseshape. However, a general qualitative analysis is also possible and may be illuminating. There are three essentially different areas in the plane of parameters (Ω, λ):

1. Weak fields and linear absorbtion for $\lambda \ll \lambda_\mathrm{c}(\Omega)$ with λ_c being the effective nonlinearity threshold, substantially dependent on the pulseshape and specified below for some typical pulse shapes. The general definition is

$$\lambda_\mathrm{c} \ f(x_{s0}) \ = 1 \,, \tag{46}$$

with x_{s0} being the root of equation (9) corresponding to $\lambda = 0$. Terms proportional to λ and λ^2 on the right-hand side of (9) can be disregarded. The exponential factor in (10) reduces to an exponentially small amplitude of high frequency harmonics, corresponding to the above-threshold quantum energy $\hbar\omega > I$, always present in the Fourier spectrum of a broadband signal.

2. Nonlinear regime: $\lambda > \lambda_\mathrm{c}$. The last term in (9) can be omitted. Then,

$$x_s = f^{\ \ 1}\left(\frac{q_\parallel \pm \mathrm{i}\sqrt{1+q_\perp^2}}{\lambda}\right). \tag{47}$$

Here, $f^{\ \ 1}(y)$ is a function, the inverse of $f(x)$. The sign of the imaginary part of its argument must be fixed so as to correspond to $x'' > 0$.

2a. High fields — $\Omega^{\ \ 1} \gg \lambda \gg 1$.

Without any loss of generality, one can always choose the point $x = 0$ to be the absolute maximum of $f'(x)$, i.e., field strength, and $f'(0) = 1$. This last condition just fixes the exact value of λ. If there are a few equivalent maxima, each of them can be treated separately. In the range of interest around this point, $f(x)$ can be approximated by a cubic parabola:

$$f(x) \approx f_0 + x \ \ \frac{1}{6} a x^3 \,, \quad 0 < a \sim 1 \,. \tag{48}$$

Then, after simple calculations,

$$\mathrm{Im}\, \Phi(x_s, \mathbf{q}) = \frac{2}{3\lambda}\left[1 + 4a \frac{(\mathbf{q} \ \ \mathbf{q}_\mathrm{m})^2}{\lambda^2}\right], \tag{49}$$

$$q_{\parallel\mathrm{m}} \approx \lambda\left[f(\infty) \ \ f_0\right]. \tag{50}$$

This corresponds in [9] to quasistatic tunneling during a short $\delta x \sim \sqrt{|\mathcal{E}|}$ time interval around the field maximum. The momentum distribution of photoelectrons is Gaussian with halfwidth $\Delta q_\parallel = (\lambda/4)\sqrt{3|\mathcal{E}|/a}$.

2b. $1 \gg \lambda \gg \lambda_\mathrm{c}(\Omega)$ **— moderate fields.** Compared to the weak and strong field cases, in this one the λ-dependence of Q^(xs, q) is more diverse, depending on the details of the pulse shape, particularly singularities of the function $f(x)$ in the upper half-plane of the complex variable x. The Gaussian shape $f(x) \sim \exp(\ \ x^2/2)$ is the particular case with the only singularity of $f(x)$ being the essential one at infinity. However the most typically pulselike function $f(x)$ has singularities (poles, branching points) in the complex plane of variable x at some x_pol with the imaginary part $x''_\mathrm{pol} \sim 1$. Then, for weak fields, just the $\exp(\ \ 2x''_\mathrm{pol}/\Omega)$ defines the amplitude of the high frequency Fourier component responsible for single quantum ionization. With the field increasing, the saddle point x_s moves from x_pol to the real axis. Let the singularity closest to the real axis be the k-th order pole, i.e.,

$$f(x) \approx \frac{A}{(x \ \ x_\mathrm{pol})^k}$$

for $|x \ \ x_\mathrm{pol}| \ll 1$. Then, as will be shown below, the ionization amplitude in the whole domain $\lambda \ll 1$, including both weak and moderate field ranges, beside the weak field factor $\exp(\ \ 2x''_\mathrm{pol}/\Omega)$, is dependent only on a single parameter,

$$z = \frac{(\lambda A)^{1/k}}{\Omega} \,, \tag{51}$$

and the moderate field range starts at $|z| \sim 1$, i.e., $\lambda_\mathrm{c} \sim \Omega^k$. Note that the first two of the above-described examples are dominated by such singularities, the first one corresponding to $k = 1$ and the second to $k = 2$. The saddle points (and possible poles of matrix element $M(\mathbf{q} + \mathbf{n}\lambda f(x))$ coincident with them in the pre-exponential factor)

$$x_s(\mathbf{q}) = x_\mathrm{pol} + \left[\frac{A\lambda}{1+q^2}\left(\pm \mathrm{i}\sqrt{1+q_\perp^2} \ \ q_\parallel\right)\right]^{1/k}, \tag{52}$$

with both signs in the argument being relevant, as all of these $2k$ points are in the close vicinity of x_pol, which itself is in the upper halfplane. However, in the moderate field strength range $|z| > 1$, only one of them dominates — that with the minimal value of x''_s; or, one pair of such points, if, depending on pole order k and $\chi \equiv \arg A$, there are in the whole set (52) such a pair of mirror symmetric relative to the imaginary axis elements, with the minimal value of the imaginary part. The second example above with $k = 2$ and $\chi = 0$ corresponds to just such a case $(x_s \;\; x_{\rm pol})_{q=0} = \sqrt{\lambda/2}\,(\pm 1 \;\; {\rm i})$. In general,

$$\operatorname{Im}\Phi(x_s, \mathbf{q}) = x''_{\rm pol} \;\; \frac{2k}{2k \;\; 1}\gamma\,(\lambda|A|)^{1/k} + \frac{q_\perp^2}{(\Delta q_\perp)^2} + \frac{(q_\parallel \;\; q_{\parallel \rm m})^2}{(\Delta q_\parallel)^2}\,, \tag{53}$$

with

$$\Delta q_\parallel^2 \approx \Delta q_\perp^2 \approx x''_{\rm pol} + o(\lambda^{1/k})\,,$$
$$q_{\parallel \rm m} = \frac{\sqrt{1 \;\; \gamma^2}}{(2k \;\; 1)\,x''_{\rm pol}}\,(\lambda|A|)^{1/k}\,,$$
$$\gamma = \max_s \frac{(x_{\rm pol} \;\; x_s)''}{|x_s \;\; x_{\rm pol}|} \tag{54}$$

with index $s = 1, 2, \ldots, 2k$ marking different elements of set (52). Thus, in the moderate field strength range, ionization probability increases as $\exp\,[4k\gamma|z|/(2k \;\; 1)]$ and the average momentum as $\lambda^{1/k}$. If there is only one dominating saddle point,

$$W_i(\Omega, \lambda) = \sqrt{\pi}\left(\frac{\Omega}{2x''_{\rm pol}}\right)^{3/2}|M_0|^2 \times \exp\left[\;\;\frac{2}{\Omega}x''_{\rm pol} + \frac{4k}{2k \;\; 1}\gamma|z|\right], \tag{55}$$

and

$$W_i(\Omega, \lambda) = \sqrt{\frac{\pi}{2}}\left(\frac{\Omega}{x''_{\rm pol}}\right)^{3/2}|M_0|^2 \times \exp\left(\;\;\frac{2}{\Omega}x''_{\rm pol} + \frac{4k}{2k \;\; 1}\gamma|z|\right) \times \left\{1 \;\; \exp\left[\;\;\frac{2}{x''_{\rm pol}\Omega}\left(\frac{\gamma|z|}{2k \;\; 1}\right)^2\right] \times \cos\left(\frac{4k}{2k \;\; 1}\sqrt{1 \;\; \gamma^2}\,|z|\right)\right\}, \tag{56}$$

if a pair of symmetric saddle points contribute. It should be noted that, in all arguments of exponential and trigonometric functions, only leading terms in $\lambda \ll 1$ are shown in these formulae.

Derived by a stationary phase asymptotic evaluation of integral in (4), formulae (55) and (56) are valid for the moderate field range $\Omega^k \ll \lambda \ll 1$. Their inapplicability for weak fields is clearly seen from the fact that they do not follow the usual $\sim \mathcal{E}^2$ dependence as $\lambda \to 0$. The reason for that was already mentioned above in discussing the second example: at $z < 1$, contributions of all $2k$ saddle points, surrounding $x_{\rm pol}$, become of the same order. It is easy to account for all of them, which would restore the correct $\sim \mathcal{E}^2$ behavior in weak fields. Still it is not the whole story. The numerical coefficient appears to be wrong. The reason is that, besides all these poles and saddle points, the point $x_{\rm pol}$ itself is the essential singularity of the integrand in (4): $\exp\,[{\rm i}(\lambda A)^2 (x \;\; x_{\rm pol})^{\;\;2k+1}/(2k \;\; 1)]$. An evaluation of the integral in (4) accounting for this whole structure in the complex plane, valid in the whole domain $|z| \ll 1$, i.e., weak and moderate field ranges, is possible in terms of a fast converging power series in z. The result again is slightly different depending on the presence or absence of the pole in the matrix element. If the matrix element is regular (no pole) and slowly varying, $M(\mathbf{q}) \approx M_0$,

$$w_{\rm i}(\Omega, \lambda, \mathbf{q}) = \lambda^2 |M_0|^2 \exp\left[\;\;\frac{2x''_{\rm pol}}{\Omega}(1 + q^2)\right] \times \left|S_k^{\rm reg}\left(\frac{z^k}{\sqrt{2k \;\; 1}}\right)\right|^2, \tag{57}$$

with

$$S_k^{\rm reg}(y) = 2\pi\sqrt{2k \;\; 1}\,y \sum_{n=0}^{\infty} \frac{(\;\;1)^{(k+1)n} y^{2n}}{n!\,[(2k \;\; 1) + k]!}\,, \tag{58}$$

and for the case of a singular matrix element,

$$w_{\rm i}(\Omega, \lambda, \mathbf{q}) = \lambda^2 |M_0|^2 \exp\left[\;\;\frac{2x''_{\rm pol}}{\Omega}(1 + q^2)\right] \left|S_k^{\rm sing}(z^k)\right|^2, \tag{59}$$

with

$$S_k^{\rm sing}(y) = 2\pi k\, y \sum_{n=0}^{\infty} (\;\;1)^{(k+1)n} a_n y^{2n}\,, \tag{60}$$

and coefficients a_n defined as

$$a_n = \sum_{m=0}^{n} \frac{(2k \;\; 1)^{\;\;m}}{m!\,[(2n + 1)k \;\; m)]!}\,. \tag{61}$$

The asymptotic form of functions $S_k^{\rm reg}$ and $S_k^{\rm sing}$ at $|z| \gg 1$ exactly coincide with the results of stationary phase calculations in the moderate field regime,

$$S_k^{\rm reg}\left(\frac{z^k}{\sqrt{2k \;\; 1}}\right) \approx \sqrt{\frac{\pi k}{|z|}}\exp\left(\frac{2k}{2k \;\; 1}\gamma|z|\right), \tag{62}$$

$$S_k^{\rm sing}(z^k) \approx \pi \exp\left(\frac{2k}{2k \;\; 1}\gamma|z|\right), \tag{63}$$

and their first terms substituted into formula (51) and (53) give the exact result for the weak field regime. Thus, for a pulse shape with a pole type of singularity, formulae (9)–(12) and (57)–(61) together describe completely the ionization probability for any field strength, restricted only from above by the atomic field, i.e., $\mathcal{E} \ll 1$. However, the weak field regime seems to be of more academic interest: for such short pulses, the effect is hardly experimentally observable.

The last of the above examples corresponds to another type of pulse shape function singularity — the logarithmic branching point ('zeroth order pole').

For long pulses and approximately monochromatic fields, the frequency dependence of true multiphoton process probability is very steep. As the whole consideration above shows, for very short pulses — HCP, OCP, and probably a few ($< 1/\Omega$) cycles-long pulses — it is much slower, though still pretty steep. Qualitatively, this slowing down can be explained as an increase, with the field increasing, of an average effective number n of photons absorbed per single ionization event. Because of a broad frequency spectrum of the pulse, the process is a single-photon one in a weak field and its multiplicity increases gradually to $n \sim \lambda^3$ [9] in the tunneling regime $\lambda \gg 1$, while in a monochromatic field it is restricted from below, $n > I/(h\omega)$.

This work was started during my visit to the Miller Institute for Basic Research of the University of California, Berkeley. I am grateful to the Miller Institute for this opportunity and especially to Professor Ron Shen for the hospitality and many valuable discussions. I am also thankful to Professor J Moloney for discussions that stimulated the beginning of this study.

References

1. McClung F J, Hellwarth R W *J. Appl. Phys.* **33** 828 (1962)
2. New G H C *Rep. Prog. Phys.* **46** 877 (1983)
3. Shank C V, in *Ultrashort Laser Pulses and Applications* (Topics in Applied Physics, Vol. 60, Ed. W Kaiser) (Berlin: Springer-Verlag, 1988) p. 5
4. Squier J et al. *Opt. Lett.* **16** 324 (1991)
5. Zhou J et al. *Opt. Lett.* **20** 64 (1995)
6. Barty C P J et al. *Opt. Lett.* **21** 668 (1996)
7. Nisoli M et al. *Opt. Lett.* **22** 522 (1997)
8. Sartania S et al. *Opt. Lett.* **22** 1562 (1997)
9. Keldysh L V *Sov. Phys. JETP* **20** 1307 (1965); *Zh. Eksp. Teor. Phys.* **47** 1945 (1964)
10. Perry M D et al. *Phys. Rev. A* **37** 747 (1988)

"Physicist history of physics", which is never correct. ...

is a sort of conventionalized myth – story that the physicists tell to their students, and those tell to their students, and is not necessarily related to the actual historical development.

R. P. Feynman, "QED: The strange theory of light and matter". Princeton University Press, Princeton, New Jersey 1986

Real-time Nonequilibrium Green's Functions

L. V. Keldysh

P. N. Lebedev Physical Institute, Moscow

Presented at the Interdisciplinary Workshop

"Progress in Nonequilibrium Green's Functions"

19 August 2002, Dresden, Germany

Abstract

A brief review of early Russian works on the Green's functions applications to the many body theory, particularly for nonequilibrium states and processes, is presented. Discussed are some general features and relations of the real-time Nonequilibrium Green's function (NGF) matrices method to some other approaches.

Application of field theoretical concepts and methods, including Green's functions, to the many body problems in the Soviet physics in the 50s and 60s of the last century was in the mainstream of the process in world science. Because of the iron curtain and the lack of our personal contacts with western colleges, it was impossible for us to participate in international meetings and conferences, many of the most important scientific journals became available in our institutes and university libraries only with a several months delay – but this only in Moscow, St. Petersburg (then Leningrad) and a few other major cities. And our main journals with a similar delay became translated into English and distributed through the world by the American Physical Society.

There were several active theoretical groups in the Soviet Union in that period – usually called "schools" – and labeled by the name of their leader: L. D. Landau, I. E. Tamm, N. N. Bogolyubov. The most active in the quantum many body theory was the Landau school. I myself belonged to the theoretical group of the P. N. Lebedev Institutes, and so to the Tamm school. However from the very beginning of my scientific activity I participated regularely (like my superviser V. L. Ginzburg) also in the Landau seminar. Several of my friends were from the closest L. D. Landau circle. So I shared many of their views and preferences. In particular, the fascination by the Feynman-Dyson diagram technique (we used the word diagrams instead of graphs). Certainly, we knew about its equivalence to the alternative approach due to J. Schwinger. Landau himself was perfect, even virtuoso in using the whole arsenal of mathematical physics methods, however, primarily as a tool for analyzing and solving real physical problems. I suppose he appreciated the diagram technique – he also used this term – as a highly general, regular and logical way, perfectly adapted, besides all that, for describing real physical phenomena and processes. In the whole collection of volumes of the famous Landau and Lifshitz course of Theoretical Physics just that approach is used to present both – quantum field theory and quantum many body theory, as in the books [1] and [2] as well. For me, as probably for many others, the diagram technique is more than just a method for doing calculations. Because of its symbolical but very spectacular presentation in terms of graphs it is more like the way of thinking about physical processes and theoretical approximations.

This rather personal remark is meant to say that the following considerations should not be regarded as a comprehensive review of the subject. Inevitably I shall speak mainly about the activity related to the diagram technique in the many body theory, which I knew better. Some alternative approaches, including the one based on Bogolyubov's idea of decoupling the infinite set of equations for GF's of successively increasing particle number by approximating higher order GFs in terms of those of lower order, were reviewed in [3-7]. Also, as our Workshop is about the NGF's, only a very short list of some previous works on the many body ground state and the thermodynamic equilibrium state (Matsubara) GF's is presented below (and only Soviet – the Western, I suppose, are known much better). I mainly intend to give an impression of the starting level at the beginning of 1960's.

Probably the first paper on GF's in many body systems in Russian journals was that of Bonch-Bruevich [7], discussing the general concept of GF in the ground state, including the relation of the zeros of $G^{-1}(\mathbf{p},\varepsilon)$ to the particle energy levels. The systematic studies of the interacting electron-phonon system in metals at $T=0$ in terms of GF's was started by Migdal and Galitskii [8, 9] resulting, in particular, in the Migdal theorem about the absense of the coupling constant renormalization by the *e-ph* interaction. Galitskii [10] calculated also the energy spectrum and ground state energy of a low density degenerate Fermi gas with short range interaction. Essential was the work of Belyaev [11] who has modified the standard diagram technique in order to describe degenerate systems of bosons. Here for the first time "anomalous" GF's were introduced, accounting for the macroscopic coherency in the system. Similar pair coherency functions for Fermi systems, introduced by Gor'kov [12], became crucial for describing superconductivity within the diagram technique. Using this Gor'kov's technique Abrikosov, Gor'kov and Khalatnikov were able to develop a theory for the majority of superconductivity related phenomena, including the electromagnetic response [1,

13]. Important for the whole quantum many body theory is the result of Landau himself about the analytical properties of GF's in the complex frequency plane, including spectral representations, valid for any system in thermodynamic equilibrium [14]. He has applied also the diagram technique to justify his Fermi liquid theory and clarify the microscopic nature of its basic notions, including the quasiparticle scattering amplitude [15].

In the following year (1959) the breakthrough occured in the theory of Matsubara's (temperature) GF's. Abrikosov, Gor'kov and Dzyaloshinskii [16] and Fradkin [17] derived the periodic boundary conditions and were able to introduce the Fourier representation in terms of discrete frequencies, which has made the diagram technique on the imaginary time (temperature) axis an effective tool for calculating any thermodynamic equilibrium parameter of many body systems. These authors have also pointed out the possibility of finding the real time GF's under thermodynamic equilibrium conditions by means of analytical continuation from the discrete set of Matsubara frequencies to the real frequency axis. In all the above mentioned papers mainly the Feynman causal function $G_c(x,x')$ was discussed, as the standard diagram technique exists just for this function. In the same year Bogolyubov and Tyablikov [18] discussed the possibility of describing the many body system in terms of retarded $G^{(r)}(x,x')$ and advanced $G^{(a)}(x,x')$ func-tions, using the infinite hierarchy of coupled equations for the GF's with increasing number of particles.

In the beginning of 1960's the interest in the field shifted to kinetic (transport) phenomena. Initially the focus was on linear response. In classical many body theory kinetics is described usually in the framework of Boltzmann's equation for the particle distribution function in the phase space. In quantum theory the description is based on evolution of the density matrix according to the Schrödinger equation. The closest quantum analog to the distribution function is the density matrix in the so called Wigner representation. Under quasiclassical conditions the Wigner density matrix can be reduced to the classical distribution function and, corresponding quantum equation, to the usual Boltzmann equation. This is the standard way of introducing quantum corrections to the Boltzmann equation (see e.g. [19]). Real time GF's are a kind of generalization of the density matrix, reducing to the latter at coinciding time arguments. So, already in [9] the relation of the two-particle GF and the Bethe-Salpeter equation to the linearized (weak field) Boltzmann equation was indicated. The next important step in that direction was done by Konstantinov and Perel' [20]. Expressing the linear correction to the single-particle density matrix in terms of the retarded density-density correlator and introducing for its calculation some special version of the diagram technique on the complex time contour they were able to derive the generalized Boltzmann equation, capable in principle to account for quantum corrections up to any desired order in the interaction. Another approach to the same linear response problem was based on the suggested in [16, 17] analytic continuation of Matsubara's GF's onto the real frequency axis. Formally the response function is the retarded self-energy of the field propagator in the medium. Eliashberg [22, 23] under quasiclassical conditions – external field, slowly varying in space and time – analyzed the analytical structure of the two particle scattering vertex in the complex frequency plane and so was able to continue the Bethe-Salpeter type equation to real frequencies, resulting in the generalized quantum Boltzmann equation for the Fermi liquid. Later Gor'kov and Eliashberg [24, 25] extended that method to the nonlinear regime for the particular case of superconductors in strong space-time dependent fields. Dzyaloshinskii [21] has found a regular way of continuing each Matsubara-Feynman graph to the real frequency in terms of spectral functions. Graphs become different depending on the time sequence ordering of different interaction vertices, resulting in an exponential increase of the number of graphs in higher orders of the perturbation theory. Generally the linear response problem was solved by these works.

My interest in the NGFs problem started in the very beginning of 1964 and was motivated purely aesthetically. The Feynman graph technique seemed to me so natural and logical that it was hard to believe that its applicability is restricted to an important but very special class of states – ground state or thermodynamic equilibrium. So the program was most simple and straightforward: to follow step by step the original Feynman-Dyson derivation, checking at what step it becomes invalid for an arbitrary state and then trying to overcome the arising difficulties, staying as close as possible to the original formulation. The first evident difference is that an arbitrary state of a many body system is described in terms of the density matrix $\hat{\rho}$ while it was a pure ground state (vacuum) $|0\rangle$ in the original procedure. However the time evolution for both is defined by the Schrödinger equation, which for the density matrix is

$$i\frac{\partial\hat{\rho}}{\partial t}=\left[\hat{H},\hat{\rho}\right] \qquad \hat{H}=\hat{H}_0(t)+\hat{H}_{\text{int}}.$$

Unlike the original case, $\hat{H}_0$ may be time dependent because of the presence of external fields. That, however, did not create any essential difficulties. Following the usual procedure of adiabatic switching of the interaction (not external fields, which are switched on realistically, not adiabatically!), in the interaction representation one gets

$$\hat{\rho}(t)=\hat{S}(t,-\infty)\hat{\rho}(-\infty)\hat{S}^{+}(t,-\infty) \quad \textit{instead of usual} \quad \Psi(t)=\hat{S}(t,-\infty)\Psi_0$$

with the standard definition of the S-matrix

$$\hat{S}(t,t')=\hat{T}\exp\left\{-i\int_{t'}^{t}\hat{H}_{\text{int}}(t_1)dt_1\right\} \textit{ and } \hat{S}^{+}(t,t')=\hat{T}\exp\left\{-i\int_{t}^{t'}\hat{H}_{\text{int}}(t_1)dt_1\right\}.$$

Here, $\hat{T}$ is the usual time ordering operator along the integration path in the direction of integration, i. e. from lower to the upper limit. The average value of any physical quantity operator $\hat{L}_0(t)$ (in the interaction representation) is then

$$\bar{L}(t)=Tr\left\{\hat{L}_0(t)\hat{\rho}(t)\right\}=Tr\left\{\hat{S}^{+}(t,-\infty)\hat{L}_0(t)\hat{S}(t,-\infty)\hat{\rho}(-\infty)\right\} \tag{1}$$

(cyclic permutation under the Tr symbol is used). This formula means the transition to the Heisenberg representation of time-dependent operators averaged over a time-independent density matrix of non-interacting (because of adiabatic switching of the interaction) fields. In the absence of interaction and external fields the density matrix was taken to be $\hat{\rho}(-\infty)=\hat{\rho}_{0T}$ - corresponding to the thermodynamic equilibrium for free particles.

At the next step the original procedure breaks down. This step for the vacuum (ground) state usually is substituting $\hat{S}(\infty,t)$ instead of $\hat{S}^{+}(t,-\infty)$, justified by the so called "vacuum stability condition", which reads as: under adiabatic transformation the non-degenerate ground state can transform only into itself, possibly multiplied by an (unessential) phase factor. Then

$$\hat{S}^{+}(t,-\infty)=\hat{S}(-\infty,t)=\hat{S}(-\infty,t)\hat{S}(t,\infty)\hat{S}(\infty,t)=\hat{S}(-\infty,\infty)\hat{S}(\infty,t)$$

Acting on the vacuum, the first factor on the r.h.s. of this relation is exactly the inverse of that phase factor – a c-number $\langle S\rangle_0^{-1}=\langle 0|\hat{S}(\infty,-\infty)|0\rangle^{-1}$. So the well-known formula appears

$$\overline{L}(t)=\langle S\rangle_0^{-1}\cdot\langle 0|\hat{T}\left(\hat{S}(\infty,-\infty)\cdot\hat{L}_0(t)\right)|0\rangle .$$

However, for an arbitrary non-equilibrium state created by an external (possibly time-dependent) field, the stability condition cannot be generally valid. So one is forced to proceed with the unchanged formula (1). But then the contour ordering comes automatically. Accounting for the opposite time ordering in $\hat{S}$ and $\hat{S}^{+}$ the formula (1) can be written as

$$\overline{L}(t)=Tr\left\{\hat{T}_C\left(\hat{S}_C\cdot L_0(t)\right)\cdot\rho_{0T}\right\}, \tag{2}$$

where, C is the contour propagating from $-\infty$ to time t and then back to $-\infty$, $\hat{T}_C$ is ordering operator and $\hat{S}_C$ - the S-matrix along this contour. If not under the time ordering symbol together with some other operator $\hat{L}_0(t)$, then $\hat{S}_C\equiv\hat{1}$, which means identical absence of all vacuum loops in this technique[1]. In order to extend all the integrals over the whole time axis, one can insert into (2) one more

[1] Strictly speaking the contour is optional in this derivation. As Pitaevskii has shown, presenting this technique in the volume "Physical Kinetics" of the Landau and Lifshitz course on Theoretical Physics, one can get all the same results, considering a perturbation series directly in formula (1) and taking into account opposite ordering in S and S^{+}. Then these two factors play the role of two contour branches. The contour, however, seems more spectacular.

factor - the identically equal to unity operator - $\hat{S}(t,\infty)\hat{S}(\infty,t)$. It does not change anything. However, the contour C propagates now from $-\infty$ to $+\infty$ and back to $-\infty$, which is much more convenient. It should be noted that both branches of the contour propagate along the real time axis. Any references to the complex time plane were eliminated from this consideration because of usually non-analytical time dependence of external fields (switching). Formulas similar to (2) hold for averaged products of any number of operators, including field operators related, in general, to different times. These are real time NGF's. Depending on the positions of the times on different branches of the contour they correspond to different time ordering of operators, as times on the reverse branch are oppositely ordered and are always "later" then any time on the direct branch; e. g. for the single particle GF four different functions exist:

$$
\begin{aligned}
&G(t_+,t'_+) = -i\left\langle \hat{T}\left(\psi_0(t),\psi_0{}^+(t')\right)\right\rangle_{0T} \qquad G(t_+,t'_-) = -i\left\langle \psi_0{}^+(t')\psi_0(t)\right\rangle_{0T} \\
&G(t_-,t'_+) = -i\left\langle \psi_0(t)\psi_0{}^+(t')\right\rangle_{0T} \qquad G(t_-,t'_-) = -i\left\langle \hat{\tilde{T}}\left(\psi_0(t),\psi_0{}^+(t')\right)\right\rangle_{0T}
\end{aligned} \tag{3}
$$

Here $\hat{\tilde{T}}$ is the reverse time ordering operator, and $\langle\ldots\rangle_{0T}$ denotes averaging over $\hat{\rho}_{0T}$. Together these four functions compose into the single contour ordered GF $G_C(x,x')$.

Now, evidently, the usual diagram technique follows from perturbative representation of (2) and all Feynman rules are valid with only one exception: GF's are defined along the whole contour and all time integrals become extended along the whole contour. We can consider $\pm$ indices of the contour branches as matrix indices. Then the 4 GF's (3) are components of a 2×2 matrix and multiplication of those matrices accounts for summing up contributions of different contour branches. All time integrals extend only along one real time axis from $-\infty$ to $+\infty$. However, those corresponding to the matrix index "– ", should be taken with a "–" sign to account for the reverse direction of integration. Now the difference to the usual Feynman diagram technique is that to each line of the graph corresponds the GF's matrix (3) and to connect 3 or 4 such matrices in each interaction point, the elementary vertex $\gamma^{k}{}_{ij}$ or $\gamma_{ij}{}^{kl}$ is introduced being equal to +1, if all indices are +, -1, if all indices are –, and 0 otherwise. Similar contour ordered GF matrices were used also by J. Schwinger in his earlier paper [26] about the Brownian motion of the harmonic quantum oscillator driven by two external forces – one regular (arbitrary function of time) and another, stochastic, defined in terms of its correlators ("thermostat").

Starting from that point the contour may be forgotten: the whole information, which it carried, is now accounted for by the structure of GF matrices. Only one real physical time remains, propagating from $-\infty$ to $+\infty$. Moreover, matrices themselves may be transformed by any canonical transformation, resulting in another equivalent description of the same system. Then the matrix indices cease to be related to contour branches. Therefore instead of ± indices usual 1,2 should be used. The canonical transformation freedom can be used to reduce the number of acting GFs, as only two of them are linearly independent. In particular, the simple transformation was found transforming (3) into the usually called "triangular" representation

$$\hat{G}_0(x,x') = \begin{pmatrix} 0 & G_0^{(a)}(x,x') \\ G_0^{(r)}(x,x') & F_0(x,x') \end{pmatrix}.$$

Here, the retarded $G_0^{(r)}(x,x')$ and advanced $G_0^{(a)}(x,x')$ GF's are defined as

$$G_0^{(r)}(x,x') = -i\cdot\theta(t-t')\cdot\left\langle\left[\psi_0(x),\psi_0^{+}(x')\right]_{\mp}\right\rangle_{0T} = \left[G_0^{(a)}(x',x)\right]^{*}$$

and

$$F_0(x,x') = -i\left\langle\left[\psi_0(x),\psi_0^{+}(x')\right]_{\pm}\right\rangle_{0T},$$

with the upper sign for Bosons and the lower for Fermions. In these formulae, $\psi_0(x)$ denotes the field operators in the interaction representation,

$$\left[\psi_0(x),\psi_0^+(x')\right]_\pm = \psi_0(x)\psi_0^+(x') \pm \psi_0^+(x')\psi_0(x).$$

The triangular representation has few evident advantages. It is minimal – the number of nonzero matrix elements cannot be reduced further by *canonical* transformation.

1. It is explicitly time symmetric (like Feynman's G_{causal} in vacuum) – retarded and advanced functions enter symmetrically. This is despite the existence of the time arrow, which is accounted for by the relative positions of elements of the GF matrix.
2. It is symmetric in emission and absorption processes. For fermions it is explicitly "charge symmetric", i.e. (anti)symmetric in electrons and holes - $iF_0(\mathbf{p},t'=t)=(1-2n_{\mathbf{p}})$ is positive for empty states and negative for occupied.
3. The functions $G^{(r,a)}(x,x')$ satisfy the universal initial condition
$$G^{(r,a)}(x,x')\big|_{t'=t} = \mp i\cdot\delta(\mathbf{r}-\mathbf{r}')$$
even after complete renormalization, then they are defined in terms of Heisenberg operators. The renormalized function F at coincident time arguments reduces to the single particle density matrix
$$i\cdot F(x,x')\big|_{t'=t} = \delta(\mathbf{r}-\mathbf{r}') \pm 2\rho(\mathbf{r},\mathbf{r}';t)$$

In that sense one can say, that $F(x,x')$ is the generalized distribution function while $G^{(r,a)}(x,x')$ describe essentially the renormalized particle dynamics. After renormalization the GF matrix is defined by the matrix Dyson equation

$$\hat{G}(x,x') = \hat{G}_0(x,x') + \iint \hat{G}_0(x,y)\cdot\hat{\Sigma}(y,y')\cdot\hat{G}(y',x')\cdot dydy'$$

which in triangular representation reduces to two equations

$$G^{(r)}(x,x') = G^{(r)}{}_0(x,x') + \iint G^{(r)}{}_0(x,y)\cdot\Sigma_r(y,y')\cdot G^{(r)}(y',x')\cdot dydy'$$

and

$$F(x,x') = \iint G^{(r)}(x,y)\cdot\Omega(y,y')\cdot G^{(a)}(y',x').$$

Note that the self energy matrix $\hat{\Sigma}$ is defined in terms of GF matrices as the sum of all exactly the same graphs as in the vacuum field theory or the ground state many body theory.

In the triangular representation

$$\hat{\Sigma} = \begin{pmatrix} \Omega & \Sigma_r \\ \Sigma_a & 0 \end{pmatrix}.$$

Those were the main results in [27], published in 1964.

In the year 1999, G. Baym in his talk at the opening session of the Conference "Kadanoff-Baym Equations – Progress and Perspectives for Many-body Physics" in my absence mentioned that paper. "The method"(round trip contour. *L.K*) "was then used by Leonid Keldysh in the Soviet Union, described first in his paper [27]. Our book was translated into Russian in the same year [29] but Keldysh did not refer to it…"[2]. Evidently, such a sentence is intended to create the impression that some method and may be also results of [27] were adopted from [29]. This statement, extremely unfriendly, being published in [28] needs a reply, especially now, after 38 years, when hardly anybody would check carefully, what was really written in so old papers. First of all, on the last page of the Russian edition of [29] among other typographical information two dates are indicated: "submitted for production 04.07.1964" and "signed for printing 05.10.1964" (still not printed). The paper [27] was published in the October 1964 issue of JETP and was

[2] To be correct that whole text (pp. 28-29) is reproduced below, except for a few lines about J. Schwinger's style:
"A crucial ingredient in the derivation of Boltzmann equations was the use of Green's functions defined on the round-trip contour along the real axis. The method was invented by Schwinger and presented in his lectures on Brownian motion at the Brandeis summer school in 1960, where I became familiar with it. Although the lectures were unpublished, Schwinger did write up his ideas in his paper *Brownian motion of a Quantum Oscillator*. ….
The round trip technique was also employed in the context of quantum electrodynamics in 1961-62 by Kalayana T. Mahanthappa, a fellow Schwinger graduate student at Harvard and Pradip Bakshi, a slightly later student of Schwinger's. Actually, Robert Mills (of Yang-Mills), while at the University of Birmingham in 1962, wrote but did not publish a lovely set of notes on round-trip Greens functions techniques, which formed the basis for later book. He refers in these notes to Schwinger's 1961 paper and remarks that, "The present work, some of which has, I believe, been duplicated independently by Baym and Kadanoff, following the methods of Martin and Schwinger, makes use of the thermodynamic Wick's theorem of Matsubara and Thouless, and others, with the integration contour in the complex time plane, distorted to include the real axes." The method then was used by Leonid Keldysh in the Soviet Union, described first in his 1964 paper. Our book was translated to Russian the same year, but Keldysh did not refer to it, writing rather, "Our technique will be close to Mills technique for equilibrium system", citing Mills' notes. Schwinger influence was widely felt."

submitted in April 1964. It is hard to imagine how I could refer to this translated version. Sure, however, all that about that Russian edition does not matter. The original version of [29] was published two years earlier and, if it in fact contained all or important parts of the results of [27], then [27] would be deprived of any significance, no matter have I seen [29] or not. Therefore, one should look what is the overlap, if any, of [27] and [29]. That corresponds more closely to the subject of our Workshop – about NGF's. So about the method. I was following, as explained above, the standard Feynman-Dyson method with perturbative expansion, S-matrix, time ordering etc. The authors of [29] used directly Heisenberg's equations of motion, which the authors themselves oppose to "... an alternative scheme, based upon an expansion of G in a power series..." (Ref. [29] p.191, and further about perturbative series, but *only* equilibrium Matsubara's.). More important, however, is that their method is based completely on the analytic continuation of Matsubara's equations from the imaginary time axis to the real one, while in [27] all the derivation is done on the real time axis in order to get results applicable to experimentally realistic external fields, which are always exactly zero before some switch-on time. So it was impossible for me to use any detail of the method of [29]. However, may be speaking about "the method" G. Baym meant contour ordering and 2×2 GF matrices. In that case the story becomes even more amusing. There is no contour ordering, no 2×2 GF matrices and even no reference (!) to Schwinger's paper [26] or to his 1960 lectures in the book [29], neither in text nor in the reference list. In the year 1999, G. Baym considers Schwinger's idea about GFs contour ordering[3] as "The crucial ingredient in the

[3] The crucial step is just the contour ordering of GF's. Not the contour itself, which was used already in Ref. [20]. To map two time arguments of GF each onto one of contour branches and later to arrange them into a 2×2 matrix, at least 4 functions are necessary, as it was done in both [26] and [27]. No matter that only 2 functions are linearly independent – one needs to know which place should be ascribed to each of the linear combinations. That could not be done, and was not done, in [29] operating with only two functions - $G^{>}$ and $G^{<}$. The Fig. 5 in [28], p.26 illustrates well "the method" of G. Baym. It is announced as "The succession of contours in deriving the generalized Boltzman equation". Plots a) and b) with all integrals along the Matsubara's imaginary time segment correspond indeed to the procedure, described in [29]. Then in the plot c) the deformation to the "real time round trip contour" is presented. But there is nothing like that in [29]. Any idea about the contour, even the word "contour" is absent. Equations (8-27) – the Kadanoff-Baym equations – contain usual (not contour) integrals along the real time axis. All integrals in the preceding formula are along imaginary Matsubara's segment. Analytic continuation is made without any contour. Performing an analytic continuation does not require the contour. Using the contour does not require the analytic continuation, as it was done in both [26] and [27].

derivation of Boltzmann equation…". However, in the year 1962, two years after his listening to Schwinger's lectures and a year after publishing [26], he considered it as not worth mentioning in the book entitled *Quantum Statistical Mechanics*.

The major difference between the Kadanoff-Baym equations [29] and the matrix Dyson equation derived in [27] is the algorithm of self energy calculation in terms of GF's. In [27] it is the standard and absolutely regular Feynman diagram technique with GF matrices. In [29], self energies are defined as two analytic in the complex time plane components of Matsubara's self energy used, however, on the real time axis. No other regular algorithm of self energy calculation is presented. So, strictly speaking, corresponding to that definition procedure must be the analytic continuation of Matsubara's self energy functions from the imaginary time axis to the real times for solving Kadanoff-Baym equations and then analytic continuation of GF's from the real time axis to imaginary times for calculating self energies. In terms of analytic functions $G^<$ and $G^>$ each of Matsubara's graphs of the n-th order transforms into $\sim n!$ graphs, differing by the time sequence of different interaction vertices. That makes the perturbation theory much more cumbersome, much like the old Schrödinger perturbation theory, strictly speaking non-renormalizable. Probably that is the reason why in [29] self energies are calculated only up to the second order in the interaction. Calculation in this approximation does not contain time integrals – self energies are proportional to GF products and do not need separate analytic continuation. The only attempt to go beyond that approximation and introduce the scattering amplitude instead of the interaction potential (chapter 13) is restricted to the equilibrium case. Sure, now there exists the possibility to calculate self energies directly on the real time axis – in terms of real time GF matrices, which calculates the whole $\hat{\Sigma}$-matrix including $\Sigma^<$ and $\Sigma^>$. However, there was nothing like that in the book [29].

To summarize that part of my talk, I believe that the paper [27] solved that problem which was announced in the title of that article: "Diagram Technique for Nonequilibrium Processes", which is applicable also to many body systems in equilibrium or in the ground state. And nothing of the methods developed in [29] was used.[4]

Another old problem, only mentioned in my paper is that of initial conditions, i.e. the description of a many body system, which at some fixed time t_0 was in some arbitrary fixed state $\hat{\rho}_0$. I shall say only a few words about some early papers in that direction. To my knowledge the first solution was presented by Hall [30]. He has shown that the diagram technique becomes modified by including new elements – all the initial correlators contained in $\hat{\rho}_0$, become build in graphs as independent (multi-particle) vertexes. Hall's technique was essentially developed by Kukharenko and Tikhodeev [31] by a renormalization procedure accounting for the decay of initial correlations and slow time evolution of basic parameters, like distribution functions and its higher self correlators, from their initial values at t_0 to following current time t, and so derived generalized Boltzmann equation and a set of transport equations for fluctuation correlations.

My point of view is that in many, may be the majority, realistic cases initial conditions result from some evolution or external conditions, which both – prehistory and external conditions – can, or should, be included in the consideration, reducing the problem to the evolution from the far past, which is already forgotten. Indeed, as Fanchenko [32] has shown, the problem of arbitrary initial conditions can be always reduced to the evolution, starting from equilibrium at $t \to -\infty$ by including in the Hamiltonian some perturbation $\hat{H}'$, resulting in the

[4] As the only overlap of [27] and [29] one may regard the derivation of the Bolzmann equation. However, it was done for different relaxation mechanisms – phonon scattering in [27] and particle collisions in [29]. Moreover, unlike [29], where generalization of the Bolzmann equation was derived, in [27] it was just an illustration without any pretension for any new result. It was done only to the lowest approximation presented in many textbooks and along the well known lines: GF at coincident time arguments - density matrix, Wigner function, imaginary parts of self energies – scattering rates etc. As a generalization of the Boltzmann equation the matrix Dyson equation was called.

evolution from $\hat{\rho}_{0T}$ at $t \to -\infty$ to $\hat{\rho}_0$ at $t = t_0$. So the problem reduces to the diagram technique in terms of usual GF matrices, with substitution, however, of complicated vertexes from $\hat{H}'$ instead of initial correlation blocks of Hall's technique. In physically realistic cases $\hat{H}'$ hardly can be very complicated.

References:

1. A. A. Abrikosov, L. P. Gor'kov and I. E. Dzyaloshinski, *Methods of the* Quantum Field Theory in Statistical Physics, Fizmatgiz, Moscow, 1962; Pergamon, Oxford, 1965.
2. D. A. Kirzhnits, *Field Theoretical Metods in the Many Body Theory*, Atomizdat, Moscow, 1963; Pergamon, Oxford, 1967
3. D. N. Zubarev, Uspekhi Fiz. Nauk **71**,71,(1960); Sov.Phys (Uspekhi) **3**, 320 (1961)
4. V. L. Bonch-Bruevich and S. V. Tyablikov, *Metod Funktsii Grina v Statisticheskoi Mekhanike,* Fizmatgiz, Moscow, 1961; Green's Function Method in the Statistical Mechanics, North Holland, Amsterdam, 1962
5. V. L. Bonch-Bruevich and Sh. M. Kogan, Ann. Physics **9**, 125 (1960)
6. D. N. Zubarev, *Nonequilibrium Statistical Thermodynamics*, Nauka, Moscow, 1971; Plenum Press, 1974.
7. V. L. Bonch-Bruevich, JETP **28**, 121 (1955); *ibid.* **30**, 343 (1956); Soviet Phys. JETP **1**, 169 (1955); *ibid.* **3**, 278 (1956)
8. A. B. Migdal, JETP **32**, 399 (1957); *ibid.* **35**, 1438 (1958); Soviet Phys. JETP**,** **5,** 333 (1957); ibid. **7**, 996 (1958).
9. V. M. Galitskii and A. B. Migdal, JETP **34**, 139 (1957); Soviet Phys. JETP **7,** 96 (1958).
10. V. M. Galitskii, JETP **34**, 151 (1957); Soviet Phys. JETP **7,** 104 (1958).
11. S. T. Belyaev, JETP **34**, 417, 433 (1958); Soviet Phys. JETP **7,** 289, 299 (1958).
12. L. P. Gor'kov, JETP **34**, 735 (1958); Soviet Phys. JETP **7,** 505 (1958).

13. A. A. Abrikosov, L. P. Gor'kov and Khalatnikov, JETP, ? , (1958); Soviet Phys. JETP **8,** 182 (1959).

14. L. D. Landau, JETP **34**, 262 (1958); Soviet Phys. JETP **7,** 505 (1958).

15. L. D. Landau, JETP **35**, 97 (1958); Soviet Phys. JETP **8,** 70 (1959).

16. A. A. Abrikosov, L. P. Gor'kov and I. E. Dzyaloshinskii , JETP **36**, 900 (1959); Soviet Phys. JETP **9,** 636 (1959).

17. E. S. Fradkin, JETP **34**, 1286 (1959); Soviet Phys. JETP **9,** 912 (1958); Nuclear Phys. **12**, 465 (1959)

18. N. N. Bogolyubov and S. V. Tyablikov, DAN **126**, 53 (1959); Sov. Phys. (Doklady) **4,** 604 (1959)

19. Yu. L. Klimontovich and V. P. Silin, Uspekhi Fizich. Nauk **70**, 247 (1960); Soviet Phys. Uspekhi **3**, 84 (1960)

20. O. V. Konstantinov and V. I. Perel, , JETP, **39**, 197, (1960); Soviet Phys. JETP **12,** 142 (1961).

21. I. E. Dzyaloshinskii, JETP **42**, 1126 (1962); Soviet Phys. JETP **15,** 778 (1962)

22. G. M. Eliashberg, JETP **41,** 1241 (1961); Soviet Phys. JETP **14,** 886 (1961).

23. G. M. Eliashberg, JETP, **42**, 1658, (1962); Soviet Phys. JETP **15,** 1151 (1962).

24. L. P. Gor'kov and G. M. Eliashberg, JETP **54**, 612 (1968); Soviet Phys. JETP **27,** 328 (1968).

25. G. M. Eliashberg, JETP **61**, 1254 (1971); Soviet Phys. JETP **34,** 668 (1972).

26. J. Schwinger, J. Math. Phys. **2**, 407 (1961)

27. L. V. Keldysh, JETP **47**, 1515 (1964); Soviet Phys. JETP **20,** 1018 (1965).

28. G. Baym, in *Progress in Nonequilibrium Green's Functions*, M. Bonitz (Ed.), World Scientific Publ., Singapore 2000, p.17.

29. L. P. Kadanoff and G. *Baym, Quantum Statistical Mechanics*, Benjamin, New York, (1962); *Kvantovaya Statisticheskaya Mekhanika*, Mir, Moskva, (1964)

30. A. G. Hall, Physica, **80**A, 369, (1975); J. Phys. A**8**, 214, (1976)

31. Yu. A. Kukharenko and S. G. Tikhodeev, JETP **83**, 1444, (1982); Soviet Phys. JETP **56,** 831, (1982).

32. C. C. Fanchenko, Teor. i Mat. Fiz. **55**, 137, (1983)

LIST OF RESEARCH PAPERS BY L.V. KELDYSH

[1] L. V. Keldysh, Behavior of Non-Metallic Crystals in Strong Electric Fields, Soviet Phys. JETP-USSR 1958, **6**, 763.
[2] L. V. Keldysh, Influence of the Lattice Vibrations of a Crystal on the Production of Electron-Hole Pairs in a Strong Electrical Field, Soviet Phys. JETP-USSR 1958, **7**, 665.
[3] L. V. Keldysh, The Effect of a Strong Electric Field on the Optical Properties of Insulating Crystals, Soviet Phys. JETP-USSR 1958, **7**, 788.
[4] B. Vul, E. Zavaritskaia, L. V. Keldysh, Impurity Conductivity of Germanium at Low Temperatures, Dokl. Akad. Nauk SSSR 1960, **135**, 1361.
[5] L. V. Keldysh, Kinetic Theory of Impact Ionization in Semiconductors, Soviet Phys. JETP-USSR 1960, **10**, 509.
[6] L. V. Keldysh, Optical Characteristics of Electrons with a Band Energy Spectrum in a Strong Electric Field, Soviet Phys. JETP-USSR 1963, **16**, 471.
[7] L. V. Keldysh, Effect of Ultrasonics on the Electron Spectrum of Crystals, Soviet Phys.-Solid State 1963, **4**, 1658.
[8] L. V. Keldysh, Y. Kopaev, The Energy Spectrum of a Degenerate Semiconductor with an Ionic Lattice, Soviet Phys.-Solid State 1963, **5,** 1026.
[9] L. V. Keldysh, Deep Levels in Semiconductors, Soviet Phys. JETP-USSR 1964, **18**, 253.
[10] L. V. Keldysh, G. Proshko, Infrared Absorption in Highly Doped Germanium, Soviet Phys.-Solid State 1964, **5**, 2481.
[11] V. Bagaev, Y. Berozashvili, B. Vul, E. Zavaritskaya, L. V. Keldysh, A. Shotov, Concerning the Energy Level Spectrum of Heavily Doped Gallium Arsenide, Soviet Phys.-Solid State 1964, **6**, 1093.
[12] L. V. Keldysh, Diagram Technique for Nonequilibrium Processes, Soviet Phys. JETP-USSR 1965, **20**, 1018. [Zh. Eksp. Teor. Fiz. 1964, **47**, 1515].
[13] L. V. Keldysh, Ionization in Field of a Strong Electromagnetic Wave, Soviet Phys. JETP-USSR 1965, 20, 1307. [Zh. Eksp. Teor. Fiz. 1964, **47**, 1945].
[14] L. V. Keldysh, Concerning Theory of Impact Ionization in Semiconductors, Soviet Phys. JETP-USSR 1965, **21**, 1135.
[15] L. V. Keldysh, Y. Kopaev, Possible Instability of Semimetallic State Toward Coulomb Interaction, Soviet Physics Solid State, Soviet Phys. Solid State, USSR 1965, **6**, 2219.
[16] L. V. Keldysh, Superconductivity In Nonmetallic Systems, Soviet Phys. Uspekhi-USSR 1965, **8**, 496.
[17] V. Bagaev, Y. Berozashvili, L. V. Keldysh, Electrooptical Effect in GaAs. JETP Lett. - USSR 1966, **4**, 246.

[18] L. V. Keldysh, T. Tratas, Dynamic Narrowing of Paramagnetic Resonance Lines in a Compensated Semiconductor, Soviet Phys. Solid State, USSR 1966, **8**, 64.
[19] L. V. Keldysh, A. Kozlov, Collective Properties of Large-Radius Excitons, JETP Lett. - USSR 1967, **5**, 190.
[20] L. V. Keldysh, A. Kozlov, Collective Properties of Excitons in Semiconductors, Soviet Phys. JETP-USSR 1968, **27**, 521. [Zh. Eksp. Teor. Fiz. 1968, **54**, 978].
[21] V. Bagaev, Y. Berozashvili, L. A. Keldysh, Anisotropy of Polarized Light Absorption Produced in GaAs and CdTe Crystals by a Strong Electric Field, JETP Lett. - USSR 1969, **9**, 108.
[22] L. V. Keldysh, M. Pkhakadze, Conductivity of Semiconductors Under Pinch-Effect Conditions, JETP Lett. - USSR 1969, **10**, 169.
[23] V. Bagaev, T. Galkina, O. Gogolin, L. V. Keldysh, Motion of Electron-Hole Drops in Germanium, JETP Lett. - USSR 1969, **10**, 195.
[24] L. V. Keldysh, O. Konstantinov, V. Perel, Polarization Effects in Interband Absorption of Light in Semiconductors Subjected to a Strong Electric Field, Soviet Phys. Semicond. - USSR 1970, **3**, 876.
[25] L. V. Keldysh, Electron-Hole Drops in Semiconductors, Soviet Phys. Uspekhi-USSR 1970, 13, 292.
[26] B. Kadomtsev, R. Sagdeev, L. V. Keldysh, I. Kobzarev, On A.A. Tyapkin's article "Expression of General Properties of Physical Processes in Space-and-Time Metric of Special Theory of Relativity", Uspekhi Fizicheskikh Nauk 1972, **106**, 660.
[27] L. V. Keldysh. Collective Properties of Excitons in Semiconductors. In Excitons in Semicondutors (Ed. B.M. Vul), Nauka, Moscow, 1971 (in Russian).
[28] R. Guseinov, L. V. Keldysh, Nature of Phase-Transitions under Excitonic Instability Conditions of a Crystal Electron Spectrum, Zh. Eksp. Teor. Fiz. 1972, **63**, 2255.
[29] L. V. Keldysh. Coherent state of excitons, Problems of Theoretical Physics. In memory of I.E. Tamm (Ed. by V.I. Ritus), Moscow Nauka 1972 (in Russian) p. 433, Physics Uspekhi **60**, 1180 (2017)
[30] L. V. Keldysh, A. Silin, Electron-Hole Liquids in Semiconductors in Magnetic-Field, Fiz. Tverdovo Tela 1973, **15**, 1532.
[31] L. V. Keldysh, A. Manenkov, V. Milyaev, G. Mikhailova, Microwave Breakdown and Exciton Condensation in Germanium, Zh. Eksp. Teor. Fiz. 1974, **66**, 2178.
[32] L. V. Keldysh, S. Tikhodeev, Absorption of Ultrasound by Electron-Hole Drops in a Semiconductor, JETP Lett. 1975, **21**, 273.
[33] L. V. Keldysh, A. Silin, Electron-Hole Fluid in Polar Semiconductors, Zh. Eksp. Teor. Fiz. 1975, **69**, 1053.
[34] L. V. Keldysh, Phonon Wind and Dimensions of Electron-Hole Drops in Semiconductors, JETP Lett. 1976, **23**, 86.
[35] L. V. Keldysh, T. Onishchenko, Electron Liquid in a Superstrong Magnetic-Field, JETP Lett. 1976, **24**, 59.
[36] E. Andryushin, V. Babichenko, L. V. Keldysh, T. Onishchenko, A. Silin, Electron-Hole Liquid in Strongly Anisotropic Semiconductors and Semimetals, JETP Lett. 1976, **24**, 185.
[37] V. Bagaev, L. V. Keldysh, N. Sibeldin, V. Tsvetkov, Phonon Wind Drag of Excitons and Electron-Hole Drops, Zh. Eksp. Teor. Fiz. 1976, **70**, 702.
[38] V. Bagaev, N. Zamkovets, L. V. Keldysh, N. Sibeldin, V. Tsvetkov, Kinetics of Exciton Condensation in Germanium, Zh. Eksp. Teor. Fiz. 1976, **70**, 1500.
[39] L. V. Keldysh, S. Tikhodeev, Ultrasound Absorption by Electron-Hole Drops in Semiconductor, Fiz. Tverdogo Tela 1977, **19**, 111.

[40] E. Andryushin, L. V. Keldysh, A. Silin, Electron-Hole Liquid and Metal-Dielectric Phase-Transition in Layer Systems, Zh. Eksp. Teor. Fiz. 1977, **73**, 1163.

[41] L. V. Keldysh, Metal-Dielectric Transformation Under Light Action, Vestnik Moskovskovo Univ. Ser. 3, Fizika Astronomia 1978, **19**, 86 (in Russian).

[42] L. V. Keldysh, Coulomb Interaction in Thin Semiconductor and Semimetal Films, JETP Lett. 1979, **29**, 658.

[43] L. V. Keldysh, Polaritons in Thin Semiconducting-Films, JETP Lett. 1979, **30**, 224. [Pis'ma v ZhETF 1979, 29, 658].

[44] V. Bagaev, M. Bonchosmolovskii, T. Galkina, L. V. Keldysh, A. Poyarkov, Entrainment of Electron-Hole Drops by a Strain Pulse Produced as a Result of Laser Irradiation of Germanium, JETP Lett. 1980, **32**, 332.

[45] E. Andriushyn, L. V. Keldysh, V. Sanina, A. Silin, Electron-Hole Liquid in Thin Semiconducting-Films, Zh. Eksp. Teor. Fiz. 1980, **79**, 1509.

[46] L. V. Keldysh, A. Kechek, On the Dielectric-Constant of The Non-Polar Fluid, Dokl. Akad. Nauk SSSR 1981, **259**, 575.

[47] A. Ivanov, L. V. Keldysh, The Propagation of Powerful Electromagnetic-Waves in Semiconductors under the Resonant Excitation of Excitons, Dokl. Akad. Nauk SSSR 1982, **264**, 1363.

[48] A. Ivanov, L. V. Keldysh, Modification of the Polariton and Phonon Spectra of a Semiconductor in the Presence of an Intense Electromagnetic-Wave, Zh. Eksp. Teor. Fiz. 1983, **84**, 404.

[49] N. Gippius, V. Zavaritskaya, L. V. Keldysh, V. Milyaev, S. Tikhodeev, Quantum Nature of the Reflection of an Exciton from the Surface of an Electron-Hole Drop, JETP Lett. 1984, **40**, 1235.

[50] P. Elyutin, L. V. Keldysh, A. Kechek, The Resonance Dielectric Permittivity of Nonpolar Liquids, Optika i Spektroskopia 1984, **57**, 282.

[51] L. V. Keldysh, S. Tikhodeev, High-Intensity Polariton Wave Near the Stimulated Scattering Threshold, Zh. Eksp. Teor. Fiz. 1986, **90**, 1852.

[52] L. V. Keldysh, S. Tikhodeev, Nonstationary Mandelstam-Brillouin Scattering of an Intense Polariton Wave, Zh. Eksp. Teor. Fiz. 1986, **91**, 78.

[53] N. Gippius, L. V. Keldysh, S. Tikhodeev, Mandelstam-Brilloiun Scattering of an Incoherent Polariton Wave, Zh. Eksp. Teor. Fiz. 1986, **91**, 2263.

[54] L. V. Keldysh. The Electron-Hole Liquid in Semiconductors. Contemporary Physics **27**, No. 5, 395 (1986)

[55] L. V. Keldysh, Excitons and Polaritons in Semiconductor Insulator Quantum Wells and Superlattices, Superlattices Microstruct. 1988, **4**, 637.

[56] A. Ivanov, L. V. Keldysh, V. Panashchenko, Low-Threshold Exciton-Biexciton Optical Stark-Effect in Direct-Gap Semiconductors, Zh. Eksp. Teor. Fiz. 1991, **99**, 641.

[57] A. Ivanov, L. V. Keldysh, V. Panashchenko, Nonlinear Optical Response of Interacting Excitons, Inst. Phys.: Conf. Ser. 1992, **126**, 431.

[58] L. V. Keldysh, Excitonic Molecules in Nonlinear Optical-Response, Phys. Status Solidi B 1992, **173**, 119.

[59] L. V. Keldysh, Coherent Excitonic Molecules, Solid State Commun. 1992, **84**, 37.

[60] N. Gippius, T. Ishihara, L. V. Keldysh, E. Muljarov, S. Tikhodeev, Dielectrically Confined Excitons and Polaritons in Natural Superlattices – Perovskite Lead Iodide Semiconductors, J. Phys. IV 1993, **3**, 437. 3rd International Conference on Optics of Excitons in Confined Systems, Univ Montpellier II, Montpellier, France, Aug 30–Sep 02, 1993.

[61] N. Gippius, S. Tikhodeev, L. V. Keldysh, Polaritons in Semiconductor-Insulator Superlattices with Nonlocal Excitonic Response, Superlattices Microstruct. 1994, **15**, 479.

[62] L. V. Keldysh, Correlations in the Coherent Transient Electron-Hole System, Phys. Status Solidi B 1995, **188**, 11. 4th International Workshop on Nonlinear Optics and Excitation Kinetics in Semiconductors (NOEKS IV), Gosen, Germany, Nov 06–10, 1994.

[63] L. V. Keldysh. Macroscopic coherent states of excitons in semiconductors. In Bose–Einstein Condensation (Eds. A. Griffin, D.W. Snoke, S. Stringari). Cambridge University Press, 1995, p. 246

[64] A. Ivanov, H. Wang, J. Shah, T. Damen, L. V. Keldysh, H. Haug, L. Pfeiffer, Coherent transient in photoluminescence of excitonic molecules in GaAs quantum wells, Phys. Rev. B 1997, **56**, 3941.

[65] L. V. Keldysh, Excitons in Semiconductor-Dielectric Nanostructures, Phys. Status Solidi A 1997, **164**, 3. 5th International Meeting on Optics of Excitons in Confined Systems (OECS 5), Göttingen, Germany, Aug 10–14, 1997.

[66] A. Ivanov, H. Haug, L. V. Keldysh, Optics of Excitonic Molecules in Semiconductors and Semiconductor Microstructures, Phys. Rep. 1998, **296**, 237.

[67] Q. Vu, H. Hang, L. V. Keldysh, Dynamics of the Electron-Hole Correlation in Femtosecond Pulse Excited Semiconductors, Solid State Commun. 2000, **115**, 63.

[68] L. V. Keldysh, Biexcitons at high densities, Phys. Status Solidi B 2002, **234**, 17.

[69] F. Klappenberger, K. Renk, R. Summer, L. V. Keldysh, B. Rieder, W. Wegscheider, Electric-field-induced reversible avalanche breakdown in a GaAs microcrystal due to cross band gap impact ionization, Appl. Phys. Lett. 2003, **83**, 704.

[70] L. V. Keldysh. Real-Time Nonequilibrium Green's Functions. In Progress in Nonequilibrium Green's functions II. (Eds: M. Bonitz, D. Semkat). World Scientific Publ, Singapore 2003, pp. 4–17.

[71] J. Reithmaier, G. Sek, A. Loffler, C. Hofmann, S. Kuhn, S. Reitzenstein, L. V. Keldysh, V. Kulakovskii, T. Reinecke, A. Forchel, Strong Coupling in a Single Quantum Dot-Semiconductor Microcavity System, Nature 2004, **432**, 197.

[72] N. Gippius, S. Tikhodeev, L. V. Keldysh, V. Kulakovskii, Hard Excitation of Stimulated Polariton-Polariton Scattering in Semiconductor Microcavities, Physics Uspekhi 2005, **48**, 306.

[73] G. Sek, C. Hofmann, J. Reithmaier, A. Loffler, S. Reitzenstein, M. Kamp, L. V. Keldysh, V. Kulakovskii, T. Reinecke, A. Forchel, Investigation of Strong Coupling Between Single Quantum Dot Excitons and Single Photons in Pillar Microcavities, Physica E 2006, 32, 471. [12th International Conference on Modulated Semiconductor Structures (MSS12), Albuquerque, NM, July 10–15, 2005.]

[74] S. Reitzenstein, A. Loffler, C. Hofmann, A. Kubanek, M. Kamp, J. Reithmaier, A. Forchel, V. Kulakovskii, L. V. Keldysh, I. Ponomarev, T. Reinecke, Coherent Photonic Coupling of Semiconductor Quantum Dots, Opt. Lett. 2006, **31**, 1738.

[75] S. Reitzenstein, C. Hofmann, A. Loeffler, A. Kubanek, J. P. Reithmaier, M. Kamp, V. D. Kulakovskii, L. V. Keldysh, T. L. Reinecke, A. Forchel, Strong and Weak Coupling of Single Quantum Dot Excitons in Pillar Microcavities, Phys. Status Solidi B 2006, **243**, 2224. [8th International Workshop on Nonlinear Optics and Excitation Kinetics In Semiconductors (NOEKS 8), Münster, Germany, February 20–24, 2006.]

[76] L. V. Keldysh, V. D. Kulakovskii, S. Reitzenstein, M. N. Makhonin, A. Forchel, Interference Effects in the Emission Spectra of Quantum Dots in High-Quality Cavities, JETP Lett. 2006, **84**, 494.

[77] S. Reitzenstein, A. Loffler, A. Kubanek, C. Hofmann, M. Kamp, J. P. Reithmaier, A. Forchel, V. D. Kulakovskii, L. V. Keldysh, I. V. Ponomarev, T. L. Reinecke, Coherent Photonic Coupling of Semiconductor Quantum Dots, Opt. Lett. 2006, **31**, 1738; Erratum: Opt. Lett. 2006, 31, 3507.
[78] L. V. Keldysh, Dynamic Tunneling, Her. Russ. Acad. Sci. 2016, **86**, 413.
[79] L. V. Keldysh, Coherent States of Excitons, Physics Uspekhi 2017, **60**, 1180.
[80] L. V. Keldysh, Multiphoton Ionization by a Very Short Pulse, Physics Uspekhi 2017, **60**, 1187.

www.ingramcontent.com/pod-product-compliance
Lightning Source LLC
Chambersburg PA
CBHW081344190326
41458CB00018B/6083
* 9 7 8 9 8 1 1 2 7 9 4 5 4 *